The Real Number System

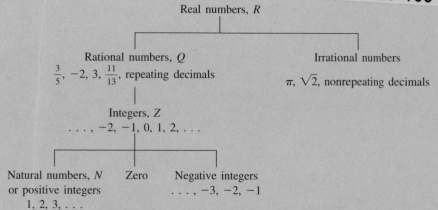

Real numbers, R

Rational numbers, Q
$\frac{3}{5}$, -2, 3, $\frac{11}{13}$, repeating decimals

Irrational numbers
π, $\sqrt{2}$, nonrepeating decimals

Integers, Z
$\ldots$, -2, -1, 0, 1, 2, $\ldots$

Natural numbers, N
or positive integers
1, 2, 3, $\ldots$

Zero

Negative integers
$\ldots$, -3, -2, -1

Some Important Properties For real numbers a, b, c, and k:

COMMUTATIVE PROPERTIES (1-4)

$a + b = b + a$
$ab = ba$

ASSOCIATIVE PROPERTIES (1-4)

$(a + b) + c = a + (b + c)$
$(ab)c = a(bc)$

DISTRIBUTIVE PROPERTIES (1-4, 1-5)

$a(b + c) = ab + ac$
$(a + b)c = ac + bc$

FUNDAMENTAL PRINCIPLE OF FRACTIONS (3-1, 3-2, 7-1)

$\dfrac{a}{b} = \dfrac{ak}{bk} \qquad b, k \neq 0$

EXPONENT PROPERTIES (1-5, 8-1, 8-2)

For integers m and n:

$a^m a^n = a^{m+n}$
$(a^n)^m = a^{mn}$
$(ab)^m = a^m b^m$
$\left(\dfrac{a}{b}\right)^m = \dfrac{a^m}{b^m} \qquad b \neq 0$
$\dfrac{a^n}{b^m} = a^{m-n} = \dfrac{1}{a^{n-m}} \qquad a \neq 0$

PROPERTIES OF EQUALITY (1-3, 2-7)

Reflexive property: $a = a$
Symmetric property: If $a = b$, then $b = a$
Transitive property: If $a = b$ and $b = c$, then $a = c$
Substitution principle: If $a = b$, then either may be substituted for the other

If $a = b$, then
Addition property: $a + c = b + c$
Subtraction property: $a - c = b - c$
Multiplication property: $a \cdot c = b \cdot c \qquad c \neq 0$
Division property: $\dfrac{a}{c} = \dfrac{b}{c} \qquad c \neq 0$

INEQUALITY PROPERTIES (5-1)

If $a < b$, then
Addition property: $a + c < b + c$
Subtraction property: $a - c < b - c$
Multiplication property: $ac < bc \qquad$ for positive c
$\qquad\qquad\qquad\qquad\quad ac > bc \qquad$ for negative c
Division property: $\dfrac{a}{c} < \dfrac{b}{c} \qquad$ for positive c
$\qquad\qquad\qquad \dfrac{a}{c} > \dfrac{b}{c} \qquad$ for negative c

ZERO PROPERTY (6-8)

$ab = 0 \qquad$ if and only if $\qquad a = 0 \qquad$ or $\qquad b = 0$

ELEMENTARY ALGEBRA

Structure and Use

Books by Barnett—Kearns—Ziegler

Barnett—Kearns: *Elementary Algebra: Structure and Use, 6th edition*
Barnett—Kearns: *Algebra: An Elementary Course, 2d edition*
Barnett—Kearns: *Intermediate Algebra: Structure and Use, 5th edition*
Barnett—Kearns: *Algebra: An Intermediate Course, 2d edition*
Barnett—Ziegler: *College Algebra, 5th edition*
Barnett—Ziegler: *College Algebra with Trigonometry, 5th edition*
Barnett—Ziegler: *Precalculus: Functions and Graphs, 3d edition*

Also available from McGraw-Hill

Schaum's Outline Series in Mathematics & Statistics

Most outlines include basic theory, definitions, hundreds of example problems solved in step-by-step detail, and supplementary problems with answers. Titles on the current list include:

Advanced Calculus
Advanced Mathematics
Analytic Geometry
Basic Mathematics for Electricity & Electronics, 2d edition
Basic Mathematics with Applications to Science and Technology
Beginning Calculus
Boolean Algebra & Switching Circuits
Calculus, 3d edition
Calculus for Business, Economics, & the Social Sciences
College Algebra
College Mathematics, 2d edition
Complex Variables
Descriptive Geometry

Differential Equations
Differential Geometry
Discrete Mathematics
Elementary Algebra, 2d edition
Essential Computer Mathematics
Finite Differences & Difference Equations
Finite Mathematics
Fourier Analysis
General Topology
Geometry, 2d edition
Group Theory
Laplace Transforms
Linear Algebra, 2d edition
Mathematical Handbook of Formulas & Tables

Matrix Operations
Modern Abstract Algebra
Numerical Analysis, 2d edition
Partial Differential Equations
Probability
Probability & Statistics
Real Variables
Review of Elementary Mathematics
Set Theory & Related Topics
Statistics, 2d edition
Technical Mathematics
Tensor Calculus
Trigonometry, 2d edition
Vector Analysis

Schaum's Solved Problems Books

Each title in this series is a complete and expert source of solved problems containing thousands of problems with worked out solutions. Titles on the current list include:

3000 Solved Problems in Calculus
2500 Solved Problems in College Algebra and Trigonometry
2500 Solved Problems in Differential Equations
2000 Solved Problems in Discrete Mathematics

3000 Solved Problems in Linear Algebra
2000 Solved Problems in Numerical Analysis
3000 Solved Problems in Precalculus

Bob Miller's Math Helpers

Bob Miller's Calc I Helper

Bob Miller's Calc II Helper

Bob Miller's Precalc Helper

Available at most college bookstores, or for a complete list of titles and prices, write to: Schaum Division, McGraw-Hill, Inc., Princeton Road, S-1, Hightstown, NJ 08520.

6TH EDITION

ELEMENTARY ALGEBRA

Structure and Use

Raymond A. Barnett
Merritt College

Thomas J. Kearns
Northern Kentucky University

McGraw-Hill, Inc.

New York   St. Louis   San Francisco   Auckland   Bogotá   Caracas
Lisbon   London   Madrid   Mexico City   Milan   Montreal
New Delhi   San Juan   Singapore   Sydney   Tokyo   Toronto

ELEMENTARY ALGEBRA: Structure and Use

This book is printed on acid-free paper.

2 3 4 5 6 7 8 9 0 VNH   VNH 9 0 9 8 7 6 5 4

ISBN 0-07-004566-6

This book was set in Times Roman by York Graphic Services, Inc.
The editors were Michael Johnson, Karen M. Minette, and David A. Damstra;
the design was done by A Good Thing, Inc.;
the production supervisor was Leroy A. Young.
The photo editor was Debra Hershkowitz.
Von Hoffmann Press, Inc., was printer and binder.

Chapter-Opening Photo Credits
1: Lawrence Migdale/Stock, Boston   *2:* Charles Krebs/The Stock Market   *3:* J. Patrick Phelan/The Stock Market
4: Guy Sauvage/Photo Researchers   *5:* Stephen Frink/The Stock Market   *6:* Bruce Hands/Comstock   *7:* Steve Elmore/
The Stock Market   *9:* Stuart Cohen/Comstock

Library of Congress Cataloging-in-Publication Data

Barnett, Raymond A.
    Elementary algebra: structure and use / Raymond A. Barnett,
Thomas J. Kearns.
        p.   cm.
    Includes index.
    ISBN 0-07-004566-6
    1. Algebra.   I. Kearns, Thomas J.   II. Title.
QA152.2.B37   1994
512'.9—dc20   93-7179

About the Authors

Raymond A. Barnett is an experienced teacher and author. He received his B.A. in mathematical statistics from the University of California at Berkeley and his M.A. in mathematics from the University of Southern California. He then went on to become a member of the Department of Mathematics at Merritt College and head of that department for 4 years. He is a member of the Mathematical Association of America (MAA), the National Council of Teachers of Mathematics (NCTM), and the American Association for the Advancement of Science. He is the author or coauthor of seventeen books in mathematics that are still in print—all with a reputation for extremely readable prose and high-quality mathematics.

Thomas J. Kearns received his B.S. from Santa Clara University and his M.S. and Ph.D. from the University of Illinois at Urbana-Champaign. After several years of teaching at the University of Delaware, he was appointed to the faculty at Northern Kentucky University, where he served as chairman of the Department of Mathematical Sciences for ten years. He is a member of the MAA, the NCTM, the American Mathematical Society, and the Operations Research Society of America. He has coauthored four texts with Raymond A. Barnett, as well as texts in college algebra and elementary statistics.

Contents

Preface

This is an introductory text in algebra for students with no background in algebra and for students who need review before proceeding further. The improvements in this sixth edition have resulted from generous responses from users and reviewers of the fifth edition. The changes in this edition have been made to make the text even more accessible to students with minimal background and to provide a better transition to material covered in subsequent courses—intermediate algebra courses in particular.

♦ NEW DESIGN

A new, accessible, **full-color** design and an expanded art program make the book more visually appealing, easier to read, and reinforces mathematical concepts.

Type style for exponents and fractions has been improved for increased clarity, and all examples and most application problems now have **captions.**

There is more **boxed material** for emphasis. More **schematics** have been added for clarity.

♦ GRAPHING

A greater emphasis is placed on **graphing** and, in particular, on the relationship between the graph of an equation

$$y = (\text{An expression in } x)$$

and the solution of the related equation

$$(\text{An expression in } x) = 0$$

(See Sections 6-8, 7-4, 8-7, 9-4.)

♦ CALCULATORS

Calculators are assumed to be available to students. Many of the problems in the text lend themselves naturally to calculator use. However, with the exception of a very few exercises, which are noted in the instructions, the problems do not *require* use of a calculator. For those instructors who want to emphasize calculators more, a **graphing calculator supplement** is available.

♦ EXAMPLES

Illustrative examples of some of the more challenging problems and applications have been added at the request of users. There are more **worked examples** and **matched problems.**

◆ EXERCISES

Exercise sets in almost every section have been considerably expanded. **Applications** have been added throughout the book and existing applications have been brought up to date.

◆ SYSTEMATIC REVIEW

Chapter review exercise sets have been expanded. Sample **Chapter Tests** have been added after every chapter, and comprehensive **Cumulative Review Exercise Sets** have been introduced following Chapters 3, 5, 7, and 9.

◆ CAUTIONS

Common student errors are identified by a special **caution** symbol at places where they naturally occur. The number of these have been increased and they are more prominently displayed.

◆ LEARNING SYSTEM

The text is **written for student comprehension.** Each concept or technique is illustrated with an example, followed by a parallel problem **(matched problem).** Answers to the matched problems are provided at the end of each section so students can immediately check their understanding. This example–matched problem structure encourages active learning rather than passive reading.

The order of topics has been chosen to provide a **smooth transition from arithmetic to algebra.** The beginning chapters develop algebraic concepts, manipulations, and applications gradually. Initially only natural numbers are used, then integers, rational numbers, and, finally (in Chapter 4), real numbers. This **gradual introduction** keeps the material accessible to students with minimal background. Subsequent chapters then extend this development.

An **informal style** is used for exposition. Definitions are illustrated with simple examples. There are **no formal statements of theorems** in this text.

◆ REVISED SECTIONS

Several sections have been substantially rewritten. In particular, see the sections on **factoring second-degree equations** (Sections 6-4, 6-5, 6-6), **square roots and radicals** (Section 8-4), and **graphing quadratic polynomials** (Section 9-4).

Review material on **fractions, decimals,** and **percent** has been combined into one optional section (see Section 3-1).

◆ IMPORTANT FEATURES RETAINED AND IMPROVED IN THE SIXTH EDITION

Graded Exercise Sets   **Graded exercise sets** are divided into A, B, and C groupings. The **A problems** are straightforward and representative of the easier examples in the section. The **B problems** represent the more challenging examples in the section, but still emphasize mechanics. The **C problems** provide a mixture of harder mechanics, theory, and extension of the material in the section. The C problems may include some challenging problems that do not match worked examples in the section. In short, the exercises are designed so that an average or below-average student will be able to experience success and the more capable students will still be challenged.

Applications   The subject matter is related to the real world through numerous **realistic applications** from the physical sciences, business and economics, the life sciences, and the social sciences. Even the most skeptical student should be convinced that algebra is really useful.

Major Topic Development   The text continues to use a **spiral technique** for major topics wherever possible; that is, a topic is introduced in a relatively simple framework and then returned to and developed further in the later sections. For example, consider these topics:

Factoring: Sections 1-6, 6-3 to 6-8, 7-1 to 7-3, 9-1

Solving equations: Sections 2-7, 3-5, 6-8, 7-4, 8-7, 9-1 to 9-3

Graphing: Sections 4-1 to 4-4, 5-3 to 5-4, 6-8, 7-4, 8-7, 9-4

Fractional forms: Chapters 3 and 7

Order and inequality: Sections 1-3, 2-4, 3-4, and Chapter 5

Word problems: 1-2, 1-3, 2-8, 3-6 to 3-8, 4-7 to 4-9, 5-2, 5-4, 7-4, 8-3, 8-7, 9-5

The use of this spiral technique continues into the companion text: *Intermediate Algebra: Structure and Use,* fifth edition.

History   **Historical comments** are included for interest and perspective.

Chapter Summaries   **Chapter summary sections** include a review of the chapter with all important terms and symbols. Also included are a **comprehensive review exercise set** for the chapter and a short **chapter practice test. Cumulative review** exercises are included after Chapters 3, 5, 7, and 9.

Answers   **Answers** to all chapter review exercises, chapter tests, cumulative review exercises, and all odd-numbered problems from the section exercises are provided in the back of the book. Answers to the exercises in the chapter summaries (review exercises, chapter tests, and cumulative review exercises) are keyed, by numbers in italics, to the corresponding sections in the text.

♦   **ADDITIONAL STUDENT AIDS**

Think Boxes   **Think Boxes** (dashed boxes) are used to enclose steps that are usually done mentally (see Sections 1-5, 1-6, 2-2, 2-3, 2-4, 2-6).

Annotation   **Annotation** of examples and development is found throughout the text to help students through critical steps (see Sections 1-1 to 1-6).

Color   **Functional use of color** guides students through critical steps (see Sections 1-6, 2-6, 2-7). Further use of color is described on page xix.

Formula Summary   **Summaries** of algebraic formulas, symbols, and real-number properties, all keyed to the sections in which they are introduced, are included inside the front cover of the book for convenient reference. Summaries of geometric and other common formulas, and the metric system, are provided inside the back cover of the book.

♦ STUDENT SUPPLEMENTS

1. A STUDENT'S SOLUTIONS MANUAL is available through your bookstore. The manual includes key ideas and formulas, solutions to odd end-of-section exercises, all solutions to the end-of-chapter exercises and chapter tests, and an appendix on setting up word problems.

2. MATHWORKS is a self-paced interactive tutorial specifically linked to the text. It reinforces selected topics and provides unlimited opportunities to review concepts and to practice problem solving. It requires virtually *no* computer training and is available for IBM, IBM compatible, and Macintosh computers.

3. Course VIDEOTAPES are available for use through your institutions learning center.
4. THE GRAPHING CALCULATOR ENHANCEMENT MANUAL presents an integrated approach that utilizes calculator-based graphing to enhance understanding and development. It includes calculator exercises and examples as well as appendices on how to use the most popular calculators.

♦ INSTRUCTOR'S SUPPLEMENTS

1. An INSTRUCTOR'S RESOURCE MANUAL provides sample tests, transparency masters, an applications index, and additional teaching suggestions and assistance.
2. An INSTRUCTOR'S SOLUTIONS MANUAL contains detailed solutions to even end-of-section exercises and all cumulative review exercises as well as the answers to all problems.
3. THE PROFESSOR'S ASSISTANT is a unique computerized test generator available to instructors. This system allows the instructor to create tests using algorithmically generated test questions and those from a standard testbank. This testing system enables the instructor to choose questions either manually or randomly by section, question type, difficulty level, and other criteria. This system is available for IBM, IBM compatible, and Macintosh computers.
4. A PRINTED AND BOUND TESTBANK is also available. This is a hard-copy listing of the questions found in the standard testbank.

♦ ERROR CHECK

This text has been carefully and independently checked and proofread by a number of people. Because of this, the authors and publisher believe it to be substantially error-free. However, if errors remain, the authors and publisher would be grateful to be notified and receive corrections. Corrections should be sent to Mathematics Editor, College Division, McGraw-Hill, Inc., 1221 Avenue of the Americas, New York, NY 10020.

♦ ACKNOWLEDGMENTS

The authors have benefited from a great deal of help in preparing this edition of the book. We thank the many users of the fifth edition for their kind remarks and helpful suggestions. We appreciate the detailed reviews provided by the following:

Greg Banks, Northwestern Connecticut Community College
Cathy Barkley, Mesa State College
Kathleen Bavelas, Manchester Community College
Edward Beardslee, Millersville University
John Bibbo, Southwestern College
Barbara Sausen, Fresno City College
Norman Cornish, University of Detroit
Virginia Carson, DeKalb College
Jean Christenson, Linn-Benton Community College
Charles Goodall, Florida College
Thomas Green, Contra Costa College
Roseanne Hofmann, Montgomery County Community College
Billie James, University of South Dakota
Richard Langlie, North Hennepin Community College
Larry Runyan, Shoreline Community College
Richard Semmler, Northern Virginia Community College
Mark Serebransky, Camden County College
Jonathan Wilkin, Northern Virginia Community College
George Witt, Glendale Community College
Cathleen Zucco, LeMoyne College

We are grateful to Steven Blasberg of West Valley College and Firmin Widmer of Cincinnati (OH) Public Schools, for checking examples, exercises, and answers. We also wish to thank Fred Safier of City College of San Francisco, for his careful preparation of the Student's Solutions Manual accompanying this book.

Finally, our thanks to the staff at McGraw-Hill for their continued support and encouragement: Michael Johnson, Mathematics Editor, for the services he provided the authors and his guidance of the project from beginning to end; Karen Minette, Associate Mathematics Editor, for guiding the preparation of the supplements; William O'Neal for his expert manuscript editing; David Damstra, Senior Editing Supervisor, for his conscientious supervision of production; Nancy Evans, Marketing Manager, for her tireless efforts in promoting the book; and Ed Millman for his helpful review.

Raymond A. Barnett
Thomas J. Kearns

To the Student

Mastery of the material in this text is important. It will be critical for your success in all college mathematics courses and quantitative courses in other disciplines. There are several things you can do that will help you reach the goal of mastering this material.

First, approach the subject with a **positive attitude.** This text has been written to help you to understand elementary algebra and to use it to solve problems. Over the past 25 years tens of thousands of students have succeeded in algebra with the help of previous editions of the book. **You, too, can succeed!** But you must want to and believe that you can.

Second, do your **work on a regular basis.** Mathematics is not a spectator sport. You cannot learn to swim, or draw, or speak a foreign language simply by watching someone else. Similarly, you cannot learn mathematics by just reading worked examples or watching your instructor work problems. You must **work problems.** This takes time and effort. Moreover, mathematical learning is cumulative—as you progress through a subject like algebra, you continually need what you have already learned. Thus, it is very important that you keep up with assigned work. **Don't fall behind.**

Third, try the following **study process.** It will help you use this text effectively:

1. **Read** the mathematical development. Keep pencil and paper at hand while you read. Make notes. Check any details that aren't provided. Try examples of your own.

2. **Work** through the illustrative example. Try to understand each step. There will be a similar problem, called a "Matched Problem," after the example.

3. **Try** the matched problem following the example. The answer to the Matched Problem can be found at the end of the section.

4. **Review** the main ideas and any new terminology in the section. Pay particular attention to boxed material and any terms in bold type.

5. **Work** the assigned exercises at the end of the section. This is the most important part of the learning process!

There are more than enough problems in this text for you to work. Use your assignments as a guide. However, if you are having trouble, you may have to do more of the A exercises to get started. If you continue to have trouble with the problems, see your instructor. If you find assignments too easy, try more of the C exercises and check with your instructor—you may be ready for the next course, Intermediate Algebra.

Good luck!

Raymond A. Barnett
Thomas J. Kearns

Pedagogical Use of Color

Color in the text figures is used to improve clarity and understanding. Various colors are used in those graphs where different lines are being plotted simultaneously and need to be distinguished.

In addition to the figures, the text has been enhanced with color as well. We have used the following colors to distinguish the various boxes:

RULES/DEFINITIONS

PROPERTIES

STRATEGIES

1

Natural Numbers

Arithmetic involves numbers, certain operations on numbers, and problems in which these operations are used. The numbers you have probably used are whole numbers 0, 1, 2, . . . ; fractions such as $\frac{1}{2}, \frac{1}{3}, \frac{2}{3}$, . . . ; maybe the negatives of these; and perhaps other numbers such as $\sqrt{2}$ or π. The operations are the familiar operations of addition, subtraction, multiplication, division, and possibly the taking of square roots.

Algebra extends the concepts of arithmetic. In addition to specific numbers, algebra involves symbols that represent unspecified or unknown numbers. These numbers and symbols are manipulated by the same basic operations used in arithmetic.

When arithmetic is extended to algebra, a wider range of problems can be attacked. For example, you need only arithmetic to solve the following problem:

A rectangular field is 110 yards long and 65 yards wide. What is its area?

The area is 7,150 square yards and is found by multiplying 110 by 65. A problem only slightly different, however, can make good use of an algebraic approach:

A rectangular field is twice as long as it is wide and has an area of 6,962 square yards. What are its dimensions?

The answer, that the field is 59 yards wide by 118 yards long, is unlikely to be found quickly by guessing and arithmetic. It is, however, easily found by using simple algebra.

In this text you will encounter not only the basic ideas of algebra, but also many practical problems that are solved using these ideas. Since algebra uses many kinds of numbers, and symbols for these numbers, it is important that we go back and

Photo reference: see Exercise 1-3, Problems 65, 66.

take a careful look at some of the properties of numbers that you may have previously encountered. We begin in this chapter using only the counting numbers 1, 2, 3, ... Algebraic ideas and methods will be extended to the integers . . . , −2, −1, 0, 1, 2, . . . in Chapter 2 and to fractions and the rational numbers in Chapter 3.

1-1   The Natural Numbers

♦   The Set of Natural Numbers
♦   Important Subsets of the Set of Natural Numbers
♦   Least Common Multiple

We begin our development of algebra using only the simplest set of numbers, the **counting numbers** 1, 2, 3, ... These numbers are also referred to as the set of **natural numbers,** and the two names can be used interchangeably.

♦   THE SET OF NATURAL NUMBERS

The word "set" here and throughout the text will be used, as it is used in everyday language, to mean a "collection." We want the collection to have the property that for any given object, it is either in the set or it is not. The word "subset" will also be used informally to refer to part, or possibly all, of a set, much like the relationship of a subcommittee to a committee. We will often represent a set by listing its **elements** between braces { } or by giving it a capital letter name. The set of natural numbers will be represented by the letter N:

$$N = \{1, 2, 3, \ldots\}$$ Natural or counting numbers

The three dots tell us that the numbers go on without end, following the pattern indicated by the first three numbers. This is a useful way to represent certain infinite sets. A set is called a **finite set** if its elements can be counted and the counting process ends; otherwise, it is an **infinite set.**

Example 1 **Recognizing** **Natural Numbers**	Select the natural numbers out of the following list: $\frac{2}{3}$, 1, $\sqrt{2}$, π, 5, 7.63, 17, $83\frac{7}{8}$, 610
Solution	1, 5, 17, and 610 are natural numbers.

Matched Problems

Each example in this text is followed by a similar problem, called a *matched problem,* for you to work. The answers to the matched problems are found at the end of the section, just before the exercise set.

Matched Problem 1   Select the natural numbers out of the following list:

$$4, \tfrac{3}{4}, 19, 305, 4\tfrac{2}{3}, 7.32, \sqrt{3}$$

Assumption

We assume that you know what natural numbers are, how to add and multiply them, and how to subtract and divide them when the result is a natural number.

You may recall that the results of addition, multiplication, subtraction, and division of numbers are called the **sum, product, difference,** and **quotient,** respectively.

♦ **IMPORTANT SUBSETS OF THE SET OF NATURAL NUMBERS**

The set of natural numbers can be separated into two subsets called even numbers and odd numbers. A natural number is an **even number** if it is exactly divisible by 2, that is, divisible without a remainder. A natural number is an **odd number** if it is not exactly divisible by 2.

Example 2
Even and Odd Numbers

Separate the set of natural numbers $\{1, 2, 3, \ldots\}$ into even and odd numbers.

Solution   The set of even numbers: $\{2, 4, 6, \ldots\}$

The set of odd numbers: $\{1, 3, 5, \ldots\}$

Matched Problem 2   Separate the following set into even and odd numbers:

$$\{8, 13, 7, 32, 57, 625, 532\}$$

Two or more numbers that are to be added or subtracted are called **terms;** two or more numbers that are to be multiplied are called **factors.** The product is a **multiple** of its factors.

Terms			**Factors**		
↓	↓	↓	↓	↓	↓
3	+ 5	+ 8	3	× 5	× 8

In algebra and higher mathematics, parentheses ( ) or the dot "·" are usually used in place of the times sign ×, since the times sign is easily confused with the letter x. In algebra x is frequently used to represent an unknown number. Thus,

$$3 \times 5 \times 8 \qquad (3)(5)(8) \qquad 3 \cdot 5 \cdot 8$$

all represent the product of 3, 5, and 8.

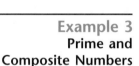

The natural numbers, excluding 1, can also be separated into two other important subsets called composite numbers and prime numbers. A natural number is a **composite number** if it can be rewritten as a product of two or more natural numbers other than itself and 1 (8 is a composite number, since $8 = 2 \cdot 4$). Stated in a different but equivalent way, a natural number is a composite number if it can be divided exactly (no remainder) by a natural number other than itself and 1. The natural number 9 is a composite number, for example, since it is exactly divisible by 3. A natural number, excluding 1, is a **prime number** if it is not a composite number. The natural number 11 is a prime number, for example, since it cannot be divided exactly by any natural number other than itself or 1. Equivalently, a number is prime if its only factors are 1 and itself. The number 1 is defined to be neither prime nor composite. The natural number 2 is the only even prime number. It can be proved that there are infinitely many odd prime numbers.

Example 3 **Prime and** **Composite Numbers**	Separate the set $\{2, 3, 4, \ldots, 18, 19\}$ into prime and composite numbers.

Solution Check each number for factors other than 1 and itself. We find that the numbers 4, 6, 8, 9, 10, 12, 14, 15, 16, and 18 are composite, because they can be factored:

$$4 = 2 \cdot 2 \qquad 6 = 2 \cdot 3 \qquad 8 = 2 \cdot 4 \qquad 9 = 3 \cdot 3$$
$$10 = 2 \cdot 5 \qquad 14 = 2 \cdot 7 \qquad 15 = 3 \cdot 5$$
$$12 = 2 \cdot 6 \qquad 16 = 2 \cdot 8 \qquad 18 = 2 \cdot 9$$
$$ = 3 \cdot 4 \qquad = 4 \cdot 4 \qquad = 3 \cdot 6$$

The remaining numbers 2, 3, 5, 7, 11, 13, 17, and 19 are prime.

Matched Problem 3 Separate the set $\{6, 9, 11, 21, 23, 25, 27, 29\}$ into prime and composite numbers.

A fundamental theorem of arithmetic states that every composite number has a unique set of prime factors. These factors may be arranged in any order. A natural number written as a product solely of prime factors is said to be **completely factored.**

Example 4 **Completely** **Factored Form**	Write each number in a completely factored form: **(A)** 8 **(B)** 36 **(C)** 60

Solution **(A)** $8 = 2 \cdot 4 = 2 \cdot 2 \cdot 2$ We continue factoring until we can go no further.

(B) $36 = \underbrace{6 \cdot 6}_{\text{36 factored}} = \underbrace{2 \cdot 3 \cdot 2 \cdot 3}_{\text{36 completely factored}}$

or $\qquad\qquad\qquad 36 = 4 \cdot 9 = 2 \cdot 2 \cdot 3 \cdot 3$

or $\qquad\qquad 36 = 3 \cdot 12 = 3 \cdot 4 \cdot 3 = 3 \cdot 2 \cdot 2 \cdot 3$

or $\qquad\qquad 36 = 2 \cdot 18 = 2 \cdot 2 \cdot 9 = 2 \cdot 2 \cdot 3 \cdot 3$

All four ways in which we factored 36 originally lead to the same set of prime factors: two 2's and two 3's. The order in which the factors are written makes no difference; for convenience we usually write them in numerical order.

(C) $60 = 10 \cdot 6 = 2 \cdot 5 \cdot 2 \cdot 3 = 2 \cdot 2 \cdot 3 \cdot 5$

It may be easier to see the factorizations in (A), (B), and (C) schematically:

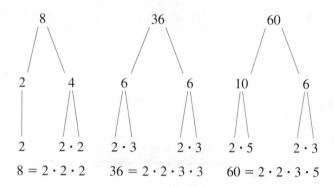

$$8 = 2 \cdot 2 \cdot 2 \qquad 36 = 2 \cdot 2 \cdot 3 \cdot 3 \qquad 60 = 2 \cdot 2 \cdot 3 \cdot 5$$

Matched Problem 4   Write each number in completely factored form:

(A) 12   **(B)** 26   **(C)** 72

In factoring natural numbers it is easiest to look for small prime factors first. Numbers with 2 or 5 as factors are easily recognized: numbers that are divisible by 2 are those that end in the digit 0, 2, 4, 6, or 8; numbers that end in 0 or 5 are the only ones divisible by 5. Thus, 46 is divisible by 2 and 65 is divisible by 5. A number is exactly divisible by 3 when the sum of its digits is divisible by 3. For example, 51 and 177 are each divisible by 3, since $5 + 1 = 6$ and $1 + 7 + 7 = 15$ are each divisible by 3:

$$51 = 3 \cdot 17 \qquad \text{and} \qquad 177 = 3 \cdot 59$$

♦ **LEAST COMMON MULTIPLE**

We can use the prime factors of numbers to aid us in finding the least common multiple (LCM) of two or more natural numbers, a process you will need to know later when dealing with fractions and certain types of equations.

The **least common multiple** of two or more natural numbers is defined to be the smallest natural number exactly divisible by each of the numbers. Often one can find the LCM by inspection. For example, the LCM of 3 and 4 is 12, since 12 is the smallest natural number exactly divisible by 3 and 4. But what is the LCM of 15 and 18?

One way to proceed is to list the multiples of 15 and the multiples of 18, and then take the smallest number that occurs on both lists:

Multiples of 15: 15, 30, 45, 60, 75, 90, 105, 120, . . .

Multiples of 18: 18, 36, 54, 72, 90, 108, 126, . . .

We see that 90 is the LCM of 15 and 18. For larger numbers, however, this process is impractical. The following method is more useful in many algebraic problems.

To find the LCM of 15 and 18, we start by writing 15 and 18 in completely factored forms:

$$15 = 3 \cdot 5 \qquad 18 = 2 \cdot 9 = 2 \cdot 3 \cdot 3$$

The different prime factors are 2, 3, and 5. The most times that 2 appears in either factorization is once; the most that 3 appears in either factorization is twice; and the most that 5 appears is once. The LCM will contain one 2, two 3's, and one 5:

$$\text{LCM of 15 and 18} = 2 \cdot 3 \cdot 3 \cdot 5 = 90$$

so 90 is the smallest natural number exactly divisible by 15 and 18.

The method is summarized as follows:

Finding the LCM

1. Factor each number completely.
2. Identify the different prime factors.
3. The LCM contains each different prime factor the most number of times it appears in any factorization of the original numbers.

Example 5
**Finding
the LCM**

Find the LCM for 8, 6, and 9.

Solution   First, write each number as a product of prime factors:

$$8 = 2 \cdot 2 \cdot 2 \qquad 6 = 2 \cdot 3 \qquad 9 = 3 \cdot 3$$

The different prime factors are 2 and 3. The most that 2 appears in any one factorization is three times, and the most that 3 appears in any one factorization is twice. The LCM will therefore contain three 2's and two 3's:

$$\text{LCM} = 2 \cdot 2 \cdot 2 \cdot 3 \cdot 3 = 72$$

so 72 is the smallest natural number exactly divisible by 8, 6, and 9.

Matched Problem 5   Find the LCM for 10, 12, and 15.

Answers to   **1.** 4, 19, 305
Matched Problems   **2.** Even: 8, 32, 532; odd: 13, 7, 57, 625
    3. Composite: 6, 9, 21, 25, 27; prime: 11, 23, 29
    4. !2 = 2 · 2 · 3; 26 = 2 · 13; 72 = 2 · 2 · 2 · 3 · 3
    5. LCM = 2 · 2 · 3 · 5 = 60

EXERCISE 1-1

The exercise sets in this text are divided according to diffi-culty into A, B, and C groupings. The A problems are mostly easy and routine. The B problems are less routine but still emphasize the mechanics of the material in the section. The C problems are more challenging and present a mixture of more complicated mechanics and a few prob-lems of a more theoretical nature.

A *Select the natural numbers out of each list:*

1. 6, 13, 3.5, $\frac{2}{3}$ **2.** 4, $\frac{1}{8}$, 22, 6.5

3. $3\frac{1}{2}$, 67, 402, 22.35   **4.** 203.17, 63, $\frac{33}{5}$, 999

Separate each list into even and odd integers:

5. 9, 14, 28, 33   **6.** 8, 24, 1, 41

7. 23, 105, 77, 426   **8.** 68, 530, 421, 72

Separate each list into composite and prime numbers:

9. 2, 6, 9, 11   **10.** 3, 4, 7, 15

11. 12, 17, 23, 27   **12.** 16, 19, 25, 39

13. 5, 13, 21, 29, 34   **14.** 29, 39, 49, 59

15. 37, 47, 57, 67   **16.** 31, 41, 51, 61

B *Let M be the set of natural numbers from 20 to 30 and N be the set of natural numbers from 40 to 50. List the fol-lowing:*

17. Even numbers in *M*   **18.** Even numbers in *N*

19. Odd numbers in *M*   **20.** Odd numbers in *N*

21. Composite numbers in *M*

22. Composite numbers in *N*

23. Prime numbers in *M*   **24.** Prime numbers in *N*

Write each of the following composite numbers as a prod-uct of prime factors:

25. 10   **26.** 21   **27.** 30   **28.** 90

29. 40   **30.** 42   **31.** 56   **32.** 63

33. 84   **34.** 72   **35.** 60   **36.** 120

37. 108   **38.** 112   **39.** 210   **40.** 252

Find the LCM for each group of numbers:

41. 9, 12   **42.** 9, 15   **43.** 6, 16

44. 12, 16   **45.** 3, 8, 12   **46.** 4, 6, 18

47. 4, 10, 15   **48.** 10, 12, 9   **49.** 10, 15, 18

50. 35, 66   **51.** 98, 110   **52.** 20, 33, 75

53. 42, 63, 90   **54.** 16, 40, 130

C *In Problems 55–68, write each of the numbers as a prod-uct of prime factors:*

55. 273   **56.** 286   **57.** 560   **58.** 910

59. 180   **60.** 200   **61.** 300   **62.** 450

63. 630   **64.** 1,050   **65.** 2,200   **66.** 1,309

67. 1,386   **68.** 1,708

69. Is every even number a prime number? Is every odd number a prime number? Is every prime number an odd number? Is every prime number except 2 an odd number?

70. Is every even number a composite number? Is every odd number a composite number? Is every even number except 2 a composite number?

Tell whether the set is finite or infinite:

71. The set of natural numbers between 1 and 1 million

72. The set of even numbers between 1 and 1 million

73. The set of all natural numbers

74. The set of all even numbers

75. The set of all the grains of sand on all the beaches in the world

76. The set of all U.S. citizens

Find the LCM for each group of numbers:

77. 35, 56, 100, 140   **78.** 10, 44, 110, 242

79. 30, 105, 385, 1155   **80.** 70, 165, 210, 231

81. 154, 220, 330, 462   **82.** 78, 105, 195, 420

*The **greatest common factor** (GCF) of two or more num-bers is the largest number that is a factor of each number. For example, the GCF of 18 and 24 is 6. The GCF can be*

found in almost the same way as the LCM: factor each number completely and identify the prime factors; the GCF contains each prime factor the fewest (rather than the most) number of times it appears in the factorizations of the original numbers. If a given prime is not a factor of each number, it is not included in the GCF. If no prime number is a common factor of each, we say the GCF is 1. Since $18 = 2 \cdot 3 \cdot 3$ and $24 = 2 \cdot 2 \cdot 2 \cdot 3$, the GCF contains one 2 and one 3. That is, the GCF is 6. Find the GCF for each group of numbers:

83. 28, 49   **84.** 56, 85   **85.** 60, 96

86. 45, 63   **87.** 21, 50   **88.** 70, 84

89. 39, 65   **90.** 63, 75   **91.** 18, 45, 66

92. 12, 30, 56

For each group of numbers, find the product, the LCM, the GCF, and the product of the LCM and GCF. Do you notice a pattern when there are only two numbers in the group? Does the pattern hold true when there are more than two?

93. 12, 16   **94.** 6, 16   **95.** 35, 66

96. 98, 110   **97.** 70, 84   **98.** 63, 75

99. 9, 10, 12   **100.** 4, 10, 15

1-2   The Transition from Arithmetic to Algebra

- ♦ Variables and Constants
- ♦ Algebraic Expressions
- ♦ Evaluating Algebraic Expressions
- ♦ From English to Algebra

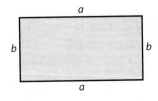

Figure 1

Consider the statement: "The perimeter of a rectangle is twice its length plus twice its width" and look at the accompanying figure.

 If we let

$$P = \text{Perimeter} \qquad a = \text{Length} \qquad b = \text{Width}$$

then the familiar formula

$$P = 2 \cdot a + 2 \cdot b \tag{1}$$

has the same meaning as the original statement, but with increased clarity and a substantial reduction in the number of symbols used. In the perimeter formula [Equation (1)], the three letters P, a, and b can be replaced with an unlimited number of different numbers, depending on the size of the rectangle. Hence, these letters are called *variables*. The symbol "2" names only one number and is consequently called a *constant*.

♦ VARIABLES AND CONSTANTS

In general, a **constant** is any symbol that names one particular number. A **variable** is a symbol that can be replaced by a number from a set containing more than one number. This latter set is called the **replacement set** for the variable. It is important to remember that the variable represents an unspecified number in the replacement set. Therefore, any operations that we can perform with an arbitrary number in the replacement set can also be done with the variable.

When variables are involved in multiplication, the dot "$\cdot$" is usually omitted. Thus "$2a$" means "2 times a," "$2b$" means "2 times b," and the perimeter formula, Equation (1), is usually written in the form $P = 2a + 2b$.

Example 1
Variables and Constants

List the constants and variables in the formula

$$F = \tfrac{9}{5}C + 32$$

for the conversion of Celsius degrees (C) to Fahrenheit degrees (F).

Solution   The constants are $\tfrac{9}{5}$ and 32, the variables C and F.

Matched Problem 1   List the constants and variables in each formula:

(A) $P = 4s$ Perimeter of a square
(B) $A = s^2$ Area of a square [*Note:* $s^2 = s \cdot s$]

A number of useful formulas such as the ones used above are listed inside the back cover of the text for convenient reference.

The introduction of variables into mathematics occurred about A.D. 1600. A French mathematician, François Vièta (1540–1603), is credited as being mainly responsible for this new idea. Many mark this point as the beginning of modern mathematics.

◆ **ALGEBRAIC EXPRESSIONS**

An **algebraic expression** is an expression built up from constants; variables; mathematical operations such as addition, subtraction, multiplication, and division (other operations will be added later); and grouping symbols such as parentheses ( ), brackets [ ], and braces { }. For example,

$$8 + 7 \qquad 3 \cdot 5 - 6 \qquad 12 - 2(8 - 5)$$
$$3x - 5y \qquad 8(x - 3y) \qquad 3\{x - 2[x + 4(x + 3)]\}$$

are all algebraic expressions.

Just as with numbers themselves, two or more algebraic expressions joined by plus (+) or minus (−) signs are called **terms**. Two or more algebraic expressions joined by multiplication are called **factors**. Here are some examples:

$5 - 2 \cdot 3$ Two terms: 5 and $2 \cdot 3$; second term has two factors: 2 and 3

$10 + 2(6 - 3)$ Two terms: 10 and $2(6 - 3)$; second term has two factors: 2 and $(6 - 3)$

$x + 3$ Two terms: x and 3

$2x + 3y - 6z$ Three terms: $2x$, $3y$, and $6z$; each term has two factors

$5[x - 3(x + 5)]$ One term: the whole thing; two factors: 5 and $[x - 3(x + 5)]$; second factor has two terms: x and $3(x + 5)$

◆ EVALUATING ALGEBRAIC EXPRESSIONS

When evaluating numerical expressions involving various operations and grouping symbols, we follow this convention:

Order of Operations

1. Simplify inside the innermost grouping first, then the next innermost, and so on.
2. Unless grouping symbols indicate otherwise, perform multiplication and division before addition and subtraction. In either case, we proceed from left to right.

For example, to illustrate rule 1:

$$2[3 + (5 - 1)]$$ First subtract 1 from 5.
$$= 2[3 + 4]$$ Then add 3 to the result.
$$= 2[7]$$ Multiply this result by 2.
$$= 14$$

To illustrate rule 2:

$$2 + 3 \cdot 5$$ First multiply 3 and 5.
$$= 2 + 15$$ Add 2 to the result.
$$= 17$$
$$2 + 3 \cdot 5 \neq 5 \cdot 5$$ $2 + 3 \cdot 5 = 2 + 15 = 17$

CAUTION

To further illustrate rule 2, we can add parentheses as grouping symbols and thereby change what the expression means:

$$(2 + 3) \cdot 5 = 5 \cdot 5 = 25$$ Here the ( ) tell us to do the addition $2 + 3$ before multiplying.

The grouping symbols [ ] and { } may be used in place of parentheses ( ). They mean the same thing.

Example 2
Order of Operations

Evaluate each expression:

(A) $8 - 2 \cdot 3$ (B) $9 - 2(5 - 3)$
(C) $(9 - 2)(5 - 3)$ (D) $2[12 - 3(8 - 5)]$

Solution

(A) $8 - 2 \cdot 3 = 8 - 6 = 2$ Multiplication precedes subtraction.
(B) $9 - 2(5 - 3) = 9 - 2 \cdot 2$ Perform operation inside parentheses first, then multiply.
$$= 9 - 4$$ Subtract.
$$= 5$$
(C) $(9 - 2)(5 - 3) = 7 \cdot 2$ Parentheses first, then multiply.
$$= 14$$ Note how parts (B) and (C) differ.

(D) $2[12 - 3(8 - 5)] = 2[12 - 3 \cdot 3]$ Parentheses first.
$\qquad\qquad\qquad\quad = 2(12 - 9)$ Parentheses ( ) can always replace
$\qquad\qquad\qquad\quad = 2 \cdot 3$ brackets [ ] and braces { } when
$\qquad\qquad\qquad\quad = 6$ the latter are used as symbols of
$\qquad\qquad\qquad\qquad\qquad\qquad\qquad$ grouping.

Matched Problem 2   Evaluate each expression:

     (A) $2 \cdot 10 - 3 \cdot 5$ **(B)** $11 - 3(7 - 5)$
     (C) $(11 - 3)(7 - 5)$ **(D)** $6[13 - 2(14 - 8)]$

     To evaluate an algebraic expression when number values of the variables are given means to replace the variables by the given values and evaluate the resulting numerical expression. Replacing the variables by numbers is often called **substituting** the numbers for the variables. The rules on order of operation given for numerical expressions apply also to algebraic expressions, since the variables represent numbers.

Example 3
Evaluating Algebraic
Expressions

Evaluate each algebraic expression for $x = 10$ and $y = 3$:

     (A) $2x - 3y$ **(B)** $x - 3(2y - 4)$
     (C) $(x - 3)(2y - 4)$ **(D)** $5[32 - x(x - 7)]$

Solution   Substitute $x = 10$ and $y = 3$ into each expression and then evaluate, following the order of operations stated above.

     (A) $2x - 3y$
$\qquad\quad 2(10) - 3(3) = 20 - 9 = 11$
     (B) $x - 3(2y - 4)$
$\qquad\quad 10 - 3(2 \cdot 3 - 4) = 10 - 3(6 - 4)$
$\qquad\qquad\qquad\qquad\qquad = 10 - 3 \cdot 2$
$\qquad\qquad\qquad\qquad\qquad = 10 - 6$
$\qquad\qquad\qquad\qquad\qquad = 4$
     (C) $(x - 3)(2y - 4)$
$\qquad\quad (10 - 3)(2 \cdot 3 - 4) = 7(6 - 4)$ Note how parts (B) and (C) differ.
$\qquad\qquad\qquad\qquad\qquad\quad = 7 \cdot 2$
$\qquad\qquad\qquad\qquad\qquad\quad = 14$
     (D) $5[32 - x(x - 7)]$
$\qquad\quad 5[32 - 10(10 - 7)] = 5(32 - 10 \cdot 3)$
$\qquad\qquad\qquad\qquad\qquad\quad = 5(32 - 30)$
$\qquad\qquad\qquad\qquad\qquad\quad = 5 \cdot 2$
$\qquad\qquad\qquad\qquad\qquad\quad = 10$

Matched Problem 3   Evaluate each algebraic expression for $x = 12$ and $y = 3$:

     (A) $x - 3y$ **(B)** $x - 4(y - 1)$
     (C) $(x - 4)(y - 1)$ **(D)** $3[x - 2(x - 9)]$

Example 4 **A Rate-Time Problem**	How far can you travel in 13 hours at 37 kilometers per hour? Use the formula "distance equals rate times time," $d = rt$.

Solution   Here we understand:

"How far" asks for distance d.

Thirteen hours gives time t.

Thirty-seven kilometers per hour gives the rate (or speed) r.

We use the formula $d = rt$ with the variable r replaced by 37 and the variable t replaced by 13 to obtain

$$d = 37 \cdot 13 = 481 \text{ kilometers}$$

Matched Problem 4   How many gallons can a water pump pump in 8 minutes if it pumps at a rate of 15 gallons per minute? Use $Q = rt$, where Q represents quantity, r rate, and t time.

♦   **FROM ENGLISH TO ALGEBRA**

English is a complex language, and it is impossible to summarize briefly all the English phrases that can be translated into algebraic or arithmetic expressions. Table 1 lists some of the more common ones that you will find in problems throughout your study of algebra.

Table 1

Phrase	Example	Algebraic Expression
Addition		
"The sum of"	The sum of 3 and 5	$3 + 5$
"Plus"	7 plus 4	$7 + 4$
"Added to"	2 added to 6	$6 + 2$
"More than"	8 more than 4	$4 + 8$
Subtraction		
"The difference of"	The difference of 5 and 3	$5 - 3$
"Minus"	7 minus 4	$7 - 4$
"Subtracted from"	2 subtracted from 6	$6 - 2$
"Less than"	4 less than 8	$8 - 4$
Multiplication		
"The product of"	The product of 5 and 3	$5 \cdot 3$
"Times"	7 times 4	$7 \cdot 4$
"Multiplied by"	2 multiplied by 6	$2 \cdot 6$
"Of"	Half of 10	$\frac{1}{2} \cdot 10$
"Times the quantity"	2 times the quantity 3 plus 4	$2(3 + 4)$
Division		
"The quotient of"	The quotient of 6 and 3	$6 \div 3$ or $6/3$ or $\frac{6}{3}$
"Over"	8 over 4	$8 \div 4$ or $8/4$ or $\frac{8}{4}$
"Divided by"	10 divided by 2	$10 \div 2$ or $10/2$ or $\frac{10}{2}$

Example 5
Writing Algebraic Expressions

If x represents a natural number, write an algebraic expression that represents the numbers described:

(A) A number 3 times as large as x
(B) A number 3 more than x
(C) A number 7 less than the product of 4 and x
(D) A number 3 times the quantity 2 less than x

Solution

(A) $3x$ "Times" corresponds to "multiply."
(B) $x + 3$ "More than" corresponds to "added to."
(C) $4x - 7$ (not $7 - 4x$)   "Less than" corresponds to "subtracted from."
(D) $3(x - 2)$ [not $3(2 - x)$ and also not $3x - 2$]

Matched Problem 5

If y represents a natural number, write an algebraic expression that represents the numbers described:

(A) A number 7 times as large as y
(B) A number 7 less than y
(C) A number 9 more than the product of 4 and y
(D) A number 5 times the quantity 4 less than y

Answers to Matched Problems

1. (A) Constant: 4; variables: P, s **(B)** Constant: 2; variables: A, s
2. (A) 5   **(B)** 5   **(C)** 16   **(D)** 6
3. (A) 3   **(B)** 4   **(C)** 16   **(D)** 18
4. 120 gallons   **5. (A)** $7y$ **(B)** $y - 7$ **(C)** $4y + 9$ **(D)** $5(y - 4)$

EXERCISE 1-2

A *Evaluate each expression:*

1. $7 + 3 \cdot 2$

2. $5 + 6 \cdot 3$

3. $8 - 2 \cdot 3$

4. $20 - 5 \cdot 3$

5. $7 \cdot 6 - 5 \cdot 5$

6. $8 \cdot 9 - 6 \cdot 11$

7. $(2 + 9) - (3 + 6)$

8. $(8 - 3) + (7 - 2)$

9. $8 + 2(7 + 1)$

10. $3 + 8(2 + 5)$

11. $(8 + 2)(7 + 1)$

12. $(3 + 8)(2 + 5)$

13. $10 - 3(7 - 4)$

14. $20 - 5(12 - 9)$

15. $(10 - 3)(7 - 4)$

16. $(20 - 5)(12 - 9)$

17. $12 - 2(7 - 5)$

18. $15 - 3(9 - 5)$

Evaluate each algebraic expression for $x = 8$ and $y = 3$:

19. $x + 2$

20. $y + 5$

21. $x - y$

22. $22 - x$

23. $x - 2y$

24. $6y - x$

25. $3x - 2y$

26. $9y - xy$

27. $y + 3(x - 5)$

28. $5 + y(x - y)$

29. $x - 2(y - 1)$

30. $x - y(x - 7)$

If x and y represent natural numbers, write an algebraic expression that represents each of the numbers described:

31. A number 5 times as large as x

32. A number 7 times as large as y

33. A number 5 more than x

34. A number 12 more than y

35. A number 5 less than x

36. A number 8 less than y

37. A number x less than 5

38. A number y less than 8

B *Identify the constants and variables in each algebraic expression:*

39. $A = \frac{1}{2}bh$ *Area of a triangle*

40. $A = ab$ *Area of a rectangle*

41. $d = rt$ *Distance-rate-time formula*

42. $C = \frac{5}{9}(F - 32)$ *Fahrenheit-Celsius formula*

43. $I = Prt$ *Simple interest*

44. $A = P(1 + rt)$ *Simple interest*

45. $y = 2x + 3$ **46.** $3x + 2y = 5$

47. $3(u + v) + 2u$ **48.** $2(x + 1) + 3(w + 5z)$

Find the area and perimeter for each rectangle ($A = ab$ and $P = 2a + 2b$). Example: If $a = 5$ meters and $b = 3$ meters, then $A = 5 \cdot 3 = 15$ square meters and $P = 2 \cdot 5 + 2 \cdot 3 = 10 + 6 = 16$ meters.

49. $a = 6$ centimeters, $b = 3$ centimeters

50. $a = 12$ feet, $b = 4$ feet

51. $a = 10$ kilometers, $b = 8$ kilometers

52. $a = 9$ meters, $b = 6$ meters

Evaluate each expression:

53. $4[15 - 10(9 - 8)]$ **54.** $6[22 - 3(13 - 7)]$

55. $7 \cdot 9 - 6(8 - 3)$ **56.** $5(8 - 3) - 3 \cdot 6$

57. $2[(7 + 2) - (5 - 3)]$ **58.** $6[(8 - 3) + (4 - 2)]$

In Problems 59–66, evaluate each expression for $w = 2$, $x = 5$, $y = 1$, and $z = 3$:

59. $w(y + z)$ **60.** $wy + wx$

61. $wy + z$ **62.** $y + wz$

63. $(z - y) + (z - w)$ **64.** $4(y + w) - 2z$

65. $2[x + 3(z - y)]$ **66.** $6[(x + z) - 3(z - w)]$

67. How far can you travel in 12 hours at 57 kilometers per hour? ($d = rt$)

68. How far can you travel in 9 hours at 43 kilometers per hour? ($d = rt$)

69. How many words can a typist type in 10 minutes if he or she can type 60 words per minute? ($Q = rt$)

70. How many gallons can a pipe fill in 20 minutes if it fills at the rate of 10 gallons per minute? ($Q = rt$)

If x represents a given natural number, write an algebraic expression that represents each of the numbers described:

71. A number 3 more than twice the given number

72. A number 3 more than the product of 12 and the given number

73. A number 3 less than the product of 12 and the given number

74. A number 3 less than twice the given number

75. A number 3 times the quantity 8 less than the given number

76. A number 6 times the quantity 4 less than the given number

77. A number that is twice as large as 4 more than a given number

78. A number that is twice the quantity that is 2 less than a given number

C *Evaluate each expression:*

79. $3[(6 - 4) + 4 \cdot 3 + 3(1 + 6)]$

80. $2[(3 + 2) + 2(7 - 4) + 6 \cdot 2]$

81. $2\{26 - 3[12 - 2(8 - 5)]\}$

82. $5\{32 - 5[(10 - 2) - 2 \cdot 3]\}$

83. $6 - \{4 - [3 - (7 - 5)]\}$

84. $1 + (4 - 3\{4 - 3[5 - (1 + 3)]\})$

85. $3[14 - 4(5 - 2)] - \{4 - [8 - 2(5 - 2)]\}$

86. $10 + 9\{8 - 7[6 - 5(4 - 3)] + 2\} - 1$

Evaluate for $u = 2$, $v = 3$, $w = 4$, and $x = 5$:

87. $2\{w + 2[7 - (u + v)]\}$

88. $3\{(u + v) + 3[x - 2(w - u)] + uv\}$

89. $(u + v)(w + x) + \{x - [w - (v - u)]\}$

90. $(v - u)(x - w) + \{x - [u + (w - v)]\}$

91. $2uv[x + (w - u)](u + v)$

92. $(x - u)\{w + v[x - (v - u)]\}$

If t represents an even number, write an algebraic expression that represents each of the numbers described:

93. A number 3 times the first even number larger than t

94. A number 5 times the first even number smaller than t

95. The sum of three consecutive natural numbers starting with t

96. The product of three consecutive natural numbers starting with t

97. The sum of three consecutive even numbers starting with t

98. The sum of three consecutive odd numbers following t

99. The product of three consecutive odd integers following t

100. The product of three consecutive even numbers starting with t

1-3 Equality and Inequality

♦ Equality Sign
♦ Algebraic Equations
♦ Properties of Equality
♦ Inequality

In the preceding sections the equality or equal sign (=) was used in a number of places. You are probably most familiar with its use in formulas such as

$$d = rt \qquad A = ab \qquad I = Prt$$

The equal sign is very important in mathematics, and you will be using it frequently. Its mathematical meaning, however, is not so obvious as it at first might seem. It is important you learn to use it correctly from the beginning.

♦ EQUALITY SIGN

An **equality sign (=)** between two expressions asserts that the two expressions are names or descriptions of exactly the same object. The **inequality sign ($\neq$)** means "is not equal to." Statements involving the use of an equality or inequality sign may be true or they may be false.

$18 - 3 = 5 \cdot 3$ True statement $7 \cdot 8 = 15$ False statement

$11 - 7 \neq \dfrac{8}{2}$ False statement $7 + 8 \neq 16$ True statement

The equality sign was not used until rather late in history—the sixteenth century. It was introduced by the English mathematician Robert Recorde (1510–1558).

♦ ALGEBRAIC EQUATIONS

If two algebraic expressions involving at least one variable are joined with an equal sign, the result is called an **algebraic equation.** The following are algebraic equations in one or more variables:

$$x + 3 = 8 \qquad x + 2 = 2 + x \qquad 2x + 3y = 12$$

Formulas such as

$$P = 2a + 2b \qquad F = \tfrac{9}{5}C + 32 \qquad P = 4s$$

are also algebraic equations. Since variables represent unspecified numbers, an algebraic equation is neither true nor false as it stands. It does not become true or false until the variables have been replaced by numbers. The set of all numbers that make an equation become true when they replace the variables is called the **solution set** of the equation.

Formulating algebraic equations is an important step in solving certain types of practical problems using algebraic methods. We address this topic now.

Example 1
Writing Equations

Translate each statement into an algebraic equation using only one variable:

(A) Fifteen is 9 more than a certain number.
(B) Three times a certain number is 7 less than twice the number.

Solution   **(A)** We let the variable x represent the unknown "certain number." The expression "9 more than a certain number" becomes $9 + x$. We translate "is" to an equal sign and obtain

$$15 \underbrace{\text{is}}\ 9 \underbrace{\text{more than}}\ \underbrace{\text{a certain number}}$$
$$15 = 9 \quad\quad + \quad\quad\quad\quad x$$

Thus, $\qquad\qquad 15 = 9 + x \qquad$ or $\qquad 15 = x + 9$

Note that $9 + x$ and $x + 9$ represent the same number. This property of addition will be discussed further in the next section.

(B)

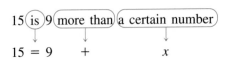

$$3 \underbrace{\text{times}}\ \underbrace{\text{a certain number}}\ \underbrace{\text{is}}\ 7 \underbrace{\text{less than}}\ \underbrace{\text{twice the number}}$$
$$3 \quad\cdot\quad\quad\quad x \quad\quad\quad = 7 \text{ subtracted from } 2x$$
$$3x = 2x - 7 \qquad (\text{not } 7 - 2x)$$

Compare the symbolic and verbal forms on the right side carefully and note how 7 and $2x$ reverse. Note that $2x - 7$ and $7 - 2x$ in general do not represent the same number.

Matched Problem 1   Translate each statement into an algebraic equation using only one variable:

(A) Eight is 5 more than a certain number. $8 = 5 + x$
(B) Six is 4 less than a certain number. $6 = x - 4$
(C) Four times a certain number is 3 less than twice the number. $4 \cdot x = 2x - 3$
(D) If 12 is added to a certain number, the sum is twice a number that is 3 less than the certain number. $x + 12 = 2(x - 3)$

Example 2
Writing Equations

If x represents an even number, write an algebraic equation that is equivalent to "the sum of three consecutive even numbers is 84."

Solution   The set of even numbers is

$$\{2, 4, 6, \ldots, x, \ldots\}$$

where x is an unspecified even number. We note that even numbers increase by twos; hence,

$$x \qquad x + 2 \qquad \text{and} \qquad x + 4$$

represent three consecutive (one following the other) even numbers, starting with x. Note that $x + 4 = (x + 2) + 2$. Thus,

$$\left(\begin{array}{c}\text{The sum of three}\\\text{consecutive even numbers}\end{array}\right)\text{ is } 84$$

$$x + (x + 2) + (x + 4) = 84$$

Matched Problem 2 If x represents an odd number, write an algebraic equation that is equivalent to "the sum of four consecutive odd numbers is 160." [*Hint:* Odd numbers increase by twos, just as even numbers do.] $x + (x+3) + (x+5) = 160$

♦ **PROPERTIES OF EQUALITY**

From the logical meaning of the equality sign, we can establish a number of rules or properties for its use. We state these properties below and will use them later when we discuss and solve equations.

Properties of Equality

If a, b, and c are names of objects, then

1. $a = a$ Reflexive property
2. If $a = b$, then $b = a$.   Symmetric property
3. If $a = b$ and $b = c$, then $a = c$.   Transitive property
4. If $a = b$, then either may replace the other in any
   statement without changing the truth or falsity of
   the statement.   Substitution principle

The symmetric property allows, for example,

$$5 = x$$

to be rewritten as $$x = 5$$

and $$A = P + Prt$$

to be rewritten as $$P + Prt = A$$

The transitive property allows us to conclude, for example, that if

$$P = 2a + 2b \quad\text{and}\quad 2a + 2b = 208$$

then $$P = 208$$

The substitution property allows us to conclude, for example, that if

$$x = 3 + y \quad\text{and}\quad y = 8$$

then $$x = 3 + 8$$

The importance of the four properties of equality will become more apparent when we start solving equations and simplifying algebraic expressions.

♦   **INEQUALITY**

If two numbers are not equal, then one of the two is larger or greater than the other. We assume that you know what it means to say that 117 is greater than 68, or that 14 is less than 53. We will use the **inequality symbols** $<$ to represent "is less than" and $>$ to represent "is greater than." The symbol $\leq$ means "is less than or equal to." Similarly, $\geq$ means "is greater than or equal to." The following are all true statements:

$$117 > 68 \qquad 117 \geq 68 \qquad 117 \geq 117$$
$$14 < 53 \qquad 14 \leq 53 \qquad 14 \leq 14$$

The statement $14 \leq 53$ is true, since $14 < 53$: a statement involving $\leq$ is true if either the "$<$" part or the "$=$" part is true; the same reasoning applies for $\geq$. The notation is summarized:

Inequality Symbols

$a < b$ a is less than b

$a > b$ a is greater than b

$a \leq b$ a is less than or equal to b

$a \geq b$ a is greater than or equal to b

The use of the words "less" and "more" can be a source of confusion, so they must be used carefully. It is important to distinguish between the statement "2 is less than 7" and the expression "2 less than 7." The first is a true statement; the second is simply a way of representing the number 5. The word "is" makes a big difference. Similarly "10 is more than 8" is a statement, whereas "10 more than 8" is an expression representing the number 18. To reduce confusion we would usually say "10 is greater than 8" rather than "10 is more than 8." Appropriate mathematical notation also reduces confusion:

$$2 < 7 \text{ is a statement} \qquad \text{and} \qquad 10 > 8 \text{ is a statement}$$
$$7 - 2 \text{ is a number} \qquad \text{and} \qquad 10 + 8 \text{ is a number}$$

While it may be clear that $117 > 68$ and $14 < 53$, similar statements in the number systems to be considered in the next chapters will be less obvious. There, it will be necessary to define the two inequality symbols $<$ and $>$ more carefully.

Answers to Matched Problems

1. (A) $8 = 5 + x$ or $8 = x + 5$ (B) $6 = x - 4$ (not $6 = 4 - x$)
   (C) $4x = 2x - 3$ (D) $x + 12 = 2(x - 3)$
2. $x + (x + 2) + (x + 4) + (x + 6) = 160$

EXERCISE 1-3

A *Indicate which of the following are true (T) and which are false (F):*

1. $10 - 6 = 4$

2. $12 + 7 = 19$

3. $6 \cdot 9 = 56$

4. $8 \cdot 7 = 54$

5. $12 - 9 \neq 2$

6. $12 - 8 \neq 6$

7. $9 \cdot 14 = 6 \cdot 21$

8. $8 \cdot 16 = 2 \cdot 64$

9. $9 - 2 \cdot 3 \neq 21$

10. $4 + 2 \cdot 5 \neq 30$

11. $32 > 19$

12. $47 \geq 43$

13. $15 \leq 10$

14. $22 < 32$

15. $4 + 5 \cdot 6 \leq 56$

16. $6 \cdot 5 - 4 < 3$

17. $3 \cdot 8 - 2(9 - 3) > 11$

18. $(8 + 4) \cdot 5 \leq 28$

19. $(9 - 5)(6 - 1) \geq 20$

20. $54 \leq (3 + 6)(8 - 3)$

21. $2 + (3 + 4)(5 - 1) > 36$

22. $9 + 8(7 - 2) < (6 + 5) \cdot 4 + 3$

In Problems 23–58, translate each statement into an algebraic equation using x as the only variable:

23. Five is 3 more than a certain number.

24. Ten is 7 more than a certain number.

25. Eight is 3 less than a certain number.

26. Fourteen is 6 less than a certain number.

27. Eighteen is 3 times a certain number.

28. Twenty-five is 5 times a certain number.

B 29. Forty-nine is 7 more than twice a certain number.

30. Twenty-seven is 3 more than 6 times a certain number.

31. Fifty-two is 8 less than 5 times a certain number.

32. Seven less than 10 times a certain number is 103.

33. Four times a given number is 3 more than 3 times that number.

34. Eight times a number is 20 more than 4 times the number.

35. Three times a certain number is 2 more than 44.

36. Six times a certain number is 4 less than 35.

37. Twice a certain number is equal to 108 less than 3 times the number.

38. Four times a certain number is 14 less than 6 times the number.

39. Two more than twice a certain number is 6 less than the number plus 41.

40. Three more than the product of 5 times a certain number is 1 greater than 6 times the number.

41. Two times a certain number is 5 greater than the number plus 38.

42. Five times a certain number is 14 less than the number plus 76.

43. Four more than twice a certain number is 16 less than 5 times the number.

44. Six less than the product 5 times a certain number is 7 more than 4 times the number.

45. Seven more than the product 4 times a certain number is 5 less than 5 times the number.

46. Eight less than the product 6 times a certain number is 11 more than 5 times the number.

47. Eleven times a certain number is 110 more than the number times 8.

48. Twelve times a certain number is 156 less than 9 times the number.

49. Four more than twice a certain number is 3 times the number.

50. Six less than twice a certain number is 14 more than the number.

51. Five more than a certain number is 3 times the quantity that is 4 less than the certain number.

52. Six less than a certain number is 5 times the quantity that is 7 more than the certain number.

C 53. The sum of three consecutive natural numbers is 90.

54. The sum of four consecutive natural numbers is 54.

55. The sum of two consecutive even numbers is 54.

56. The sum of three consecutive odd numbers is 105.

57. The sum of three consecutive multiples of 5 is 75.

58. The sum of two consecutive multiples of 7 is 119.

59. Does the phrase ''7 more than 4 times a certain number'' represent an expression or a statement?

60. Does the phrase ''3 is more than 5 times a certain number'' represent an expression or a statement?

APPLICATIONS

61. Pythagoras found that the octave chord could be produced by placing the movable bridge so that a taut string is divided into two parts with the longer piece twice the length of the shorter piece (see the figure). If the total string is 54 centimeters long and we let x represent the length of the shorter piece, write an equation relating the lengths of the two pieces and the total length of the string.

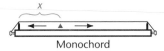

Monochord

62. A steel rod 7 meters long is cut into two pieces so that the longer piece is 1 meter less than twice the length of the shorter piece. Write an equation relating the lengths of the two pieces with the total length.

63. In a rectangle of area 50 square centimeters, the length is 10 centimeters more than the width. Write an equation relating the area with the length and the width.

64. In a rectangle of area 75 square yards, the length is 5 more than 3 times the width. Write an equation relating the area with the length and the width.

65. A rectangular field with length equal to twice the width has area of 1,440 square yards. Write an equation in one variable relating the area to the length and width.

66. A rectangular field with length equal to three times the width has perimeter of 200 meters. Write an equation in one variable relating the perimeter to the length and width.

67. A rectangular solid with length and height both equal to twice the width has a volume of 864 cubic inches. Write an equation in one variable relating the volume to the width, length, and height.

68. A rectangular solid with a height equal to twice the width and with a length equal to three times the width has a volume of 384 cubic centimeters. Write an equation in one variable relating the volume to the width, height, and length.

69. The *girth* of a rectangular solid is defined to be twice the sum of the width and height. (See the figure.) Postal regulations restrict the size of a rectangular box to be shipped parcel post. The girth plus the length must not exceed 108 inches. Express this requirement for the largest box that can be sent as an equation using h, l, and w, for height, length, and width, respectively.

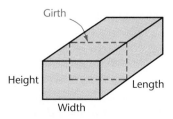

70. A rectangular box to be shipped parcel post (see Problem 69) has height equal to twice the width, and length equal to four times the width. The box exactly meets the postal restriction on its size. Write an equation relating the height of the box to the postal limit of 108 inches.

  ## 1-4 Properties of Addition and Multiplication

♦ Commutative Properties
♦ Associative Properties
♦ Simplifying Algebraic Expressions

Algebra in many ways can be thought of as a game—a game that requires the manipulation of symbols to change algebraic expressions from one form to another. As is the case in any game, one must know the rules to play.

In this section we discuss some of the basic rules in the ''game of algebra.'' These rules depend on properties of the arithmetic operations in the underlying number systems.

◆ COMMUTATIVE PROPERTIES

Assume we are given the set of natural numbers

$$N = \{1, 2, 3, \ldots\}$$

and the operations

$$+ \quad - \quad \cdot \quad \div$$

When we apply these operations on two natural numbers, does it matter what order the numbers are written in? For example, which of the following are true?

$8 + 4 = 4 + 8$ True

$8 - 4 = 4 - 8$ False ($8 - 4 = 4$; $4 - 8$ hasn't been defined yet, but it is -4, not 4)

$8 \cdot 4 = 4 \cdot 8$ True

$8 \div 4 = 4 \div 8$ False ($8 \div 4 = 2$; $4 \div 8$ has not been defined yet, but it is $\frac{1}{2}$, not 2)

The first and third equations are true, and the second and fourth are false. In fact, we can add or multiply any two natural numbers in either order. We therefore say that the natural numbers are **commutative** with respect to addition and multiplication. By this we simply mean that the order of operation in addition or in multiplication doesn't matter ($2 + 3 = 3 + 2$ and $2 \cdot 3 = 3 \cdot 2$). More formally, we state:

Commutative Properties

For all natural numbers a and b

$a + b = b + a$ The order in addition doesn't matter: $3 + 5 = 5 + 3$.

$ab = ba$ The order in multiplication doesn't matter: $3 \cdot 5 = 5 \cdot 3$.

On the other hand, subtraction and division are not commutative. The order in subtraction and division does matter.

The commutative properties are also referred to as the *commutativity* of addition and multiplication. The properties can be used to rewrite algebraic expressions in alternative forms that represent the same numbers.

Example 1
Using Commutativity

If x and y are natural numbers, use the commutative properties for addition and multiplication to write each of the following expressions in an equivalent form:

(A) $x + 7$ **(B)** $y5$ **(C)** yx **(D)** $3 + 5x$

Solution   Using the commutative properties simply means reversing the order in sums and products:

(A) $x + 7 = 7 + x$ Each example illustrates the use of a commutative property. Notice
(B) $y5 = 5y$ that the order is reversed in each case.
(C) $yx = xy$
(D) $3 + 5x = 5x + 3$ and also $3 + 5x = 3 + x5$

Matched Problem 1

If a and b are natural numbers, use the commutative properties for addition and multiplication to write each of the following expressions in an equivalent form:

(A) $a + 3$ (B) $b5$ (C) ba (D) $b + a$

♦ ASSOCIATIVE PROPERTIES

Suppose you are given the following four problems:

$$8 + 4 + 2 \qquad 8 - 4 - 2 \qquad 8 \cdot 4 \cdot 2 \qquad 8 \div 4 \div 2$$

Notice in the addition problem that if we add 4 to 8 first and then add 2, we get the same result as adding 2 to 4 first and then adding this sum to 8. That is,

$$(8 + 4) + 2 = 8 + (4 + 2) \qquad \begin{cases} (8 + 4) + 2 = 12 + 2 = 14 & \text{Left side} \\ 8 + (4 + 2) = 8 + 6 = 14 & \text{Right side} \end{cases}$$

It appears that grouping the terms in addition doesn't seem to make a difference. Does grouping make a difference for any of the other three operations? That is,

$$\text{Does } (8 - 4) - 2 = 8 - (4 - 2)?$$
$$\text{Does } (8 \cdot 4) \cdot 2 = 8 \cdot (4 \cdot 2)?$$
$$\text{Does } (8 \div 4) \div 2 = 8 \div (4 \div 2)?$$

Doing the arithmetic on both sides of each equal sign, and you should do this, we see that the only equation of the three that is true is the one involving multiplication. So we can also change the grouping in multiplication without changing the result. Thus, we say that the natural numbers are **associative** with respect to both addition and multiplication. This means that we may insert or remove parentheses at will in addition and insert or remove parentheses at will in multiplication. More formally, we state:

Associative Properties

For all natural numbers a, b, and c

$$(a + b) + c = a + (b + c)$$ Grouping doesn't matter in addition:
$(3 + 2) + 5 = 3 + (2 + 5)$

$$(ab)c = a(bc)$$ Grouping doesn't matter in multiplication:
$(4 \cdot 3)2 = 4(3 \cdot 2)$

Because of the associative properties, we can write the sum of three natural numbers, a, b, and c, simply as $a + b + c$, since

$$a + (b + c) = (a + b) + c = a + b + c$$

Similarly, the product can be written as abc, since

$$a(bc) = (ab)c = abc$$

On the other hand, subtraction and division are not associative. Grouping for both of these operations does matter:

$$(8 - 4) - 2 \neq 8 - (4 - 2)$$

and $$(8 \div 4) \div 2 \neq 8 \div (4 \div 2)$$

The associative properties are also referred to as the *associativity* of addition and multiplication. They, too, can be used to transform algebraic expressions into other equivalent forms.

Example 2
Using Associativity

If x, y, and z are natural numbers, use the associative properties for addition and multiplication to write each of the following expressions in an equivalent form:

(A) $(x + 3) + 5$ **(B)** $2(3x)$ **(C)** $(x + y) + z$ **(D)** $(xy)z$

Solution **(A)** Using the associative property means to regroup by shifting parentheses.

$$(x + 3) + 5 \boxed{= (x + (3) + 5)}$$
$$= x + (3 + 5) = x + 8$$

Notice how the parentheses are moved in each case. No order is changed.

The dashed boxes are used here to indicate steps that are usually done mentally. This notation will be used throughout the text.

(B) $2(3x) \boxed{= (2 \cdot (3)x)} = (2 \cdot 3)x = 6x$

(C) $(x + y) + z \boxed{= (x + (y) + z)} = x + (y + z)$

(D) $(xy)z \boxed{= (x(y)z)} = x(yz)$

Matched Problem 2 If a, b, and c are natural numbers, use the associative properties for addition and multiplication to write each of the following expressions in an equivalent form:

(A) $(a + 5) + 7 = a + (?)$ **(B)** $5(9b) = (?)b$
(C) $(a + b) + c = a + (?)$ **(D)** $(ab)c = a(?)$

♦ **SIMPLIFYING ALGEBRAIC EXPRESSIONS**

We can use the associative and commutative properties together to rewrite expressions like

$$(x + 3) + (y + 5) \quad \text{and} \quad (3x)(5y)$$

in simpler form. Since only addition is involved in the first expression, we can drop the parentheses. Then we can rearrange and regroup the terms to obtain

$$(x + 3) + (y + 5) = x + 3 + y + 5$$
$$= x + y + 3 + 5$$
$$= x + y + 8$$

Similarly, since only multiplication is involved in the second expression, we can drop the parentheses. Then we can rearrange and regroup the factors to obtain

$$(3x)(5y) = 3x5y$$
$$= 3 \cdot 5xy$$
$$= 15xy$$

The process of rewriting expressions in forms that are simpler is called **simplifying** the expression.

It is important to remember that the associative and commutative properties are behind the operations performed in both the above examples. In general, we can state the following:

Use of Commutative and Associative Properties—Summary

In addition or multiplication, the commutative and associative properties permit us to change the order of the operation at will and insert or remove parentheses as we please. The same thing is not true for subtraction or division.

Example 3
Simplifying

Remove parentheses and simplify, using the commutative and associative properties mentally.

(A) $(a + 5) + (b + 2) + (c + 4)$ **(B)** $(2x)(3y)(4z)$

Solution

(A) $(a + 5) + (b + 2) + (c + 4)$
$$= a + 5 + b + 2 + c + 4$$
$$= a + b + c + 5 + 2 + 4$$
$$= a + b + c + 11$$

Since only addition is involved, we can (mentally) rearrange and regroup the variables and the constants.

(B) $(2x)(3y)(4z) = 2x3y4z$
$$= (2 \cdot 3 \cdot 4)(xyz)$$
$$= 24xyz$$

Since only multiplication is involved, we can (mentally) rearrange and regroup the variables and the constants.

Matched Problem 3

Remove parentheses and simplify, using commutative and associative properties mentally.

(A) $(u + 4) + (v + 5) + (w + 3)$ **(B)** $(4m)(8n)(2p)$

Answers to
Matched Problems

1. **(A)** $3 + a$ **(B)** $5b$ **(C)** ab **(D)** $a + b$
2. **(A)** $5 + 7$ **(B)** $5 \cdot 9$ **(C)** $b + c$ **(D)** bc
3. **(A)** $u + v + w + 12$ **(B)** $64mnp$

EXERCISE 1-4

A *Remove parentheses and simplify:*

1. $(7 + x) + 3$
2. $(5 + z) + 12$
3. $4 + (5 + x)$
4. $3 + (x + 8)$
5. $3(5x)$
6. $5(8x)$
7. $(2x) \cdot 12$
8. $(8x) \cdot 7$
9. $(7a)(4b)$
10. $(3x)(4y)$
11. $(7 + a) + (9 + b)$
12. $(x + 7) + (y + 8)$
13. $(6 + x) + (x + 7)$
14. $(x + 2) + (x + 13)$
15. $(x + 3) + (8 + x)$
16. $(5 + x) + (x + 3)$
17. $3 + (4 + 5x)$
18. $2 + (4x + 7)$
19. $3(4 \cdot 5x)$
20. $2(4x \cdot 7)$
21. $3(4x \cdot 5y)$
22. $2(4x \cdot 7y)$

B *Remove parentheses and simplify:*

23. $(3x)(4 \cdot 5y)$
24. $(2x)(4 \cdot 7y)$
25. $(x + 2) + 10 + (x + 13)$
26. $(6 + x) + 5 + (x + 7)$
27. $(5 + x) + 11 + (x + 3)$
28. $(x + 3) + 8 + (8 + x)$
29. $(3a)(5b)(2c)$
30. $(2x)(8y)(3z)$
31. $(4u)(5v)(3w)$
32. $(2x)(3y)(4z)$
33. $(x + 2) + (y + 4) + (z + 8)$
34. $(a + 3) + (b + 5) + (c + 2)$
35. $(u + 5) + (v + 10) + (w + 4)$
36. $(r + 6) + (s + 8) + (t + 10)$
37. $x + (y + 1) + (z + 2) + (w + 3)$
38. $(a + 8) + (b + 6) + (c + 4) + d$
39. $a(2b)(3c)(4d)$
40. $(5x)(4y)(3z)w$
41. $(x + 2) + (y + 4) + (z + 6) + (w + 8)$
42. $(a + 5) + (b + 1) + (c + 3) + (d + 7)$

C *Each statement illustrates either a commutative property or an associative property. State which.*

43. $5 + z = z + 5$
44. $bc = cb$
45. $(5x)y = 5(xy)$
46. $(a + 5) + 7 = a + (5 + 7)$
47. $3x + x5 = 3x + 5x$
48. $5(x8) = 5(8x)$
49. $3 + (x + 2) = 3 + (2 + x)$
50. $(5x)y = y(5x)$
51. $5 + (x + 3) = (x + 3) + 5$
52. $(x + 2) + (y + 3) = (x + 2) + (3 + y)$
53. $(x + 3) + (y + 2) = (y + 2) + (x + 3)$
54. $(x + 3) + (y + 2) = x + [3 + (y + 2)]$
55. $(x + 3) + (y + 2) = [(x + 3) + y] + 2$
56. $x + [3 + (y + 2)] = (x + 3) + (y + 2)$

In Problems 57–68, remove grouping symbols and simplify:

57. $1 + [x + (3 + y)]$
58. $4 + [(5 + a) + 6]$
59. $(3 + a) + [6 + (b + 4)]$
60. $[x + (y + 3)] + 9$
61. $\{[(x + 4) + 5] + y\}$
62. $\{a + [2 + (y + 5)]\}$
63. $\{[(a + 3) + b] + 8\} + 6$
64. $\{[x + (7 + y)] + 3\} + 2$
65. $\{[(x + 1) + (y + 2)] + (z + 5)\} + 9$
66. $\{(6 + a) + [(b + 3) + (c + 5)]\} + 8$
67. $\{[2 + (a + 3)] + 4(5b)\} + 2c$
68. $x + \{3(7y) + [4 + (z + 8)]\}$

69. If a statement is not true for all natural numbers a and b, find replacements for a and b that show that the statement is false.
   (A) $a + b = b + a$ **(B)** $ab = ba$
   (C) $a - b = b - a$ **(D)** $a \div b = b \div a$

70. Repeat the preceding problem for:
   (A) $(a + b) + c = a + (b + c)$
   (B) $(ab)c = a(bc)$
   (C) $(a - b) - c = a - (b - c)$
   (D) $(a \div b) \div c = a \div (b \div c)$

1-5 Exponents and Order of Operations

♦ Exponents
♦ An Exponent Property
♦ Order of Operations

Natural number exponents such as the 2 in $x^2 = x \cdot x$ provide a useful shorthand notation for repeated factors in products. In this section, we define natural number exponents and begin to look at their properties. We also incorporate exponents into the order-of-operations rules that were introduced in Section 1-2.

♦ EXPONENTS

We start by defining

$$b^2 = bb \qquad 2^2 = 2 \cdot 2 = 4$$

Thus, $3^2 = 3 \cdot 3 = 9$, $6^2 = 6 \cdot 6 = 36$, and so on. There is obviously no reason to stop here: you no doubt can guess how b^3, 2^3, b^4, and 2^4 should be defined. If you guessed

$$b^3 = bbb \qquad 2^3 = 2 \cdot 2 \cdot 2 = 8$$
$$b^4 = bbbb \qquad 2^4 = 2 \cdot 2 \cdot 2 \cdot 2 = 16$$

then you have anticipated the following general definition of b^n, where n is any natural number and b is any number:

$$b^n = \underbrace{bbb \cdots b}_{n \text{ factors of } b} \qquad b^n \text{ is read ``}b\text{ to the }n\text{th power.''}$$

Here b is called the **base** and n the **exponent.** In addition, we define b^1 as

$$b^1 = b$$

and usually use b in place of b^1. Finally, we emphasize that $3b$ and b^3 are quite different. Compare

$$3b = 3 \cdot b \qquad \text{Three times } b$$
$$b^3 = b \cdot b \cdot b \qquad \text{3 factors of } b$$

The quantity b^2 is read as "b squared" or "the square of b" or "b to the second power." Similarly b^3 is read "b cubed" or "the cube of b" or "b to the third power." Higher powers, such as b^4 and b^{12}, are read "b to the fourth power" and "b to the twelfth power," respectively.

Example 1 **Exponent Forms**	**(A)** Write without exponents: x^2, t^1, 3^4, $5x^3y^5$ **(B)** Write in exponent form: xxx, $2xxy$, $2 \cdot 2 \cdot 2 \cdot 2$, $3xxxyy$

Solution   **(A)** From exponent forms to forms without exponents:

$$x^2 = xx \qquad t^1 = t \qquad 3^4 = 3 \cdot 3 \cdot 3 \cdot 3 \qquad 5x^3y^5 = 5xxxyyyyy$$

(B) From forms without exponents to exponent forms:

$$xxx = x^3 \qquad 2xxy = 2x^2y \qquad 2 \cdot 2 \cdot 2 \cdot 2 = 2^4 \qquad 3xxxyy = 3x^3y^2$$

Matched Problem 1   **(A)** Write without exponents: y^3, 2^4, $3x^3y^4$
(B) Write in exponent form: uu, $5 \cdot 5 \cdot 5 \cdot 5$, $7xxxxyyy$

♦ **AN EXPONENT PROPERTY**

Something interesting happens if we multiply two exponent forms with the same base:

$$x^3x^5 = (xxx)(xxxxx)$$
$$= xxxxxxxx$$
$$= x^8$$

which we could get by simply adding the exponents in x^3x^5. This example suggests the following general property of exponents:

> ### First Law of Exponents
>
> For any natural numbers m and n and any number b
>
> $$b^mb^n = b^{m+n} \qquad b^2b^3 = b^{2+3} = b^5$$

Expressed in words, this says: the product of exponent forms with a common base is the base raised to the sum of the exponents. This first law of exponents is one of five very important exponent laws you will encounter in this book. The remaining four laws are discussed in Chapter 8.

Example 2 **First Law of Exponents**	Apply the first law of exponents to simplify the following expressions: **(A)** x^3x^4 **(B)** $5^{10} \cdot 5^{23}$ **(C)** $(2y^2)(3y^5)$ **(D)** $(3x^2y)(4x^4y^5)$

Solution   **(A)** $x^3x^4 = x^{3+4} = x^7$ Not x^{12}

(B) $5^{10} \cdot 5^{23} = 5^{10+23} = 5^{33}$ Not 25^{33}

(C) $(2y^2)(3y^5) \boxed{= (2 \cdot 3)(y^2 y^5)} = 6y^7$

(D) $(3x^2 y)(4x^4 y^5) \boxed{= (3 \cdot 4)(x^2 x^4)(y y^5)} = 12x^6 y^6$

Note how commutative and associative properties are used in parts (C) and (D) where we rearranged the factors and regrouped them.

Matched Problem 2   Apply the first law of exponents to simplify the following expressions:

(A) $y^5 y^3$ **(B)** $3^{17} \cdot 3^{20}$ **(C)** $(3a^6)(5a^3)$ **(D)** $(2x^2 y^4)(3xy^2)$

♦ **ORDER OF OPERATIONS**

Unless grouping symbols dictate otherwise, exponents are to be applied before multiplication or division. For example, $3 \cdot 2^2$ means first raise 2 to the second power, then multiply by 3, to obtain 12. On the other hand, because of the parentheses, $(3 \cdot 2)^2$ means first multiply 3 times 2 and then raise the result to the second power, to obtain 36. Similarly, we can compare $3x^5$ and $(3x)^5$:

$3x^5 = 3 \cdot x \cdot x \cdot x \cdot x \cdot x$ Right (3 times five factors of x)

$3x^5 \neq 3x \cdot 3x \cdot 3x \cdot 3x \cdot 3x$ The exponent 5 in $3x^5$ applies only to the base x.

$(3x)^5 = 3x \cdot 3x \cdot 3x \cdot 3x \cdot 3x$ Right (five factors of $3x$)

The order of operations introduced in Section 1-2 is thus extended to include exponents:

Order of Operations

1. Simplify inside the innermost grouping symbols first, then the next innermost, and so on.
2. Unless grouping symbols indicate otherwise, apply exponents before multiplication or division is performed.
3. Unless grouping symbols indicate otherwise, perform multiplication and division before addition and subtraction. In either case, proceed from left to right.

Example 3
Order of Operations

Evaluate the following expressions for $x = 2$ and $y = 3$.

(A) $(x + y)^2$ **(B)** $x + y^2$ **(C)** $(xy)^2$ **(D)** xy^2

Solution   In each case, we substitute and simplify.

(A) $(2 + 3)^2 = 5^2 = 25$ **(B)** $2 + 3^2 = 2 + 9 = 11$
(C) $(2 \cdot 3)^2 = 6^2 = 36$ **(D)** $2 \cdot 3^2 = 2 \cdot 9 = 18$

Matched Problem 3   Evaluate the following expressions for $x = 12$ and $y = 2$:

      (A) $(x - y)^2$ **(B)** $x - y^2$ **(C)** $(x/y)^2$ **(D)** x/y^2

Example 4
Order of Operations

Evaluate the following expressions for $x = 7$, $y = 8$, and $z = 2$:

      (A) $x + 3y^2$ **(B)** $xy + yz^2$ **(C)** $x - y \div z$ **(D)** $xy - z$

Solution   **(A)** $7 + 3 \cdot 8^2 = 7 + 3 \cdot 64 = 7 + 192 = 199$
          (B) $7 \cdot 8 + 8 \cdot 2^2 = 56 + 8 \cdot 4 = 56 + 32 = 88$
          (C) $7 - 8 \div 2 = 7 - 4 = 3$
          (D) $7 \cdot 8 - 2 = 56 - 2 = 54$

Matched Problem 4   Evaluate the following expressions for $x = 50$, $y = 4$, and $z = 2$:

      (A) $x - 3y^2$ **(B)** $xy - yz^2$ **(C)** $x + y \div z$ **(D)** $xy + z$

Answers to
Matched Problems

1. **(A)** yyy, $2 \cdot 2 \cdot 2 \cdot 2$, $3xxxyyyy$ **(B)** u^2, 5^4, $7x^4y^3$
2. **(A)** y^8 **(B)** 3^{37} (not 9^{37})   **(C)** $15a^9$ **(D)** $6x^3y^6$
3. **(A)** 100   **(B)** 8   **(C)** 36   **(D)** 3
4. **(A)** 2   **(B)** 184   **(C)** 52   **(D)** 202

EXERCISE 1-5

A *Write without exponents:*

  1. x^3 **2.** y^4 **3.** $2x^3y^2$

  4. $5a^2b^3$ **5.** $3w^2xy^3$ **6.** $7ab^3c^2$

Write in exponent form:

  7. xxx **8.** $yyyy$ **9.** $2xxxyy$

10. $7uuvvvv$ **11.** $3xyyzzz$ **12.** $9aabccc$

Multiply, using the first law of exponents:

13. $u^{10}u^4$ **14.** m^8m^7 **15.** aa^5

16. b^7b **17.** $w^{12}w^7$ **18.** $n^{23}n^{10}$

19. $y^{12}y^4$ **20.** u^4u^{44} **21.** $3^{10} \cdot 3^{20}$

22. $7^8 \cdot 7^5$ **23.** $9^5 \cdot 9^6$ **24.** $2^5 \cdot 2^{12}$

Evaluate the following for $x = 3$ and $y = 5$:

25. $(x + y)^2$ **26.** $x^2 + y$

27. $(xy)^2$ **28.** x^2y

Evaluate the following for $x = 7$ and $y = 4$:

29. $x^2 + y$ **30.** $(x + y)^2$

31. x^2y **32.** $(xy)^2$

B *Multiply, using the first law of exponents:*

33. x^2xx^4 **34.** mm^3m^4

35. yyy^6y^2 **36.** uu^2uu^4

37. $(2x^3)(3x)(4x^5)$ **38.** $(3u^4)(2u^5)(u^7)$

39. $(a^2b)(ab^2)$ **40.** $(cd^2)(c^2d^3)$

41. $(4x)(3xy^2)$ **42.** $(5b)(2a^2b^3)$

43. $(2xy)(3x^3y)$ **44.** $(3xy^2z^3)(5xyz^2)$

45. $(3x^2y^2)(4xy^3)$ **46.** $(2x^3y^3)(5x^2y)$

47. $(xy)(yz)(xz)$ **48.** $(x^2y)(y^2z)(z^2x)$

49. $(xy^2z^3)(xyz)(xyz^3)$ **50.** $(x^2yz)(xy^2z)(xyz^3)$

51. $(2x^3yz)(3y^3z)(4xz^3)$ **52.** $(3xy^3z)(2xyz^2)(5x^2yz)$

Evaluate the following expressions for $a = 16$ and $b = 2$:

53. $(a + b)^3$ **54.** $(a - b)^3$

55. $a + b^3$ **56.** $a - b^3$

57. $(ab)^3$ **58.** $(a/b)^3$

59. $a \cdot b^3$ **60.** a/b^3

Evaluate the following expressions for $a = 54$ and $b = 3$:

61. $(a - b)^3$ **62.** $(a + b)^3$

63. $a - b^3$ **64.** $a + b^3$

65. $(a/b)^3$ **66.** $(a \cdot b)^3$

67. a/b^3 **68.** $a \cdot b^3$

C *Evaluate the following expressions for* $x = 24$, $y = 12$, *and* $z = 2$:

69. $5x + 4y^2z$ **70.** $x^2y - xy^2$

71. $x + y \div z$ **72.** $xy - y \div z^2$

73. $x \div y + z$ **74.** $x - y \div z$

75. $x - y - z$ **76.** $x \div y \div z$

In Problems 77–84, evaluate the expressions for $x = 40$, $y = 20$, *and* $z = 2$:

77. $x^2y - xy^2$ **78.** $5x + 4y^2z$

79. $xy - y \div z^2$ **80.** $x + y \div z$

81. $x - y \div z$ **82.** $x \div y + z$

83. $x \div y \div z$ **84.** $x - y - z$

APPLICATIONS

85. The distance s in feet that an object falls in a time of t seconds is 16 times the square of the time (see the figure).
   (A) Write a formula that indicates the distance s that the object falls in t seconds; use exponents if possible.
   (B) Identify the constants and variables.

(C) How far will the object fall in the first 8 seconds?

86. The Pythagorean theorem in geometry states that for a right triangle, the square of the length of the hypotenuse is equal to the sum of the squares of the lengths of the sides. Write this statement as an algebraic formula using a, b, and c as shown in the figure.

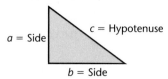

1-6 **Relating Addition and Multiplication**

♦ Distributive Property
♦ Common Factors
♦ Extended Distributive Property

We now introduce another important property of the natural numbers. This property involves both multiplication and addition and is called the **distributive property.**

♦ **DISTRIBUTIVE PROPERTY**

The distributive property relates expressions like $3(5 + 2)$ and $3 \cdot 5 + 3 \cdot 2$. If we compute each of these, we get

$$3(5 + 2) = 3 \cdot 7 \qquad 3 \cdot 5 + 3 \cdot 2 = 15 + 6$$
$$= 21 \qquad\qquad\qquad\qquad = 21$$

Remember that multiplications are done before additions.

Thus $\qquad\qquad\qquad\qquad\qquad 3(5 + 2) = 3 \cdot 5 + 3 \cdot 2$

Notice that the right-hand side of the last equation is obtained by multiplying each term in the parentheses by 3 and then adding the results. This holds in general and is stated formally as follows:

Distributive Property

For all natural numbers a, b, and c

$$a(b + c) = ab + ac \qquad 5(2 + 3) = 5 \cdot 2 + 5 \cdot 3$$

Notice in the distributive property that one side of the equation is a product $a(b + c)$ and the other side is a sum $ab + ac$. The property allows us to rewrite certain products as sums, and certain sums as products. We sometimes say that *multiplication distributes over addition*, since we can rewrite $a(b + c)$ by multiplying "over" the parentheses to obtain $ab + ac$. We call the process **multiplying out.**

Example 1
Multiplying Out

Multiply out using the distributive property:

(A) $3(x + y)$ **(B)** $4(w + 2)$ **(C)** $x(x + 1)$ **(D)** $2x^2(3x + 2y)$

Solution

(A) $3(x + y) = 3x + 3y$

(B) $4(w + 2) = 4w + 4 \cdot 2 = 4w + 8$

(C) $x(x + 1) = x \cdot x + x \cdot 1 = x^2 + x$

(D) $2x^2(3x + 2y) = 2x^2 \cdot 3x + 2x^2 \cdot 2y$
$$= (2 \cdot 3)(x^2 x) + (2 \cdot 2)(x^2 y)$$
$$= 6x^3 + 4x^2 y$$

Matched Problem 1

Multiply out using the distributive property:

(A) $2(a + b)$ **(B)** $5(x + 3)$ $2a + 2b$ $5x + 15$
(C) $u(u^2 + 1)$ **(D)** $3n^2(2m^2 + 3n)$

$u^3 + 1u$ $6m^2 n^2 + 9n^3$

The symmetric property of equality and the commutative properties of addition and multiplication allow us to write the distributive property in several equivalent forms. The following are true for all natural numbers a, b, and c:

$$a(b + c) = ab + ac \tag{1}$$

$$(b + c)a = ba + ca \tag{2}$$

$$ab + ac = a(b + c) = (b + c)a \tag{3}$$

$$ba + ca = (b + c)a = a(b + c) \tag{4}$$

◆ COMMON FACTORS

Multiplying out $a(b + c)$ to $ab + ac$ [Equation (1)] is sometimes just called multiplying. In doing so, we are changing a product into a sum. The reverse process of changing from $ab + ac$ to $a(b + c)$ is called **factoring out** or **taking out the common factor.** In particular, in Equation (3) we are taking out the common factor a:

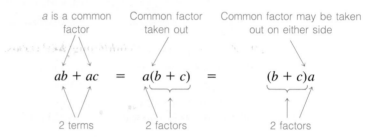

In this case, we are changing a sum to a product.

<table>
<tr><td rowspan="2">**Example 2**
Taking Out
Common Factors</td></tr>
</table>

Take out all factors common to all terms:

(A) $2x + 2y$ **(B)** $3w + 6$ **(C)** $x^2 + x$ **(D)** $6y + 4y^2$

Solution   **(A)** $2x + 2y = 2(x + y)$

(B) $3w + 6 = 3w + 3 \cdot 2 = 3(w + 2)$

(C) $x^2 + x = x \cdot x + x \cdot 1 = x(x + 1)$ The number 1 is a factor of every number.

(D) $6y + 4y^2 = 2y \cdot 3 + 2y \cdot 2y = 2y(3 + 2y)$

Multiplying the expression on the right should take you back to the expression with which you started on the left. This provides a way to check your work when taking out common factors. For instance, in Example 2(D), if we multiply out the answer, we obtain the original expression:

$$2y(3 + 2y) = 2y \cdot 3 + 2y \cdot 2y = 6y + 4y^2$$

Matched Problem 2   Take out all factors common to all terms:

(A) $5x + 5y$ **(B)** $2w + 8$ **(C)** $u^2 + u$ **(D)** $6y^2 + 4y^3$

$5(x+y)$ $2(w+4)$ $u(u+1)$ $2y^2(3+2y)$

In Example 2, the common factors were taken out on the left. It is just as valid to take them out on the right, as was shown in Equations (3) and (4) above. The next example illustrates this process, which results in useful simplifications.

Example 3
Taking Out
Common Factors

Apply the distributive property to simplify:

(A) $3x + 5x$ (B) $7y + 2y$ (C) $3x^2y^2 + 4x^2y^2$

Solution (A) $3x + 5x = (3 + 5)x = 8x$

(B) $7y + 2y = (7 + 2)y = 9y$

(C) $3x^2y^2 + 4x^2y^2 = (3 + 4)x^2y^2 = 7x^2y^2$

Notice how much simpler the result is in each case. Using the distributive property in this way is called **combining like terms.** Combining like terms, which we discuss in detail in Section 1-7, will be very useful in solving equations.

Matched Problem 3

Apply the distributive property to simplify:

(A) $2x + 3x$ (B) $8u + 3u$ (C) $5uv^2 + 7uv^2$
    $(2 + 3)x$ $(8 + 3)u$ $(5 + 7)uv^2$

♦ **EXTENDED DISTRIBUTIVE PROPERTY**

By repeated use of the distributive property we can extend the sum to any number of terms. Thus:

Extended Distributive Property

$$a(b + c + d) = ab + ac + ad \qquad 2(x + y + z) = 2x + 2y + 2z$$
$$a(b + c + d + e) = ab + ac + ad + ae \qquad \begin{aligned} & 3(w + x + y + z) \\ & = 3w + 3x + 3y + 3z \end{aligned}$$

and so on.

We can use the extended distributive property to multiply out and take out common factors. The processes work in exactly the same way as when only two terms are involved in the sums.

Example 4
Multiplying Out

Apply the extended distributive law to multiply out:

(A) $3(x + y + z)$ (B) $2x(x + 3y + 2)$
(C) $3x^2y(2xy^3 + 3xy + x^2y^2)$ (D) $(x + 2)(x + 3)$

Solution (A) $3(x + y + z) = 3x + 3y + 3z$
(B) $2x(x + 3y + 2) = 2x^2 + 6xy + 4x$

(C) $3x^2y(2xy^3 + 3xy + x^2y^2) = 6x^3y^4 + 9x^3y^2 + 3x^4y^3$

(D) $(x + 2)(x + 3)$ Think of $x + 2$ as a single quantity. Treat this product just like $a(x + 3) = ax + a3$.

$= (x + 2)x + (x + 2)3$ Multiply out each term.

$= x^2 + 2x + 3x + 6$ Combine like terms.

$= x^2 + (2 + 3)x + 6$

$= x^2 + 5x + 6$

Matched Problem 4   Apply the extended distributive law to multiply out:

(A) $5(a + b + c)$ **(B)** $3x(2x + 3y + 5)$ $6x^2 + 9xy + 15x$
(C) $2u^2v^3(4u^2v + 2uv + uv^2)$ **(D)** $(y + 1)(y + 5)$ $(y+1)y + (y+1)5$

$5a + 5b + 5c$ $8u^4v^4 + 4u^3v^4 + 2u^3v^5$ $y^2 + y + 5y + 5$
$y^2 + 6y + 5$

Example 5
Taking Out
Common Factors

Apply the extended distributive law to take out common factors:

(A) $ma + mb + mc$ **(B)** $4x^2 + 2xy + xz$

Solution   **(A)** $ma + mb + mc = m(a + b + c)$
(B) $4x^2 + 2xy + xz = x(4x + 2y + z)$

Matched Problem 5   Apply the extended distributive law to take out common factors:

(A) $ax + ay + az$ **(B)** $4x^3 + 2xy + 6x^2$

$2x(2x^2 + y + 3x)$

In the introduction to Section 1-4, an analogy was drawn between algebra and games. The properties we have considered in Sections 1-4 to 1-6 are, so to speak, some of the rules of the game of algebra. However, as is true for most games, one must practice to become good at this game of algebra.

Answers to
Matched Problems

1. **(A)** $2a + 2b$ **(B)** $5x + 15$ **(C)** $u^3 + u$ **(D)** $6m^2n^2 + 9n^3$
2. **(A)** $5(x + y)$ **(B)** $2(w + 4)$ **(C)** $u(u + 1)$ **(D)** $2y^2(3 + 2y)$
3. **(A)** $5x$ **(B)** $11u$ **(C)** $12uv^2$
4. **(A)** $5a + 5b + 5c$ **(B)** $6x^2 + 9xy + 15x$
   (C) $8u^4v^4 + 4u^3v^4 + 2u^3v^5$ **(D)** $y^2 + 6y + 5$
5. **(A)** $a(x + y + z)$ **(B)** $2x(2x^2 + y + 3x)$

EXERCISE 1-6

A *Compute:*   *Multiply, using the distributive property:*

1. $2(1 + 5)$ and $2 \cdot 1 + 2 \cdot 5$ 5. $4(x + y)$ 6. $5(a + b)$ 7. $7(m + n)$

2. $3(4 + 2)$ and $3 \cdot 4 + 3 \cdot 2$ 8. $9(u + v)$ 9. $6(x + 2)$ 10. $3(y + 7)$

3. $5(2 + 7)$ and $5 \cdot 2 + 5 \cdot 7$ 11. $(2 + m)5$ 12. $(3 + n)8$ 13. $(x + 4)12$

4. $7(3 + 2)$ and $7 \cdot 3 + 7 \cdot 2$ 14. $(a + 5)7$ 15. $(x + y)3$ 16. $(a + b)4$

Take out factors common to all terms:

17. $3x + 3y$ **18.** $2a + 2b$ **19.** $5m + 5n$

20. $7u + 7v$ **21.** $xa + ya$ **22.** $um + vm$

23. $2x + 4$ **24.** $3y + 9$ **25.** $6x + 15$

26. $4x + 14$ **27.** $9a + 24$ **28.** $10a + 25$

29. $12x + 42$ **30.** $24x + 54$

Multiply, using the distributive property:

31. $2(x + y + z)$ **32.** $4(a + b + c)$

33. $(x + y + z)3$ **34.** $(a + b + 3)4$

Take out factors common to all terms:

35. $7x + 7y + 7z$ **36.** $9a + 9b + 9c$

37. $2m + 2n + 6$ **38.** $3x + 3y + 12$

B *Multiply, using the distributive property:*

39. $x(1 + x)$ **40.** $y(y + 7)$

41. $(1 + y^2)y$ **42.** $(x^2 + 3)x$

43. $3x(2x + 5)$ **44.** $5y(2y + 7)$

45. $(m^2 + 3m)2m^2$ **46.** $(a^3 + 2a^2)3a^2$

47. $3x(2x^2 + 3x + 1)$ **48.** $2y(y^2 + 2y + 3)$

49. $5(2x^3 + 3x^2 + x + 2)$

50. $4(y^4 + 2y^3 + y^2 + 3y + 1)$

51. $3x^2(2x^3 + 3x^2 + x + 2)$

52. $7m^3(m^3 + 2m^2 + m + 4)$

Apply the distributive property to simplify:

53. $3x + 7x$ **54.** $4y + 5y$

55. $2u + 9u$ **56.** $8m + 5m$

57. $2xy + 3xy$ **58.** $5mn + 7mn$

59. $2x^2y + 8x^2y$ **60.** $3uv^2 + 5uv^2$

61. $7x + 2x + 5x$ **62.** $8y + 3y + 4y$

Take out factors common to all terms:

63. $36x + 90$ **64.** $30x + 100$

65. $18a + 42$ **66.** $24a + 108$

67. $x^2 + 2x$ **68.** $y^2 + 3y$

69. $u^2 + u$ **70.** $m^2 + m$

71. $2x^3 + 4x$ **72.** $3u^5 + 6u^3$

73. $x^2y + xy^2$ **74.** $xy^3 + x^3y$

75. $a^2bc + ab^2c$ **76.** $ab^2c^2 + a^2bc^2$

77. $x^2 + xy + xz$ **78.** $y^3 + y^2 + y$

79. $3m^3 + 6m^2 + 9m$ **80.** $12x^3 + 9x^2 + 3x$

81. $5u^2v + 15uv^2$ **82.** $2x^3y^2 + 4x^2y^3$

C *Multiply, using the distributive property:*

83. $4m^2n^3(2m^3n + mn^2)$ **84.** $5uv^2(2u^3v + 3uv^2)$

85. $(2xy^3 + 4x + y^2)3x^2y$

86. $(c^2d + 2cd + 4c^3d^2)2cd^3$

87. $4x^2yz^3(3x^2z + yz)$

88. $3u^2v^3w(5uw^4 + vw^3)$

Take out factors common to all terms:

89. $a^2bc + ab^2c + abc^2$ **90.** $m^3n + mn^2 + m^2n^2$

91. $16x^3yz^2 + 4x^2y^2z + 12xy^2z^3$

92. $27u^5v^2w^2 + 9u^2v^3w^4 + 12u^3v^2w^5$

Multiply, using the distributive property:

93. $(x + 2)(x + 5) = x(x + 5) + 2(x + 5) = ?$
    (finish the multiplication)

94. $(x + 4)(x + 1) = x(x + 1) + 4(x + 1) = ?$
    (finish the multiplication)

95. $(a + 6)(a + 2)$ **96.** $(a + 2)(a + 7)$

97. $(x + 3)(x + 2)$ **98.** $(m + 5)(m + 3)$

99. $(u + v)(c + d) = (u + v)c + (u + v)d = ?$
    (finish the multiplication)

100. $(m + n)(x + y) = (m + n)x + (m + n)y = ?$
    (finish the multiplication)

1-7 Rewriting Algebraic Expressions

♦ Coefficient
♦ Like Terms
♦ Combining Like Terms
♦ Multiplying and Combining Like Terms

In Example 3 in the last section we introduced an important procedure that will be formalized in this section. The procedure is called *combining like terms*. Our first task will be to make the notion of ''like terms'' more precise.

♦ COEFFICIENT

The factor 3 in $3x^2y$ is called the numerical coefficient of the term and of the variable factors x^2y. In general, the constant factor in a term is called the **numerical coefficient** (or simply the **coefficient**) of the term and of the variables. If no constant factor appears in the term, then the coefficient is understood to be 1. For example, x^2y^2 has a coefficient of 1, since $x^2y^2 = 1 \cdot x^2y^2$. If no variable occurs in the term, the coefficient is the entire term. In this case the term is referred to as a **constant term.**

Example 1
Identifying Coefficients

What are the coefficients of each term in the algebraic expression

$$2x^3 + x^2y + 3xy^2 + y^3$$

Solution The coefficient of x^3 is 2, that of x^2y is 1, that of xy^2 is 3, and that of y^3 is 1.

Matched Problem 1 What are the coefficients of each term in the algebraic expression

$$5x^4 + 2x^3y + x^2y^2 + 4xy^3 + y^4$$

♦ LIKE TERMS

If two or more terms are exactly alike, except possibly for their numerical coefficients or the order in which the factors are multiplied, then they are called **like terms.** Thus, like terms must have the same variables to the same powers, but their coefficients do not have to be the same.

Example 2
Identifying Like Terms

List the like terms in

(A) $4x + 2y + 3x$ **(B)** $9x^2y + 3xy + 2x^2y + x^2y$

Solution **(A)** In $4x + 2y + 3x$, $4x$ and $3x$ are like terms.
(B) In $9x^2y + 3xy + 2x^2y + x^2y$, the first, third, and fourth terms, counting from the left, are like terms.

Matched Problem 2 List the like terms in

(A) $5m + 6n + 2n$ (B) $2xy + 3xy^3 + xy + 2xy^3$

                                 6n + 2n   ✗

♦ **COMBINING LIKE TERMS**

If an algebraic expression contains two or more like terms, these terms can always be combined into a single term using the distributive property.

Example 3
Combining Like Terms

Combine like terms:

(A) $4x^2 + 5x^2$ (B) $5t + 4s + 7t + s$

Solution (A) $4x^2 + 5x^2 = (4 + 5)x^2 = 9x^2$

(B) $5t + 4s + 7t + s = 5t + 7t + 4s + s$
$$= (5 + 7)t + (4 + 1)s$$
$$= 12t + 5s$$

Matched Problem 3 Combine like terms:

(A) $6y + 5y$ (B) $4x + 7y^2 + x + 2y^2$

                     11y   5x + 9y²

The rule for combining like terms can be stated simply as follows:

Rule for Combining Like Terms

To combine like terms, add their numerical coefficients.

As indicated in the last example, most of the work in combining like terms can be done mentally.

Example 4
Combining Like Terms Mentally

Combine like terms mentally:

(A) $7x + 2y + 3x + y$ (B) $2u^2 + 3u + 4u^2$
(C) $(3x^2 + x + 2) + (4x^2 + 2x + 1)$

Solution (A) $7x + 2y + 3x + y = 7x + 3x + 2y + 1y$
$$= 10x + 3y$$

(B) $2u^2 + 3u + 4u^2 = 2u^2 + 4u^2 + 3u$
$$= 6u^2 + 3u$$

(C) $(3x^2 + x + 2) + (4x^2 + 2x + 1)$
$$= 3x^2 + 1x + 2 + 4x^2 + 2x + 1$$
$$= 3x^2 + 4x^2 + 1x + 2x + 2 + 1$$
$$= 7x^2 + 3x + 3$$

Matched Problem 4   Combine like terms mentally:

(A) $4x + 7y + 9x$ (B) $3x^2 + y^2 + 2x^2 + 3y^2$
(C) $(2m^2 + 3m + 5) + (m^2 + 4m + 2)$

$13x + 7y$ $5x^2 + 4y^2$ $3m^2 + 7m + 7$

♦ MULTIPLYING AND COMBINING LIKE TERMS

By multiplying out and combining like terms, an algebraic expression can be rewritten in a different but equivalent form. Quite often the result is simpler and more useful than the original.

Example 5
Simplifying

Multiply out and combine like terms:

(A) $3x(x + 5) + 4x(2x + 3)$
(B) $2x(x^2 + 2x + 1) + x(3x^2 + x + 2)$
(C) $3x(2x + 4y) + 2y(3x + y) + 2x^2 + 3y^2$
(D) $(4x + 3)(3x + 2) = (4x + 3)3x + (4x + 3)2$

Solution

(A) $3x(x + 5) + 4x(2x + 3) = 3x^2 + 15x + 8x^2 + 12x$
$= 11x^2 + 27x$

(B) $2x(x^2 + 2x + 1) + x(3x^2 + x + 2)$
$= 2x^3 + 4x^2 + 2x + 3x^3 + x^2 + 2x$
$= 5x^3 + 5x^2 + 4x$

(C) $3x(2x + 4y) + 2y(3x + y) + 2x^2 + 3y^2$
$= 6x^2 + 12xy + 6xy + 2y^2 + 2x^2 + 3y^2$ Note: $6yx = 6xy$
$= 8x^2 + 18xy + 5y^2$

(D) $(4x + 3)(3x + 2) = (4x + 3)3x + (4x + 3)2$
$= 12x^2 + 9x + 8x + 6$
$= 12x^2 + 17x + 6$

Matched Problem 5   Multiply out and combine like terms:

(A) $4m(m + 3) + m(6m + 1)$ $4m^2 + 12m + 6m^2 + m = 10m^2 + 13m$
(B) $3x(2x^3 + x + 1) + 2x(x^3 + 3x^2 + 2)$ $6x^4 + 3x^2 + 3x + 2x^4 + 6x^3 + 4x$
(C) $4x^2 + 3y(2x + y) + 2x(x + 3y) + y^2$
$8x^4 + 6x^3 + 3x^2 + 7x$
(D) $(3x + 2)(2x + 1)$

Answers to
Matched Problems

1. 5, 2, 1, 4, and 1   2. (A) $6n, 2n$ (B) $2xy, xy; 3xy^3, 2xy^3$
3. (A) $11y$ (B) $5x + 9y^2$
4. (A) $13x + 7y$ (B) $5x^2 + 4y^2$ (C) $3m^2 + 7m + 7$
5. (A) $10m^2 + 13m$ (B) $8x^4 + 6x^3 + 3x^2 + 7x$
(C) $6x^2 + 12xy + 4y^2$ (D) $6x^2 + 7x + 2$

EXERCISE 1-7

A *Indicate the numerical coefficient of each term:*

1. $4x$ 2. $7ab$ 3. $8x^2y$ 4. $9uv^2$

5. x^3 6. y^5 7. u^2v^3 8. m^3n^5

Given the algebraic expression $2x^4 + 3x^3 + x^2 + 5x$, indicate each of the following:

9. The coefficient of x^4 10. The coefficient of x^2

11. The coefficient of x 12. The coefficient of x^3

Select like terms in each group of terms:

13. $8x$, $5y$, x **14.** x, y, $2x$

15. $3x$, $2y$, $4x$, $5y$ **16.** $3m$, $2n$, $5m$, $7n$

17. $6x^2$, x^3, $3x^2$, x^2, $4x^3$ **18.** $2y^2$, $3y^4$, $5y^4$, y^2, y^4

19. $2u^2v$, $3uv^2$, u^2v, $5uv^2$

20. $5mn^2$, m^2n, $2m^2n$, $3mn^2$

Combine like terms:

21. $5x + 4x$ **22.** $2m + 3m$

23. $3u + u$ **24.** $x + 7x$

25. $7x^2 + 2x^2$ **26.** $4y^3 + 6y^3$

27. $4xy + 9xy$ **28.** $3ab + 8ab$

29. $2x + 3x + 5x$ **30.** $4u + 5u + u$

31. $2a + 5a + 9a$ **32.** $6x + 3x + 11x$

33. $2x + 3y + 5x + y$ **34.** $m + 2n + 3m + 4n$

35. $2x + 3y + 5 + x + 2y + 1$

36. $3a + b + 1 + a + 4b + 2$

B *Select like terms in each group:*

37. m^2n, $4mn^2$, $2mn$, $3mn$, $5m^2n$, mn^2

38. $3u^2v$, $2uv$, u^2v, $2uv^2$, $4uv$, uv^2

Combine like terms:

39. $2t^2 + t^2 + 3t^2$ **40.** $6x^3 + 3x^3 + x^3$

41. $a^4 + 2a^4 + 3a^4$ **42.** $7x^2 + x^2 + 5x^2$

43. $3xy^2 + 6xy^2 + 7xy^2$ **44.** $3a^2b + 8a^2b + 6a^2b$

45. $3x + 5y + x + 4z + 2y + 3z$

46. $2r + 7t + r + 4s + r + 3t + s$

47. $9x^3 + 4x^2 + 3x + 2x^3 + x$

48. $y^3 + 2y + 3y^2 + 4y^3 + 2y^2 + y + 5$

49. $x^2 + xy + y^2 + 3x^2 + 2xy + y^2$

50. $3x^2 + 2x + 1 + x^2 + 3x + 4$

51. $(2x + 1) + (2x + 3) + (2x + 5)$

52. $(4x + 1) + (3x + 2) + (2x + 5)$

53. $(t^2 + 5t + 3) + (3t^2 + t) + (2t + 7)$

54. $(4x^4 + 2x^2 + 3) + (x^4 + 3x^2 + 1)$

55. $(x^3 + 3x^2y + xy^2 + y^3) + (2x^3 + 3xy^2 + y^3)$

56. $(2u^3 + uv^2 + v^3) + (u^3 + v^3) + (u^3 + 3u^2v)$

Multiply, using the distributive property, and combine like terms:

57. $2(x + 5) + 3(2x + 7)$

58. $5(m + 7) + 2(3m + 6)$

59. $x(x + 1) + x(2x + 3)$

60. $2t(3t + 5) + 3t(4t + 1)$

61. $5(t^2 + 2t + 1) + 3(2t^2 + t + 4)$

62. $4(u^2 + 3u + 2) + 2(2u^2 + u + 1)$

63. $y(y^2 + 2y + 3) + (y^3 + y) + y^2(y + 1)$

64. $2y(y^2 + 2y + 5) + 7y(3y + 2) + y(y^2 + 1)$

65. $2x(3x + y) + 3y(x + 2y)$

66. $3m(2m + n) + 2n(3m + 2n)$

67. $2x^2(2x^2 + y^2) + y^2(x^2 + 3y^2)$

68. $3u^2(u^2 + 2v^2) + v^2(2u^2 + v^2)$

69. $3m^4(m^2 + 2m + 1) + m^3(m^3 + 3m^2 + m)$

70. $4x^3(x^2 + 3x) + 2x^4(3x + 1)$

C *In Problems 71–84, multiply using the distributive property, and combine like terms.*

71. $2xy^2(3x + x^2y) + 3x^2y(y + xy^2)$

72. $3s^2t^3(2s^3t + s^2t^2) + 2s^3t^2(3s^2t^2 + st^3)$

73. $3u^2v(2uv^2 + u^2v) + 2uv^2(u^2v + 2u^3)$

74. $4m^3n^2(3mn^2 + n) + 2mn^2(2m^3n^2 + m^2n)$

75. $(x + 1)(x + 2) = x(x + 2) + 1(x + 2) = ?$
(finish the multiplication)

76. $(x + 3)(x + 5) = x(x + 5) + 3(x + 5) = ?$
(finish the multiplication)

77. $(x + 6)(x + 2)$ **78.** $(x + 4)(x + 1)$

79. $(2x + 3)(3x + 2) = (2x + 3)3x + (2x + 3)2 = ?$
(finish the multiplication)

80. $(x + 2)(2x + 3) = (x + 2)2x + (x + 2)3 = ?$
(finish the multiplication)

81. $(x + 2y)(2x + y)$ **82.** $(3x + y)(x + 3y)$

83. $(x + 3)(x^2 + 2x + 5)$ **84.** $(r + s + t)(r + s + t)$

85. If y represents an odd number, write an algebraic expression for the product of y and the next odd number. Write as the sum of two terms.

86. If y represents the first of four consecutive even numbers, write an algebraic expression that would represent the product of the first two added to the product of the last two. Simplify.

87. An even number plus the product of it and the next even number is 180. Introduce a variable and write as an algebraic equation. Simplify the left and right sides of the equation where possible.

88. There exist at least two consecutive odd numbers such that 5 times the first plus twice the second is equal to twice the first plus 3 times the second. Introduce a variable and write as an algebraic equation. Simplify the left and right sides of the equation where possible.

$$5x + 2y = 2x + 3y$$

89. If x represents a natural number, write an algebraic expression for the sum of four consecutive natural numbers starting with x. Simplify the expression by combining like terms.

90. If t represents an even number, write an algebraic expression for the sum of three consecutive even numbers starting with t. Simplify.

CHAPTER SUMMARY

1-1 THE NATURAL NUMBERS

A **set** is a collection of objects. A set is called **finite** if its elements can be counted and the counting ends; otherwise it is **infinite.** The infinite set $\{1, 2, 3, \ldots\}$ is called the set of **counting numbers** or **natural numbers.** The **even numbers,** those natural numbers exactly divisible by 2, and the **odd numbers,** those not divisible by 2, are subsets of the natural numbers. Two or more numbers being added or subtracted are called **terms;** when multiplied, they are called **factors.** A **composite number** is a natural number that can be written as the product of two natural numbers, neither being 1. A **prime number** is a natural number, not 1, that is not a composite number. A natural number is **completely factored** when it is written as a product of primes. The **least common multiple** (LCM) of two or more numbers is the smallest natural number exactly divisible by each. The LCM can be found by taking each prime as a factor the most number of times it occurs in the factorizations of the given numbers.

1-2 THE TRANSITION FROM ARITHMETIC TO ALGEBRA

A **variable** is a symbol that is used to represent any of the numbers in a set, called the **replacement set** for the variable. A **constant** is a symbol for one object in a set. An **algebraic expression** is an expression built up from constants, variables, mathematical operations (addition, subtraction, multiplication, division), and grouping symbols. Algebraic expressions joined by addition or subtraction are called **terms;** when joined by multiplication, they are called **factors.**

1-3 EQUALITY AND INEQUALITY

The **equality sign** $(=)$ between two expressions means that they are names or descriptions for the same thing. The symbol $\neq$ means "is not equal to." The symbols $<$ and $>$ mean "less than" and "greater than," respectively. The symbols $\leq$ and $\geq$ mean "less than or equal to" and "greater than or equal to," respectively. An **algebraic equation** is a form joining two algebraic expressions, involving at least one variable, by an equality sign. The set of replacements for variables that make an algebraic equation true is called the **solution set** of the equation. Equality satisfies these properties:

1. Reflexive property: $a = a$.
2. Symmetric property: If $a = b$, then $b = a$.
3. Transitive property: If $a = b$ and $b = c$, then $a = c$.
4. Substitution principle: If $a = b$, then either may be substituted for the other.

1-4 PROPERTIES OF ADDITION AND MULTIPLICATION

Addition and multiplication are **commutative,** that is,

$$a + b = b + a \qquad ab = ba$$

and **associative,** that is,

$$(a + b) + c = a + (b + c) \qquad (ab)c = a(bc)$$

1-5 EXPONENTS AND ORDER OF OPERATIONS

The product $bbb \cdots \cdot b$ with n factors of b can be written as b^n; b is called the **base,** n the **exponent.** For any natural numbers m and n, $b^m b^n = b^{m+n}$; this is the **first law of exponents.** The **order of operations** for arithmetic operations, unless grouping symbols indicate otherwise, is first to apply exponents, then do multiplications and divisions left to right, and then additions and subtractions left to right. Simplification within grouping symbols begins with the innermost symbols and works outward.

1-6 RELATING ADDITION AND MULTIPLICATION

The **distributive property** states that

$$a(b + c) = ab + ac$$

That is, multiplication distributes over addition. Rewriting $a(b + c)$ as $ab + ac$ is referred to as **multiplying out.** Rewriting $ab + ac$ as $a(b + c)$ is referred to as **factoring out the common factor** a.

1-7 REWRITING ALGEBRAIC EXPRESSIONS

The constant factor in a term is called the **coefficient.** Two terms identical except possibly for their coefficients are called **like terms** and may be combined using the distributive property.

CHAPTER REVIEW EXERCISE

Work through all the problems in this chapter review and check the answers in the back of the book. Answers to all review problems are there, and following each answer is a number in italics indicating the section in which that type of problem is discussed. Where weaknesses show up, review appropriate sections in the text.

All variables represent natural numbers:

A **1.** Given $G = \{10, 11, 12, 13, 14, 15\}$:
   (A) Write the set of odd numbers in G.
   (B) Write the set of prime numbers in G.

2. Given the expression $x^3 + 3x^2 + 7x + 4$:
   (A) What is the coefficient of x^2?
   (B) What is the coefficient of x^3?

Find the LCM:

3. 6, 15   **4.** 10, 18   **5.** 12, 21   **6.** 14, 20

Write as a product of prime numbers:

7. 42   **8.** 56   **9.** 84   **10.** 98

Evaluate:

11. $12 - 5 \cdot 2$ **12.** $5 + 3(7 - 5)$

13. $x - 4(x - 7)$ for $x = 9$

14. $(x + 4)(x - 4)$ for $x = 6$

Multiply:

15. $x^{12} x^{13}$ **16.** $(2x^3)(3x^5)$ **17.** $2^5 \cdot 2^{20}$

18. $x(x + 1)$ **19.** $5(2x + 3y + z)$ **20.** $3u(2u^2 + u)$

Combine like terms:

21. $3y + 6y$ **22.** $2m + 5n + 3m$

23. $3x^2 + 2x + 4x^2 + x$ **24.** $3x^2y + 2xy^2 + 5x^2y$

Write in a factored form by taking out factors common to all terms:

25. $3m + 3n$

26. $8u + 8v + 8w$

27. $xy + xw$

28. $4x + 8w$

29. $xz + yz$

30. $x^2 + x$

31. $abc + adc$

32. $3ab + 12ac$

In Problems 33–36, let x represent a natural number and write an algebraic expression that represents each of the following:

33. A number 12 times as large as x

34. A number 3 more than 3 times x

35. A number 5 less than twice x

36. A number 3 times as large as the number 4 less than x

B 37. Let A be the set of natural numbers starting at 21 and ending at 31:
(A) List the elements in the set A that are primes.
(B) Is A finite or infinite?

38. Given $5x^3 + 3x^2 + x + 7$:
(A) What is the coefficient of the second term?
(B) What is the coefficient of the third term?
(C) What is the exponent of the variable in the third term?

Write as a product of prime factors:

39. 120   **40.** 96   **41.** 110   **42.** 140

Find the LCM:

43. 3, 4, 9

44. 6, 5, 9

45. 15, 18, 30

46. 12, 18, 10

47. 12, 30, 42

Evaluate:

48. $2 + 3^2$

49. $(8 - 5) \cdot 4$

50. $1 + 5^2 - 3^2$

51. $(8 + 10) - 3(7 - 3)$

52. $2 \cdot 9 - 6(8 - 2 \cdot 3)$

53. $2[12 - 2(6 - 3)]$

54. $2[(8 + 4) - (7 - 5)]$

55. $2[x + 3(x - 4)]$ for $x = 6$

56. $6[(x + y) - 3(x - y)]$ for $x = 7$ and $y = 5$

57. $x^2 + x + 7$ for $x = 5$

58. $x^2y + xy^2$ for $x = 3$ and $y = 5$

Multiply as indicated and combine like terms where possible:

59. $(2x^3)(3x)(3x^4)$

60. $(3xy^2z)(4x^2y^3z^3)$

61. $3y^3(2y^2 + y + 5)$

62. $2(5u^2 + 2u + 1) + 3(3u^2 + u + 5)$

63. $3x(x + 5) + 2x(2x + 3) + x(x + 1)$

64. $(x + 2y)3x + (x + 2y)y$

Write in a factored form by taking out factors common to all terms:

65. $u^3 + u^2 + u$

66. $6x^2y + 3xy^2$

67. $3m^5 + 6m^4 + 15m^2$

Translate each statement into an algebraic equation using only the variable x:

68. Twenty-four is 6 less than twice a certain number.

69. Three times a given number is 12 more than that number.

70. The sum of four consecutive natural numbers is 138. (Let x be the first of the four consecutive natural numbers.)

71. The sum of three consecutive even numbers is 78. (Let x be the first of the three consecutive even numbers.)

C *Find the LCM:*

72. 132, 270

73. 90, 108, 132

74. 12, 30, 50, 80

Write as a product of prime numbers:

75. 169   **76.** 540   **77.** 1485   **78.** 2520

Evaluate:

79. $4\{20 - 4[(11 - 3) - 3 \cdot 2]\}$

80. $3\{x + 2[8 - 2(x - y)]\}$ for $x = 5$ and $y = 3$

81. $2[x^2 + 3(7 + x)]$ for $x = 3$

82. $5[xy + 2x(y + 5)]$ for $x = 4$ and $y = 2$

Multiply and combine like terms where possible:

83. $5u^3v^2(2u^2v^2 + uv + 2)$

84. $2x^3(2x^2 + 1) + 3x^2(x^3 + 3x + 2)$

85. $(4x + 3)(2x + 1)$

Write in factored form by taking out factors common to all terms.

86. $12x^3yz^2 + 9x^2yz$

87. $20x^3y^2 + 5x^2y^3 + 15x^2y^2$

88. $3abc + 9a^2bc + 18a^3b^2c$

89. $8x^4y + 4yz^2 + 12x^2z^4$

90. $3xyz + 15xy^2 + 18yz$

91. $21ab^3c + 12bc^3 + 4a^2$

In Problems 92–98, each statement is justified by either a commutative or an associative property, or the distributive property. State which.

92. $x3 = 3x$

93. $(x + 3) + 2 = x + (3 + 2)$

94. $(3 + x) + 5 = (x + 3) + 5$

95. $(x + 3) + (x + 5) = x + [3 + (x + 5)]$

96. $(x + 1)(x + 2) = (x + 2)(x + 1)$

97. $(x + 1)(x + 2) = (x + 1)x + (x + 1)2$

98. $3(a + b) = 3(b + a)$

99. If x represents a natural number, write an algebraic equation that represents the fact that the square of this number is the product of 2 less than the number and 1 less than the number.

100. If x represents the first of three consecutive odd numbers, write an algebraic equation that represents the fact that 4 times the first is equal to the sum of the second and third.

CHAPTER PRACTICE TEST

The following practice test is provided for you to test your knowledge of the material in this chapter. You should try to complete it in 50 minutes or less. The answers in the back of the book indicate the section in the text that covers the material in the question. Actual tests in your class may vary from this practice test in difficulty, length, or emphasis, depending on the goals of your course or instructor.

Find the quantity requested:

1. The sum of the prime numbers among 20, 21, 22, 23, 24, 25, 26, 27, 28, 29, 30.

2. The LCM of 48 and 64.

Evaluate the expression:

3. $3 \cdot 5^2 - 8 \cdot 9$

4. $[(8 - 2^2)^2 - 3 \cdot 5]$

5. $x^3 + 3x^2 + 2x + 1$ for $x = 2$

6. $(xy - x)z + x$ for $x = 2$, $y = 5$, and $z = 4$

7. $xy - x + xz^2$ for $x = 2$, $y = 5$, and $z = 4$

Rewrite the expression in the form requested:

8. 150 as a product of primes

9. $xxxyyyzz$ in exponent form

10. x^5x^8 as a power of x

11. $ab + ac + ad$ as a product with two factors

12. $x(y + z + w)$ as a sum of three terms

13. $2x(3x + 5y + 8z)$ by multiplying out

14. $3xy^2 + 9x^2y^2 + 15x^2y$ by taking out all common factors

15. $4x^2 + 7x^2 + 12x^2$ by combining like terms

16. $2ab + 4c + 11ab + 5ab + 6c$ by combining like terms

17. $(x + 3)(x + 7)$ by multiplying out

Simplify the given expression and write the final answer in the form indicated:

18. $(3x^4)(5x^7)$ [*product*]

19. $3x + 7xy + 15x + 8xy$ [*product*]

20. $x(x + 1)(x + 2)$ [*sum*]

Identify the property used in each step:

21. $a(b + c) + ad = (ab + ac) + ad$
$= ab + (ac + ad)$

22. $ab + (ac + ad) = ab + a(c + d)$
$= a[b + (c + d)]$

23. $a(bc) + ab = (ab)c + ab$
$= ab(c + 1)$

Translate the given statement into an algebraic equation using x as the only variable:

24. 11 more than a number is 7 less than 3 times the number.

25. Twice the square of a number is 1 more than the square of the quantity 2 more than the given number.

2 Integers

In the first chapter we limited ourselves to the simplest number system within your experience, the natural numbers. We were able to develop many basic algebraic processes without the distracting influence of more complicated numbers such as negatives, decimals, fractions, and radicals. We will find that most of these processes carry on without change to the more involved number systems to be presented in this and the next chapter. In this chapter, we will extend basic algebraic ideas to the set of integers.

Photo reference: see Exercise 2-1, Problem 29.

2-1   **The Integers**

♦ Natural Numbers, or Positive Integers
♦ Integers and a Number Line

The set of natural numbers has limitations. For example, neither

$$3 \div 5 \quad \text{nor} \quad 2 - 5$$

has an answer that is a natural number. As a step toward overcoming these limitations and at the same time increasing our problem-solving power, we now extend the natural numbers to the integers.

♦ NATURAL NUMBERS, OR POSITIVE INTEGERS

We start by giving the natural numbers another name. From now on they will also be called **positive integers.** To help us emphasize the difference between the positive integers (natural numbers) and the negative integers that are to be introduced shortly, we will sometimes place a plus sign in front of a numeral used to name a natural number. Thus, we may use either

$$+3 \text{ or } 3 \qquad +25 \text{ or } 25 \qquad +372 \text{ or } 372$$

and so on.

♦ INTEGERS AND A NUMBER LINE

If we form a **number line** (a line with numbers associated with points on the line) using the positive integers, and divide the line to the left into line segments equal to those used on the right, how should the points on the left be labeled?

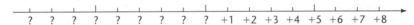

As you might guess, we label the first point to the left of $+1$ with *zero*

$$0$$

and the other points in succession with

$$-1, -2, -3, \ldots$$

These last numbers are called **negative integers.**

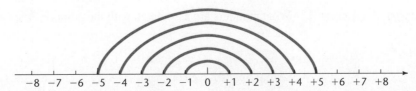

Figure 1

In general, to each positive integer there corresponds a unique (one and only one) number called a negative integer: -1 to $+1$, -2 to $+2$, -3 to $+3$, and so on. The minus sign is part of the number symbol. The elements in the integer pairs -1 and $+1$, -2 and $+2$, and so on are often referred to as **opposites** of each other. Zero has no sign attached.

In historical terms, zero and the negative integers are relatively recent concepts. Both concepts were introduced as numbers in their own right between A.D. 600 and 700. Hindu mathematicians in India are given credit for their invention. The growing importance of commercial activities seemed to be the stimulus. Since business transactions involve decreases as well as increases, it was found that both transactions could be treated at once if the positive integers represented amounts received and the negative integers represented amounts paid out. Since then, negative numbers have been put to many other uses, such as recording temperatures below 0, indicating altitudes below sea level, and representing deficits in financial statements.

By collecting the positive integers, 0, and the negative integers into one set, we obtain the set of **integers,** *Z*, the subject of this chapter.

The Set of Integers, *Z*

$$\{\,\ldots,\ -4,\ -3,\ -2,\ -1,\ 0,\ +1,\ +2,\ +3,\ +4,\ \ldots\,\}$$

The set $\{0,\ +1,\ +2,\ +3,\ \ldots\}$ consisting of the positive integers together with zero is referred to as the set of **whole numbers** or the set of **nonnegative integers.** We do not attempt to give a precise definition of each integer. We do, however, assume the existence of this set of numbers, and we will learn to manipulate them according to certain rules.

Example 1
Associating Numbers with Points

What numbers are associated with the points *a*, *b*, *c*, and *d* on the following number line?

Solution

The point *a* is eight units to the left of 0, so the associated number is -8. The points *b*, *c*, and *d* on the number line are associated with -6, 0, and $+7$, respectively.

Matched Problem 1

What numbers are associated with the points *a*, *b*, *c*, and *d* on the following number line?

Example 2 **Associating Points** **with Numbers**	Locate the set of numbers $\{-8, -3, 0, +5, +9\}$ on a number line.

Solution Draw a number line and locate each number with a solid dot:

Matched Problem 2 Locate the set of numbers $\{-12, -6, +2, +10\}$ on a number line.

Example 3
Using Signed Numbers

Express each of the following by means of an appropriate positive or negative integer:

(A) A bank deposit of $37   **(B)** A bank withdrawal of $52

Solution Associate positive integers with increases in the bank balance and negative integers with decreases in the balance:

(A) A deposit represents an increase, so we associate the increase with $+37$.
(B) A withdrawal represents a decrease in the balance, so we associate this with -52.

Matched Problem 3 Express each of the following by means of an appropriate positive or negative integer:

(A) A 36-yard gain in football
(B) A 5-yard loss in football

Without negative numbers it is not possible to perform the operation

$$7 - 12$$

or to solve the equation

$$8 + x = 2$$

Before this course is over, many more uses of negative numbers will be considered. In the next several sections we will learn how to add, subtract, multiply, and divide integers—operations essential to many uses.

Answers to
Matched Problems

1. $a = -13$; $b = -6$; $c = -1$; $d = +9$

2.
$$-12 \quad -6 \quad\quad 2 \quad\quad 10$$

3. **(A)** $+36$ **(B)** -5

EXERCISE 2-1

A   **1.** What numbers are associated with points *a*, *b*, *c*, and *d*?

2. What numbers are associated with points *a*, *b*, *c*, *d*, and *e*?

3. What numbers are associated with the points *a*, *b*, *c*, *d*, and *e*?

4. What numbers are associated with the points *a*, *b*, *c*, *d*, and *e*?

Locate each set of numbers on a number line:

5. $\{-4, -2, 0, +2, +4\}$

6. $\{-7, -4, 0, +4, +8\}$

7. $\{-25, -20, -15, +5, +15\}$

8. $\{-30, -20, -5, +10, +15\}$

Using the figure for Problem 2, write down the number associated with the following points:

9. 3 units left of *d*   **10.** 4 units right of *e*

11. 4 units right of *a*   **12.** 2 units left of *b*

13. 10 units left of *d*   **14.** 20 units right of *a*

Using the figure for Problem 3, write down the number associated with the following points:

15. 3 units right of *d*   **16.** 4 units left of *e*

17. 5 units left of *b*   **18.** 6 units right of *a*

19. 11 units right of *c*   **20.** 14 units left of *d*

B *Let A be the set $\{-2, 4, -\frac{3}{5}, 3.14, 17, 6{,}035, -21, 0, \sqrt{13}, \frac{2}{9}, 1\}$ and B be the set $\{23, 0, -5, 1.4142, \frac{1}{3}, -\sqrt{2}, 6, -1, 33, 712, -8\}$:*

21. List the positive integers in *A*.

22. List the positive integers in *B*.

23. List the negative integers in *A*.

24. List the negative integers in *B*.

25. List the integers in *A*.

26. List the integers in *B*.

27. List the nonintegers in *A*.

28. List the nonintegers in *B*.

Referring to the figure, express each of the following quantities by means of an appropriate integer:

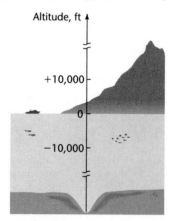

29. A mountain height of 20,270 feet (Mount McKinley, highest mountain in the United States)

30. A mountain peak 29,141 feet high (Mount Everest, highest point on earth)

31. A valley depth of 280 feet below sea level (Death Valley, the lowest point below sea level in the western hemisphere)

32. An ocean depth of 35,800 feet (Marianas Trench in the western Pacific, greatest known depth in the world)

33. A mountain height of 6,960 meters (Aconcagua in Argentina, the highest mountain in the western hemisphere)

34. An ocean depth of 25,197 feet (the greatest depth in the Caribbean Sea)

35. An ocean depth of 4,020 meters (the greatest depth in the Gulf of Mexico)

36. A mountain height of 6,684 feet (Mount Mitchell in North Carolina, the highest mountain in the eastern United States)

Referring to the figure, express each of the following quantities by means of an appropriate integer:

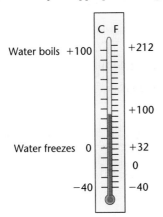

Water boils +100 … +212

+100

Water freezes 0 … +32

0

−40 … −40

C F

37. 5° below freezing on the Celsius scale

38. 35° below freezing on the Celsius scale

39. 5° below freezing on the Fahrenheit scale

40. 100° below boiling on the Fahrenheit scale

41. 35° below freezing on the Fahrenheit scale

42. 220° below boiling on the Fahrenheit scale

In Problems 43–60, express each quantity by means of an appropriate integer:

43. A bank deposit of $25

44. A bank balance of $237

45. A bank withdrawal of $10

46. An overdrawn checking account of $17

47. A 9-yard loss in football

48. A 23-yard gain in football

49. A rise in the Dow Jones Industrial Average of 12 points

50. A decline in the Dow Jones Industrial Average of 13 points.

51. The year 200 B.C.

52. The year A.D. 1995

53. A time of 110 seconds before the lift-off of the space shuttle

54. A time of 24 seconds after the lift-off of the space shuttle

55. An increase of 6 points in the Consumer Price Index

56. A decrease of 1 point in the Consumer Price Index

57. A golf score of 6 under par

58. A golf score of 4 over par

59. A golf score of even par

60. No change in the Consumer Price Index

61. Enter each of the following numbers in your calculator.* If necessary, read the instruction book to find how negative numbers are entered and displayed.
   (A) +23   **(B)** −6   **(C)** −78   **(D)** +31
   (E) −103

62. Enter each of the following numbers in your calculator:
   (A) −38   **(B)** +9   **(C)** −110   **(D)** +234
   (E) −87

C *In each problem start at 0 on a number line and give the number associated with the final position:*

63. Move 7 units in the positive direction, 4 units in the negative direction, 5 more units in the negative direction, and finally 3 units in the positive direction.

64. Move 4 units in the negative direction, 7 units in the positive direction, and 13 units in the negative direction.

65. Move 5 units in the negative direction, 2 units in the positive direction, 4 units in the negative direction, and 6 units in the positive direction.

66. Move 3 units in the positive direction, 5 units in the negative direction, 7 units in the positive direction, and 2 units in the negative direction.

67. Move 6 units right, 3 left, 4 left, 5 right, 2 left.

68. Move 4 units left, 7 right, 8 left, 3 right, 6 right.

Express the net gain or loss as an appropriate integer:

69. In banking: a $23 deposit, a $20 withdrawal, a $14 deposit

70. In banking: a $32 deposit, a $15 withdrawal, an $18 withdrawal

71. In football: a 5-yard gain, a 3-yard loss, a 4-yard loss, an 8-yard gain, a 9-yard loss

72. In an elevator: up 2 floors, down 7 floors, up 3 floors, down 5 floors, down 2 floors

73. In an elevator: down 3 floors, up 6 floors, down 2 floors, up 5 floors

74. In football: a 3-yard loss, a 6-yard gain, a 10-yard gain, an 8-yard loss, a 22-yard gain

* Most problems in this text do not *require* the use of a calculator. However, many of the problems, especially those in subsequent chapters that involve decimals, are made easier by using one. It is assumed that students have access to basic calculators.

75. In any sport: a 4-game winning streak, a 2-game losing streak, a 3-game winning streak, a 1-game losing streak, a 2-game winning streak

76. In any sport: a 2-game losing streak, a 5-game winning streak, a 3-game losing streak, a 1-game winning streak, a 2-game losing streak

77. In sales: 5 units above quota, 4 units above quota, 2 units below quota, 3 units above quota

78. In sales: 1 unit below quota, 4 units above quota, 3 units below quota, exactly at quota, 1 unit below quota

79. In gambling: a $2 win, a $4 loss, a $2 loss, a $6 win, a $2 win

80. In gambling: a $5 loss, a $10 win, a $20 win, a $5 loss, a $5 loss

---

2-2   Opposites and Absolute Values

♦   The Opposite of a Number
♦   The Absolute Value of a Number
♦   Combined Operations

An algebraic expression is like a recipe in that it contains instructions on how to proceed in its evaluation. For example, if we were to evaluate

$$5(x + 2y)$$

for $x = 10$ and $y = 3$, we would write

$$5(10 + 2 \cdot 3)$$

Using our order of operation rules, we would multiply 2 and 3, add the product to 10, and then multiply the sum by 5. Thus, the instructions expressed symbolically involve the operations "multiply" and "add."

In this section we are going to define two more operations on numbers called "the opposite of" and "the absolute value of," and you will get further practice in following symbolic instructions. These two new operations are widely used in mathematics and its applications. We will use them in the following sections to help us define addition, subtraction, multiplication, and division for integers. We start by defining the opposite of a number.

♦   THE OPPOSITE OF A NUMBER

By the **opposite of a number** x, we mean an operation on x, symbolized by

$$-x \qquad \text{Opposite of } x$$

that produces another number. What number? It changes the sign of x if x is not 0, and if x is 0, it leaves it alone.

Example 1
Finding Opposites

Find:

(A) $-(+3)$ **(B)** $-(-5)$ **(C)** $-(0)$ **(D)** $-[-(+3)]$

Solution   **(A)** To find the opposite of $+3$, change the sign. That is, $-(+3) = -3$.
(B) $-(-5) = +5$ **(C)** $-(0) = 0$ **(D)** $-[-(+3)] = -[-3] = +3$

Matched Problem 1   Find:

(A) $-(+7)$ **(B)** $-(-6)$ **(C)** $-(0)$ **(D)** $-[-(-4)]$

Graphically, the opposite of a number is its "mirror image" relative to 0 (see the figure).

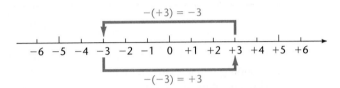

$-(+3) = -3$

$$-6 \quad -5 \quad -4 \quad -3 \quad -2 \quad -1 \quad 0 \quad +1 \quad +2 \quad +3 \quad +4 \quad +5 \quad +6$$

$-(-3) = +3$

As a consequence of the definition of the opposite of a number, we note the following important properties:

1. The opposite of a positive number is a negative number.
2. The opposite of a negative number is a positive number.
3. The opposite of 0 is 0.

CAUTION

Thus, we see that $-x$ is not necessarily a negative number: **$-x$ represents a positive number if x is negative and a negative number if x is positive** (see the figure: if x is -3, then $-x$ is 3; if x is 3, then $-x$ is -3).
The negative sign $(-)$ is used in three distinct ways:

1. As the operation "subtract": $7 \overset{\downarrow}{-} 5 = 2$ 7 minus 5
2. As the operation "the opposite of":

$$\overset{\downarrow}{-}(-6) = +6 \quad \text{Opposite of negative 6}$$
$$\overset{\downarrow}{-}(+3) = -3 \quad \text{Opposite of 3}$$

3. As part of a number symbol: $\overset{\downarrow}{-}8$ Negative 8

The opposite of a number x, $-x$, is sometimes referred to as "the negative of x." Since this terminology often causes confusion at this stage, we will avoid its use until later.

♦ **THE ABSOLUTE VALUE OF A NUMBER**

The **absolute value of a number** x is an operation on x, denoted symbolically by

$$|x| \quad \text{Absolute value of } x$$

that produces another number. What number? If x is positive or 0, it leaves it alone; if x is negative, it makes it positive. Thus, the absolute value of a number may be thought of as the number "without the sign."

Don't confuse square brackets [x] with the absolute value symbol |x|.

Geometrically, |x| represents the distance—a positive quantity—that x is from 0 (see the figure).

$$\overset{\longleftarrow |-5| = +5 \longrightarrow}{\underset{-6\ -5\ -4\ -3\ -2\ -1\ \ 0\ +1\ +2\ +3\ +4\ +5\ +6}{|\ \ \ \ \ \ \ \ \ \ \ \ \ \ |}}\overset{\longleftarrow |+5| = +5 \longrightarrow}{|\ \ \ \ \ \ \ \ \ \ \ \ \ \ |}$$

Symbolically, and more formally,

$$|x| = \begin{cases} x & \text{if } x \text{ is a positive number or } 0 \\ -x & \text{if } x \text{ is a negative number} \end{cases}$$

Do not be afraid of this symbolic form of the definition. It represents a first exposure to more precise mathematical representations, and it will take on more meaning with repeated exposure.

Example 2
Finding Absolute Values

Evaluate:

(A) |-7| **(B)** |$+7$| **(C)** |0|

Solution   **(A)** Since -7 is 7 units to the *left* of 0, |-7| = $+7$.
(B) |$+7$| = $+7$ **(C)** |0| = 0

Matched Problem 2

Evaluate:

(A) |$+5$| **(B)** |-5| **(C)** |0|

We see that:

1. The absolute value of a positive number is a positive number.
2. The absolute value of a negative number is a positive number.
3. The absolute value of 0 is 0.

We therefore conclude that:

The absolute value of a number is never negative.

♦   **COMBINED OPERATIONS**

"The absolute value of" and "the opposite of" operations are often used in combination, and it is important to perform the operations in the right order—generally, from the inside out.

Example 3
Evaluating Opposites and Absolute Values

Evaluate:

(A) |$-(-3)$| **(B)** $-|-3|$ **(C)** $\{-(|-5| - |-2|)\}$

Solution   **(A)** $|-(-3)| = |+3| = +3$ **(B)** $-|-3| = -(+3) = -3$
(C) $\{-(|-5| - |-2|)\} = -(5 - 2) = -(+3) = -3$

Matched Problem 3   Evaluate:

(A) $|-(+5)|$ **(B)** $-|+5|$ **(C)** $-(|-7| + |-3|)$
        5   -5   -10

Example 4
**Recognizing Opposites
and Absolute Values**

Replace each question mark with an appropriate integer:

(A) $-(?) = +7$ **(B)** $-(?) = -7$ **(C)** $|?| = +3$ **(D)** $|?| = -3$
        -7   7

Solution   **(A)** -7 since $-(-7) = +7$ **(B)** $+7$ since $-(+7) = -7$
(C) $+3$ or -3 since $|+3| = +3$ and $|-3| = +3$
(D) No integer will work, since the absolute value of a number is never negative.

Matched Problem 4   Replace each question mark with an appropriate integer:

(A) $-(?) = -6$ **(B)** $-(?) = +6$ **(C)** $|?| = -1$ **(D)** $|?| = +5$
        6   -6 ± 5

Example 5
**Evaluating Opposites
and Absolute Values**

Evaluate for $x = -3$ and $y = +2$:

(A) $-x$ **(B)** $|x|$ **(C)** $-(-y)$ **(D)** $-|-y|$
        3   3   2   -2

Solution   **(A)** $-x \boxed{= -(\ )} = -(-3) = +3$ Remember that the dashed boxes are "think"
steps and usually are not written down as a
separate step.

(B) $|x| \boxed{= |\ \ |} = |-3| = +3$

(C) $-(-y) \boxed{= -[-(\ )]} = -[-(+2)] = -(-2) = +2$

(D) $-|-y| \boxed{= -|-(\ )|} = -|-(+2)| = -|-2| = -(+2) = -2$

Matched Problem 5   Evaluate for $x = +3$ and $y = -2$:

(A) $-y$ **(B)** $|y|$ **(C)** $-(-x)$ **(D)** $-|-x|$
        2   2   3   -3

**Answers to
Matched Problems**

1. (A) -7 **(B)** $+6$ **(C)** 0 **(D)** -4
2. (A) $+5$ **(B)** $+5$ **(C)** 0
3. (A) $+5$ **(B)** -5 **(C)** -10
4. (A) 6 **(B)** -6 **(C)** No value   **(D)** $+5$ or -5
5. (A) $+2$ **(B)** $+2$ **(C)** $+3$ **(D)** -3

EXERCISE 2-2

A *Evaluate the expression:*

1. $-(+9)$ 2. $-(+14)$ 3. $-(-2)$

4. $-(-3)$ 5. $|+4|$ 6. $|+10|$

7. $|-6|$ 8. $|-7|$ 9. $-(0)$

10. $|0|$

Choose the italicized word in parentheses that correctly completes the sentence.

11. The opposite of a number is *(always, sometimes, never)* a negative number.

12. The opposite of a number is *(always, sometimes, never)* a positive number.

13. The absolute value of a number is *(always, sometimes, never)* a negative number.

14. The absolute value of a number is *(always, sometimes, never)* a positive number.

Replace each question mark with an appropriate integer:

15. $-(+11) = ?$ 16. $-(-15) = ?$

17. $-(?) = +5$ 18. $+(?) = -8$

19. $|-13| = ?$ 20. $|+17| = ?$

21. $|+11| = ?$ 22. $|-22| = ?$

23. $+(?) = -10$ 24. $-(?) = +8$

25. $|?| = +2$ 26. $|?| = +8$

27. $|?| = -4$ 28. $|?| = 0$

B *Evaluate:*

29. $-[-(+6)]$ 30. $-[-(-11)]$

31. $|-(-5)|$ 32. $|-(+7)|$

33. $-|-5|$ 34. $-|+7|$

35. $-|+15|$ 36. $-|-12|$

37. $|-(+16)|$ 38. $|-(-13)|$

39. $-[(-7)]$ 40. $-[-(+9)]$

41. $(|-3| + |-2|)$ 42. $(|-7| - |+3|)$

43. $-(|-12| - |-4|)$ 44. $-(|-6| + |-2|)$

45. $-(|-5| + |-8|)$ 46. $-(|-12| - |-7|)$

47. $-(|-12| - |-8|)$ 48. $-(|-14| - |-6|)$

Evaluate for $x = +7$ and $y = -5$:

49. $-x$ 50. $|x|$ 51. $|y|$

52. $-y$ 53. $-|x|$ 54. $-|y|$

55. $-(-y)$ 56. $-(-x)$ 57. $|-y|$

58. $|-x|$ 59. $|x| - |y|$ 60. $-(|x| + |y|)$

Find the solution set of each equation from the set of integers:

61. $|+5| = x$ 62. $|-7| = x$

63. $-x = -3$ 64. $-x = +8$

65. $|x| = +6$ 66. $|x| = +9$

67. $|x| = -4$ 68. $-|x| = +4$

69. $-|-x| = +11$ 70. $-|-x| = -11$

71. $|-x| = +15$ 72. $-|-x| = +9$

73. $-|-x| = -21$ 74. $|-x| = +21$

75. $-|-18| = -x$ 76. $-|+19| = -x$

77. $-|+13| = -x$ 78. $-|-15| = -x$

C *Solve each equation by describing the set of integers for which the equation is true:*

79. $|x| = 0$ 80. $-x = 0$

81. $-x = |x|$ 82. $x = |x|$

83. $-(-x) = x$ 84. $|-x| = |x|$

85. $|-x| = x$ 86. $-x = x$

87. $-|x| = |x|$ 88. $|-x| = -|x|$

     # 2-3 **Addition**

♦   Definition of Addition
♦   Adding Several Integers

How should addition in the integers be defined so that we can assign numbers to each of the following sums?

$$(+2) + (+5) = ? \qquad (-2) + (+5) = ? \qquad (+5) + 0 = ?$$
$$(+2) + (-5) = ? \qquad (-2) + (-5) = ? \qquad 0 + (-5) = ?$$

To give us an idea, let us think of addition of integers in terms of deposits and withdrawals in a checking account, starting with a 0 balance. If we do this, then a deposit of $2 followed by another deposit of $5 would provide us with a balance of $7. Thus, as we would expect from addition in the natural numbers,

$$(+2) + (+5) = (+7)$$

Similarly, a deposit of $2 followed by a withdrawal of $5 would yield an over-drawn account of $3, and we would write

$$(+2) + (-5) = -3$$

Continuing in the same way, we can assign to each sum the value that indicates the final status of our account after the two transactions have been completed. Hence,

$$(-2) + (+5) = +3 \qquad (+5) + 0 = +5$$
$$(-2) + (-5) = -7 \qquad 0 + (-5) = -5$$

We can visualize these activities on a number line.

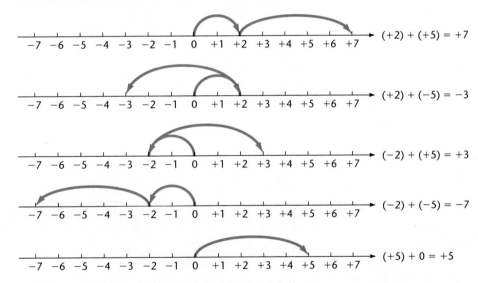

In each case, adding a positive integer corresponds to moving to the *right*, adding a negative integer corresponds to moving to the *left*.

◆ DEFINITION OF ADDITION

We want to define addition of integers so that we get the same results as above, as well as satisfy the rearrangement (commutative) and grouping (associative) proper-ties we had with the natural numbers. Before we state the definition of addition, we point out that **two numbers of like sign** are either two positive numbers or two negative numbers and that **two numbers of unlike sign** are numbers such that one is positive and the other is negative. Now we are ready to define addition by considering three possibilities for the two numbers to be added: zero is one of the numbers, the two numbers are of like sign, or the two numbers are of unlike sign.

Addition of Integers

1. **Adding zero** to any integer does not change that integer.
   $$0 + (-5) = -5 \qquad +5 + 0 = +5 \qquad 0 + 0 = 0$$

2. To **add two integers with the same sign,** ignore the sign, add, and then attach the common sign to this sum.
   $$(+5) + (+2) = +(5 + 2) = +7$$
   $$(-5) + (-2) = -(5 + 2) = -7$$

3. To **add two integers with unlike signs,** ignore the signs, subtract the smaller unsigned number from the larger, then attach the sign associated with the larger unsigned number to this difference.
   $$(+2) + (-5) = -(5 - 2) = -3$$
   $$(-2) + (+5) = +(5 - 2) = +3$$

More formally, to add two integers with like signs, we add their absolute values, and attach their common sign to this sum. To add two integers with unlike signs, we subtract the smaller absolute value from the larger and attach the sign of the number with the larger absolute value to this difference.

If you think of each example in the definition in terms of deposits and withdrawals in a bank account or moving back and forth on a number line, then the processes and results will seem more reasonable. You will need to be able to do addition of integers mentally as in Example 1 below. It is important that you practice integer addition enough so that you can perform the operation with speed and accuracy.

Example 1
Adding Integers

Add:

(A) $(+4) + (+6)$ (B) $(-4) + (-6)$ (C) $(-4) + (+6)$
(D) $(+4) + (-6)$ (E) $0 + (-1)$

Solution

(A) $(+4) + (+6) = +(4 + 6) = +10$

(B) $(-4) + (-6) = -(4 + 6) = -10$

(C) $(-4) + (+6) = +(6 - 4) = +2$

(D) $(+4) + (-6) = -(6 - 4) = -2$

(E) $0 + (-1) \qquad = -1$

Matched Problem 1

Add:

(A) $(-2) + (+7)$ (B) $(+2) + (-7)$
(C) $(-2) + (-7)$ (D) $0 + (-5)$

Addition problems are often written vertically instead of horizontally. Thus

$$\begin{array}{r} -8 \\ +3 \\ \hline -5 \end{array} \quad \text{and} \quad (-8) + (+3) = -5$$

mean the same thing.

The commutative and associative properties of addition were introduced for positive integers in Section 1-4. These properties are also true for addition of integers in general.

Addition Properties

For all integers *a*, *b*, and *c*:

(A) $a + b = b + a$ COMMUTATIVE PROPERTY
$(-2) + (+3) = (+3) + (-2)$

(B) $(a + b) + c = a + (b + c)$ ASSOCIATIVE PROPERTY
$[(-1) + (+3)] + (-2) = (-1) + [(+3) + (-2)]$

As a consequence of these properties, we will have essentially the same freedom that we had with the natural numbers in rearranging terms and inserting or removing parentheses relative to addition.

♦ ADDING SEVERAL INTEGERS

Now let us turn to the problem of adding three or more integers. The commutative and associative properties of the integers allow this procedure:

Steps in Adding Several Integers

1. Add all positive integers.
2. Add all negative integers.
3. Add the two resulting sums.

Example 2
Adding Several Integers

Add: $(+3) + (-6) + (+8) + (-4) + (-5)$.

Solution

$(+3) + (-6) + (+8) + (-4) + (-5)$ List the positive integers first.

$= [(+3) + (+8)] + [(-6) + (-4) + (-5)]$ Add the positive integers. Add the negative integers.

$= (+11) + (-15)$ Add the resulting sums.
$= -4$

The addition can also be done vertically:

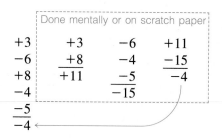

Matched Problem 2   Add: $(-8) + (-4) + (+6) + (-3) + (+10) + (+1)$.

The number zero plays a special role in addition. Opposites add up to 0. Conversely, if two numbers add to 0, then the numbers must be opposites. More formally we can state:

Addition of Opposites

1. For each integer a, the sum of a and its opposite is 0; that is, $a + (-a) = 0$. $5 + (-5) = 0$ $-5 + [-(-5)] = 0$
2. If the sum of two numbers is 0, then each must be the opposite of the other; that is, if $a + b = 0$, then $a = -b$ and $b = -a$.
   If $a + 5 = 0$, then $a = -5$ and $5 = -a$

Example 3
Adding Opposites   Replace the question marks with appropriate symbols:

(A) $(+5) + (-5) = ?$ (B) $(-y) + y = ?$
(C) $(+7) + (?) = 0$ (D) $m + (?) = 0$

Solution   (A) $(+5) + (-5) = 0$ (B) $(-y) + y = 0$
(C) $(+7) + (-7) = 0$ (D) $m + (-m) = 0$

Matched Problem 3   Replace the question marks with appropriate symbols:

(A) $(-3) + (+3) = ?$ (B) $d + (-d) = ?$
(C) $(-4) + (?) = 0$ (D) $(-w) + (?) = 0$

Answers to   **1. (A)** $+5$ **(B)** -5 **(C)** -9 **(D)** -5 **2.** $+2$
Matched Problems   **3. (A)** 0 **(B)** 0 **(C)** $+4$ **(D)** w

---

EXERCISE 2-3

6. -3 7. -7 8. $+8$ 9. 0 10. -4
   $\underline{-4}$ $\underline{-1}$ $\underline{+2}$ $\underline{+3}$ $\underline{0}$

11. $(+5) + (+4)$ 12. $(-7) + (-3)$

A *Add:*

13. $(-8) + (+2)$ 14. $(+3) + (-7)$

1. $+5$ 2. $+7$ 3. -9 4. -6 5. $+6$
   $\underline{+6}$ $\underline{-4}$ $\underline{+6}$ $\underline{+8}$ $\underline{-8}$

15. $(-6) + (-3)$ 16. $(+2) + (+3)$

17. $0 + (-9)$ 18. $(+2) + 0$

19. $\begin{array}{r} +4 \\ -3 \\ -5 \\ -7 \\ +9 \\ \hline \end{array}$ **20.** $\begin{array}{r} -6 \\ -4 \\ +8 \\ +3 \\ -5 \\ \hline \end{array}$ **21.** $\begin{array}{r} -7 \\ +2 \\ -3 \\ -1 \\ +5 \\ \hline \end{array}$ **22.** $\begin{array}{r} +6 \\ -4 \\ -8 \\ -2 \\ +9 \\ \hline \end{array}$

23. $(+5) + (-8) + (-9) + (+7)$

24. $(-8) + (-7) + (+3) + (+9)$

25. $(-6) + 0 + (+5) + (-2) + (-1)$

26. $(+9) + (-3) + 0 + (-8)$

27. $\begin{array}{r} +11 \\ -23 \\ \hline \end{array}$ **28.** $\begin{array}{r} -12 \\ -21 \\ \hline \end{array}$ **29.** $\begin{array}{r} -403 \\ -219 \\ \hline \end{array}$ **30.** $\begin{array}{r} -307 \\ +231 \\ \hline \end{array}$

31. $(-63) + (+25)$ **32.** $(-45) + (-73)$

33. $(-237) + (-431)$ **34.** $(-197) + (+364)$

35. $\begin{array}{r} +12 \\ -18 \\ -23 \\ +4 \\ -11 \\ \hline \end{array}$ **36.** $\begin{array}{r} -63 \\ +45 \\ -3 \\ +17 \\ +12 \\ \hline \end{array}$

37. $(+12) + (+7) + (-37) + (+14)$

38. $(-23) + (-35) + (+43) + (-33)$

B *Replace each question mark with an appropriate integer:*

39. $(-3) + ? = -7$ **40.** $? + (-9) = -13$

41. $(+8) + ? = +3$ **42.** $(-12) + ? = +4$

43. $? + (-12) = -7$ **44.** $(+54) + ? = -33$

45. $(+33) + ? = -44$ **46.** $? + (-14) = +20$

47. $(-8) + (+8) = ?$ **48.** $(-8) + ? = 0$

49. $? + (+8) = 0$ **50.** $(+12) + ? = 0$

51. $? + (-12) = 0$ **52.** $(+12) + (-12) = ?$

53. $0 + ? = 0$ **54.** $0 + 0 = ?$

Add:

55. $(+3,462) + (-5,237) + (-1,304) + (-7,064)$

56. $(+2,062) + (-3,896) + (+6,438) + (-7,064)$

57. $(-2,345) + (+4,567) + (+1,001) + (-6,243)$

58. $(-1,357) + (-2,468) + (+9,753) + (-4,826)$

Evaluate:

59. $|-8| + |+6|$ **60.** $|(-8) + (+6)|$

61. $(-|3|) + (-|+3|)$ **62.** $|-5| + [-(-8)]$

63. $(-8) + |-7| + (+9) + (-4)$

64. $(-|-7|) + (-5) + |-10| + (-3)$

65. $[-(-9)] + |-9| + (-9) + (-|-9|)$

66. $|-(-8)| + (-8) + |-8| + [-(-8)]$

Evaluate for $x = -5$, $y = +3$, and $z = -2$:

67. $x + y$ **68.** $y + z$

69. $|(-x) + z|$ **70.** $-(|x| + |z|)$

71. $-|(-x) + y|$ **72.** $|-x + |-z||$

73. $-|x + |-y||$ **74.** $-|x + (-z)|$

C *In Problems 75–80, use addition of integers to find the net gain or loss by an appropriate integer:*

75. In banking: a $32 deposit, a $15 withdrawal, an $18 withdrawal

76. In an elevator: down 3 floors, up 6 floors, down 2 floors, up 5 floors

77. In football: a 3-yard loss, a 6-yard gain, a 10-yard gain, an 8-yard loss, a 22-yard gain

78. In any sport: a 4-game winning streak, a 2-game losing streak, a 3-game winning steak, a 1-game losing streak, a 2-game winning streak

79. In sales: 1 unit below quota, 4 units above quota, 3 units below quota, exactly at quota, 1 unit below quota

80. In gambling: a $2 win, a $4 loss, a $2 loss, a $6 win, a $2 win

81. You own a stock that is traded on a stock exchange. On Monday it closed at $23 per share, it fell $3 on Tuesday and another $6 on Wednesday, it rose $2 on Thursday, and finished strongly on Friday by rising $7. Use addition of signed numbers to determine the closing price of the stock on Friday.

82. Your football team is on the opponent's 10-yard line and in four downs gains 8 yards, loses 4 yards, loses another 8 yards, and gains 13 yards. Use addition of signed numbers to determine whether a touchdown was made.

83. A spelunker (cave explorer) descended 2,340 (vertical) feet into the 3,300-foot Gouffre Berger, the world's deepest pothole cave, located in the Isere province of France. On his ascent he climbed 732 feet, slipped back 25 feet and then another 60 feet, climbed 232 feet, and finally slipped back 32 feet. Use addition of signed numbers, starting with $-2,340$, to find his final position.

84. In a card game (such as rummy, where cards held in your hand after someone goes out are counted against you) the following scores were recorded after four hands of play. Who was ahead at this time and what was his or her score?

Russ	Jan	Paul	Meg
+35	+80	−5	+15
+45	+5	+40	−10
−15	−35	+25	+105
−5	+15	+35	−5

85. Your charge account begins with a balance of $28 owed and then has the following activity: a charge of $36, a charge of $21, a charge of $48, a return for credit of $33, a charge of $42, and a payment of $50. What is your new balance?

86. Your charge account begins with a balance of $30 owed and then has the following activity: a charge of $18, a charge of $45, a return for credit of $22, a charge of $38, a charge of $75, and a payment of $150. What is your new balance?

87. Utility companies make use of *degree-days* to estimate power usage throughout the year. *Cooling degree-days* are used in the summer, and *heating degree-days* are used in the winter. A **cooling degree-day** is defined to be the number of degrees by which the average temperature for a day exceeds some standard, usually 65°F. Days for which the average temperature is 65° or less are counted as 0 cooling degree-days. Thus average daily temperatures of 73° and 62° result in cooling degree-days of +8 and 0, respectively. Cooling degree-days for each day are added to give some idea of power usage. What is the weekly total of cooling degree-days for a week in which the average daily temperatures were 68°, 71°, 72°, 65°, 63°, 62°, and 66°?

88. Repeat Problem 87 for a week with average daily temperatures of 60°, 65°, 70°, 73°, 68°, 64°, and 67°.

89. A **heating degree-day** is the same as a cooling degree-day (see Problem 87) except that the number of degrees *below* 65°F is recorded. Thus a temperature of 57° results in a degree-day of +8, and a temperature of 70° in a degree-day of 0. What is the total number of heating degree-days for a week with average daily temperatures of 68°, 59°, 58°, 63°, 65°, 62°, and 67°?

90. Repeat Problem 89 for a week with average daily temperatures of 60°, 68°, 59°, 73°, 64°, 63°, and 70°.

In Problems 91–94, replace each question mark with an appropriate symbol (variables represent integers):

91. $a + (-a) = ?$

92. $(-x) + x = ?$

93. $m + ? = 0$

94. $(-x) + ? = 0$

95. Give a reason for each of the following steps:

$$
\begin{aligned}
(a + b) + (-a) &= (-a) + (a + b) \\
&= [(-a) + a] + b \\
&= 0 + b \\
&= b
\end{aligned}
$$

96. Give a reason for each of the following steps:

$$
\begin{aligned}
(-b) + (a + b) &= (a + b) + (-b) \\
&= a + [b + (-b)] \\
&= a + 0 \\
&= a
\end{aligned}
$$

  # 2-4 Subtraction

♦ Definition of Subtraction
♦ Order in the Integers

Given two natural numbers, we know how to subtract the smaller from the larger:

$$(+8) - (+5) = +3$$

Notice that we obtain the same result by changing the sign of the number being subtracted and then adding:

$$(+8) - (+5) = (+8) + (-5) = +3$$

This process will allow us to subtract any integer from any other, and we will define subtraction this way.

♦ **DEFINITION OF SUBTRACTION**

Subtraction of Integers

To **subtract one integer from another** change the sign of the integer being subtracted and add.

Change sign.

$$(-3) - (-4) = (-3) + (+4) = +1$$

Subtraction becomes addition.

More formally, to subtract one integer from another, add the opposite of the integer being subtracted. That is, if a and b are integers, then

$$a - b = a + (-b)$$

Thus, any subtraction problem can be changed to an equivalent addition problem. For example,

$$(+3) - (-4) = (+3) + (+4) = +7$$
$$(+3) - (+4) = (+3) + (-4) = -1$$
$$(-3) - (+4) = (-3) + (-4) = -7$$

Recall that addition of integers is related to moving back and forth on the number line. To add a positive integer to a given one, start at the given integer and move to the *right* the number of units being added; to add a negative integer, move to the *left:*

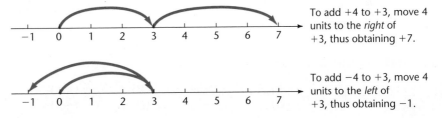

To add +4 to +3, move 4 units to the *right* of +3, thus obtaining +7.

To add −4 to +3, move 4 units to the *left* of +3, thus obtaining −1.

Figure 1

Since we subtract by changing the sign and adding, these rules are reversed for subtraction. To subtract a positive integer we move that many units to the *left;* to subtract a negative integer we move to the *right:*

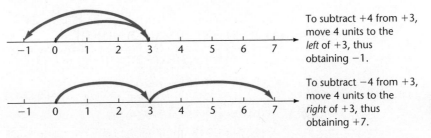

To subtract +4 from +3, move 4 units to the *left* of +3, thus obtaining −1.

To subtract −4 from +3, move 4 units to the *right* of +3, thus obtaining +7.

Figure 2

Example 1
Subtracting Integers

Subtract:

(A) $(+7) - (-8)$ **(B)** $(-4) - (+5)$
(C) $(-9) - (-4)$ **(D)** $0 - (+8)$

Solution **(A)** $(+7) - (-8) = (+7) + (+8) = +15$

Change the sign of (-8) and add

(B) $(-4) - (+5) = (-4) + (-5) = -9$

Change the sign of $(+5)$ and add

(C) $(-9) - (-4) = (-9) + (+4) = -5$ Change the sign of (-4) and add.
(D) $0 - (+8) = 0 + (-8) = -8$ Change the sign of $(+8)$ and add.

An equivalent definition of subtraction says that the difference $a - b$ is what must be added to b to get a. That is,

$$a - b = c \qquad \text{means} \qquad a = b + c$$

This way of looking at subtraction provides a useful check on calculations. For example, you can check the work in Example 1:

$$(+7) - (-8) \stackrel{?}{=} +15$$

Check that $+7 = (-8) + (+15)$, and this is true.

$$(-9) - (-4) \stackrel{?}{=} -5$$

Check that $-9 = (-4) + (-5)$, and this is true.

Matched Problem 1

Subtract:

(A) $(+6) - (-9)$ **(B)** $(-3) - (-5)$
(C) $(-7) - (-2)$ **(D)** $(-3) - (+8)$

Example 2
Subtracting Integers

Evaluate:

(A) $[(-3) - (+2)] - [(+2) - (+5)]$
(B) $(-x) - y$ for $x = -3$ and $y = +5$

Solution **(A)** $[(-3) - (+2)] - [(+2) - (+5)]$ Evaluate inside the brackets first.

$$= [(-3) + (-2)] - [(+2) + (-5)]$$

$$= (-5) - (-3)$$ Now subtract.

$$= (-5) + (+3)$$

$$= -2$$

(B) $(-x) - y$ Substitute $x = -3$ and $y = +5$.

$[-(-3)] - (+5)$ Evaluate inside the brackets first.

$= (+3) - (+5)$ Subtract.

$= (+3) + (-5)$

$= -2$

Matched Problem 2   Evaluate:

(A) $[(+3) + (-8)] - [(-2) - (+3)]$
(B) $x - (-y)$ for $x = -3$ and $y = +8$

♦ **ORDER IN THE INTEGERS**

In Chapter 1, we introduced the inequality symbols $<$, $>$, $\leq$, and $\geq$ to deal with the natural order of the positive integers and assumed that it was obvious to you that the statement 3 is less than 7, that is, that

$$3 < 7$$

is a true statement. But does it seem equally obvious that the statements

$$-8 < -1 \qquad -15 < 1 \qquad -320 > -10,000$$

are also true? To make the inequality relation precise so that we can interpret it relative to *all* integers, we need a careful definition of the concept.

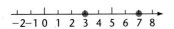

Intuitively, one integer is greater than a second if it is located to the right of the second on the number line. Thus, from their locations on the number line, $7 > 3$. Since moving to the right means adding a positive integer and moving to the left means subtracting a positive integer, we are led to a formal definition of $<$ and $>$:

Definition of $<$ and $>$

If a and b are integers, then we write $a < b$ if there exists a positive integer p such that $a + p = b$. Equivalently, $b > a$ if the difference $b - a$ is positive.

> $3 < 7$ because there is a positive integer, namely, 4, such
> that $3 + 4 = 7$, or equivalently because $7 - 3 = +4$ is positive.

The definition is not so complicated as it might seem and will prove useful in establishing properties of inequalities later. Certainly, one would expect that if a positive number were added to *any* integer it would make it larger and if it were subtracted from *any* integer it would make it smaller. That is essentially what the definition states. Note that if $b > a$, then it follows that $a < b$, and vice versa. That is, $b > a$ and $a < b$ mean the same thing.

___

Example 3
Inequalities in the Integers

Replace each question mark with $<$ or $>$:

(A) $2\,?\,3$ **(B)** $-8\,?\,-1$ **(C)** $0\,?\,-5$ **(D)** $-2\,?\,-10{,}000$

Solution

(A) $2 < 3$ Since $2 + 1 = 3$ or since $3 - 2 = +1$ is positive
(B) $-8 < -1$ Since $-8 + 7 = -1$ or $(-1) - (-8) = +7$ is positive
(C) $0 > -5$ Since $0 - (-5) = +5$ is positive
(D) $-2 > -10{,}000$ Since $-2 - (-10{,}000) = +9{,}998$ is positive

Matched Problem 3

Replace each question mark with $<$ or $>$:

(A) $4\,?\,6$ **(B)** $6\,?\,4$ **(C)** $-6\,?\,-4$
(D) $-9\,?\,9$ **(E)** $3\,?\,-9$ **(F)** $0\,?\,-14$

As in Chapter 1, the inequality symbols $<$ and $>$ are often combined with the equality symbol $=$ to form statements such as $a \le b$ or $c \ge d$:

$$a \le b \qquad \text{means} \qquad a \text{ is less than or equal to } b$$

$$c \ge d \qquad \text{means} \qquad c \text{ is greater than or equal to } d$$

Answers to Matched Problems

1. **(A)** $+15$ **(B)** $+2$ **(C)** -5 **(D)** -11
2. **(A)** 0 **(B)** $+5$
3. **(A)** $<$ **(B)** $>$ **(C)** $<$ **(D)** $<$ **(E)** $>$ **(F)** $>$

___

EXERCISE 2-4

A *Subtract as indicated:*

1. $(+9) - (+4)$ **2.** $(+10) - (+7)$

3. $(+9) - (-4)$ **4.** $(+10) - (-7)$

5. $(+4) - (+9)$ **6.** $(+7) - (+10)$

7. $(-9) - (-4)$ **8.** $(-10) - (-7)$

9. $(-4) - (-9)$ **10.** $(-7) - (-10)$

Replace each question mark with $<$ or $>$:

11. $7\,?\,5$ **12.** $3\,?\,6$

13. $5\,?\,7$ **14.** $6\,?\,3$

15. $-7\,?\,-5$ **16.** $-3\,?\,-6$

17. $-5\,?\,-7$ **18.** $-6\,?\,-3$

19. $0\,?\,8$ **20.** $5\,?\,0$

21. $0\,?\,-8$ **22.** $-5\,?\,0$

23. $-7\,?\,5$ **24.** $-6\,?\,3$

25. $-842\,?\,0$ **26.** $-905\,?\,-10$

27. $900\,?\,-1{,}000$ **28.** $505\,?\,-55$

Subtract as indicated:

29. $(+6) - (-8)$ **30.** $(-4) - (+7)$

31. $(+6) - (+10)$ **32.** $(+6) - (-10)$

33. $(-9) - (-3)$ **34.** $0 - (-7)$

35. $0 - (+5)$ **36.** $(-1) - (+6)$

B **37.** $(-12) - (-27)$ **38.** $(+57) - (+92)$

39. $0 - (-87)$ **40.** $0 - (+101)$

41. $(-271) - (+44)$ **42.** $(+327) - (-73)$

43. $(-245) - 0$ **44.** $(+732) - 0$

45. $(+8{,}063) - (-9{,}810)$

46. $(-6{,}024) - (-5{,}321)$

47. $(-1{,}203) - (+4{,}027)$

48. $(+2{,}539) - (+7{,}681)$

49. $(-984) - (-1{,}803)$

50. $(+4{,}398) - (-1{,}661)$

51. $(+3{,}899) - (+5{,}604)$

52. $(-609) - (+4{,}477)$

(C) $\dfrac{-27}{+9} = -3$

(D) $\dfrac{+27}{-9} = -3$

Quotients of numbers with unlike signs are negative.

(E) $\dfrac{0}{-4} = 0$

Zero divided by a nonzero number is zero.

(F) $\dfrac{+7}{0}$

(G) $\dfrac{0}{0}$

These are not defined, since 0 cannot be a divisor.

Matched Problem 3   Divide:

(A) $\dfrac{+18}{+6}$ **(B)** $\dfrac{-18}{-6}$ **(C)** $\dfrac{+18}{-6}$ **(D)** $\dfrac{-18}{+6}$

(E) $\dfrac{0}{-6}$ **(F)** $\dfrac{-18}{0}$ **(G)** $\dfrac{0}{0}$

The multiplication properties

$$(-a)b = a(-b) = -(ab) \qquad (-a)(-b) = ab$$

have exact counterparts for division:

Division Properties

For any integers a and b with $b \neq 0$:

$$(-a) \div b = a \div (-b) = -(a \div b) \qquad \text{or} \qquad \dfrac{-a}{b} = \dfrac{a}{-b} = -\dfrac{a}{b}$$

$$(-a) \div (-b) = a \div b \qquad \text{or} \qquad \dfrac{-a}{-b} = \dfrac{a}{b}$$

For example,

$$\dfrac{-30}{5} = \dfrac{30}{-5} = -\dfrac{30}{5} = -6 \qquad \text{and} \qquad \dfrac{-30}{-5} = \dfrac{30}{5} = +6$$

◆ **COMBINED OPERATIONS**

Let us finish this section by considering examples involving all four arithmetic operations $(+, -, \cdot, \div)$. Recall that:

Multiplication and division precede addition and subtraction, unless grouping symbols indicate otherwise. Also, all the operations are performed from left to right.

Example 4
Operations on Integers

Evaluate:

(A) $\dfrac{-18}{-3} + (-2)(+3)$ **(B)** $(+6)(-3) - \dfrac{-20}{4}$

(C) $(-1)(-2)(-3) - \left[(-4) - \dfrac{-6}{+2}\right]$

Solution   **(A)** $\dfrac{-18}{-3} + (-2)(+3) = (+6) + (-6) = 0$

(B) $(+6)(-3) - \dfrac{-20}{4} = (-18) - (-5) = -13$

(C) $(-1)(-2)(-3) - \left[(-4) - \dfrac{-6}{+2}\right] = (-6) - [(-4) - (-3)]$

$$= (-6) - (-1)$$
$$= -5$$

Matched Problem 4

Evaluate:

(A) $(-4)(-3) + \dfrac{-16}{+2}$ **(B)** $\dfrac{(+2)(-9)}{-3} + \dfrac{(+4) - (-6)}{-5}$

(C) $(-3)[(-4) - (-2)] - \dfrac{-24}{-3}$

Example 5
Evaluating Integer Expressions

Evaluate each expression for $x = -24$, $y = +6$, $z = -3$:

(A) $\dfrac{xz}{y}$ **(B)** $yz + \dfrac{x}{y}$ **(C)** $\dfrac{y}{z} - \dfrac{x}{y}$

Solution   **(A)** $\dfrac{xz}{y} = \dfrac{(-24)(-3)}{+6} = \dfrac{+72}{+6} = +12$

(B) $yz + \dfrac{x}{y} = (+6)(-3) + \dfrac{-24}{+6} = (-18) + (-4) = -22$

(C) $\dfrac{y}{z} - \dfrac{x}{y} = \dfrac{+6}{-3} - \dfrac{-24}{+6} = (-2) - (-4) = (-2) + (+4) = +2$

Matched Problem 5

Evaluate each algebraic expression in Example 5 for $x = +18$, $y = -9$, $z = -3$.

Answers to
Matched Problems

1. **(A)** $+30$ **(B)** -42 **(C)** $+40$ **(D)** -72
2. **(A)** $+3$ **(B)** -3 **(C)** $+6$ **(D)** -6
3. **(A)** $+3$ **(B)** $+3$ **(C)** -3 **(D)** -3 **(E)** 0
    (F) Not defined   **(G)** Not defined
4. **(A)** $+4$ **(B)** $+4$ **(C)** -2
5. **(A)** $+6$ **(B)** $+25$ **(C)** $+5$

EXERCISE 2-5

All variables represent integers.

A *Multiply or divide as indicated:*

1. $(-8)(-4)$ **2.** $(+8)(+4)$ **3.** $(+8)(-4)$

4. $(-8)(+4)$ **5.** $(0)(-7)$ **6.** $(-5)(0)$

7. $\dfrac{-4}{-2}$ **8.** $\dfrac{+14}{+7}$ **9.** $\dfrac{-6}{+2}$

10. $\dfrac{+8}{-4}$ **11.** $\dfrac{-6}{0}$ **12.** $\dfrac{0}{0}$

13. $(+2)(-7)$ **14.** $(-2)(-7)$ **15.** $(-2)(+7)$

16. $(+2)(+7)$ **17.** $(+1)(0)$ **18.** $(0)(+6)$

19. $(-9)/(-3)$ **20.** $(+9)/(+3)$ **21.** $(-9)/(+3)$

22. $(+9)/(-3)$ **23.** $0/(+3)$ **24.** $(-9)/0$

B *Evaluate:*

25. $(-2) + (-1)(+3)$ **26.** $(-3)(-2) + (+4)$

27. $\dfrac{-9}{+3} + (-4)$ **28.** $(-8) + \dfrac{-12}{-2}$

29. $(+2)[(+3) + (-2)]$ **30.** $(+5)[(-4) + (+6)]$

31. $\dfrac{(-10) + (-6)}{-4}$ **32.** $\dfrac{(-12) + (+4)}{+2}$

33. $(+4) - (-2)(-4)$ **34.** $(-7) - (+4)(-3)$

35. $(-6)[(+3) - (+8)]$ **36.** $(-3)[(-2) - (-4)]$

37. $(-7)(+2) - \dfrac{-20}{+5}$ **38.** $\dfrac{+36}{-12} - (-4)(-8)$

39. $(-6)(+2) + (-2)^2$ **40.** $(-3)^2 + (-2)(+1)$

41. $(-6)(+7) - (-3)^2$ **42.** $(-6) - (-2)^2$

43. $(-4)(-3) - \left[(-8) - \dfrac{-6}{+2}\right]$

44. $(+7)(-2) - \left[(+10) - \dfrac{+9}{-3}\right]$

Evaluate for $w = +2$, $x = -3$, $y = 0$, and $z = -24$:

45. wx **46.** wz **47.** z/x **48.** z/w **49.** xyz

50. wyz **51.** y/x **52.** y/z **53.** w/y **54.** z/y

55. $z/(wx)$ **56.** xy/z **57.** $wx - \dfrac{z}{w}$ **58.** $\dfrac{z}{x} - wz$

59. $\dfrac{z}{w} - \dfrac{z}{x}$ **60.** $\dfrac{+15}{x} - \dfrac{-12}{w}$

61. $wxy - \dfrac{y}{z}$ **62.** $\dfrac{xy}{w} - xyz$

Evaluate the following expressions for the given values of x and y: (A) $(-x)y$, (B) $x(-y)$, (C) $-(xy)$.

63. $x = -4$, $y = +3$ **64.** $x = -2$, $y = -5$

65. $x = -4$, $y = -3$ **66.** $x = +2$, $y = -5$

Evaluate the following expressions for the given values of x and y: (A) $(-x)/y$, (B) $x/(-y)$, (C) $-(x/y)$.

67. $x = -6$, $y = +3$ **68.** $x = -20$, $y = -5$

69. $x = -6$, $y = -3$ **70.** $x = +20$, $y = -5$

Evaluate the following expressions for the given values of x and y: (A) $(-x)(-y)$, (B) xy.

71. $x = -4$, $y = +3$ **72.** $x = -2$, $y = -5$

73. $x = -4$, $y = -3$ **74.** $x = +2$, $y = -5$

Evaluate the following expressions for the given values of x and y: (A) $(-x)/(-y)$, (B) x/y.

75. $x = -6$, $y = +3$ **76.** $x = -20$, $y = -5$

77. $x = -6$, $y = -3$ **78.** $x = +20$, $y = -5$

Evaluate the following expressions for the given value of x: (A) $(-1)x$, (B) $-x$.

79. $x = -3$ **80.** $x = +8$

C *What integer replacements for x will make each equation true?*

81. $(-3)x = -24$ **82.** $(+5)x = -20$

83. $\dfrac{-24}{x} = -3$ **84.** $\dfrac{x}{-4} = +12$

85. $(-3)x = 0$ **86.** $\dfrac{x}{+32} = 0$

87. $\dfrac{x}{0} = +4$ **88.** $\dfrac{0}{x} = 0$

Evaluate:

89. $(-271)(-196)$ **90.** $(-37)(+166)$

91. $(+304)(-32)$ **92.** $(-831)(-104)$

93. $\dfrac{-10,881}{+403}$ **94.** $\dfrac{+22,977}{-621}$

95. $\dfrac{-12,276}{-1,023}$ **96.** $\dfrac{-7,854}{-462}$

97. $\dfrac{-2,808}{+12} - (+131)(-13)$

98. $(+17)(-29) - \dfrac{-7,000}{-14}$

2-6 **Simplifying Algebraic Expressions**

♦ Addition and Subtraction without Grouping Symbols
♦ Removing Grouping Symbols
♦ Multiplying by $+1$ or by -1
♦ Order of Removing Grouping Symbols
♦ Inserting Grouping Symbols

We are close to being able to use algebra to solve practical problems. First, however, we need to streamline our methods of representing algebraic expressions. For ease of reading and faster manipulation, it is desirable to reduce the number of grouping symbols and plus signs to a minimum. Thus, **we drop the plus sign from numerals that name positive integers unless a particular emphasis is desired,** so we will usually write

$$1, 2, 3, \ldots \quad \text{instead of} \quad +1, +2, +3, \ldots$$

♦ **ADDITION AND SUBTRACTION WITHOUT GROUPING SYMBOLS**

When three or more terms are combined by addition or subtraction and symbols of grouping are omitted, we convert (mentally) any subtraction to addition (Section 2-4) and add. Thus,

$$8 - 5 + 3 \underbrace{= 8 + (-5) + 3}_{\text{Think}} = 6$$

**Example 1
Adding and Subtracting**

Evaluate:

(A) $2 - 3 - 7 + 4$ **(B)** $-4 - 8 + 2 + 9$ **(C)** $-3 - 8$

Solution

(A) $2 - 3 - 7 + 4 \underbrace{= 2 + (-3) + (-7) + 4}_{\text{Think}} = -4$

(B) $-4 - 8 + 2 + 9 \underbrace{= (-4) + (-8) + 2 + 9}_{\text{Think}} = -1$

(C) $-3 - 8 \underbrace{= (-3) + (-8)}_{\text{Think}} = -11$ Note that $-3 - 8$ is not the same as $(-3)(-8)$, nor is it $(-3) - (-8)$.

Matched Problem 1

Evaluate:

(A) $5 - 8 + 2 - 6$ **(B)** $-6 + 12 - 2 - 1$ **(C)** $-5 - 9$

From the preceding discussion comes the following general process for rearranging terms:

Rearranging Terms

The terms in an algebraic expression may be rearranged without restriction so long as the sign preceding each term accompanies it in the process.

For example,

$$a - b + c = a + c - b \qquad x^2 + y^2 - 2xy = x^2 - 2xy + y^2$$

We now modify the earlier definition of a numerical coefficient and state a simple mechanical rule for combining like terms in more involved algebraic expressions. The **(numerical) coefficient** of a given term in an algebraic expression includes the sign that precedes it.

Example 2
Coefficients

In $3x^3 - 2x^2 - x + 3$, what is the coefficient of

(A) x^3?   (B) x^2?   (C) x?

Solution   Since $3x^3 - 2x^2 - x + 3 \boxed{= 3x^3 + (-2x^2) + (-1x) + 3}$ the coefficient of

Think

(A) x^3 is 3   (B) x^2 is -2 (C) x is -1

Matched Problem 2

In $2y^3 - y^2 - 2y + 4$, what is the coefficient of

(A) y^3?   (B) y^2?   (C) y?

To see the effect of including the sign with the coefficient, consider the following:

$$3x - 2y - 5x + 7y$$
$$= 3x + (-2y) + (-5x) + 7y$$
$$= 3x + (-5x) + (-2y) + 7y \qquad \text{Using the commutative and associative}$$
$$\text{properties mentally}$$
$$= [3 + (-5)]x + [(-2) + 7]y \qquad \text{Using the distributive property}$$
$$= -2x + 5y$$

Thus, including the sign with the coefficient does not change the mechanical rule we had before for combining like terms. That is:

Combining Like Terms

Like terms are combined by adding their numerical coefficients.

We should think of the above example as follows:

$$3x - 2y - 5x + 7y \begin{aligned} &= 3x - 5x - 2y + 7y \\ &= (3 - 5)x + (-2 + 7)y \end{aligned}$$
$$= -2x + 5y$$

Example 3
Combining Like Terms

Combine like terms:

(A) $2x - 3x$ $\quad -1x$

(B) $-4x - 7x$ $\quad -11x$

(C) $3x + 5y + 6x + 2y$

(D) $2x^2 + 3x + 5 + 5x^2 - 2x + 3$

$\quad 9y + -3y$ $\qquad\qquad 7x^2 - 5x - 2$

Solution

(A) $2x - 3x = (2 - 3)x = -x$

(B) $-4x - 7x = (-4 - 7)x = -11x$

(C) $3x - 5y + 6x + 2y = (3 + 6)x + (-5 + 2)y = 9x - 3y$

(D) $2x^2 - 3x - 5 + 5x^2 - 2x + 3 = (2 + 5)x^2 + (-3 - 2)x + (-5 + 3)$
$$= 7x^2 - 5x - 2$$

Matched Problem 3

Combine like terms:

(A) $4x - 5x$ $\quad -1x$

(B) $-8x - 2x$ $\quad -10x$

(C) $7x + 8y + 5x + 10y$

(D) $4x^2 + 5x + 8 + 3x^2 - 7x - 2$

$\quad 2x - 2y$ $\qquad\qquad x^2 - 2x - 10$

♦ **REMOVING GROUPING SYMBOLS**

How can we simplify expressions such as

$$2(3x + 5y) + 2(x + 3y)$$

$6x + -10y + -2x - 6y = 4x + 16y$

We would like to multiply and combine like terms as we did when only plus signs were involved. Since any subtraction can be converted to addition, we can proceed as follows:

$$2(3x - 5y) - 2(x + 3y) \begin{aligned} &= 2[3x + (-5)y] + (-2)(x + 3y) \\ &= 6x + (-10)y + (-2)x + (-6)y \end{aligned}$$
$$= 6x - 10y - 2x - 6y$$
$$= 4x - 16y$$

Mechanically, we usually leave out the justifying steps in the dashed box and use the following process, which is simply an application of the distributive property:

Mechanics of Removing Grouping Symbols

Parentheses can be cleared by multiplying each term within the parentheses by the coefficient outside the parentheses.

$$2(3 + 5) = 2 \cdot 3 + 2 \cdot 5 = 6 + 10 = 16$$

$$-2(3 + 5) = (-2) \cdot 3 + (-2) \cdot 5 = (-6) + (-10) = -16$$

$$2(3 - 5) = 2 \cdot 3 - 2 \cdot 5 = 6 - 10 = -4$$

$$-2(3 - 5) = (-2) \cdot 3 - (-2) \cdot 5 = -6 - (-10) = -6 + 10 = 4$$

Example 4
Removing Parentheses

Remove parentheses and simplify:

(A) $2(x - 3y) + 3(2x - y)$ **(B)** $4x(2x + y) - 3x(x - 3y)$

Solution   **(A)** $2(x - 3y) + 3(2x - y) = 2x - 6y + 6x - 3y = 8x - 9y$
(B) $4x(2x + y) - 3x(x - 3y) = 8x^2 + 4xy - 3x^2 + 9xy = 5x^2 + 13xy$

Matched Problem 4   Remove parentheses and simplify:

(A) $3(2x - 3y) + 2(x - 2y)$ **(B)** $5x(x - 2y) - 3x(2x + y)$

$6x + {}^-9y + 2x + {}^-4y = 8x - 13y$

$5x^2 + {}^-10xy + {}^-6x^2 + {}^-3xy = {}^-x^2 - 13xy$

◆ **MULTIPLYING BY +1 OR BY −1**

If parentheses do not have an accompanying coefficient, we can treat them as if the coefficient is +1 or −1. Since $-1 \cdot a = -a$, we can rewrite an expression like $-(2x - 3y)$ as $-1 \cdot (2x - 3y)$ and thus remove parentheses:

$$-(2x - 3y) = -1 \cdot (2x - 3y) = -1 \cdot 2x - (-1) \cdot 3y$$
$$= -2x - (-3y) = -2x + (+3y)$$
$$= -2x + 3y$$

Since $+1 \cdot a = a$, we can rewrite an expression like $(x + 5y)$ as $+1 \cdot (x + 5y)$ and simply remove the parentheses:

$$(x + 5y) = +1 \cdot (x + 5y)$$
$$= x + 5y$$

This leads to simple rules for removing parentheses when they are not preceded by a coefficient:

Removing Parentheses Having No Coefficient

1. If parentheses are preceded by a minus sign, the sign of each term within the parentheses is changed when the parentheses are removed.

Minus sign   Both signs are changed
↓   ↓ ↓
$$-(a - b) = -a + b$$

2. If parentheses are preceded by a plus sign, the signs of the terms within the parentheses remain unchanged when the parentheses are removed.

Plus sign   No sign change   No sign   No sign change
↓   ↓ ↓   ↓   ↓ ↓
$$+(a - b) = a - b$$ $$(a - b) = a - b$$

Using statements 1 and 2, we see that

$$(x + 7) - (x - 3) = x + 7 - x + 3 = 10$$

The first rule is particularly important. You can either apply the rule directly or remember that it is a result of thinking of parentheses preceded by a minus sign as being multiplied by -1.

Example 5

Removing Parentheses

Remove parentheses and simplify:

(A) $(x + 5y) + (x - 3y)$ **(B)** $(x + 5y) - (2x - 3y)$

$2x + 2y$ $x + 5y + {}^-2x + 3y = -x + 8y$

Solution **(A)** $(x + 5y) + (x - 3y)$ $= 1(x + 5y) + 1(x - 3y)$ Think of the coefficient of

  ⌐ - - - - - - - - - - - - - - ⌐ $(x + 5y)$ and $(x - 3y)$ as 1,
  Think   or use rule 2 above.

$$= x + 5y + x - 3y$$

$$= 2x + 2y$$

(B) $(x + 5y) - (2x - 3y)$

$= 1(x + 5y) - 1(2x - 3y)$ Think of the coefficient of $-(2x - 3y)$

  ⌐ - - - - - - - - - - - - - - ⌐   as -1, or use rule 1 above.
  Think

$$= x + 5y - 2x + 3y$$

$$= -x + 8y$$

Matched Problem 5   Remove parentheses and simplify:

(A) $(3x + 2y) + (2x - 4y)$ **(B)** $(3x + 2y) - (2x - 4y)$

$3x + 2y + 2x + {}^-4y = 5x - 2y$ $3x + 2y + {}^-2x + 4y = x + 6y$

◆ ORDER OF REMOVING GROUPING SYMBOLS

Recall from Section 1-5 that when grouping symbols exist within grouping symbols, we apply the following rule to simplify the expression:

Order of Removing Grouping Symbols

Generally, when removing grouping symbols, we work from the inside out, removing the innermost grouping symbols first.

Referring to the figure, with the size of grouping symbols exaggerated to clarify the pairing of the symbols, the rule says this:

1. Remove the inner symbols first.
2. Remove the symbols labeled 2 next.
3. Remove the outer symbols, labeled 3, last.

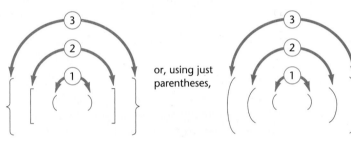

or, using just parentheses,

Figure 1

Example 6
Removing
Grouping Symbols

Remove grouping symbols and combine like terms:

(A) $4x - 3[x - 2(x + 3)]$ **(B)** $u - \{u - [u - 2(u - 1)]\}$

Handwritten annotations:
$4y - 3[x + ^-2x - 6]$ $u - (u + [u + ^-2u + 2])$
$4x + ^-3x + 6x + 18\ \ 7x + 18$ $u + ^-u + ^+u + 2u + 2 = -u + 2$

Solution We work from the inside out, removing the innermost grouping symbols first.

(A) $4x - 3[x - 2(x + 3)]$ Remove ( ) first.
$\quad = 4x - 3[x - 2x - 6]$ Combine like terms.
$\quad = 4x - 3[-x - 6]$ Remove [ ].
$\quad = 4x + 3x + 18$ Combine like terms.
$\quad = 7x + 18$

(B) $u - \{u - [u - 2(u - 1)]\}$ Work from the inside out. We can collect like
$\quad = u - \{u - [u - 2u + 2]\}$ terms as we go or wait until the end. Here
$\quad = u - \{u - u + 2u - 2\}$ we wait until the end.
$\quad = u - u + u - 2u + 2$
$\quad = -u + 2$

Collecting like terms as we go, we obtain the same result:

$$u - \{u - [u - 2(u - 1)]\} = u - \{u - [u - 2u + 2]\} = u - \{u - [-u + 2]\}$$
$$= u - \{u + u - 2\} = u - \{2u - 2\}$$
$$= u - 2u + 2 = -u + 2$$

In expressions such as those in Example 6, we could use only parentheses for grouping and write

$$4x - 3(x - 2(x + 3)) \qquad \text{in place of} \qquad 4x - 3[x - 2(x + 3)]$$

and

$$u - (u - (u - 2(u - 1))) \qquad \text{in place of} \qquad u - \{u - [u - 2(u - 1)]\}$$

but the use of brackets [ ] and braces { } makes it easier to see the groupings.

Matched Problem 6   Remove grouping symbols and combine like terms:

    (A) $3y + 2[y - 3(y - 5)]$ **(B)** $3m - \{m - [2m - (m - 1)]\}$

$3y + 2(y + ^-3y + 15)$

$3y + 2y - 6y + 30 = -y + 30$ $3m - (m - (2m - m + 1)) = 3m - \frac{(m - 2m + m - 1)}{3m + ^-m + 2m - m + 1}$

$3m + 1$

♦ **INSERTING GROUPING SYMBOLS**

Later we will find it useful to reverse the process described above and insert grouping symbols. In the following example we will replace each question mark with an appropriate algebraic expression so that if the parentheses are removed, we will obtain the expression to the left of the equal sign.

Example 7
Inserting Parentheses   Replace each question mark with an appropriate algebraic expression:

    (A) $3x + 4a - 6b = 3x + 2(?)$ **(B)** $6a - 3x + 9y = 6a - 3(?)$
    (C) $5 + 2x - 3y = 5 + (?)$ **(D)** $5 + 2x - 3y = 5 - (?)$

Solution   **(A)** We need to rewrite $4a - 6b$ as $2(?)$. That is, we need to factor out a 2 to obtain $4a - 6b = 2(2a - 3b)$. Thus $3x + 4a - 6b = 3x + 2(2a - 3b)$.
    (B) $6a - 3x + 9y = 6a - 3(x - 3y)$ **(C)** $5 + 2x - 3y = 5 + (2x - 3y)$
    (D) $5 + 2x - 3y = 5 - (-2x + 3y)$

In each part of the example, note that if parentheses are to be preceded by a plus sign, the terms placed within the parentheses remain unchanged. If parentheses are to be preceded by a minus sign, each term placed inside the parentheses undergoes a sign change.

Matched Problem 7   Replace each question mark with an appropriate algebraic expression:

    (A) $2y + 3z - 4w = 2y + (?)$ **(B)** $2y + 3z - 4w = 2y - (?)$
    (C) $a + 4b - 12c = a + 4(?)$ **(D)** $u - 6v - 9w = u - 3(?)$

Answers to
Matched Problems   **1.** **(A)** -7 **(B)** 3 **(C)** -14 **2.** **(A)** 2 **(B)** -1 **(C)** -2
    3. **(A)** $-x$ **(B)** $-10x$ **(C)** $2x - 2y$ **(D)** $x^2 - 2x - 10$
    4. **(A)** $8x - 13y$ **(B)** $-x^2 - 13xy$ **5.** **(A)** $5x - 2y$ **(B)** $x + 6y$
    6. **(A)** $-y + 30$ **(B)** $3m + 1$
    7. **(A)** $3z - 4w$ **(B)** $-3z + 4w$ **(C)** $b - 3c$ **(D)** $2v + 3w$

EXERCISE 2-6

A *Evaluate:*

1. $3 - 2 + 4$ 2. $3 + 4 - 2$

3. $4 - 8 - 9$ 4. $-8 + 4 - 9$

5. $-4 + 7 - 6$ 6. $7 - 6 - 4$

7. $2 - 3 - 6 + 5$ 8. $5 + 2 - 6 - 3$

9. $-3 - 2 + 6 - 4$ 10. $-8 + 9 - 4 - 3$

11. $-7 + 1 + 6 + 2 - 1$

12. $9 - 5 - 4 + 7 - 6 + 10$

13. $-5 - 3 - 8 + 15 - 1$

14. $1 - 12 + 5 + 7 - 1 + 6 - 8$

Remove symbols of grouping where present and combine like terms:

15. $7x - 3x$ 16. $9x - 4x$

17. $2x - 5x - x$ 18. $4t - 8t - 9t$

19. $-3y + 2y - 5y - 6y$ 20. $2y - 3y - 6y + 5y$

21. $2x - 3y - 5x$ 22. $4y - 3x - y$

23. $2x + 8y - 7x - 5y$ 24. $5m + 3n - m - 9n$

25. $2(m + 3n) + 4(m - 2n)$

26. $3(u - 2v) + 2(3u + v)$

27. $2(x - y) - 3(x - 2y)$

28. $4(m - 3n) - 3(2m + 4n)$

29. $(x + 3y) + (2x - 5y)$ 30. $(2u - v) + (3u - 5v)$

31. $x - (2x - y)$ 32. $m - (3m + n)$

33. $(x + 3y) - (2x - 5y)$ 34. $(2u - v) - (3u - 5v)$

35. $3(2x - 3y) - (2x - y)$

36. $2(3m - n) - (4m + 2n)$

B 37. $3xy + 4xy - xy$ 38. $3xy - xy + 4xy$

39. $-x^2y + 3x^2y - 5x^2y$

40. $-4r^3t^3 - 7r^3t^3 + 9r^3t^3$

41. $3x^2 - 2x + 5 - x^2 + 4x - 8$

42. $y^3 + 4y^2 - 10 + 2y^3 - y + 7$

43. $2x^2y + 3xy^2 - 5xy + 2xy^2 - xy$

44. $a^2 - 3ab + b^2 + 2a^2 + 3ab - 2b^2$

45. $x - 3(x + 2y) + 5y$ 46. $y - 2(x - y) - 3x$

47. $-3(-t + 7) - (t - 1)$

48. $-2(-3x + 1) - (2x + 4)$

49. $-2(y - 7) - 3(2y + 1) - (-5y + 7)$

50. $2(x - 1) - 3(2x - 3) - (4x - 5)$

51. $3x(2x^2 - 4) - 2(3x^3 - x)$

52. $5y(2y - 3) + 3y(-2y + 4)$

53. $3x - 2[2x - (x - 7)]$ 54. $y - [5 - 3(y - 2)]$

55. $2t - 3t[4 - 2(t - 1)]$ 56. $2u - 3u[4 - (u - 3)]$

Combine like terms:

57. $23,041 - 2,315 - 43,201 + 12,792$

58. $-623 + 328 - 34 - 512 + 402$

59. $312x + 203y - 278x - 461y$

60. $2,530u - 1,462v - 187u - 239v$

61. $89u - 106v - 137u + 47v$

62. $4,881x - 783y - 6,008x - 608y$

In Problems 63–82, replace each question mark with an appropriate algebraic expression:

63. $2 + 3x - y = 2 + (?)$

64. $5 + m - 2n = 5 + (?)$

65. $2 + 3x - y = 2 - (?)$

66. $5 + m - 2n = 5 - (?)$

67. $x - 4y - 8z = x - 4(?)$

68. $2x - 3a + 12b = 2x - 3(?)$

69. $w^2 - x + y - z = w^2 - (?)$

70. $w^2 - x + y - z = w^2 + (?)$

71. $4x + 6y - z = 2(?) - z$

72. $9a - 6b + 2z = 3(?) + 2z$

73. $4x + 6y - z = -2(?) - z$

74. $9a - 6b + 2z = -3(?) + 2z$

75. $a - b + c - d = (?) - d$

76. $-x + y - z + w = (?) + w$

77. $a - b + c - d = -(?) - d$

78. $-x + y - z + w = -(?) + w$

79. $2x + 4y - 5z - 10w = 2(?) + 5(?)$

80. $3a - 9b + 4c - 28d = 3(?) + 4(?)$

81. $2x + 4y - 5z - 10w = -2(?) - 5(?)$

82. $3a - 9b + 4c - 28d = -3(?) - 4(?)$

83. The width of a rectangle is 5 meters less than its length. If x is the length of the rectangle, write an algebraic expression that represents the perimeter of the rectangle. Then multiply and combine like terms.

84. The length of a rectangle is 8 feet more than its width. If y is the width of the rectangle, write an algebraic expression that represents its area. Change the expression to a form without parentheses.

In Problems 85–90, remove symbols of grouping and combine terms:

85. $x - \{x - [x - (x - 1)]\}$

86. $2t - 3\{t + 2[t - (t + 5)] + 1\}$

87. $2x[3x - 2(2x + 1)] - 3x[8 + (2x - 4)]$

88. $-2t\{-2t(-t - 3) - [t^2 - t(2t + 3)]\}$

89. $3x^2 - 2\{x - x[x + 4(x - 3)] - 5\}$

90. $w - \{x - [z - (w - x) - z] - (x - w)\} + x$

91. A coin purse contains dimes and quarters only. There are four more dimes than quarters. If x represents the number of quarters, write an algebraic expression that represents the value of the money in the purse in cents. Clear grouping symbols and combine like terms.

92. A pile of coins consists of nickels, dimes, and quarters. There are twice as many dimes as nickels and four fewer quarters than dimes. If x represents the number of nickels, write an algebraic expression that represents the value of the pile of coins in cents. Remove grouping symbols and combine like terms.

  ## 2-7   Solving Equations

- ♦ Solutions to Equations
- ♦ Solving Equations
- ♦ Equations with No Solution or with Infinitely Many Solutions
- ♦ Formulas and Literal Equations

We have now reached the point where we can begin solving equations other than by guessing. For example, you would not be likely to guess the solution of

$$2(2x + 5) + 2x = 52$$

This equation is related to a practical problem that we will see in the next section.

Up to now we have mainly been dealing with algebraic *expressions* such as $2(2x + 5) + 2x$, the left side of the above *equation*. Expressions can be *simplified*, or *rewritten*, in various forms, all representing the same quantity. The inclusion of the equal sign in the statement changes our emphasis: we are now interested in what values of the variable x will make the statement true. That is, we want to *solve* an equation. You can substitute 7 for x and find that the expression on the left side of the above equation *is*, in fact, 52. In this section we will develop a *systematic* procedure for finding that $x = 7$.

♦ SOLUTIONS TO EQUATIONS

A **solution** or **root** of an equation in one variable is a replacement of the variable by a number that makes the left side equal to the right. For example, -2 is a solution of

$$4 + x = 2$$

since

$$4 + (-2) = 2$$

To **solve an equation** is to find all its solutions. To **check** a solution means to substitute the solution value into the original equation. If this results in a true statement, then we know we have found a correct solution.

Knowing what we mean by a solution of an equation is one thing; finding it is another. Our objective now is to develop a systematic method of solving equations that is free from guesswork. We start by introducing the idea of equivalent equations. We say that **two equations are equivalent** if they have exactly the same solutions.

The basic idea in solving equations is to perform operations on equations that produce simpler equivalent equations and to continue the process until we reach an equation whose solution is obvious—generally, an equation such as

$$x = -3$$

With a little practice you will find the methods that we are going to develop very easy to use and very powerful. The following properties of equality produce equivalent equations when applied:

Properties of Equality

(A) ADDITION PROPERTY The same quantity may be added to each side of an equation.

$$a = b$$
$$a + c = b + c$$

(B) SUBTRACTION PROPERTY The same quantity may be subtracted from each side of an equation.

$$a = b$$
$$a - c = b - c$$

(C) MULTIPLICATION PROPERTY Each side of an equation may be multiplied by the same nonzero quantity.

$$a = b$$
$$ca = cb$$

(D) DIVISION PROPERTY Each side of an equation may be divided by the same nonzero quantity.

$$a = b$$
$$\frac{a}{c} = \frac{b}{c}$$

Here is an illustration of each of these properties:

(A) If $x - 3 = 7$, then $(x - 3) + 3 = 7 + 3$; simplifying each side yields $x = 10$.

(B) If $x + 5 = 2$, then $(x + 5) - 5 = 2 - 5$; that is, $x = -3$.

(C) If $\frac{x}{2} = 7$, then $2 \cdot \left(\frac{x}{2}\right) = 2 \cdot 7$. Remember that $\frac{x}{2}$ means a number that multiplied by 2 gives x, so $2 \cdot \left(\frac{x}{2}\right) = x$ and $x = 14$.

(D) If $3x = 15$, then $\frac{3x}{3} = \frac{15}{3}$. Now $\frac{3x}{3} = x$, since $3x = 3 \cdot x$, so $x = 5$.

In each case, simplifying the resulting equation on both sides leads to a solution.

We can also think of the process of solving equations as a game. The objective of the game is to isolate the variable with a coefficient of 1 on one side of the equation, usually the left, leaving a constant on the other side. The rules of the game include the equality properties given above as well as the simplifying processes discussed in Section 2-6.

♦ SOLVING EQUATIONS

We are now ready to solve equations. Several examples will illustrate the process.

Example 1
Solving an Equation

Solve $x - 5 = -2$ and check.

$x - 5 + 5 = -2 + 5 = 3$

Solution

$x - 5 = -2$	How can we isolate x on the left side?
$x - 5 + 5 = -2 + 5$	Add 5 to each side using the addition property of equality.
$x = 3$	Solution.

Check

$x - 5 = -2$	Replace x with 3.
$3 - 5 \overset{?}{=} -2$	Evaluate both sides.
$-2 \overset{\checkmark}{=} -2$	We have a check, since both sides are equal.

After some practice the steps in the dashed boxes should be done mentally. If in doubt, include them.

Matched Problem 1

Solve $x + 8 = -6$ and check.

$x + 8 - 8 = -6 - 8 =$
$x = -14$

Example 2
Solving an Equation

Solve $-3x = 15$ and check.

$\dfrac{-3x}{3} = \dfrac{15}{3} = -5$

Solution

$-3x = 15$	How can we make the coefficient of x positive 1. That is, how can we isolate x on the left side?
$\dfrac{-3x}{-3} = \dfrac{15}{-3}$	Divide each side by -3 using the division property of equality. Note that adding 3 to each side, a common mistake, will not isolate x.
$x = -5$	Solution.

The checking is left to you.

Matched Problem 2

Solve $5x = -20$ and check.

$\dfrac{5}{5} \quad \dfrac{5}{-4}$

Do not confuse the following two types of equations:

(A) $x + 2 = 8$ **(B)** $2x = 8$

$-2 = 8 - 2$ $\dfrac{2}{2} \quad \dfrac{2}{2}$

6 4

Right method of solution for each:

(A) $x + 2 = 8$ **(B)** $2x = 8$

$$x + 2 - 2 = 8 - 2 \qquad \frac{2x}{2} = \frac{8}{2}$$

$$x = 6 \qquad\qquad x = 4$$

To remove the 2 from the left side in either equation, we perform the operation opposite of that in which the 2 is involved. For the first equation, the opposite of addition is subtraction. For the second equation the opposite of multiplication is division.

For reasons that will be apparent in Chapters 4 and 5, an expression that can be written in the form $ax + b$, where a and b are constants and x is a variable, is called a **linear expression** in x. For example,

$$x - 5, \ x + 8, \ -3x, \ 5x, \ x + 2, \ \text{and} \ 3x - 5$$

are all linear expressions in x. The equations we are solving here equate two linear expressions. Such equations are called **linear** or **first-degree equations.** Linear equations involve no power of the variable higher than 1, and have no variables in denominators or in radicals. We can state a strategy for solving such equations.

Strategy for Solving Linear Equations

Objective: Isolate the variable with a coefficient of 1 on one side of the equation, usually the left, leaving a constant on the other side. To do this:

1. Simplify each side of the equation separately by removing any grouping symbols and combining any like terms.
2. Use addition or subtraction properties of equality to get all variable terms on one side, usually the left, and all constant terms on the other side, usually the right. Simplify both sides again.
3. Finally, if the variable has a coefficient other than 1, use the multiplication or division property of equality to make the coefficient of the variable 1.
4. Check your answer by substituting it into the original equation.

The following examples illustrate the use of this strategy.

Example 3
Solving an Equation

Solve $2x - 8 = 5x + 4$ and check.

$$2x - 8 = 5x + 4 \qquad 2x = 5x$$
$$+8 \qquad\qquad +8$$
$$\frac{-3x}{-3} = \frac{12x}{-3} = -4$$

Solution $2x - 8 = 5x + 4$ To remove -8 from the left side, add 8 to both sides using the addition property of equality.

$$2x - 8 + 8 = 5x + 4 + 8$$

$$2x = 5x + 12$$

To remove $5x$ from the right side, subtract $5x$ from both sides using the subtraction property of equality.

$$2x - 5x = 5x + 12 - 5x$$

$$-3x = 12$$

To isolate x on the left side with a coefficient of $+1$, divide both sides by -3 using the division property of equality.

$$\frac{-3x}{-3} = \frac{12}{-3}$$

$$x = -4$$

We have solved the equation!

Check

$$2x - 8 = 5x + 4$$

$$2(-4) - 8 \overset{?}{=} 5(-4) + 4$$

$$-8 - 8 \overset{?}{=} -20 + 4$$

$$-16 \overset{\checkmark}{=} -16$$

Replace x with -4 and proceed as in Examples 1 and 2.

Matched Problem 3   Solve $3x - 9 = 7x + 3$ and check.

$$3x - 9 = 7x + 3 + 9$$
$$3x - 7x \quad \frac{-4}{-4} \qquad \frac{12}{-4} \overset{x}{=} -3$$

Example 4
Solving an Equation

Solve and check:

(A) $3x - 2(2x - 5) = 2(x + 3) - 8$

$$3x - 2(2x - 5) = 2x + 6 - 8$$
$$3x + 4x + 10 = 2x - 2$$

(B) $3x^2 + 5(x - 1) = 3x(x + 2) - 3$

Solution   **(A)** We will simplify the expressions on each side of the equal sign first, and then proceed as in the preceding example:

$$3x + {}^-4x = 2x - 2 + {}^-10$$

$$\frac{-3x}{-3} = \frac{-12}{-3} = x$$

$$4$$

$$3x - 2(2x - 5) = 2(x + 3) - 8 \qquad \text{Remove grouping symbols.}$$

$$3x - 4x + 10 = 2x + 6 - 8 \qquad \text{Now combine like terms.}$$

$$-x + 10 = 2x - 2 \qquad \text{Subtract 10 from each side.}$$

$$-x = 2x - 12 \qquad \text{Subtract } 2x \text{ from each side.}$$

$$-3x = -12 \qquad \text{Divide each side by } -3.$$

$$x = 4 \qquad \text{We now have the solution.}$$

Check

$$3x - 2(2x - 5) = 2(x + 3) - 8$$

$$3(4) - 2[2(4) - 5] \overset{?}{=} 2[(4) + 3] - 8$$

$$12 - 2(8 - 5) \overset{?}{=} 2(7) - 8$$

$$12 - 2(3) \overset{?}{=} 14 - 8$$

$$12 - 6 \overset{?}{=} 6$$

$$6 \overset{\checkmark}{=} 6$$

(B) $3x^2 + 5(x - 1) = 3x(x + 2) - 3$ First clear parentheses.

$3x^2 + 5x - 5 = 3x^2 + 6x - 3$

$3x^2 + 5x - 5 - 3x^2 = 3x^2 + 6x - 3 - 3x^2$ Note how $3x^2$ drops out by subtracting it from each side. If this did not happen, we could not solve the equation at this time.

$5x - 5 = 6x - 3$ Add 5 to each side.

$5x = 6x + 2$ Subtract $6x$ from each side.

$-x = 2$ Multiply each side by -1.

$x = -2$

Check $3x^2 + 5(x - 1) = 3x(x + 2) - 3$

$3(-2)^2 + 5[(-2) - 1] \overset{?}{=} 3(-2)[(-2) + 2] - 3$

$12 + (-15) \overset{?}{=} (-6)(0) - 3$

$-3 \overset{\checkmark}{=} -3$

Matched Problem 4   Solve and check:

(A) $8x - 3(x - 4) = 3(x - 4) + 6$
(B) $x(x + 3) + 3 = x^2 + 4(x + 1)$

◆ **EQUATIONS WITH NO SOLUTION OR WITH INFINITELY MANY SOLUTIONS**

The equations studied so far all had exactly one solution. It is possible for some equations of this type to have no solution. This will be true if the equation can be changed into an equivalent form that is never true, for example, $0 = 1$. It is also possible that every number is a solution. This will be true if the equation can be changed into an equivalent form that is always true, for example, $x = x$ or $3 = 3$.

Example 5
Solving Equations   Solve:

(A) $2(x - 3) = 2(x + 5) - 5$ **(B)** $2(x - 3) = 2(x + 1) - 8$

Solution   **(A)** $2(x - 3) = 2(x + 5) - 5$ Remove grouping symbols.

$2x - 6 = 2x + 10 - 5$ Combine like terms.

$2x - 6 = 2x + 5$ Subtract $2x$ from each side.

$-6 = 5$ Impossible!

Therefore, the equation has no solution.

(B) $2(x - 3) = 2(x + 1) - 8$

$2x - 6 = 2x + 2 - 8$

$2x - 6 = 2x - 6$ This is always true.

Therefore, every number in the replacement set is a solution to the equation.

Matched Problem 5   Solve:

(A) $7 + 3(x - 5) = 4x - (x + 5)$ **(B)** $7 + 3(x - 5) = 4x - (x + 8)$

♦ **FORMULAS AND LITERAL EQUATIONS**

The familiar formula for the perimeter of a rectangle, $P = 2l + 2w$, contains three variables: P, l, and w. As written, the formula expresses the perimeter P in terms of the length l and width w. We can also think of this formula as an equation in any of the three variables. Since both sides of the equation are linear expressions in each of the variables, the equation is a linear equation in each variable. We can solve such an equation for any of the variables in terms of the other variables by treating the other variables as if they were constant. Since we are manipulating variables denoted by different letters and the letters remain in the solution, such equations are also called **literal equations.**

Example 6
Solving Literal Equations

Solve these equations, the perimeter and area formulas for rectangles, for w:

(A) $A = lw$ **(B)** $P = 2l + 2w$

Solution   **(A)** $A = lw$ The variable w is on the right side in a single term. We solve for w by dividing by its coefficient, l.

$\dfrac{A}{l} = w$

(B) $P = 2l + 2w$ We will try to isolate w on the right side.

$P - 2l = 2l + 2w - 2l$ Subtract $2l$ from both sides. The quantity $2l$ will be treated like a number.

$P - 2l = 2w$ Divide both sides by 2.

$\dfrac{P - 2l}{2} = w$

Notice that the solutions must be left in fraction form, since we cannot actually perform the division. Fractional forms will be investigated in detail in the next chapter.

Matched Problem 6   Solve the equations in Example 6 for l.

Answers to Matched Problems
1. $x = -14$ **2.** $x = -4$ **3.** $x = -3$ **4. (A)** -9 **(B)** -1

5. (A) No solution   **(B)** All numbers   **6. (A)** $l = \dfrac{A}{w}$ **(B)** $l = \dfrac{P - 2w}{2}$

EXERCISE 2-7

A *Solve and check:*

1. $x + 5 = 8$
2. $x + 2 = 7$
3. $x + 8 = 5$
4. $x + 7 = 2$
5. $x + 9 = -3$
6. $x + 4 = -6$
7. $x - 3 = 2$
8. $x - 4 = 3$
9. $x - 5 = -8$
10. $x - 7 = -9$
11. $y + 13 = 0$
12. $x - 5 = 0$
13. $4x = 32$
14. $9x = 36$
15. $6x = -24$
16. $7x = -21$
17. $-3x = 12$
18. $-2x = 18$
19. $-8x = -24$
20. $-9x = -27$
21. $3y = 0$
22. $-5m = 0$
23. $4x - 7 = 5$
24. $3y - 8 = 4$
25. $2y + 5 = 9$
26. $4x + 3 = 19$
27. $2y + 5 = -1$
28. $2w + 18 = -2$
29. $-3t + 8 = -13$
30. $-4m + 3 = -9$
31. $4m = 2m + 8$
32. $3x = x + 6$
33. $2x = 8 - 2x$
34. $3x = 10 - 7x$
35. $2n = 5n + 12$
36. $3y = 7y + 8$
37. $2x - 7 = x + 1$
38. $4x - 9 = 3x + 2$
39. $3x - 8 = x + 6$
40. $4y + 8 = 2y - 6$
41. $2t + 9 = 5t - 6$
42. $3x - 4 = 6x - 19$

B 43. $x - 3 = x + 7$
44. $2y + 8 = 2y - 6$
45. $2x + 2(x - 6) = 52$
46. $5x + 10(x + 7) = 100$
47. $x + (x + 2) + (x + 4) = 54$
48. $10x + 25(x - 3) = 275$
49. $2(x + 7) - 2 = x - 3$
50. $5 + 4(t - 2) = 2(t - 7) + 1$
51. $-3(4 - t) = 5 - (t + 1)$
52. $5x - (7x - 4) - 2 = 5 - (3x + 2)$
53. $x(x + 2) = x(x + 4) - 12$
54. $x(x - 1) + 5 = x^2 + x - 3$
55. $t(t - 6) + 8 = t^2 - 6t - 3$
56. $x(x - 4) - 2 = x^2 - 4(x + 3)$
57. $3 + 2(x - 4) = x - (5 - x)$
58. $x + 7 = 2x - (5 - x)$

59. $3x - (x + 1) = 2x + 1$
60. $3x - (1 - x) = x + 3(x - 1)$
61. $x^2 + 3x + 5 = x^2 + 3(x + 1) - 1$
62. $2(x - 5) + 8 = -2(1 - x)$
63. $2 - x = x - 2(x - 1)$
64. $x = x - (1 - x) - (x - 1)$
65. $7 - 2x - (-x + 4) = 3 + x - (x - 3)$
66. $x + 3[5 - (x - 2)] = 7 + 3x - (2 + x)$
67. $x - [1 - (x - 1)] = 2(x - 1)$
68. $1 - 3x = x + 13 - 2(x + 1)$

C *Solve each equation for the indicated variable:*

69. $A = bh;\ b$
70. $A = bh;\ h$
71. $d = rt;\ r$
72. $d = rt;\ t$
73. $I = prt;\ r$
74. $I = prt;\ t$
75. $A = P(1 + rt);\ P$
76. $A = P(1 + rt);\ t$
77. $y = mx + b;\ x$
78. $y = mx + b;\ m$
79. $P = a + b + c;\ a$
80. $P = a + b + c;\ b$

In Problems 81–90, solve and check:

81. $-307x + 2{,}132 = 4{,}281$
82. $642x - 1{,}304 = -8{,}366$
83. $391x + 2{,}312 = 161x + 1{,}162$
84. $31x - 420 = 73x - 714$
85. $u - [u - 3(u - 1)] = u - (1 - u)$
86. $3[t - (1 + t)] = t + 4$
87. $1 - [x - (2 - x)] = x - (1 - x)$
88. $1 - w = 1 - [w - (1 - w)]$
89. $3 + z - 5(z + 3) = 6 - z$
90. $7 + x - (3 - 2x) = 6 - \{1 - [2x - (1 - x)]\}$

91. Which of the following are equivalent to $3x - 6 = 6$: $3x = 12,\ 3x = 0,\ x = 4,\ x = 0$?

92. Which of the following are equivalent to $2x + 5 = x - 3$: $2x = x - 8,\ 2x = x + 2,\ 3x = -8,\ x = -8$?

93. Which of the following equations are equivalent to $-2x + 7 = 13$: $2x - 7 = 13,\ 2x + 7 = -13,\ 2x = 6,\ -2x = 6,\ 2x = -6,\ x = 3,\ x = -3$?

94. Which of the following equations are equivalent to $-5x + 10 = 30$: $5x - 10 = -30,\ 5x + 10 = -30,\ 5x = 20,\ -5x = 20,\ 5x = -20,\ -5x = -20,\ x = 4,\ x = -4$?

2-8   Word Problems and Applications

♦ The Processes of Algebra
♦ Solving Word Problems
♦ Strategy

We have now introduced a number of algebraic concepts and techniques. Before we add even more, it will be helpful to put our work in context and to see what it means to "do" algebra. Despite what might at first appear to be a large number of different things to learn, the ideas and methods that will be covered in this book all fall into a half-dozen categories that we might call the *processes* of algebra. These are summarized in this section.

Even more importantly, it is time to emphasize that algebraic manipulations are learned for a purpose, namely, to solve problems. We will begin to study word problems as an indication of the kinds of real-world applications that require algebra. Our initial examples will be in very simplified settings so that you can concentrate on the process, rather than on the details of a complicated real-world problem. Word problems will be considered frequently in the chapters ahead. This section introduces a strategy for attacking them.

♦ THE PROCESSES OF ALGEBRA

You can get a good idea of what the subject of algebra is about by considering the words in the instructions to the exercises you have already worked. These words include "evaluate," "translate," "write in a particular form," "multiply out," "combine like terms," and so forth. Additional instructions will be added as we progress in the text. However, almost everything you do will fall into one of the following categories:

1. Evaluate
2. Translate
3. Rewrite
4. Solve
5. Graph (This won't be introduced until Chapter 4.)
6. Use

You evaluate algebraic expressions like $2x^2 + y$ by substituting particular values for x and y. Such an expression might come from translating a phrase like "twice the square of one number plus another number" into a symbolic form.

The category "rewrite" includes instructions such as combining like terms, multiplying out, writing products as sums, writing sums as products, simplifying, removing grouping symbols, inserting grouping symbols, and so forth. A great deal of attention in algebra courses is focused on rewriting algebraic expressions. Some forms of algebraic expressions are more useful than others for a particular purpose. For example, in the last section, you solved equations by rewriting and simplifying each side.

The most important of these six categories is the last: "use." We learn algebra to use it. Solving word problems will *use* the algebraic processes we have discussed for a purpose. A word problem must be formulated in such a way that it can be *translated* into algebraic language. To *solve* a resulting algebraic equation, you will usually have to do some *rewriting*. You will then check your solution by *evaluating* expressions related to the original problem.

♦ **SOLVING WORD PROBLEMS**

Let us start by solving a fairly simple problem dealing with numbers. Through this problem you will learn some basic ideas about setting up and solving word problems in general.

Example 1
A Number Problem

Find three consecutive integers whose sum is 66.

Solution   Identify one of the unknowns with a variable, say, x, then write other unknowns in terms of the variable. In this case, we let x represent the first integer. Then the next integer is represented by $x + 1$, and the next by $x + 2$. Thus, x, $x + 1$, and $x + 2$ represent three consecutive integers starting with the integer x.

$$x + (x + 1) + (x + 2) = 66$$ Write an equation that relates the unknown quantities with other facts in the problem—the sum of three consecutive integers is 66.

$$x + x + 1 + x + 2 = 66$$ Solve the equation.

$$3x + 3 = 66$$

$$3x = 63$$

$$\left.\begin{array}{r} x = 21 \\ x + 1 = 22 \\ x + 2 = 23 \end{array}\right\}$$ Write all answers requested.

Check
$$\left.\begin{array}{r} 21 \\ 22 \\ \underline{23} \\ 66 \end{array}\right\}$$ Three consecutive integers

Checking your answer in the original equation is not enough, since you might have made a mistake in setting up the equation. A check is provided only if the conditions in the original problem are satisfied.

Thus we have found three consecutive integers whose sum is 66.

Matched Problem 1   Find three consecutive integers whose sum is 54.

Example 2
A Number Problem

Find three consecutive even numbers such that twice the second plus 3 times the third is 7 times the first.

Solution   Let $x =$ First even number

Then $x + 2 =$ Second even number

and $x + 4 =$ Third even number

$$\left(\begin{array}{c}\text{Twice the second}\\ \text{even number}\end{array}\right) + \left(\begin{array}{c}\text{Three times the third}\\ \text{even number}\end{array}\right) = \left(\begin{array}{c}\text{Seven times the first}\\ \text{even number}\end{array}\right)$$

$$2(x + 2) \qquad + \qquad 3(x + 4) \qquad = \qquad 7x$$

$$2x + 4 + 3x + 12 = 7x$$

$$5x + 16 = 7x$$

$$-2x = -16$$

$$x = 8 \qquad \text{First even number}$$

$$x + 2 = 10 \qquad \text{Second even number}$$

$$x + 4 = 12 \qquad \text{Third even number}$$

Check   First, it is true that 8, 10, and 12 are three consecutive even numbers. Next check that twice the second plus 3 times the third is 7 times the first:

$$2 \cdot 10 + 3 \cdot 12 \stackrel{?}{=} 7 \cdot 8$$

$$20 + 36 \stackrel{?}{=} 56$$

$$56 \stackrel{\checkmark}{=} 56$$

Matched Problem 2   Find three consecutive even numbers such that the second plus twice the third is 4 times the first.

Example 3
A Geometry Problem

Find the dimensions of a rectangle with a perimeter of 52 centimeters if its length is 5 centimeters more than twice its width.

Solution   First draw a figure, then label parts using an appropriate variable. In this case, since the length is given in terms of the width, we let x represent the width; then the length will be given by $2x + 5$.

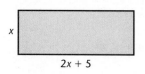

$2x + 5$

Figure 1

$$2(\text{Length}) + 2(\text{Width}) = \text{Perimeter}$$

$$2(2x + 5) + 2x = 52$$

$$4x + 10 + 2x = 52$$

$$6x = 42$$

$$x = 7 \text{ centimeters} \qquad \text{Width}$$

$$2x + 5 = 19 \text{ centimeters} \qquad \text{Length}$$

Check   First, 19 is 5 more than twice 7. The perimeter is

$$2 \cdot 19 + 2 \cdot 7 = 38 + 14 = 52$$

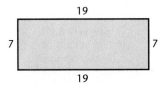

Figure 2

Matched Problem 3   Find the dimensions of a rectangle, given that the perimeter of the rectangle is 30 meters and the length is 7 meters more than its width.

Before we work our next example, we consider a few questions about collections of coins and their value.

How much are 8 nickels worth in cents?

$$\begin{pmatrix} \text{Value of one} \\ \text{nickel in cents} \end{pmatrix} \times \begin{pmatrix} \text{Number of} \\ \text{nickles} \end{pmatrix} = \begin{pmatrix} \text{Total value of all} \\ \text{nickels in cents} \end{pmatrix}$$

$$\qquad\quad 5 \qquad\quad\times \qquad 8 \qquad = \qquad 40 \text{ cents}$$

How much are x nickels worth in cents?

$$\qquad\quad 5 \qquad\quad\times \qquad x \qquad = \qquad 5x$$

How much are 6 dimes worth in cents?

$$\begin{pmatrix} \text{Value of one} \\ \text{dime in cents} \end{pmatrix} \times \begin{pmatrix} \text{Number of} \\ \text{dimes} \end{pmatrix} = \begin{pmatrix} \text{Total value of all} \\ \text{dimes in cents} \end{pmatrix}$$

$$\qquad\quad 10 \qquad\quad\times \qquad 6 \qquad = \qquad 60 \text{ cents}$$

How much are $(x + 7)$ dimes worth in cents?

$$\qquad\quad 10 \qquad\quad\times \qquad (x + 7) \qquad = \qquad 10(x + 7)$$

Example 4
A Money Problem   In a pile of coins that is composed of only dimes and nickels, there are 7 more dimes than nickels. If the total value of all the coins in the pile is $1, how many of each type of coin are in the pile?

Solution   Let x = Number of nickels in the pile   Do not confuse the number of nickels with the value of the nickels or the number of dimes with the value of the dimes.

Then $x + 7$ = Number of dimes in the pile

$$\left(\begin{array}{c}\text{Value of nickels}\\\text{in cents}\end{array}\right) + \left(\begin{array}{c}\text{Value of dimes}\\\text{in cents}\end{array}\right) = \left(\begin{array}{c}\text{Value of pile}\\\text{in cents}\end{array}\right)$$

$$5x \qquad + \qquad 10(x + 7) \qquad = \qquad 100$$

The value of the nickels in cents is 5 cents times the number of nickels. Similar logic applies to the value of the dimes.

$$5x + 10x + 70 = 100$$
$$15x = 30$$
$$x = 2 \qquad \text{Nickels}$$
$$x + 7 = 9 \qquad \text{Dimes}$$

Check　First, 9 dimes is 7 more than 2 nickels.

$$\begin{array}{r}\text{Value of 2 nickels in cents} = 10\\ \text{Value of 9 dimes in cents} = \underline{90}\\ \text{Total value} = 100 \text{ cents} = \$1\end{array}$$

Matched Problem 4　A boy has dimes and quarters worth $1.80 in his pocket. If there are twice as many dimes as quarters, how many of each does he have?

Example 5
A Rate-Time-Distance Problem

A car leaves town A and travels at 55 miles per hour toward town B at the same time that a car leaves town B and travels 45 miles per hour toward town A. If the towns are 600 miles apart, how long will it take the two cars to meet? Set up an equation and solve.

Solution　Let t = number of hours until both cars meet. Then draw a diagram and label known and unknown parts:

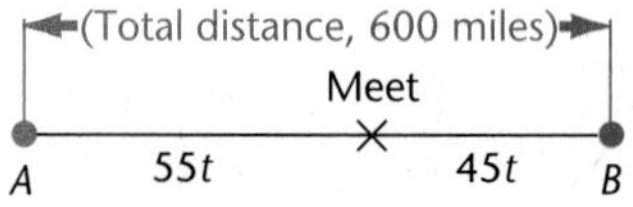

Figure 3

$$\left(\begin{array}{c}\text{Distance car from}\\A\text{ travels}\end{array}\right) + \left(\begin{array}{c}\text{Distance car from}\\B\text{ travels}\end{array}\right) = \left(\begin{array}{c}\text{Total}\\\text{distance}\end{array}\right)$$

$$55t \qquad + \qquad 45t \qquad = \qquad 600$$

$$100t = 600$$

$$t = \frac{600}{100}$$

$$t = 6 \text{ hours}$$

Check　You should check this answer yourself.

Matched Problem 5　If an older printing press can print 50 handbills per minute and a newer press can print 75, how long will it take both together to print 1,500 handbills? Set up an equation and solve.

♦ **STRATEGY**

You are now beginning to see the power of algebra. A solution to a problem that was difficult to obtain by arithmetical computation can be obtained instead by a deductive process involving a symbol that represents the solution.

There are many different types of algebraic applications—so many, in fact, that no single approach will apply to all. The following suggestions, however, may be of help to you:

Strategy for Solving Word Problems

1. Read the problem very carefully—several times if necessary.
2. Write down important facts and relationships on a piece of scratch paper. Draw figures if it is helpful. Write down any formulas that might be relevant.
3. Identify the unknown quantities in terms of a single variable if possible.
4. Look for key words and relationships in a problem that will lead to an equation involving the variables introduced in step 3.
5. Solve the equation. Write down all the solutions asked for in the original problem.
6. Check the solutions in the original problem.

The key step, and usually the most difficult one, in the above strategy is the fourth: finding the equation. There is no set procedure for carrying out this step. A variety of techniques will be illustrated in examples throughout the text.

Most importantly, you must remember that mathematics is not a spectator sport! Just reading examples is not enough. You must set up and solve problems yourself.

Answers to **1.** 17, 18, 19   **2.** 10, 12, 14   **3.** 4 meters by 11 meters
Matched Problems **4.** 4 quarters and 8 dimes   **5.** 12 minutes

---

EXERCISE 2-8

A 1. Find three consecutive integers whose sum is 78.

2. Find three consecutive integers whose sum is 96.

3. Find three consecutive even numbers whose sum is 54.

4. Find three consecutive even numbers whose sum is 42.

5. How long would it take you to drive from San Francisco to Los Angeles, a distance of about 424 miles, if you could average 53 miles per hour? (Use $d = rt$.)

6. How long would it take you to bicycle from Columbus, Ohio, to Cincinnati, a distance of 126 miles, if you could average 14 miles per hour?

7. If you bicycled from home to a town 105 miles away, and then returned, in a total riding time of 14 hours, what was your average speed?

8. If you drove from Berkeley to Lake Tahoe, a distance of 200 miles, in 4 hours, what was your average speed?

9. About 8 times the height of an iceberg of uniform cross section is under water as is above the water. If the total height of an iceberg from bottom to top is 117 feet, how much is above and how much is below the surface?

10. A chord called an octave can be produced by dividing a stretched string into two parts so that one part is twice as long as the other part. How long will each part of the string be if the total length of the string is 57 inches?

11. The sun is about 390 times as far from the earth as the moon is. If the sun is approximately 93,210,000 miles from the earth, how far is the moon from the earth?

12. A ship sailing from San Francisco to New York through the Panama Canal travels approximately 5,260 miles. If this represents a savings of 7,800 miles over the route around Cape Horn, how long is the Cape Horn route?

13. A piece of lumber 18 feet long is to be cut into three pieces, with the two smaller pieces the same length and the larger piece twice as long as the smaller pieces combined. How long should the pieces be?

14. You are asked to construct a triangle with two equal angles so that the third angle is twice the size of either of the two equal ones. How large should each angle be? [*Note:* The sum of the three angles in any triangle is 180°.]

B 15. Find three consecutive odd numbers such that the sum of the first and second is 5 more than the third.

16. Find three consecutive odd numbers such that the sum of the second and third is 1 more than 3 times the first.

17. Find the dimensions of a rectangle with perimeter 66 feet if its length is 3 feet more than twice the width.

18. Find the dimensions of a rectangle with perimeter 128 inches if its length is 6 inches less than 4 times the width.

19. An isosceles triangle, a triangle having two sides of equal length, has a perimeter of 161 inches. If the equal sides are three times as long as the third side, how long are the sides?

20. The length and width of the sides of a rectangle are two consecutive integers. If the perimeter of the rectangle is 50 feet, what are its dimensions?

21. In a pile of coins containing only quarters and dimes, there are 3 fewer quarters than dimes. If the total value of the pile is $2.75, how many of each type of coin are in the pile?

22. If you have 20 dimes and nickels in your pocket worth $1.40, how many of each do you have?

23. A basketball team scored 67 points on 2- and 3-point baskets and made 11 more 2-point baskets than 3-point baskets. How many of each type did they make?

24. A golfer's round of 18 holes consisted of scores of 4's and 5's. She had four more 4's than 5's. What was her score?

25. A toy rocket that is shot vertically upward with an initial velocity of 160 feet per second (fps) has at time

t a velocity given by the equation $v = 160 - 32t$, where air resistance is neglected. In how many seconds will the rocket reach its highest point? [*Hint:* Find *t* when $v = 0$.]

26. In the preceding problem, when will the rocket's velocity be 32 feet per second?

27. Air temperature drops approximately 5°F per 1,000 feet in altitude above the surface of the earth up to 30,000 feet. If *T* represents temperature and *A* represents altitude in thousands of feet, and if the temperature on the ground is 60°F, then we can write

$$T = 60 - 5A \qquad 0 \le A \le 30$$

If you were in a balloon, how high would you be if the thermometer registered −50°F?

28. In Problem 27, how high would you be if the thermometer registered −80°F?

29. A mechanic charges $45 per hour for his labor and $28 per hour for that of his assistant. On a repair job the bill was $390 with $309 for labor and $81 for parts. If the assistant worked 2 hours less than the mechanic, how many hours did each work?

30. If the assistant in the previous problem worked 2 hours *more* than the mechanic, and the bill was $275 for labor, how many hours did each work?

C 31. A mother and son in a kayak went up a river and back in 6 hours. If their rate up the river was 2 miles per hour and back 4 miles per hour, how far did they go up the river?

32. In a combined cycling-canoe race, a contestant can ride along the river road at 24 miles per hour and paddle her canoe back to the starting point at 8 miles per hour. If her time for the race is 8 hours, how far upstream does the race run?

33. In a swimming-cycling-running triathalon, a contestant must swim 2 miles, cycle 50 miles, and run 26 miles. A contestant who takes twice as long on the riding portion and three times as long on the running portion as he does for the swimming finishes the race in 6 hours. How long does he take cycling?

34. Two participants in a fund-raising walk-a-thon raised $61.20 between them. One participant earned $1.40 per lap of the track. The other earned $1.80 per lap and walked two more laps. How many laps did each walk? [*Hint:* Convert all dollar amounts to cents.]

35. One ship leaves England and another leaves the United States at the same time. The distance between the two ports is 3,150 miles. The ship from the United States averages 25 miles per hour, and the one from England 20 miles per hour. If they both travel the

same route, how long will it take the ships to reach a rendezvous point, and how far from the United States will they be at that time?

36. A cyclist leaves Denver heading east at the same time another leaves Kansas City, 600 miles away, heading west on the same highways. The eastbound cyclist can average 25 miles per hour, since the road is mostly downhill. The other cyclist can average only 15 miles per hour. The cyclists ride until they meet. How long does it take?

37. On one leg of a cross-country trip a cyclist leaves at 7 A.M. traveling at an average speed of 20 miles per hour. His support vehicle leaves at 10 A.M. and drives at an average speed of 50 miles per hour. When will the truck catch up to the cyclist?

38. At 8 A.M. your father left by car on a long trip. An hour later you find that he has left his wallet behind. You decide to take another car to try to catch up with

him. From past experience you know that he averages about 42 miles per hour. If you can average 54 miles per hour, how long will it take you to catch him?

39. In a copy center, two machines are being used to run 5,000 copies. If the first machine makes 75 copies per minute and the second makes 125, how long will it take to complete the job?

40. A postal bar code reader can read and sort 28,000 pieces of mail per hour; a newer model can do 30,000 per hour. If both are used simultaneously, how long will it take to sort 348,000 pieces of mail?

41. Find four consecutive integers such that the sum of the first and last is the same as the sum of the second and third. Be careful!

42. Find four consecutive even numbers such that the sum of the first and last is the same as the sum of the second and third. Be careful!

CHAPTER SUMMARY

2-1   THE INTEGERS

The natural numbers are also called **positive integers.** For each positive integer there is a corresponding **negative integer.** The positive integers and negative integers together with 0 make up the set of **integers** $Z = \{ \ldots, -2, -1, 0, +1, +2, \ldots \}$. The integers can be represented on a **number line:**

$$\xleftarrow{\qquad} \overset{|}{-6} \ \overset{|}{-5} \ \overset{|}{-4} \ \overset{|}{-3} \ \overset{|}{-2} \ \overset{|}{-1} \ \overset{|}{0} \ \overset{|}{1} \ \overset{|}{2} \ \overset{|}{3} \ \overset{|}{4} \ \overset{|}{5} \ \overset{|}{6} \xrightarrow{\qquad}$$

2-2   OPPOSITES AND ABSOLUTE VALUES

The **opposite** of a nonzero integer is produced by changing its sign; 0 is its own opposite. The **absolute value** of a number is given by

$$|x| = \begin{cases} x & \text{if } x \text{ is positive or } 0 \\ -x & \text{if } x \text{ is negative} \end{cases}$$

2-3   ADDITION

For any integer a, $a + 0 = 0 + a = a$. To **add** two nonzero integers:

1. If they are of like sign, add the numerical parts and attach the common sign.
2. If they are of opposite sign, take the difference of the numerical parts and attach the sign of the larger numerical part.

Addition is **commutative,** that is, $a + b = b + a$, and **associative,** that is, $a + (b + c) = (a + b) + c$.

2-4   SUBTRACTION

To **subtract** two integers, change the sign of the number being subtracted and add: $a - b = a + (-b)$. For integers a and b, $a < b$ if there is a positive integer p such that $a + p = b$.

2-5   MULTIPLICATION AND DIVISION

For any integer, a, $a \cdot 0 = 0 \cdot a = 0$. To **multiply** two nonzero integers:

1. If they are of like sign, multiply the numerical parts and attach a plus sign.
2. If they are of opposite sign, multiply the numerical parts and attach a minus sign.

Multiplication is **commutative,** that is, $ab = ba$; **associative,** that is, $a(bc) = (ab)c$; and **distributive over addition,** that is, $a(b + c) = ab + ac$. Moreover, $(+1)a = a$, $(-1)a = -a$, $(-a)b = a(-b) = -(ab)$, and $(-a)(-b) = ab$ for all integers a and b.

For any nonzero integer a, $0 \div a = 0$, and $a \div 0$ is not defined; $0 \div 0$ is also not defined. To **divide** two nonzero integers:

1. If they are of like sign, divide the numerical parts and attach a plus sign.
2. If they are of opposite sign, divide the numerical parts and attach a minus sign.

2-6   SIMPLIFYING ALGEBRAIC EXPRESSIONS

The plus sign may be omitted in denoting positive integers. Terms in algebraic expressions may be rearranged so long as the sign preceding each term accompanies it in the process. A grouping symbol can be removed from an expression by multiplying each term within the symbol by the coefficient attached to the symbol.

2-7   SOLVING EQUATIONS

The **solution** of an equation in one variable is a replacement of the variable that makes the equation true. To **solve** an equation means to find all its solutions. Two equations are **equivalent** if they have exactly the same solutions. The following **properties of equality** produce equivalent equations; if $a = b$, then:

$$
\begin{array}{lll}
a + c = b + c & & \text{Addition property} \\
a - c = b - c & & \text{Subtraction property} \\
a \cdot c = b \cdot c & (c \neq 0) & \text{Multiplication property} \\
a/c = b/c & (c \neq 0) & \text{Division property}
\end{array}
$$

An expression that can be written in the form $ax + b$, with a and b constant and x variable, is called a **linear expression** in x. An equation that equates two linear expressions is called a **linear** or **first-degree equation.** The basic strategy for solving linear equations is to:

1. Simplify each side.
2. Get all variable terms on one side and constant terms on the other by using the addition and subtraction properties. Simplify each side.
3. Use the multiplication or division property to make the coefficient of the variable 1.

The same strategy can be applied to solving **literal equations** that are linear, that is, equations that are linear in the variable for which we are to solve.

2-8   WORD PROBLEMS AND APPLICATIONS

A general strategy for solving word problems is, in brief:

1. Read the problem very carefully.
2. Write down the important facts and relationships.
3. Identify the unknown quantities in terms of one variable.
4. Find the equation.
5. Solve the equation and write down all solutions asked for.
6. Check the solution.

CHAPTER REVIEW EXERCISE

Work through all the problems in this chapter review and check the answers in the back of the book. Answers to all review problems are there, and following each answer is a number in italics indicating the section in which that type of problem is discussed. Where weaknesses show up, review appropriate sections in the text.

A **1.** What numbers are associated with the points *a*, *b*, *c*, *d*, *e*, and *f*?

2. Locate the following numbers on a number line: -14, -7, -4, $+1$, $+8$, and $+12$.

In Problems 3–30, evaluate as indicated:

3. $-(+4)$

4. $|-(+3)|$

5. $-(-8)$

6. $-|-15|$

7. $(-8) + (+3)$

8. $(-9) + (-4)$

9. $(-3) - (-9)$

10. $(+4) - (+7)$

11. $(-11) - (+6)$

12. $(+14) - (-5)$

13. $(-7)(-4)$

14. $(+3)(-6)$

15. $(-9)(+8)$

16. $(+18)/(-3)$

17. $(-16)/(+4)$

18. $(-12)/(-2)$

19. $0/(+2)$

20. $(-6)/0$

21. $-6 + 8 - 5 + 1$

22. $(-2)(+5) + (-6)$

23. $(-4)(-3) - (+20)$

24. $\dfrac{-8}{-2} - (-4)$

25. $(-6)(0) + \dfrac{0}{-3}$

26. $\dfrac{-9}{+3} - \dfrac{-12}{-4}$

27. $|-13| - |+9|$

28. $-8 - |(+4) - (+9)|$

29. $|7 - |6 - 15||$

30. $-\{-4 - |-6 - [-8 - (-10)]|\}$

31. For the expression $3x^4 + 7x - 18$, identify:
  (A) the coefficient of *x*
  (B) the constant term

32. Evaluate the expression $3x^4 + 7x - 18$ for $x = 2$.

Remove symbols of grouping, if present, and combine like terms:

33. $4x - 3 - 2x - 5$

34. $(2x - 3) + (3x + 1)$

35. $3(m + 2n) - (m - 3n)$

36. $2(x - 3y) - 4(2x + 3y)$

In Problems 37–42, solve and check:

37. $3x - 11 = 4$

38. $7x + 13 = -8$

39. $4x - 9 = x - 15$

40. $2x - 5 = 3x + 2$

41. $2x - 9 = 15 - (6 - x)$

42. $11 - 3(x - 2) = x - (x + 1)$

43. Express each quantity by means of an appropriate integer:
  (A) Salton Sea's surface at 245 feet below sea level
  (B) Mount Whitney's height of 14,495 feet

44. Find three consecutive integers whose sum is 159.

45. A rectangle with a perimeter of 160 yards has a length 3 times the width. How long is the rectangle?

46. A triangle has its largest angle 3 times as big as its smallest and the third angle twice as big as the smallest. What are the angles? [Recall that the sum of the angles of a triangle is 180°.]

B *Evaluate for $x = -12$, $y = -2$, and $z = +3$:*

47. $-x$

48. $-(-z)$

49. $-|-y|$

50. $x - y$

51. xy

52. $x + y$

53. x/z

54. $x - z$

55. $|-z|$

56. $|x + z|$

57. $(z - y) - x$

58. $(3y + x)/z$

59. $(x/y) - yz$

60. $(4z + x)/y$

61. $\left(yz - \dfrac{x}{z}\right) - xz$

62. $\dfrac{0}{x} + x(0)y$

Remove grouping symbols and combine like terms:

63. $(3x^2y^2 - xy) - (5x^2y^2 + 4xy)$

64. $3y(2y^2 - y + 4) - 2y(y^2 + 2y - 3)$

65. $7x - 3[(x + 7y) - (2x - y)]$

66. $2x^2y(3xy - 5) - 3xy^2(4x^2 - 1)$

Replace each question mark with an appropriate algebraic expression:

67. $a + 2b + 4c = a + 2(?)$

68. $3x - 6y + 9 = 3x - 3(?)$

69. $3(x - 2y) - x + 2y = 3(x - 2y) - (?)$

In Problems 70–75, solve and check:

70. $3(m - 2) - (m + 4) = 8$

71. $2x + 3(x - 1) = 8 - (x - 1)$

72. $3x + 7 = x - (14 - 2x)$

73. $1 - (2 - 3x) = (3x - 2) + 1$

74. $6 - 5(x - 4) = 4(x + 8) - (-4 - x)$

75. $3(x + 7) - 5(x - 9) = 4(x - 1) + 8(2x + 6)$

76. If the sum of four consecutive even numbers is 188, find the numbers.

77. A pile of coins consists of nickels and quarters. How many of each kind are there if the whole pile is worth $1.45 and there are 3 fewer quarters than twice the number of nickels?

78. Express the net gain by means of an appropriate integer: a 15° rise in temperature followed by a 30° drop, another 15° drop, a 25° rise, and, finally, a 40° drop.

79. Show that subtraction is not associative by evaluating the following expressions for $x = +7$, $y = -3$, and $z = -5$:
(A) $(x - y) - z$ (B) $x - (y - z)$

80. Show that division is not associative by evaluating the following expressions for $x = +16$, $y = -8$, and $z = -2$:
(A) $(x \div y) \div z$ (B) $x \div (y \div z)$

81. Simplify: $3x - 2\{x - 2[x - (4x + 2)]\}$.

Solve the equation:

82. $(x + 3) + (x + 5) = (x + 1) + (x + 7)$

83. $(x + 5) - (x + 7) = (x + 9) - (x + 11)$

84. $[x - (x - 3)] - 3 = [3 - (x - 3)] + x$

85. $1 + \{1 - [1 - (1 - x)]\} = x - \{1 - [x - (1 - x)]\}$

In Problems 86–89, solve for the indicated variable:

86. $ax + by = c$; x **87.** $E = mc^2$; m

88. $x + y + z = 1$; z

89. $2x + y + 2z = x + 2y + z$; x

90. Two swimmers are raising money for their team in a swim-a-thon. One raises $2.10 per lap. The other raises $2.50 per lap but swims 10 fewer laps. If they raise $205 between them, how many laps did each swim? [Convert all dollars to cents and solve.]

91. The parcel post rate for mailing packages between two zones is $2.74 for anything up to 2 pounds. For heavier packages, the rate is $2.74 plus 38 cents per pound for each pound above 2 pounds. The cost of sending two packages was $34.36. If one package weighed 6 pounds more than the other and each weighs at least 2 pounds, what were the weights? [Convert all dollars to cents and solve.]

92. A space shuttle passes over Vandenberg Air Force Base at 8 A.M. traveling at 17,000 miles per hour. Another shuttle, attempting a rendezvous, passes over the same spot at 9 A.M. traveling at 18,000 miles per hour. How long will it take the second shuttle to catch up with the first?

Identify the property of the integers that justifies each of the following statements:

93. If $x - 8 = 13$, then $x = 21$.

94. $3x + y = y + 3x$

95. $3x + 3y = 3(x + y)$

96. If $7x = 56$, then $x = 8$.

97. $7x = x \cdot 7$

98. If $x + 23 = 8$, then $x = -15$.

99. $(3x)y = 3(xy)$

100. If $18x = 45$, then $2x = 5$.

CHAPTER PRACTICE TEST

The following practice test is provided for you to test your knowledge of the material in this chapter. You should try to complete it in 50 minutes or less. The answers in the back of the book indicate the section in the text that covers the material in the question. Actual tests in your class may vary from this practice test in difficulty, length, or emphasis, depending on the goals of your course or instructor.

Evaluate the expression:

1. $(-3) - (-15)$ **2.** $|-8| - (-8)$

3. $(-3)(+6) + (+72) \div (-6)$

4. $-[-|(-5)|]$

5. $xy - |x + y|$ for $x = -4$ and $y = -9$

6. $-x^2 - (-x)^2$ for $x = -3$

7. $-1 + 2 - 3 + 4 - 5 + 6$

Rewrite the expression in the form requested:

8. $-(-3)x = (-3)(?)$

9. $-(x - 3)$ without grouping symbols

10. $x^2 - 5x + 7 - 2x^2 - 9 + 2x$ with like terms combined

11. $-(a + b - c)$ without parentheses

12. $-x^2 + x^3 - 2x^4 + 4x + 2x^4 - 5x^3 + 6x^2$ with like terms combined and the variables in the order x^4, x^3, x^2, x

13. $-5 - [x - (2x - 3)]$ without grouping symbols and with like terms combined

14. $5 - 4x + 3y - 2z = 5 - (?)$

In Problems 15–21, solve the equation:

15. $3x + 11 = 6x - 10$

16. $4x - 2(x - 5) = 17 + 3(11 - x)$

17. $x(x + 8) - 4 = -(4 - 8x - x^2)$

18. $2(9 - 3x) = -3(2x - 6)$

19. $4x + 5(x - 6) = 3(3x + 10)$

20. $3 - 4(5 - 6x) = 8x - 2(5 - x) - 7$

21. $M = l + 2(w + h)$ for w

22. The National Basketball Association championship series is a best of seven series that ends when a team wins four games. During the 1980s the Los Angeles Lakers played in the series eight times, winning by 2 games, winning again by 2, losing by 4, losing by 1, winning by 2, winning by 2, winning by 1, and finally losing by 4. How many more games did the Lakers win than lose in these eight series? Solve using addition of integers.

23. Find the dimensions of a rectangle if the length is 3 centimeters more than twice the width and the perimeter is 624 centimeters.

24. Find three consecutive integers whose sum is 177.

25. A kayaker enters a river 30 miles downstream from where her friend puts in. If she can paddle upstream at 3 miles per hour and her friend can come downstream at 7 miles per hour, how long will it take them to meet?

3

Rational Numbers

In the last chapter we formed the set of integers by extending the natural numbers to include 0 and the negative integers. With this extension came added power to perform more operations on more numbers and to solve more equations than we could using only the natural numbers. But with the integers we are not able to find

$$2 \div 5$$

or to solve

$$2x = 3$$

We need fractions. Recall that in the last chapter we said we would use

$$a \div b \qquad a/b \qquad \frac{a}{b} \qquad b\overline{)a}$$

interchangeably; hence,

$$4 \div 2 \qquad 4/2 \qquad \frac{4}{2} \qquad 2\overline{)4}$$

are different names for the number 2. However, what does

$$\frac{3}{2}$$

name? Certainly not an integer. We are going to extend the set of integers so that $\frac{3}{2}$ will name a number and division will always be defined, except by 0. The extended number system will be called the set of *rational* numbers because the numbers are related to *ratios* of integers. Section 3-1 reviews the arithmetic of fractions, decimals, and percent and **may be omitted unless a review of these topics is needed.** Section 3-2 introduces the rational number system.

Photo reference: see Exercise 3-6, Problem 41.

3-1 Fractions, Decimals, and Percent (Optional Review)

- ♦ Rewriting Fractions in Higher or Lower Terms
- ♦ Multiplication and Division
- ♦ Addition and Subtraction
- ♦ Decimals and Decimal Arithmetic
- ♦ Percent

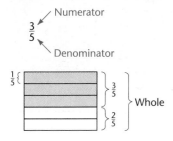

Figure 1

In this section we will review the arithmetic of fractions, decimals, and percent. **Students with a clear understanding of these topics can omit this review** and proceed directly to Section 3-2.

A **fraction** is most easily thought of as a number representing parts of a whole, as illustrated in Figure 1.

Thus, if a whole is divided into five equal parts, $\frac{3}{5}$ represents 3 of the 5 parts. An **improper fraction,** in which the numerator is larger than the denominator, as in $\frac{7}{5}$, represents 7 of these 5 parts, and thus more than the whole.

♦ REWRITING FRACTIONS IN HIGHER OR LOWER TERMS

Figure 2

The fractions $\frac{3}{5}$ and $\frac{6}{10}$ are equal, as shown in Figure 2. The fact that

$$\frac{2 \cdot 3}{2 \cdot 5} = \frac{6}{10}$$

illustrates a fundamental principle of fractions:

Fundamental Principle of Fractions

We can multiply the numerator and denominator by the same nonzero number:

$$\frac{a}{b} = \frac{ka}{kb} \qquad k \neq 0$$

Using this principle to rewrite a fraction like $\frac{6}{10}$ as $\frac{3}{5}$ is called **reducing a fraction to lower terms;** we are in effect removing the common factor 2 from each term. If we remove all the common factors of the numerator and denominator, we obtain a fraction having no common factors in the resulting numerator and denominator, as is the case for $\frac{3}{5}$. We say we have reduced the fraction to **lowest terms.** Rewriting a fraction like $\frac{3}{5}$ as $\frac{6}{10}$ is called **raising a fraction to higher terms.**

Example 1
Higher and
Lower Terms

Use the fundamental principle of fractions to rewrite each fraction as indicated:

(A) $\dfrac{18}{24} = \dfrac{?}{12}$ (B) $\dfrac{3}{4} = \dfrac{9}{?}$ (C) $\dfrac{15}{24}$ reduced to lowest terms

Solution   (A) $\dfrac{18}{24} = \dfrac{2 \cdot 9}{2 \cdot 12} = \dfrac{9}{12}$ Reducing to *lower* terms

(B) $\dfrac{3}{4} = \dfrac{3 \cdot 3}{3 \cdot 4} = \dfrac{9}{12}$ Raising to *higher* terms

(C) We must remove all common factors from the numerator and denominator. The only common factor of 15 and 24 is 3, so

$$\dfrac{15}{24} = \dfrac{3 \cdot 5}{3 \cdot 8} = \dfrac{5}{8}$$

and the fraction $\frac{5}{8}$ is in lowest terms.

Matched Problem 1   Use the fundamental principle of fractions to rewrite each fraction as indicated:

(A) $\dfrac{18}{30} = \dfrac{?6}{10}$ (B) $\dfrac{3}{5} = \dfrac{9}{?15}$ (C) $\dfrac{21}{30}$ reduced to lowest terms

---

♦ MULTIPLICATION AND DIVISION

> To multiply two fractions, multiply their numerators and place the product over the product of the denominators:
>
> $$\dfrac{a}{b} \cdot \dfrac{c}{d} = \dfrac{a \cdot c}{b \cdot d}$$

Example 2
Multiplying Fractions

Multiply: $\dfrac{3}{5} \cdot \dfrac{2}{7}$

Solution $\dfrac{3}{5} \cdot \dfrac{2}{7} = \dfrac{3 \cdot 2}{5 \cdot 7} = \dfrac{6}{35}$

Matched Problem 2   Multiply: (A) $\dfrac{7}{4} \cdot \dfrac{5}{6}$ (B) $\dfrac{3}{8} \cdot \dfrac{5}{7}$

Example 3
Multiplying Fractions

Multiply and reduce to lowest terms: $\dfrac{8}{9} \cdot \dfrac{3}{4}$

Solution $\dfrac{8}{9} \cdot \dfrac{3}{4} = \dfrac{\overset{2}{\cancel{8}}}{\underset{3}{\cancel{9}}} \cdot \dfrac{\overset{1}{\cancel{3}}}{\underset{1}{\cancel{4}}}$ Any factor in a numerator can be removed with a like factor in a denominator.

$= \dfrac{2}{3}$

Matched Problem 3 Multiply and reduce to lowest terms:

(A) $\dfrac{2}{15} \cdot \dfrac{3}{8}$ (B) $\dfrac{9}{7} \cdot \dfrac{14}{24}$

To divide two fractions, invert the divisor and multiply:

$$\frac{a}{b} \div \frac{c}{d} = \frac{a}{b} \cdot \frac{d}{c}$$

Invert divisor

Example 4
Dividing Fractions

Divide and reduce to lowest terms: $\dfrac{5}{8} \div \dfrac{3}{4}$

Solution $\dfrac{5}{8} \div \dfrac{3}{4} = \dfrac{5}{8} \cdot \dfrac{4}{3}$ Invert divisor and multiply. Do not remove common factors before inverting the divisor.

$= \dfrac{5}{\overset{}{\underset{2}{8}}} \cdot \dfrac{\overset{1}{4}}{3}$

$= \dfrac{5}{6}$

Matched Problem 4 Divide and reduce to lowest terms:

(A) $\dfrac{7}{16} \div \dfrac{14}{10}$ (B) $\dfrac{42}{7} \div \dfrac{14}{49}$

◆ ADDITION AND SUBTRACTION

If two fractions have the same denominator, we add or subtract them by adding or subtracting their numerators and placing the result over the common denominator:

$$\frac{a}{b} + \frac{c}{b} = \frac{a+c}{b} \qquad \frac{a}{b} - \frac{c}{b} = \frac{a-c}{b}$$

Example 5
Adding and Subtracting Fractions

Perform the indicated operation and reduce to lowest terms:

(A) $\dfrac{5}{6} + \dfrac{4}{6}$ (B) $\dfrac{5}{8} - \dfrac{3}{8}$

Solution (A) $\dfrac{5}{6} + \dfrac{4}{6} = \boxed{\dfrac{5+4}{6}} = \dfrac{9}{6} = \dfrac{3}{2}$ (B) $\dfrac{5}{8} - \dfrac{3}{8} = \boxed{\dfrac{5-3}{8}} = \dfrac{2}{8} = \dfrac{1}{4}$

Matched Problem 5   Perform the indicated operation and reduce to lowest terms:

(A) $\dfrac{11}{24} + \dfrac{4}{24}$ (B) $\dfrac{17}{12} - \dfrac{9}{12}$

If two fractions do not have a common denominator, we must change them so that they do before we can add or subtract. The most convenient common denominator to use is the least common multiple (LCM) of the denominators. **The LCM is the smallest number exactly divisible by each denominator.** The least common multiple of the denominators is also called the **least common denominator (LCD).**

Example 6
Adding and
Subtracting Fractions

Perform the indicated operation and reduce to lowest terms:

(A) $\dfrac{3}{4} + \dfrac{2}{3}$ (B) $\dfrac{5}{6} - \dfrac{11}{15}$

Solution   (A) $\dfrac{3}{4} + \dfrac{2}{3} = \boxed{\dfrac{3 \cdot 3}{3 \cdot 4} + \dfrac{4 \cdot 2}{4 \cdot 3}}$ The LCD is 12 (the smallest number divisible by 4 and 3). Use the fundamental principle of fractions to make each denominator 12.

$= \dfrac{9}{12} + \dfrac{8}{12}$

$= \boxed{\dfrac{9 + 8}{12}}$

$= \dfrac{17}{12}$

(B) $\dfrac{5}{6} - \dfrac{11}{15} = \boxed{\dfrac{5 \cdot 5}{5 \cdot 6} - \dfrac{2 \cdot 11}{2 \cdot 15}}$ LCD = 30

$= \dfrac{25}{30} - \dfrac{22}{30}$

$= \boxed{\dfrac{25 - 22}{30}}$

$= \dfrac{3}{30} = \dfrac{1}{10}$

Matched Problem 6   Perform the indicated operations and reduce to lowest terms:

(A) $\dfrac{5}{6} + \dfrac{4}{9}$ (B) $\dfrac{7}{10} - \dfrac{3}{25}$ $\dfrac{35-6}{50} \quad \dfrac{29}{50}$

$\dfrac{15+8}{18} = \dfrac{23}{18} \quad 1\dfrac{5}{18}$

♦ **DECIMALS AND DECIMAL ARITHMETIC**

A decimal fraction is a way of representing numbers in decimal form. The base 10 is central to the process. Recall:

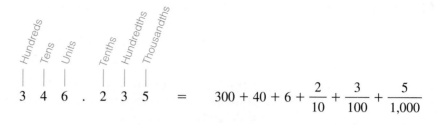

$$3\ 4\ 6\ .\ 2\ 3\ 5\ =\ 300 + 40 + 6 + \frac{2}{10} + \frac{3}{100} + \frac{5}{1{,}000}$$

Thus,

$$0.2 = \frac{2}{10} \qquad 0.03 = \frac{3}{100} \qquad 0.005 = \frac{5}{1{,}000} \qquad 0.235 = \frac{235}{1{,}000}$$

Fractions to Decimal Fractions

To convert a fraction into a decimal fraction, divide the denominator into the numerator.

Example 7
Converting Fractions to Decimals

Convert $\frac{12}{23}$ to a decimal fraction rounded to three decimal places.

Solution

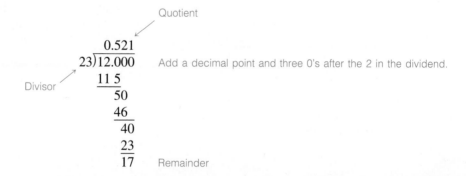

In order to **round** to three decimal places, we do the following. If the remainder after carrying out the division to three decimal places is greater than or equal to half the divisor, we add 1 to the last decimal place in the quotient. If the remainder is less than half the divisor, we leave the last decimal place alone. In this case the remainder 17 is more than half the divisor 23, so we add 1 to the third decimal place to obtain

$$\frac{12}{23} \approx 0.522 \qquad \approx \text{means "approximately equal to"}$$

We write 0.522 instead of .522 because the latter might be mistaken for the whole number 522. Placing the 0 to the left of the decimal point keeps the decimal point from getting lost.

Matched Problem 7   Convert $\frac{26}{35}$ to a decimal fraction rounded to two decimal places.

We will briefly review the arithmetic operations on decimal fractions.

> To add two or more decimal fractions, line up decimal points and add as in whole-number arithmetic. The decimal point is carried straight down to the sum.

Example 8
Adding Decimals

Add: 325.2, 62.25, 3.012

Solution

$$
\begin{array}{r}
\downarrow \\
325.2 \\
62.25 \\
3.012 \\
\hline
390.462 \\
\uparrow
\end{array}
$$

Line up decimal points.

Matched Problem 8   Add: 22.06, 204.135, 3.4

> To subtract one decimal fraction from another, line up the decimals and subtract as in whole-number arithmetic. The decimal point is carried straight down to the difference.

Example 9
Subtracting Decimals

Subtract 23.427 from 125.8.

Solution

$$
\begin{array}{r}
\downarrow \\
125.800 \\
23.427 \\
\hline
102.373 \\
\uparrow
\end{array}
$$

Add two 0's

Difference

Line up decimal points.

Matched Problem 9   Subtract 325.63 from 407.5.

> To multiply two decimal fractions, multiply as in whole-number arithmetic. The product has as many decimal places as the sum of the number of decimal places used in the two original decimal fractions.

Example 10
Multiplying Decimals

Multiply 36.24 and 13.6.

Solution

$$
\begin{array}{r}
36.\,24 \\
13.\,6 \\
\hline
21\ 744 \\
108\ 72 \\
362\ 4 \\
\hline
492.\,864
\end{array}
$$

2 decimal places
1 decimal place } Add
3 decimal places

Matched Problem 10

Multiply 103.2 and 26.72.

> To divide one decimal fraction by another, divide as in whole-number arithmetic. To locate the decimal in the quotient, move the decimal point in the dividend and divisor as many places to the right as there are decimal places in the divisor. Then move the decimal point straight up from the dividend to the quotient.

Example 11
Dividing Decimals

Divide 425.3 by 2.43, and round your answer to two decimal places.

Solution

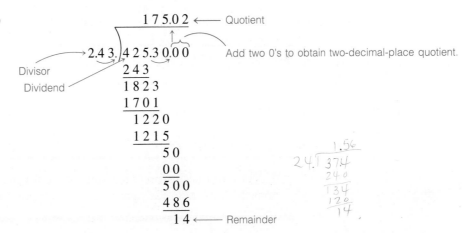

The remainder 14 is less than half the divisor 243, so we do not add 1 to the second decimal place. The quotient rounded to two decimal places is 175.02.

Matched Problem 11

Divide 3.74 by 2.4, and round your answer to two decimal places.

♦ **PERCENT**

If 1 is divided into 100 equal parts, then each part is called **1 percent.** The symbol % is read "percent." Thus,

Fraction		Decimal		Percent
$\dfrac{23}{100}$	=	0.23	=	23%
$\dfrac{4}{100}$	=	0.04	=	4%
$\dfrac{162}{100}$	=	1.62	=	162%
$\dfrac{0.3}{100}$	=	0.003	=	0.3%

It will often be necessary to convert back and forth between percent and decimal forms.

> To convert a percent to a decimal, remove the percent symbol and move the decimal point two places to the left.

Example 12
Converting Percent to Decimal

Convert each percent to a decimal:

(A) 43%   (B) 7%   (C) 234%   (D) 0.3%

Solution

(A) $43\% = 43.\% = 0.43$ (B) $7\% = 07.\% = 0.07$

(C) $234\% = 234.\% = 2.34$ (D) $0.3\% = 00.3\% = 0.003$

Matched Problem 12

Convert each percent to a decimal:

(A) 59%   (B) 2%   (C) 105%   (D) 0.7%

> To convert a decimal or whole number to a percent, shift the decimal point two places to the right and use the percent symbol.

Example 13
Converting Decimal to Percent

Convert each decimal to a percent:

(A) 0.23   (B) 0.06   (C) 4.37   (D) 0.064

Solution

(A) $0.23 = 0.23 = 23\%$ (B) $0.06 = 0.06 = 6\%$

(C) $4.37 = 4.37 = 437\%$ (D) $0.064 = 0.064 = 6.4\%$

Matched Problem 13 Convert each decimal to a percent:

(A) 0.71   (B) 0.05   (C) 1.09   (D) 0.003

Applied problems frequently require taking a percent of a given quantity. The "of" here indicates a multiplication, just as it would in finding half *of* something.

To find a percent of a given quantity, convert the percent to a decimal and multiply.

Example 14
Finding "Percent of"

(A) Find 23% of 45.   (B) Find 6.2% of 28.

Solution (A) 23% of 45 = 0.23 × 45 = 10.35
(B) 6.2% of 28 = 0.062 × 28 = 1.736

Matched Problem 14 (A) 4% of 64   (B) 108% of 22   (C) 23% of 5   (D) 0.5% of 30

Answers to
Matched Problems

1. (A) $\frac{6}{10}$ (B) $\frac{9}{15}$ (C) $\frac{7}{10}$ 2. (A) $\frac{35}{24}$ (B) $\frac{15}{56}$
3. (A) $\frac{1}{20}$ (B) $\frac{3}{4}$ 4. (A) $\frac{5}{16}$ (B) 2
5. (A) $\frac{5}{8}$ (B) $\frac{2}{3}$ 6. (A) $\frac{23}{18}$ (B) $\frac{29}{50}$
7. 0.74 8. 229.595 9. 81.87 10. 2,757.504 11. 1.56
12. (A) 0.59 (B) 0.02 (C) 1.05 (D) 0.007
13. (A) 71% (B) 5% (C) 109% (D) 0.3%
14. (A) 2.56 (B) 23.76 (C) 1.15 (D) 0.15

EXERCISE 3-1

Since this section is included only to provide a brief review of arithmetic operations on fractions, decimals, and percent, the designations A, B, and C for level of difficulty are omitted.

Rewrite the fraction in the form indicated by reducing to lower terms or raising to higher terms:

1. $\frac{5}{6} = \frac{20}{?}$ 2. $\frac{4}{9} = \frac{28}{?}$ 3. $\frac{5}{?} = \frac{20}{44}$

4. $\frac{7}{?} = \frac{28}{36}$ 5. $\frac{?}{12} = \frac{14}{24}$ 6. $\frac{?}{15} = \frac{21}{45}$

7. $\frac{2}{5} = \frac{?}{45}$ 8. $\frac{5}{8} = \frac{?}{56}$ 9. $\frac{15}{18} = \frac{45}{?}$

10. $\frac{32}{12} = \frac{8}{?}$ 11. $\frac{5}{11} = \frac{?}{44}$ 12. $\frac{15}{6} = \frac{?}{18}$

13. $\frac{25}{?} = \frac{75}{48}$ 14. $\frac{56}{?} = \frac{8}{7}$ 15. $\frac{?}{13} = \frac{30}{65}$

16. $\frac{?}{42} = \frac{11}{7}$

Reduce to lowest terms:

17. $\frac{28}{49}$ 18. $\frac{56}{84}$ 19. $\frac{60}{96}$ 20. $\frac{45}{63}$

21. $\frac{21}{56}$ 22. $\frac{70}{84}$ 23. $\frac{39}{65}$ 24. $\frac{63}{75}$

Perform the indicated operations and reduce to lowest terms:

25. $\frac{2}{3} \cdot \frac{4}{5}$ 26. $\frac{3}{4} \cdot \frac{2}{7}$ 27. $\frac{1}{2} \div \frac{2}{3}$

28. $\frac{3}{4} \div \frac{4}{3}$ 29. $\frac{4}{9} \cdot \frac{3}{12}$ 30. $\frac{5}{12} \cdot \frac{9}{10}$

31. $\dfrac{10}{12} \div \dfrac{6}{18}$ **32.** $\dfrac{18}{24} \div \dfrac{12}{9}$ **33.** $\dfrac{5}{12} + \dfrac{3}{12}$

34. $\dfrac{3}{8} + \dfrac{7}{8}$ **35.** $\dfrac{11}{9} - \dfrac{5}{9}$ **36.** $\dfrac{17}{14} - \dfrac{9}{14}$

37. $\dfrac{1}{4} + \dfrac{2}{3}$ **38.** $\dfrac{3}{5} + \dfrac{1}{2}$ **39.** $\dfrac{5}{6} - \dfrac{3}{4}$

40. $\dfrac{3}{4} - \dfrac{1}{3}$ **41.** $\dfrac{5}{12} + \dfrac{3}{8}$ **42.** $\dfrac{11}{18} + \dfrac{5}{12}$

43. $\dfrac{7}{9} - \dfrac{5}{12}$ **44.** $\dfrac{7}{20} - \dfrac{9}{30}$

45. $\dfrac{2}{3} \cdot \left(\dfrac{3}{4} \div \dfrac{9}{12} \right)$ **46.** $\dfrac{4}{5} \div \left(\dfrac{8}{10} \div \dfrac{3}{4} \right)$

47. $\dfrac{8}{9} \cdot \left(\dfrac{3}{4} - \dfrac{2}{3} \right)$ **48.** $\dfrac{7}{5} \div \left(\dfrac{5}{6} - \dfrac{1}{4} \right)$

Add:

49. 3.1, 2.5, 0.2   **50.** 6.4, 0.3, 5.6

51. 23.2, 2.45, 6.012   **52.** 405.03, 21.105, 5.2

Subtract:

53. 25.32 from 43.05   **54.** 6.09 from 13.12

55. 23.56 from 103.2   **56.** 5.69 from 41.2

Multiply:

57. 2.5 by 13   **58.** 24 by 1.6

59. 4.26 by 0.002   **60.** 3.04 by 0.006

Convert to decimal fractions rounded to two decimal places:

61. $\dfrac{5}{6}$ **62.** $\dfrac{7}{9}$ **63.** $\dfrac{34}{46}$ **64.** $\dfrac{16}{24}$

Divide and round answers to one decimal place:

65. 84 by 2.2   **66.** 68 by 4.5

67. 36.2 by 4.6   **68.** 4.02 by 6.4

Change percents to decimals:

69. 67%   **70.** 14%   **71.** 9%   **72.** 1%

73. 216%   **74.** 308%   **75.** 0.6%   **76.** 0.1%

77. 7.4%   **78.** 2.8%   **79.** 23.1%   **80.** 64.5%

Change decimals to percents:

81. 0.12   **82.** 0.21   **83.** 0.08   **84.** 0.02

85. 3.25   **86.** 6.04   **87.** 0.007   **88.** 0.004

89. 0.072   **90.** 0.069   **91.** 0.405   **92.** 0.236

Perform the calculations:

93. 12% of 403   **94.** 18% of 40

95. 6% of 4,000   **96.** 8% of 2,000

97. 125% of 200   **98.** 150% of 44

99. 6.5% of 24   **100.** 4.8% of 36

101. 0.4% of 20   **102.** 0.7% of 80

3-2 Rational Numbers and the Fundamental Principle of Fractions

- ♦ The Set of Rational Numbers
- ♦ Rational Numbers and the Number Line
- ♦ Opposite of and Absolute Value of a Rational Number
- ♦ Fundamental Principle of Fractions

This section introduces the set of rational numbers. Basic operations on the numbers are considered in Sections 3-3 and 3-4.

♦ THE SET OF RATIONAL NUMBERS

Rational Numbers

Any number that can be written in the form

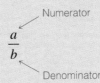

$$\frac{a}{b}$$

where a and b are integers, with $b \neq 0$, is called a **rational number.**

Thus,

$$\frac{1}{3} \quad \frac{3}{5} \quad \frac{9}{1} \quad \frac{-2}{7} \quad \frac{10}{-5} \quad \frac{-3}{-2} \quad \frac{0}{3}$$

are all rational numbers. Every fraction represents a rational number, but the rational numbers also include "fractions" with negative numerator or denominator. Every integer is a rational number, but some rational numbers are not integers. For example, the integer 9 is also a rational number, since it can be written as the quotient of two integers:

$$9 = \frac{9}{1} \quad \text{or} \quad \frac{18}{2} \quad \text{or} \quad \frac{-27}{-3}$$

and so on, but the rational number $\frac{3}{5}$ is not an integer.

On the basis of our experience with multiplying and dividing signed quantities in the preceding chapter, it seems reasonable to define the quotient of any two integers with like signs as a **positive rational number** and the quotient of any two integers with unlike signs as a **negative rational number.** Thus,

$$\frac{+2}{+3} = \frac{2}{3} \qquad \frac{-2}{-3} = \frac{2}{3} \qquad \frac{-2}{+3} = -\frac{2}{3} \qquad \frac{+2}{-3} = -\frac{2}{3}$$

Therefore, we will use $-\frac{2}{3}$, $\frac{-2}{3}$, and $-2/3$ interchangeably, since they all represent the same rational number. We could also include $\frac{2}{-3}$, but we usually try to avoid writing a final form with a negative denominator.

♦ RATIONAL NUMBERS AND THE NUMBER LINE

Locating rational numbers on a number line proceeds as one would expect: the positive numbers are located to the right of 0 and the negative numbers to the left. Where do we locate a number such as $\frac{7}{4}$? We divide each unit on the number line into four segments and identify $\frac{7}{4}$ with the endpoint of the seventh segment to the right of 0. Where is $-\frac{3}{2}$ located? Halfway between -1 and -2.

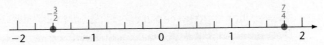

By proceeding as described, every rational number can be associated with a point on a number line. However, as you will see in Chapter 4, there are points on the number line that do not correspond to any rational number.

Example 1
Number Line

Locate $\frac{1}{2}$, $-\frac{3}{4}$, $\frac{5}{2}$, and $-\frac{9}{4}$ on a number line.

Solution

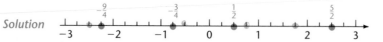

Matched Problem 1   Locate $\frac{3}{4}$, $-\frac{1}{2}$, $\frac{7}{4}$, and $-\frac{5}{2}$ on a number line.

♦ **OPPOSITE OF AND ABSOLUTE VALUE OF A RATIONAL NUMBER**

The **opposite of (negative of) a rational number** and the **absolute value of a rational number** are defined as for the integers. Thus,

$$-\left(\frac{2}{3}\right) = -\frac{2}{3} \qquad -\left(-\frac{2}{3}\right) = \frac{2}{3} \qquad \left|\frac{2}{3}\right| = \frac{2}{3} \qquad \left|-\frac{2}{3}\right| = \frac{2}{3}$$

♦ **FUNDAMENTAL PRINCIPLE OF FRACTIONS**

We now consider a very important property of rational numbers and of fractions in general. Recall from arithmetic the processes of reducing a fraction to lowest terms and that of raising a fraction to higher terms:

(A) $\dfrac{8}{12} = \dfrac{8 \div 4}{12 \div 4} = \dfrac{2}{3}$ or $\dfrac{8}{12} = \dfrac{4 \cdot 2}{4 \cdot 3} = \dfrac{2}{3}$ Lowest terms

(B) $\dfrac{3}{4} = \dfrac{5 \cdot 3}{5 \cdot 4} = \dfrac{15}{20}$ Higher terms

The **fundamental principle of fractions** is the basis for these processes.

Fundamental Principle of Fractions

For any nonzero integers b and k and any integer a:

(A) $\dfrac{ka}{kb} = \dfrac{\cancel{k}a}{\cancel{k}b} = \dfrac{a}{b}$

We may divide out a common factor k from both the numerator and denominator. This is called **reducing to lower terms.**

$$\frac{6}{4} = \frac{2 \cdot 3}{2 \cdot 2} = \frac{3}{2}$$

(B) $\dfrac{a}{b} = \dfrac{ka}{kb}$

We may multiply the numerator and denominator by the same nonzero factor k. This is called **raising to higher terms.**

$$\frac{1}{5} = \frac{1 \cdot 3}{5 \cdot 3} = \frac{3}{15}$$

The principle is stated in two parts to draw your attention to the two ways to view the equality

$$\frac{a}{b} = \frac{ka}{kb}$$

Viewing it as rewriting $\dfrac{ka}{kb}$ in the form $\dfrac{a}{b}$, as in (A), we are **reducing to lower terms.** That is, we are getting smaller numbers in the numerator and denominator. The process of dividing out a common factor from both the numerator and the denominator of a fraction is sometimes called **canceling.** This terminology is not used in this text. Viewing the fundamental principle the other way, as rewriting $\dfrac{a}{b}$ in the form $\dfrac{ka}{kb}$, we are **raising to higher terms,** that is, obtaining a larger numerator and denominator.

Property (A) also provides the basis for reducing fractions to low*est* terms. To reduce a fraction to **lowest terms,** we divide the numerator and denominator by the *largest* nonzero common divisor. That is, we remove *all* common factors from the numerator and denominator.

**Example 2
Rewriting Rational
Numbers**

Replace question marks with appropriate symbols; assume no variables are 0:

(A) $\dfrac{27}{18} = \dfrac{?\,9}{6}$ (B) $\dfrac{27}{18} = \dfrac{?\,3}{2}$ (C) $\dfrac{6x}{9x} = \dfrac{?\,2}{3}$

(D) $\dfrac{-3}{7} = \dfrac{?\,-18}{42}$ (E) $\dfrac{5x}{3} = \dfrac{10x^2}{?\,6t}$

Solution (A) $\dfrac{27}{18} = \dfrac{9 \cdot 3}{6 \cdot 3} = \dfrac{9}{6}$ Lower terms (Can be further reduced.)

(B) $\dfrac{27}{18} = \dfrac{3 \cdot 9}{2 \cdot 9} = \dfrac{3}{2}$ Lowest terms (Cannot be further reduced.)

(C) $\dfrac{6x}{9x} = \dfrac{2 \cdot 3x}{3 \cdot 3x} = \dfrac{2}{3}$ Lowest terms

(D) $\dfrac{-3}{7} = \dfrac{6 \cdot (-3)}{6 \cdot 7} = \dfrac{-18}{42}$ Higher terms

(E) $\dfrac{5x}{3} = \dfrac{2x \cdot 5x}{2x \cdot 3} = \dfrac{10x^2}{6x}$ Higher terms

Matched Problem 2

Replace question marks with appropriate symbols; assume no variables are 0:

(A) $\dfrac{24}{32} = \dfrac{?\,6}{8}$ (B) $\dfrac{24}{32} = \dfrac{?\,3}{4}$ (C) $\dfrac{8m}{12m} = \dfrac{-2}{?\,-3}$

(D) $\dfrac{2}{3} = \dfrac{?\,8y}{12y}$ (E) $\dfrac{7}{4} = \dfrac{14y^2}{?\,8y^2}$

CAUTION When reducing fractions to lower terms, it is important to keep in mind that it is only common *factors* in products that can be divided out or removed. Common *terms* in sums or differences cannot be removed. For example:

Common Factors $\dfrac{2a}{3a} = \dfrac{2}{3}$ *a* is a common factor and can be divided out from both the numerator and the denominator. Thus, $\dfrac{2 \cdot 5}{3 \cdot 5} = \dfrac{2}{3}$

Common Terms $\dfrac{2+a}{3+a} \neq \dfrac{2}{3}$ *a* is a common *term*, not a common factor, and cannot be divided out. Thus, $\dfrac{2+5}{3+5} \neq \dfrac{2}{3}$

**Example 3
Lowest Terms**

Reduce to lowest terms:

(A) $\dfrac{3xy}{15x^2}$ **(B)** $\dfrac{-33xy^2}{15x^2y}$ **(C)** $\dfrac{6(x+1)}{15(x+2)}$

Solution **(A)** $\dfrac{3xy}{15x^2} = \dfrac{3 \cdot x \cdot y}{3 \cdot 5 \cdot x \cdot x} = \dfrac{y}{5x}$

(B) $\dfrac{-33xy^2}{15x^2y} = \dfrac{3 \cdot (-11) \cdot x \cdot y \cdot y}{3 \cdot 5 \cdot x \cdot x \cdot y} = \dfrac{-11y}{5x} = -\dfrac{11y}{5x}$

(C) $\dfrac{6(x+1)}{15(x+2)} = \dfrac{3 \cdot 2(x+1)}{3 \cdot 5(x+2)} = \dfrac{2(x+1)}{5(x+2)}$ Note that *x* is not a common factor.

Matched Problem 3 Reduce to lowest terms:

(A) $\dfrac{12x^2}{4xy}$ $\dfrac{3x}{y}$ **(B)** $\dfrac{-28x^2y^2}{16xy}$ $-7xy$ **(C)** $\dfrac{8(y-3)}{20(y+3)}$ $\dfrac{2(y-3)}{5(y+3)}$

$\dfrac{}{4}$

**Answers to
Matched Problems**

1.

2. (A) 6   **(B)** 3   **(C)** -3 **(D)** $8y$ **(E)** $8y^2$

3. (A) $\dfrac{3x}{y}$ **(B)** $-\dfrac{7xy}{4}$ **(C)** $\dfrac{2(y-3)}{5(y+3)}$

EXERCISE 3-2

Do not change improper fractions to mixed fractions in your answers; that is, write $\frac{7}{2}$, not $3\frac{1}{2}$. All variables represent integers.

A *Locate the given rational numbers on a number line:*

1. What rational numbers are associated with points *a*, *b*, and *c*?

2. What rational numbers are associated with points *c*, *d*, and *e*?

3. What rational numbers are associated with points *a*, *b*, *c*, and *d*?

4. What rational numbers are associated with points c, d, e, and f?

Locate the given rational numbers on a number line:

5. $-\dfrac{3}{2},\ -\dfrac{1}{4},\ \dfrac{5}{4},\ \dfrac{3}{2}$ **6.** $-\dfrac{4}{3},\ -\dfrac{2}{3},\ \dfrac{1}{3},\ \dfrac{5}{3}$

7. $-\dfrac{5}{3},\ -\dfrac{5}{6},\ \dfrac{2}{3},\ \dfrac{7}{6}$ **8.** $-\dfrac{7}{8},\ -\dfrac{1}{2},\ \dfrac{1}{4},\ \dfrac{5}{8}$

Identify the factor that must be divided out in order to reduce each fraction to lowest terms:

9. $\dfrac{21}{56}$ **10.** $\dfrac{51}{68}$ **11.** $\dfrac{21}{40}$ **12.** $\dfrac{25}{44}$

13. $\dfrac{18}{45}$ **14.** $\dfrac{45}{66}$ **15.** $\dfrac{12}{30}$ **16.** $\dfrac{30}{56}$

Reduce to lowest terms:

17. $\dfrac{21}{56}$ **18.** $\dfrac{51}{68}$ **19.** $\dfrac{21}{40}$ **20.** $\dfrac{25}{44}$

21. $\dfrac{18}{45}$ **22.** $\dfrac{45}{66}$ **23.** $\dfrac{12}{30}$ **24.** $\dfrac{30}{56}$

Replace question marks with appropriate symbols:

25. $\dfrac{8}{12}=\dfrac{?}{3}$ **26.** $\dfrac{12}{16}=\dfrac{?}{4}$ **27.** $\dfrac{1}{5}=\dfrac{3}{?}$

28. $\dfrac{3}{4}=\dfrac{?}{20}$ **29.** $\dfrac{21x}{28x}=\dfrac{?}{4}$ **30.** $\dfrac{36y}{54y}=\dfrac{2}{?}$

31. $\dfrac{5}{6}=\dfrac{-20a}{?}$ **32.** $\dfrac{-4}{9}=\dfrac{28x}{?}$

33. $\dfrac{5}{?}=\dfrac{20z}{-44z}$ **34.** $\dfrac{7}{?}=\dfrac{-28x}{36x}$

35. $\dfrac{?}{-12}=\dfrac{14a}{24a}$ **36.** $\dfrac{?}{-15}=\dfrac{21x}{45x}$

37. $\dfrac{2k}{-5k}=\dfrac{?}{45k}$ **38.** $\dfrac{-5x}{8x}=\dfrac{?}{56x}$

B 39. $\dfrac{3}{7}=\dfrac{?}{21x^2}$ **40.** $\dfrac{4}{5}=\dfrac{28m^3}{?}$

41. $\dfrac{6x^3}{4xy}=\dfrac{?}{2y}$ **42.** $\dfrac{9xy}{12y^2}=\dfrac{3x}{?}$

43. $\dfrac{2ab}{3b^2}=\dfrac{2a}{?}$ **44.** $\dfrac{2a^2b}{5ab}=\dfrac{2a}{?}$

45. $\dfrac{10a^2b^2}{5ab}=\dfrac{?}{1}$ **46.** $\dfrac{12ab^2}{4a^2b}=\dfrac{3b}{?}$

47. $\dfrac{15x}{8x}=\dfrac{-45xy}{?}$ **48.** $\dfrac{-32x^2y}{12x^2y}=\dfrac{8y}{?}$

49. $\dfrac{5xy^2}{-11xy^2}=\dfrac{?}{44xy^3}$ **50.** $\dfrac{15a}{6a}=\dfrac{?}{-18ab}$

51. $\dfrac{-25}{?}=\dfrac{75ab}{48ab}$ **52.** $\dfrac{-56}{?}=\dfrac{8x^2}{7x^2}$

53. $\dfrac{?}{13ab}=\dfrac{-30abc}{65abc}$ **54.** $\dfrac{?}{42z}=\dfrac{-11xyz}{7xyz}$

Reduce to lowest terms:

55. $\dfrac{9x}{6x}$ **56.** $\dfrac{27y}{15y}$ **57.** $\dfrac{-3x}{12x}$ **58.** $\dfrac{18y}{-8y}$

C 59. $\dfrac{2y^2}{8y^3}$ **60.** $\dfrac{6x^3}{15x}$ **61.** $\dfrac{12a^2b}{3ab^2}$

62. $\dfrac{21x^2y^3}{35x^3y}$ **63.** $\dfrac{60xy^2}{-96y}$ **64.** $\dfrac{-45a^2b}{63a}$

65. $\dfrac{-21a^2b^2}{56ab}$ **66.** $\dfrac{-70ab}{84a^2b^2}$ **67.** $\dfrac{-39x^3y}{65x}$

68. $\dfrac{-63xy^3}{75xy}$ **69.** $\dfrac{-28x^2yz^2}{49xy^2}$ **70.** $\dfrac{-56ab^2c}{85abc^2}$

71. $\dfrac{-2xy^2}{8x^2}$ **72.** $\dfrac{25mn^3}{-15m^2n^2}$ **73.** $\dfrac{60x^3y^4}{25x^4y^3}$

74. $\dfrac{60x^3y^4}{24x^2y^5}$ **75.** $\dfrac{35x^2y^2}{56xy}$ **76.** $\dfrac{30a^3b^2}{105a^2b}$

77. $\dfrac{44a^2b^3}{110ab}$ **78.** $\dfrac{165x^2y}{210x}$ **79.** $\dfrac{-210ab}{231a^2b^3}$

80. $\dfrac{78a^2b}{-105a^3b^3}$ **81.** $\dfrac{105x^2y^2}{-195x^3y^4}$ **82.** $\dfrac{-70xy^2}{231x^2y}$

83. $\dfrac{4(x^2+y)}{6(x^2+y)}$ **84.** $\dfrac{3(a+b^2)}{9(a+b^2)}$ **85.** $\dfrac{4(x^2+y)}{12(x+y^2)}$

86. $\dfrac{3(a+b^2)}{12(a^2+b)}$ **87.** $\dfrac{12(x-y^2)}{20(x-y^2)}$ **88.** $\dfrac{32(x^2-y)}{10(x^2-y)}$

89. $\dfrac{-15(x-y)}{40(x-y)}$ **90.** $\dfrac{-35(xz-y)}{90(xz-y)}$ **91.** $\dfrac{22(x+y)}{121(x+y)}$

92. $\dfrac{225(x-yz)}{100(x-yz)}$ **93.** $\dfrac{3x(y+z)}{15(xy+xz)}$ **94.** $\dfrac{4(xy+xz)}{14x(y+z)}$

95. $\dfrac{5(x+y)^2}{30(x+y)^3}$ **96.** $\dfrac{8(x+y)^3}{42(x+y)}$

3-3   Multiplication and Division

♦ Multiplication
♦ Sign Properties
♦ Division

In this section we consider multiplication and division of algebraic forms representing rational numbers. In the next section we will consider addition and subtraction.

♦ MULTIPLICATION

In arithmetic you multiply fractions by multiplying their numerators (tops) and multiplying their denominators (bottoms). This is exactly what we do with rational numbers in general.

Definition of Multiplication for Rational Numbers

If a, b, c, and d are integers, with b and d different from 0, then

$$\frac{a}{b} \cdot \frac{c}{d} = \frac{a \cdot c}{b \cdot d} \qquad \frac{3}{4} \cdot \frac{5}{7} = \frac{3 \cdot 5}{4 \cdot 7} = \frac{15}{28}$$

Example 1
Multiplying Rational Number Expressions

Multiply:

(A) $\dfrac{2}{5} \cdot \dfrac{3}{7}$ **(B)** $(-8) \cdot \dfrac{9}{5}$ **(C)** $\dfrac{2x}{3y^2} \cdot \dfrac{x^2}{5y}$

Solution

(A) $\dfrac{2}{5} \cdot \dfrac{3}{7} = \dfrac{2 \cdot 3}{5 \cdot 7} = \dfrac{6}{35}$

(B) $(-8) \cdot \dfrac{9}{5} = \dfrac{-8}{1} \cdot \dfrac{9}{5} = \dfrac{(-8)(9)}{(1)(5)} = \dfrac{-72}{5}$ or $-\dfrac{72}{5}$ or $-14\dfrac{2}{5}$ or -14.4

(C) $\dfrac{2x}{3y^2} \cdot \dfrac{x^2}{5y} = \dfrac{(2x)(x^2)}{(3y^2)(5y)} = \dfrac{2x^3}{15y^3}$

Matched Problem 1   Multiply:

(A) $\dfrac{3}{4} \cdot \dfrac{3}{5}$ $\dfrac{9}{20}$ **(B)** $(-5) \cdot \dfrac{3}{4}$ $\dfrac{-15}{4}$ **(C)** $\dfrac{3x^2}{2y} \cdot \dfrac{x}{4y^2}$ $\dfrac{3x^3}{8y^3}$

It follows from the definition of multiplication that we can continue to rearrange and regroup factors that represent rational numbers in the same way we rearrange and regroup factors that represent integers or natural numbers. That is, **multiplication of rational numbers is commutative and associative.**

♦ SIGN PROPERTIES

The following sign properties of rational numbers, and of fractions in general, are used with great frequency in mathematics. Their misuse accounts for many algebraic errors. These properties are a consequence of the fundamental principle of fractions and the definition of opposites. Justification of the properties will be given in Section 3-4.

Sign Properties

For each integer a and each nonzero integer b:

(A) $\dfrac{-a}{-b} = \dfrac{a}{b}$ $\qquad\qquad$ $\dfrac{-2}{-3} = \dfrac{2}{3}$

(B) $\dfrac{-a}{b} = \dfrac{a}{-b} = -\dfrac{a}{b}$ $\qquad$ $\dfrac{-2}{3} = \dfrac{2}{-3} = -\dfrac{2}{3}$

(C) $(-1)\dfrac{a}{b} = -\dfrac{a}{b}$ $\qquad\qquad$ $(-1)\dfrac{2}{3} = -\dfrac{2}{3}$ or $\dfrac{-2}{3}$ or $\dfrac{2}{-3}$

In choosing among $\dfrac{-2}{3}$, $\dfrac{2}{-3}$, or $-\dfrac{2}{3}$ for a final answer, we would generally choose either $\dfrac{-2}{3}$ or $-\dfrac{2}{3}$, forms with a positive denominator, thus avoiding the use of a negative denominator.

We now consider multiplication examples of a more general type. The following example illustrates one use of the sign properties.

Example 2
Sign Properties in Multiplication

Multiply and write your answer in lowest terms:

(A) $\dfrac{5x^2}{9y^2} \cdot \dfrac{-6y}{10x}$ $\qquad$ **(B)** $\dfrac{-3x}{2y} \cdot \dfrac{6y^2}{9x^2}$

Solution **(A)** $\dfrac{5x^2}{9y^2} \cdot \dfrac{-6y}{10x}$ $= \dfrac{5xx \cdot 2 \cdot (-3)y}{3 \cdot 3yy \cdot 2 \cdot 5x}$ Factor numerator and denominator.

$\qquad\qquad = \dfrac{5xx \cdot 2 \cdot \overset{-1}{(-3)} \cdot y}{3 \cdot 3yy \cdot 2 \cdot 5x}$ Remove common factors.

$\qquad\qquad = \dfrac{-x}{3y} = -\dfrac{x}{3y}$ Answer is in lowest terms.

After a little experience you will probably proceed by repeated division of the numerator and denominator by common factors until all common factors are eliminated. That is, you will proceed something like this:

$$\dfrac{5x^2}{9y^2} \cdot \dfrac{-6y}{10x} = \dfrac{(5x^2)(-6y)}{(9y^2)(10x)} = \dfrac{-x}{3y}$$

(B) $\dfrac{-3x}{2y} \cdot \dfrac{6y^2}{9x^2} = \dfrac{\overset{-1}{\cancel{(-3x)}}\overset{3y}{\cancel{(6y^2)}}}{\underset{1}{\cancel{(2y)}}\,\underset{\underset{x}{3x}}{\cancel{(9x^2)}}} = \dfrac{-y}{x}$ or $-\dfrac{y}{x}$

We could also proceed as in the first part of part (A) by factoring the numerator and denominator and removing common factors.

Matched Problem 2   Multiply and write your answer in lowest terms:

(A) $\dfrac{10x}{6y^2} \cdot \dfrac{12y}{5x^2}$ **(B)** $\dfrac{-7x^2}{3y^2} \cdot \dfrac{12y}{14x}$

♦ **DIVISION**

The rule for dividing fractions is "To divide one fraction by another, invert the divisor and multiply." The same rule will hold for dividing rational numbers. It is not difficult to see why this mechanical rule is valid. To start, we define division for rational numbers as in the integers.

Definition of Division

If a/b and c/d are any two rational numbers with $c/d \neq 0$, then

$$\frac{a}{b} \div \frac{c}{d} = Q \qquad \text{if and only if} \qquad \frac{c}{d} \cdot Q = \frac{a}{b}$$

and Q is unique.

That is, the *quotient* Q is the unique number that the *divisor* c/d must be multiplied by to produce the *dividend* a/b.

$$\frac{2}{3} \div \frac{1}{5} = \frac{10}{3} \quad \text{because} \quad \frac{1}{5} \cdot \frac{10}{3} = \frac{2}{3}$$

As a result of this definition, we can establish the following mechanical rule for carrying out division.

Mechanical Rule for Division

To divide one rational number by another rational number different from 0, invert the divisor and multiply. That is,

$$\frac{a}{b} \div \frac{c}{d} = \frac{a}{b} \cdot \frac{d}{c} \qquad\qquad \frac{2}{3} \div \frac{1}{5} = \frac{2}{3} \cdot \frac{5}{1} = \frac{10}{3}$$

Divisor   Inverted divisor

To establish this rule, we have to show that the product of the divisor (c/d) and the quotient $[(a/b) \cdot (d/c)]$ is equal to the dividend (a/b):

$$\frac{c}{d} \cdot \left(\frac{a}{b} \cdot \frac{d}{c}\right) = \frac{c}{d} \cdot \frac{ad}{bc}$$

$$= \frac{\overset{1}{\cancel{c}}a\overset{1}{\cancel{d}}}{\underset{1}{\cancel{d}}b\underset{1}{\cancel{c}}} = \frac{a}{b}$$

Example 3
Dividing Rational Number Expressions

Divide and reduce to lowest terms:

(A) $\dfrac{6}{14} \div \dfrac{21}{2}$ (B) $\dfrac{12x}{5y} \div \dfrac{9y}{8x}$ (C) $\dfrac{18a^2b}{15c} \div \dfrac{12ab^2}{5c}$

(D) $\dfrac{-3x}{yz} \div 12x$

Solution

(A) $\dfrac{6}{14} \div \dfrac{21}{2} = \dfrac{\overset{2}{\cancel{6}}}{\underset{7}{\cancel{14}}} \cdot \dfrac{\overset{1}{\cancel{2}}}{\underset{7}{\cancel{21}}} = \dfrac{2}{49}$ Do not remove common factors in division before inverting the divisor.

(B) $\dfrac{12x}{5y} \div \dfrac{9y}{8x} = \dfrac{\overset{4x}{\cancel{12x}}}{5y} \cdot \dfrac{8x}{\underset{3y}{\cancel{9y}}} = \dfrac{32x^2}{15y^2}$

(C) $\dfrac{18a^2b}{15c} \div \dfrac{12ab^2}{5c} = \dfrac{\overset{3a}{\cancel{18a^2b}}}{\underset{1}{\cancel{15c}}} \cdot \dfrac{\overset{1}{\cancel{5c}}}{\underset{2b}{\cancel{12ab^2}}} = \dfrac{a}{2b}$

(D) $\dfrac{-3x}{yz} \div 12x = \boxed{\dfrac{-3x}{yz} \div \dfrac{12x}{1}} = \dfrac{-3x}{yz} \cdot \dfrac{1}{\underset{4}{\cancel{12x}}} = \dfrac{-1}{4yz}$ or $-\dfrac{1}{4yz}$

Matched Problem 3

Divide and reduce to lowest terms:

(A) $\dfrac{8}{9} \div \dfrac{4}{3}$ (B) $\dfrac{8x}{3y} \div \dfrac{6x}{9y}$ (C) $\dfrac{15mn^2}{8x} \div \dfrac{9m^2n}{8x}$ (D) $\dfrac{6x}{wz} \div (-3x)$

Answers to Matched Problems

1. (A) $\dfrac{9}{20}$ (B) $\dfrac{-15}{4}$ or $-\dfrac{15}{4}$ (C) $\dfrac{3x^3}{8y^3}$

2. (A) $\dfrac{4}{xy}$ (B) $\dfrac{-2x}{y}$ or $-\dfrac{2x}{y}$

3. (A) $\dfrac{2}{3}$ (B) 4 (C) $\dfrac{5n}{3m}$ (D) $\dfrac{-2}{wz}$ or $-\dfrac{2}{wz}$

EXERCISE 3-3

Do not change improper fractions to mixed fractions in your answers; that is, write $\frac{7}{2}$, not $3\frac{1}{2}$. All variables represent integers.

A *Multiply and reduce to lowest terms:*

1. $\dfrac{2}{5} \cdot \dfrac{3}{7}$ **2.** $\dfrac{3}{8} \cdot \dfrac{3}{5}$ **3.** $\dfrac{14}{9} \cdot \dfrac{-3}{2}$

4. $\dfrac{15}{4} \cdot \dfrac{-3}{11}$ **5.** $\dfrac{4}{5} \cdot \dfrac{7x}{3y}$ **6.** $\dfrac{5}{7} \cdot \dfrac{2x}{3y}$

7. $\dfrac{3a}{5} \cdot \dfrac{4b}{9}$ **8.** $\dfrac{4x}{7y} \cdot \dfrac{4}{3}$ **9.** $\dfrac{x}{2y} \cdot \dfrac{3x}{y^2}$

10. $\dfrac{2m}{n^2} \cdot \dfrac{3m^2}{5n^2}$ **11.** $\dfrac{-3x}{4y} \cdot \dfrac{2x}{5y}$ **12.** $\dfrac{5a}{6b} \cdot \dfrac{-5a^2}{4b^2}$

13. $\dfrac{3}{7} \cdot \dfrac{-2}{11}$ **14.** $\dfrac{-2}{5} \cdot \dfrac{4}{3}$ **15.** $\dfrac{-5}{3} \cdot \dfrac{2}{-7}$

16. $\dfrac{2}{-5} \cdot \dfrac{-3}{7}$ **17.** $\dfrac{4ab}{3} \cdot \dfrac{1}{2c}$ **18.** $\dfrac{a}{3b} \cdot \dfrac{2a}{b}$

19. $\dfrac{-6ab^2}{5} \cdot \dfrac{-a}{7}$ **20.** $\dfrac{-3a}{2} \cdot \dfrac{-5}{b}$

Divide:

21. $\dfrac{3}{5} \div \dfrac{5}{7}$ **22.** $\dfrac{3}{4} \div \dfrac{4}{5}$ **23.** $\dfrac{-3}{4} \div \dfrac{4}{5}$

24. $\dfrac{-2}{3} \div \dfrac{5}{6}$ **25.** $\dfrac{2x}{3} \div \dfrac{5}{7y}$ **26.** $\dfrac{3}{2u} \div \dfrac{5y}{11}$

27. $\dfrac{-3x}{y} \div \dfrac{2}{3x}$ **28.** $\dfrac{4a}{3} \div \dfrac{-b}{4}$ **29.** $\dfrac{3}{7} \div \dfrac{-2}{3}$

30. $\dfrac{-4}{5} \div \dfrac{3}{7}$ **31.** $\dfrac{-2a}{3b^2} \div \dfrac{-3b}{2}$

32. $\dfrac{-ab}{3} \div \dfrac{-3}{b}$ **33.** $\dfrac{5x}{-2y} \div \dfrac{-4y}{x}$

34. $\dfrac{3xy}{-5} \div \dfrac{-10}{7x}$

B *Multiply or divide as indicated and reduce to lowest terms:*

35. $\dfrac{3}{4} \cdot \dfrac{8}{9}$ **36.** $\dfrac{2}{9} \cdot \dfrac{3}{10}$ **37.** $\dfrac{1}{25} \div \dfrac{15}{4}$

38. $\dfrac{7}{3} \div \dfrac{2}{3}$ **39.** $3 \cdot \dfrac{5}{3}$ **40.** $\dfrac{5}{7} \cdot 7$

41. $\dfrac{2}{3} \div \dfrac{4}{9}$ **42.** $\dfrac{5}{11} \div \dfrac{55}{44}$ **43.** $\dfrac{4}{-5} \cdot \dfrac{15}{16}$

44. $\dfrac{8}{3} \cdot \dfrac{-12}{24}$ **45.** $\dfrac{-3}{8} \cdot \dfrac{-18}{7}$ **46.** $\dfrac{-5}{6} \div \dfrac{15}{-7}$

47. $\dfrac{-3}{8} \div \dfrac{-18}{7}$ **48.** $\dfrac{-5}{6} \cdot \dfrac{15}{-7}$

49. $\dfrac{-15}{8} \div \dfrac{5}{-12}$ **50.** $\dfrac{-15}{8} \cdot \dfrac{5}{-12}$

51. $\dfrac{2x}{3yz} \cdot \dfrac{6y}{4x}$ **52.** $\dfrac{2a}{3bc} \cdot \dfrac{9c}{a}$

53. $\dfrac{6x}{5y} \div \dfrac{3x}{10y}$ **54.** $\dfrac{9m}{8n} \div \dfrac{3m}{4n}$

55. $2xy \div \dfrac{x}{y}$ **56.** $\dfrac{x}{3y} \div 3y$

57. $-x \cdot \dfrac{-y}{x}$ **58.** $-y \div \dfrac{y}{-x}$

59. $\dfrac{-x}{2y} \div \dfrac{3x^2}{-4y}$ **60.** $\dfrac{-5x}{14y} \cdot \dfrac{21y^2}{-20x^2}$

61. $\dfrac{2x^2}{3y^2} \cdot \dfrac{9y}{4x}$ **62.** $\dfrac{3x^2}{4} \cdot \dfrac{16y}{12x^3}$

63. $\dfrac{2x}{3y} \div \dfrac{4x}{6y^2}$ **64.** $\dfrac{a}{4c} \div \dfrac{a^2}{12c^2}$

65. $\dfrac{6a^2}{7c} \cdot \dfrac{21cd}{12ac}$ **66.** $\dfrac{8x^2}{3xy} \cdot \dfrac{12y^3}{6y}$

67. $\dfrac{3uv^2}{5w} \div \dfrac{6u^2v}{15w}$ **68.** $\dfrac{21x^2y^2}{12cd} \div \dfrac{14xy}{9d}$

69. $\dfrac{-6x^3}{5y^2} \div \dfrac{18x}{10y}$ **70.** $\dfrac{9u^4}{4v^3} \div \dfrac{-12u^2}{15v}$

C *Perform the operations as indicated and reduce to lowest terms:*

71. $\left(\dfrac{9}{10} \div \dfrac{4}{6}\right) \cdot \dfrac{3}{5}$ **72.** $\dfrac{9}{10} \div \left(\dfrac{4}{6} \cdot \dfrac{3}{5}\right)$

73. $\dfrac{-21}{16} \cdot \dfrac{12}{-14} \cdot \dfrac{8}{9}$ **74.** $\dfrac{18}{15} \cdot \dfrac{-10}{21} \cdot \dfrac{3}{-1}$

75. $\dfrac{2x^2}{3y^2} \cdot \dfrac{6yz}{2x} \cdot \dfrac{y}{-xz}$ **76.** $\dfrac{-a}{-b} \cdot \dfrac{12b^2}{15ac} \cdot \dfrac{-10}{4b}$

77. $\left(\dfrac{-3}{4} \cdot \dfrac{8}{15}\right) \div \dfrac{-10}{3}$ **78.** $\left(\dfrac{-3}{4} \div \dfrac{8}{15}\right) \cdot \dfrac{-10}{3}$

79. $\dfrac{-3}{4} \cdot \left(\dfrac{8}{15} \div \dfrac{-10}{3}\right)$ **80.** $\dfrac{-3}{4} \div \left(\dfrac{8}{15} \cdot \dfrac{-10}{3}\right)$

81. $\dfrac{4xy}{3z} \cdot \left(\dfrac{9xz}{10y} \div \dfrac{-12zy}{5x}\right)$

82. $\dfrac{4xy}{3z} \div \left(\dfrac{xz}{10y} \cdot \dfrac{-12zy}{5x} \right)$

83. $\left(\dfrac{4xy}{3z} \cdot \dfrac{9xz}{10y} \right) \div \dfrac{-12zy}{5x}$

84. $\left(\dfrac{4xy}{3z} \div \dfrac{xz}{10y} \right) \cdot \dfrac{-12zy}{5x}$

85. $\left(\dfrac{a}{b} \div \dfrac{c}{d} \right) \div \dfrac{e}{f}$

86. $\dfrac{a}{b} \div \left(\dfrac{c}{d} \div \dfrac{e}{f} \right)$

87. $\dfrac{ab}{cd} \cdot \left(\dfrac{ac}{bd} \div \dfrac{ad}{bc} \right)$

88. $\dfrac{ab}{cd} \div \left(\dfrac{ac}{bd} \cdot \dfrac{ad}{bc} \right)$

89. $\left(\dfrac{ab}{cd} \cdot \dfrac{-ac}{bd} \right) \div \dfrac{-ad}{bc}$

90. $\left(\dfrac{ab}{cd} \div \dfrac{-ac}{bd} \right) \cdot \dfrac{-ad}{bc}$

3-4 Addition and Subtraction

- ♦ Addition and Subtraction with Common Denominators
- ♦ Addition and Subtraction with Different Denominators
- ♦ Averages of Numbers
- ♦ Order in the Rational Numbers

In this section we consider the operations of addition and subtraction on rational numbers. We look at the problem of finding *averages,* which will lead to many applications, and conclude with some observations about the "greater than" relationship. A more detailed consideration of inequalities involving rational numbers is, however, deferred until Chapter 5.

♦ ADDITION AND SUBTRACTION WITH COMMON DENOMINATORS

As in the preceding sections, we will generalize from arithmetic. You would probably add $\frac{1}{2}$ and $\frac{2}{3}$ as follows:

$$\dfrac{1}{2} = \dfrac{3}{6}$$

$$\dfrac{2}{3} = \dfrac{4}{6}$$

$$\overline{\phantom{\dfrac{2}{3}} \dfrac{7}{6}}$$

In algebra we often leave fractions in what is called "improper form"; that is, we write $\frac{7}{6}$ rather than $1\frac{1}{6}$, since the latter might be confused with the product $(1)(\frac{1}{6})$. However, if $\frac{7}{6}$ were the final answer in an applied problem, we would generally write it in the form $1\frac{1}{6}$.

To perform the addition, you changed each fraction to an equivalent form having a common denominator, then added the numerators and placed the sum over the common denominator. We will proceed in the same way with rational numbers, but we will find it more convenient to work horizontally. We start by defining addition and subtraction of rational numbers with common denominators.

Addition and Subtraction of Rational Numbers—Common Denominators

If a, b, and c are integers with $b \neq 0$, then

$$\frac{a}{b} + \frac{c}{b} = \frac{a+c}{b} \qquad \frac{a}{b} - \frac{c}{b} = \frac{a-c}{b}$$

That is, two rational numbers with common denominators are added or subtracted by adding or subtracting the numerators and placing the result over the common denominator.

$$\frac{2}{5} + \frac{4}{5} = \frac{2+4}{5} = \frac{6}{5} \qquad \frac{3}{5} - \frac{4}{5} = \frac{3-4}{5} = \frac{-1}{5}$$

An immediate consequence of the definitions of addition and multiplication is that **addition and multiplication of rational numbers are commutative and associative.** Moreover, **multiplication distributes over addition.**

Example 1
Adding and Subtracting Rational Number Expressions

Add or subtract as indicated:

(A) $\dfrac{1}{8} + \dfrac{3}{8}$ **(B)** $\dfrac{5x}{5x} - \dfrac{2}{5x}$ **(C)** $\dfrac{2x}{6x^2} + \dfrac{5}{6x^2}$

Solution   **(A)** $\dfrac{1}{8} + \dfrac{3}{8} = \dfrac{1+3}{8} = \dfrac{4}{8} = \dfrac{1}{2}$ **(B)** $\dfrac{5x}{5x} - \dfrac{2}{5x} = \dfrac{5x-2}{5x}$

(C) $\dfrac{2x}{6x^2} + \dfrac{5}{6x^2} = \dfrac{2x+5}{6x^2}$

CAUTION

Note that in Example 1(B)

$$\frac{5x-2}{5x} \neq -2 \qquad \text{and} \qquad \frac{5x-2}{5x} \neq \frac{1-2}{1}$$

Only common *factors* can be divided out. $5x$ is a term in the numerator, not a factor. For example, if $x = 3$:

$$\underbrace{\frac{15-2}{15} \neq -2}_{\substack{\text{15 cannot be} \\ \text{divided out.}}} \qquad \underbrace{\frac{15-2}{15} = \frac{13}{15}}_{\text{Right}}$$

Matched Problem 1

Add or subtract as indicated:

(A) $\dfrac{7}{3} - \dfrac{5}{3}$ **(B)** $\dfrac{3}{2u} + \dfrac{2u}{2u}$ **(C)** $\dfrac{3m^2}{12m^3} - \dfrac{2m}{12m^3}$

♦ ADDITION AND SUBTRACTION WITH DIFFERENT DENOMINATORS

How do we add or subtract rational numbers, or algebraic expressions representing rational numbers, when the denominators are not the same? We use the fundamental principle of fractions

$$\frac{a}{b} = \frac{ka}{kb} \qquad k, b \neq 0 \qquad \text{Fundamental principle of fractions}$$

to obtain equivalent forms having common denominators and then add or subtract the resulting numbers as above.

Addition and Subtraction of Rational Numbers—Different Denominators

1. Find a common denominator of the numbers to be added or subtracted.
2. Convert each of the numbers to an equivalent form with this common denominator by using the fundamental principle of fractions.
3. Add or subtract using the rule for adding or subtracting with common denominators.

In step 1, the product of the denominators will always work as a common denominator. However, the common denominator that generally results in the least amount of computation is the least common multiple (LCM) of all the denominators. The LCM (see Section 1-1) of the denominators is the "smallest" quantity exactly divisible by each denominator and is called the **least common denominator (LCD)**.

Finding the LCD

1. Determine the LCD (the "smallest" quantity exactly divisible by each denominator) by inspection, if possible.
2. If the LCD is not obvious, then it can always be found as follows:
   (A) Factor each denominator completely, representing multiple factors as powers.
   (B) The LCD must contain each *different* factor from these factorizations to the highest power it occurs in any one factorization.

Note that step 2 is the same process that was used in Section 1-1 to find the LCM of natural numbers. However, the process of factoring algebraic denominators as required in step 2(A) is not always routine. We will return to this problem in Chapter 7. For now we will restrict ourselves to denominators that can be readily factored.

Example 2
Adding and Subtracting Rational Number Expressions

Add or subtract and reduce to lowest terms:

(A) $\dfrac{1}{3} + \dfrac{5}{12}$ **(B)** $\dfrac{5}{24y} - \dfrac{7}{10y}$ **(C)** $\dfrac{3}{8x} + \dfrac{7}{12x^2}$

$\dfrac{4}{12} + \dfrac{5}{12} \quad \dfrac{9}{12} \quad \dfrac{3}{4} \qquad \dfrac{25}{120y} - \dfrac{84}{120y} \quad -\dfrac{59}{120y} \qquad \dfrac{9x}{24x^2} + \dfrac{14}{24x^2} \qquad \dfrac{9x+14}{24x^2}$

Solution **(A)** *Step 1* Find the LCD. The smallest number exactly divisible by 3 and 12 is 12. Therefore,

$$\text{LCD} = 12$$

Step 2 Use the fundamental principle of fractions to convert each fraction into a form having the LCD as a denominator. Then combine into a single fraction and reduce to lowest terms:

$$\dfrac{1}{3} + \dfrac{5}{12} = \dfrac{4 \cdot 1}{4 \cdot 3} + \dfrac{5}{12} \qquad \text{Convert to forms having the same LCD.}$$

$$= \dfrac{4}{12} + \dfrac{5}{12} \qquad \text{Combine into a single fraction.}$$

$$= \dfrac{4 + 5}{12} \qquad \text{Reduce to lowest terms.}$$

$$= \dfrac{9}{12}$$

$$= \dfrac{3}{4}$$

(B) In this case the LCD is not obvious, so we factor each denominator completely:

$$24y = 8 \cdot 3y = 2 \cdot 2 \cdot 2 \cdot 3 \cdot y = 2^3 \cdot 3 \cdot y$$
$$10y = 2 \cdot 5 \cdot y$$

Since the LCD must contain each *different* factor from these factorizations to the highest power that it occurs in any one, we first write down all the different factors:

$$2, 3, 5, y$$

Then we observe the highest power to which each occurs in any one factorization, to obtain

$$\text{LCD} = 2^3 \cdot 3 \cdot 5 \cdot y = 120y$$

Now multiply numerators and denominators in the original problem by appropriate quantities to obtain $120y$ as a common denominator:

$$\frac{5}{24y} - \frac{7}{10y} = \frac{5 \cdot 5}{5 \cdot 24y} - \frac{12 \cdot 7}{12 \cdot 10y}$$

$$= \frac{25}{120y} - \frac{84}{120y}$$

$$= \frac{25 - 84}{120y}$$

$$= \frac{-59}{120y}$$

(C) Find the LCD:

$$8x = 2^3 \cdot x$$

$$12x^2 = 2^2 \cdot 3 \cdot x^2$$

$$\text{LCD} = 2^3 \cdot 3 \cdot x^2 = 24x^2$$

Use the LCD to add:

$$\frac{3}{8x} + \frac{7}{12x^2} = \frac{(3x)(3)}{(3x)(8x)} + \frac{2 \cdot 7}{2(12x^2)}$$

$$= \frac{9x}{24x^2} + \frac{14}{24x^2}$$

$$= \frac{9x + 14}{24x^2}$$

Matched Problem 2   Add or subtract and reduce to lowest terms:

(A) $\dfrac{4}{5} - \dfrac{2}{15}$ **(B)** $\dfrac{5}{18xy} + \dfrac{3}{4xy}$ **(C)** $\dfrac{2}{9x^2y} - \dfrac{7}{12xy^2}$

Adding or subtracting several rational numbers follows the same rules as for two. If it is necessary to find a common denominator, we look for a common denominator for all the numbers involved.

Example 3
Adding and Subtracting Several Rational Number Expressions

Add or subtract as indicated:

(A) $\dfrac{-3}{4} - \dfrac{-1}{3} + \dfrac{5}{6}$ **(B)** $\dfrac{3}{2x^2} - \dfrac{-5}{x} + 1$

Solution   **(A)** $\dfrac{-3}{4} - \dfrac{-1}{3} + \dfrac{5}{6} = \dfrac{3(-3)}{3(4)} - \dfrac{4(-1)}{4(3)} + \dfrac{2(5)}{2(6)}$ *Note:* LCD = 12

$$= \frac{-9}{12} - \frac{-4}{12} + \frac{10}{12}$$

$$= \frac{-9 - (-4) + 10}{12}$$

$$= \frac{-9 + 4 + 10}{12}$$

$$= \frac{5}{12}$$

(B) $\dfrac{3}{2x^2} - \dfrac{-5}{x} + 1 = \dfrac{3}{2x^2} - \dfrac{2x(-5)}{2x(x)} + \dfrac{2x^2}{2x^2}$ *Note:* LCD = $2x^2$

$$= \frac{3}{2x^2} - \frac{-10x}{2x^2} + \frac{2x^2}{2x^2}$$

$$= \frac{3 - (-10x) + 2x^2}{2x^2}$$

$$= \frac{3 + 10x + 2x^2}{2x^2}$$

Matched Problem 3   Add or subtract as indicated:

(A) $\dfrac{-1}{2} - \dfrac{-3}{4} + \dfrac{2}{3}$ **(B)** $2 - \dfrac{-3}{x} + \dfrac{4}{3x^2}$

Adding a number to its opposite should yield 0. This observation, together with the definition of addition of rational numbers, justifies some of the sign properties given in the sign properties box in Section 3-3. For example,

$$\frac{-a}{b} = -\frac{a}{b}$$

since

$$\frac{a}{b} + \frac{-a}{b} = \frac{a + (-a)}{b} = \frac{0}{b} = 0$$

Thus, adding $\dfrac{-a}{b}$ to $\dfrac{a}{b}$ yields 0, so $\dfrac{-a}{b}$ must be the opposite of $\dfrac{a}{b}$. The other parts of sign properties (B) and (C) in the box in Section 3-3 are explained similarly.

♦   **AVERAGES OF NUMBERS**

Rational numbers occur naturally when one takes the average of several numbers, either integers or rational numbers. The **average** of a set of numbers is the sum of the numbers divided by the number of numbers in the set. For example, the average of 3, 12, −9, 5, 0, −4, 2, and −6 (a set of eight numbers) is

$$\frac{3 + 12 - 9 + 5 + 0 - 4 + 2 - 6}{8} = \frac{3}{8}$$

We have divided by 8 since there are eight numbers to be averaged. Observe that the average can also be written as

$$\frac{1}{8} \cdot (3 + 12 - 9 + 5 + 0 - 4 + 2 - 6)$$

since

$$\frac{1}{8} \cdot S = \frac{1}{8} \cdot \frac{S}{1} = \frac{S}{8}$$

Example 4
Finding an Average

Find the average of the following numbers:

(A) 11, −7, 22, −15, 31, 13, −8, 10, −18
(B) 3x, −6x, 4y, −7y, 2x, 6x, 5y
(C) 3/5, 3/8, 1/10, 1/4

Solution

(A) The average is the sum of the numbers divided by 9, since there are 9 numbers to be averaged:

$$\frac{11 - 7 + 22 - 15 + 31 + 13 - 8 + 10 - 18}{9} = \frac{39}{9} = 4\frac{1}{3}$$

(B) $\dfrac{3x - 6x + 4y - 7y + 2x + 6x + 5y}{7} = \dfrac{5x + 2y}{7}$

(C) Find the sum of the four numbers using the LCD of 40:

$$\frac{3}{5} + \frac{3}{8} + \frac{1}{10} + \frac{1}{4} = \frac{24}{40} + \frac{15}{40} + \frac{4}{40} + \frac{10}{40} = \frac{53}{40}$$

The average is the sum divided by 4, that is,

$$\frac{53}{40} \div 4 = \frac{53}{40} \cdot \frac{1}{4} = \frac{53}{160}$$

Matched Problem 4

Find the average of the following numbers:

(A) 25, −16, 40, −9, 36, −49, 4, 64, −50, 18
(B) 4a, 3b, −6b, 7a, −5a, 8b, −2b, 2a
(C) 1/5, 5/2, 3/4, 3/10, 1/20

♦ **ORDER IN THE RATIONAL NUMBERS**

There is a natural order on the set of rational numbers, just as on the set of integers. One number is greater than a second if the first lies to the right of the second on the number line. However, in contrast to the integers, it is not always obvious which of two rational numbers is greater. Compare, for example, $\frac{4}{7}$ and $\frac{6}{11}$. One way to compare them is to convert both to decimal fractions

$$\frac{4}{7} = 0.5714\ldots \qquad \frac{6}{11} = 0.5454\ldots$$

where it becomes clear that $\frac{4}{7}$ is the greater. Another approach follows that given in Section 2-4:

Definition of $<$ and $>$

If a and b are rational numbers, then we write $a < b$ if there exists a positive number p such that $a + p = b$. Equivalently, $b > a$ if the difference $b - a$ is positive.

> $\frac{1}{4} < \frac{3}{4}$ because there is a positive number, namely, $\frac{2}{4}$, such that $\frac{1}{4} + \frac{2}{4} = \frac{3}{4}$, or equivalently because $\frac{3}{4} - \frac{1}{4} = +\frac{2}{4}$ is positive.

Thus, since

$$\frac{4}{7} - \frac{6}{11} = \frac{44}{77} - \frac{42}{77} = \frac{2}{77}$$

and $\frac{2}{77}$ is positive, $\frac{4}{7} > \frac{6}{11}$.

Example 5
Rational Number Inequalities

Select the larger rational number:

(A) $\frac{3}{13}, \frac{4}{17}$ **(B)** $\frac{5}{7}, \frac{14}{19}$

Solution **(A)** $\frac{3}{13} = 0.2307\ldots$ and $\frac{4}{17} = 0.2352\ldots$

Thus $\frac{4}{17} > \frac{3}{13}$

(B) $\frac{14}{19} - \frac{5}{7} = \frac{14 \cdot 7}{19 \cdot 7} - \frac{5 \cdot 19}{7 \cdot 19} = \frac{98}{133} - \frac{95}{133} = \frac{3}{133} > 0$

Since $\frac{14}{19} - \frac{5}{7}$ is positive, $\frac{14}{19} > \frac{5}{7}$.

Matched Problem 5 Select the larger rational number:

(A) $\frac{5}{21}, \frac{7}{31}$ **(B)** $\frac{8}{15}, \frac{14}{27}$

Answers to Matched Problems

1. (A) $\dfrac{2}{3}$ (B) $\dfrac{3 + 2u}{2u}$ (C) $\dfrac{3m^2 - 2m}{12m^3}$

2. (A) $\dfrac{2}{3}$ (B) $\dfrac{37}{36xy}$ (C) $\dfrac{8y - 21x}{36x^2y^2}$

3. (A) $\dfrac{11}{12}$ (B) $\dfrac{6x^2 + 9x + 4}{3x^2}$

4. (A) $\dfrac{63}{10} = 6\dfrac{3}{10} = 6.3$ (B) $\dfrac{8a + 3b}{8}$ (C) $\dfrac{76}{100} = \dfrac{19}{25} = 0.76$

5. (A) $\dfrac{5}{21}$ (B) $\dfrac{8}{15}$

EXERCISE 3-4

A *Combine into single fractions and reduce to lowest terms. Work horizontally.*

1. $\dfrac{2}{3} + \dfrac{4}{3}$

2. $\dfrac{3}{4} + \dfrac{5}{4}$

3. $\dfrac{-3}{5} + \dfrac{7}{5}$

4. $\dfrac{2}{3} + \dfrac{-5}{3}$

5. $\dfrac{3}{8} + \dfrac{1}{2}$

6. $\dfrac{2}{5} + \dfrac{3}{10}$

7. $\dfrac{2}{3} + \dfrac{3}{5}$

8. $\dfrac{1}{2} + \dfrac{4}{7}$

9. $\dfrac{7}{11} - \dfrac{3}{11}$

10. $\dfrac{5}{3} - \dfrac{2}{3}$

11. $\dfrac{7}{11} - \dfrac{-3}{11}$

12. $\dfrac{5}{3} - \dfrac{-2}{3}$

13. $\dfrac{1}{2} - \dfrac{3}{8}$

14. $\dfrac{2}{5} - \dfrac{3}{10}$

15. $\dfrac{3}{5} - \dfrac{2}{3}$

16. $\dfrac{1}{2} - \dfrac{4}{7}$

17. $\dfrac{3}{5xy} + \dfrac{-6}{5xy}$

18. $\dfrac{-6}{5x^2} + \dfrac{4}{5x^2}$

19. $\dfrac{3y}{x} + \dfrac{2y}{x}$

20. $\dfrac{x}{5y} + \dfrac{2x}{5y}$

21. $\dfrac{3}{7y} - \dfrac{-3}{7y}$

22. $\dfrac{-2}{3x} - \dfrac{2}{3x}$

23. $\dfrac{1}{2x} + \dfrac{2}{3x}$

24. $\dfrac{3}{5m} + \dfrac{5}{2m}$

25. $\dfrac{3x}{2} + \dfrac{2x}{3}$

26. $\dfrac{4m}{3} + \dfrac{m}{7}$

27. $\dfrac{3}{5x} - \dfrac{2}{3}$

28. $\dfrac{2}{3} - \dfrac{3}{4y}$

29. $\dfrac{2}{3} + \left(\dfrac{1}{4} \cdot \dfrac{3}{7}\right)$

30. $\left(\dfrac{2}{3} + \dfrac{1}{4}\right)\dfrac{3}{7}$

31. $\left(\dfrac{8}{3} - \dfrac{1}{2}\right)\left(\dfrac{4}{7} \div \dfrac{3}{5}\right)$

32. $\dfrac{8}{3} - \dfrac{1}{2}\left(\dfrac{4}{7} \div \dfrac{3}{5}\right)$

33. $\left(\dfrac{1}{3} - \dfrac{1}{4}\right) - \left(\dfrac{1}{5} - \dfrac{1}{6}\right)$

34. $\left(\dfrac{1}{3} - \dfrac{1}{5}\right) \div \left(\dfrac{1}{4} - \dfrac{1}{6}\right)$

B *Find the average of the listed numbers:*

35. $3, -7, -5, 14, 22, -13, 7, 11$

36. $-2, -12, 21, -8, 6, 9, -4, 10$

37. $1/2, 1/4, 2/5, 1/8, 3/4$

38. $3/5, 7/10, 3/2, 6/5, 1/4$

Find the LCM:

39. $2x^2, 3x$

40. $3y, 4y^2$

41. $8m^3, 12m$

42. $9n^2, 12n^4$

43. $4x, 3y, 8xy$

44. xy, yz, xz

45. $y^3, 3y^2, 2y$

46. $5x^3, 2x^2, 3x$

47. $9, 28, 42$

48. $6, 8, 15, 20$

49. $12, 18, 30$

50. $50, 15, 6$

Combine into single fractions and reduce to lowest terms. Work horizontally.

51. $\dfrac{x}{y} - \dfrac{y}{x}$

52. $\dfrac{a}{b} + \dfrac{b}{a}$

53. $\dfrac{x}{y} - 2$

54. $1 - \dfrac{1}{x}$

55. $5 - \dfrac{-3}{x}$

56. $\dfrac{-2}{m} - 4$

57. $\dfrac{1}{xy} - \dfrac{3}{y}$

58. $\dfrac{2a}{b} + \dfrac{-1}{ab}$

59. $\dfrac{3}{2x^2} + \dfrac{4}{3x}$

60. $\dfrac{5}{3y} + \dfrac{3}{4y^2}$

61. $\dfrac{5}{8m^3} - \dfrac{1}{12m}$

62. $\dfrac{2}{9n^2} - \dfrac{5}{12n^4}$

63. $\dfrac{1}{3} - \dfrac{-1}{2} + \dfrac{5}{6}$

64. $\dfrac{-3}{4} + \dfrac{2}{5} - \dfrac{-3}{2}$

65. $\dfrac{x^2}{4} - \dfrac{x}{3} + \dfrac{-1}{2}$

66. $\dfrac{2}{5} - \dfrac{x}{2} - \dfrac{-x^2}{3}$

67. $\dfrac{3}{4x} - \dfrac{2}{3y} + \dfrac{1}{8xy}$ **68.** $\dfrac{1}{xy} - \dfrac{1}{yz} + \dfrac{-1}{xz}$

69. $\dfrac{3}{y^3} - \dfrac{-2}{3y^2} + \dfrac{1}{2y} - 3$ **70.** $\dfrac{1}{5x^3} + \dfrac{-3}{x^2} - \dfrac{-2}{3x} - 1$

C 71. $\dfrac{y}{9} - \dfrac{-1}{28} - \dfrac{y}{42}$ **72.** $\dfrac{5x}{6} - \dfrac{3}{8} - \dfrac{x}{15} - \dfrac{3}{20}$

73. $\dfrac{x^2}{12} + \dfrac{x}{18} - \dfrac{1}{30}$ **74.** $\dfrac{3x}{50} - \dfrac{x}{15} - \dfrac{-2}{6}$

75. $\dfrac{x}{3}\left(\dfrac{1}{4} - \dfrac{x}{2}\right)$ **76.** $\left(\dfrac{x}{5} - \dfrac{x}{2}\right) \div \dfrac{x}{10}$

77. $x\left(\dfrac{1}{2} + \dfrac{x}{3}\right) - \dfrac{x^2}{6}$ **78.** $\dfrac{x^2}{2} \div \dfrac{x}{3} - \dfrac{x}{4}$

Select the larger rational number:

79. $\dfrac{2}{7}, \dfrac{7}{25}$ **80.** $\dfrac{6}{13}, \dfrac{5}{11}$

81. $\dfrac{8}{15}, \dfrac{10}{19}$ **82.** $\dfrac{5}{11}, \dfrac{16}{35}$

83. $\dfrac{3}{7}, \dfrac{8}{19}$ **84.** $\dfrac{5}{13}, \dfrac{7}{18}$

85. $\dfrac{-3}{8}, \dfrac{-10}{27}$ **86.** $\dfrac{-5}{6}, \dfrac{-11}{13}$

87. $-\dfrac{5}{9}, -\dfrac{6}{11}$ **88.** $-\dfrac{2}{13}, -\dfrac{5}{33}$

The concept of heating and cooling degree-days was introduced in Problems 87 and 89 in Section 2-3. A cooling degree-day is the number of degrees by which the average temperature for a particular day exceeds 65°F. If the average is 65° or less, the day contributes 0 to the cooling degree-day total. In the same way, heating degree-days measure how much the average temperature is below

65°F. The average temperature for a given day is defined to be the average of the highest temperature and the lowest temperature for the day. Problems 89–96 refer to the following data, which summarize daily highs and lows for a 2-week period:

Day	Su	M	T	W	Th	F	Sa
High	68	64	69	71	70	73	70
Low	53	56	58	62	64	65	61

Day	Su	M	T	W	Th	F	Sa
High	68	67	69	71	73	75	72
Low	58	59	55	59	57	58	54

89. What is the average of the high temperatures for the first week?

90. What is the average of the high temperatures for the second week?

91. What are the average daily temperatures for the first week?

92. What are the average daily temperatures for the second week?

93. How many heating degree-days are accumulated in the first week?

94. How many heating degree-days are accumulated in the second week?

95. How many cooling degree-days are accumulated in the first week?

96. How many cooling degree-days are accumulated in the second week?

Find the averages of the listed numbers.

97. $3x, 6y, -4y, 10x, -3x, 2y, 13x$

98. $a, -5b, 7b, 3a, 10b, -5a, -3b, 2a$

3-5 Solving Equations Involving Fractions and Decimals

- ♦ Solving Equations by Clearing Fractions
- ♦ Solving Equations without Clearing Fractions
- ♦ Equations versus Expressions
- ♦ Solving Equations Involving Decimals

The equations we solved earlier had integer coefficients. In most practical applications, rational number coefficients occur more frequently than integers. We now have the tools to convert any equation with rational coefficients in either fraction or decimal form into one with integer coefficients. This converts the equation to a form that can be solved using the method outlined in Section 2-7. That method is summarized briefly as follows:

Solving Equations

1. Simplify each side.
2. Get all variable terms on one side and constant terms on the other by using the addition and subtraction properties of equality.
3. Use the multiplication or division property of equality to make the coefficient of the variable equal to 1.

If we are restricted to working only in the integers, step 3 may not always be possible. For example, we cannot isolate the variable in $3x = 2$ without introducing the fraction 2/3. However, using the rational numbers we will always be able to perform this final step for an equation with rational coefficients.

◆ **SOLVING EQUATIONS BY CLEARING FRACTIONS**

We can solve equations involving fractional coefficients by first eliminating the fractions. We do so by simply multiplying each side of the equation by an integer that is a common multiple of all the denominators. This process is called **clearing fractions.** Any multiple, including the product of all denominators, will work but using the LCD will usually result in less calculation.

Example 1
Solving Equations with Rational Coefficients

Solve $\dfrac{x}{4} = \dfrac{3}{8}$.

Solution   To clear fractions, we multiply both sides of the equation by 8, the LCM of 4 and 8.

$$8\left(\frac{x}{4}\right) = 8\left(\frac{3}{8}\right)$$

$$\frac{8x}{4} = \frac{24}{8}$$

$$2x = 3 \qquad \text{We could also have divided 8 by 4 on the left and 8 by 8 on the right, before multiplying, to obtain } 2x = 3 \text{ directly.}$$

$$x = \frac{3}{2} \qquad \text{This step could not have been done if } x \text{ were restricted to the integers.}$$

The fractions in this example could also have been cleared by multiplying both sides of the equation by the product of the denominators to obtain

$$8x = 12$$

This is the same result that we would obtain by so-called **cross-multiplying.**

$$\frac{x}{4} \bowtie \frac{3}{8}$$

that is, multiplying the numerator of each side by the denominator of the other.

 Cross-multiplying applies *only* to clearing fractions from an equation of the form

$$\frac{A}{B} = \frac{C}{D}$$

Matched Problem 1   Solve $\dfrac{x}{9} = \dfrac{2}{3}$.

Example 2
Solving Equations

Solve $\dfrac{x}{3} - \dfrac{1}{2} = \dfrac{5}{6}$.

Solution   The LCM of 3, 2, and 6 is 6. Thus, we multiply both sides of the equation by 6 (or $\frac{6}{1}$) to clear fractions:

$$\frac{6}{1} \cdot \left(\frac{x}{3} - \frac{1}{2} \right) = \frac{6}{1} \cdot \frac{5}{6}$$ Next clear ( ) before dividing out common factors.

$$\frac{\overset{2}{\cancel{6}}}{1} \cdot \frac{x}{\underset{1}{\cancel{3}}} - \frac{\overset{3}{\cancel{6}}}{1} \cdot \frac{1}{\underset{1}{\cancel{2}}} = \frac{\overset{1}{\cancel{6}}}{1} \cdot \frac{5}{\underset{1}{\cancel{6}}}$$ All denominators should divide out, resulting in an equation with integer coefficients.

$$2x - 3 = 5$$ Add 3 to each side.

$$2x = 8$$ Divide each side by 2.

$$x = 4$$

With a little experience you will be able to do the dashed box steps mentally. Notice that each term on each side of the equation is multiplied by 6, and since 6 is the LCM of the denominators, each denominator will divide into 6 exactly.

Before you try the matched problem, observe that

$$\frac{5}{12}x$$

can also be written in the form

$$\frac{5x}{12}$$

since

$$\frac{5}{12}x = \frac{5}{12} \cdot \frac{x}{1} = \frac{5x}{12}$$

You should be able to shift easily from one form to the other.

Matched Problem 2 Solve $\frac{1}{4}x - \frac{2}{3} = \frac{5}{12}x$.

$\left(\textit{Hint:}\quad \text{First, write each term as a single fraction: } \frac{1}{4}x = \frac{x}{4} \text{ and } \frac{5}{12}x = \frac{5x}{12}.\right)$

Example 3
Solving Equations

Solve $5 - \dfrac{2x - 1}{4} = \dfrac{x + 2}{3}$.

Solution It is a good idea to enclose any numerator with more than one term in parentheses first before multiplying both sides by the LCM of the denominators. The LCM of 4 and 3 is 12.

$$5 - \frac{(2x - 1)}{4} = \frac{(x + 2)}{3}$$

$$12 \cdot 5 - \overset{3}{\cancel{12}} \cdot \frac{(2x - 1)}{\underset{1}{\cancel{4}}} = \overset{4}{\cancel{12}} \cdot \frac{(x + 2)}{\underset{1}{\cancel{3}}} \qquad \text{Eliminate denominators.}$$

$$60 - 3(2x - 1) = 4(x + 2) \qquad \text{Clear ( ).}$$

$$60 - 6x + 3 = 4x + 8 \qquad \text{Combine like terms.}$$

$$-6x + 63 = 4x + 8 \qquad \text{Subtract 63 from each side.}$$

$$-6x = 4x - 55 \qquad \text{Subtract } 4x \text{ from each side.}$$

$$-10x = -55 \qquad \text{Divide both sides by } -10.$$

$$x = \frac{-55}{-10} \qquad \text{Reduce your answer.}$$

$$x = \frac{11}{2} \text{ or } 5.5$$

Matched Problem 3 Solve $\dfrac{x + 3}{4} - \dfrac{x - 4}{2} = \dfrac{3}{8}$.

♦ **SOLVING EQUATIONS WITHOUT CLEARING FRACTIONS**

In some problems it is just as easy to work with the fractional coefficients directly as to clear them. As an example, we reconsider Example 1 and resolve $x/4 = 3/8$. Instead of multiplying both sides by 8, we multiply by 4:

$$4\left(\frac{x}{4}\right) = 4\left(\frac{3}{8}\right)$$

$$x = \frac{3}{2}$$

In this case, we isolated x more quickly by working with the fractions rather than clearing them.

Example 4
Solving without Clearing Fractions

Solve without clearing fractions: $\dfrac{3x}{5} + \dfrac{3}{10} = \dfrac{4}{5}$

Solution   We subtract 3/10 from each side to obtain

$$\frac{3x}{5} = \frac{4}{5} - \frac{3}{10} = \frac{8}{10} - \frac{3}{10} = \frac{5}{10} = \frac{1}{2}$$

Now multiply each side by 5/3:

$$x = \frac{5}{3} \cdot \frac{1}{2} = \frac{5}{6}$$

Matched Problem 4   Solve without clearing fractions: $\dfrac{2x}{3} - \dfrac{4}{9} = \dfrac{1}{3}$

♦ EQUATIONS VERSUS EXPRESSIONS

A very common error occurs about now—expressions involving fractions tend to be confused with equations involving fractions. Consider the two problems:

(A) Solve: $\dfrac{x}{2} + \dfrac{x}{3} = 10.$ **(B)** Add: $\dfrac{x}{2} + \dfrac{x}{3} + 10.$

The problems look very much alike, but they are actually quite different. To solve the *equation* in part (A) we multiply both sides by 6 to clear the fractions. This works so well for equations, students want to do the same thing for problems like (B). The only catch is that part (B) is not an equation and the multiplication property of equality does not apply. If we multiply the *expression* in part (B) by 6, we obtain an expression 6 times as large as the original. To add in part (B) we find the LCD and proceed as in Section 3-4. Thus, we obtain

(A) $\dfrac{x}{2} + \dfrac{x}{3} = 10$ **(B)** $\dfrac{x}{2} + \dfrac{x}{3} + 10$

$$6 \cdot \frac{x}{2} + 6 \cdot \frac{x}{3} = 6 \cdot 10$$

$$= \frac{3 \cdot x}{3 \cdot 2} + \frac{2 \cdot x}{2 \cdot 3} + \frac{6 \cdot 10}{6 \cdot 1}$$

$$3x + 2x = 60$$

$$= \frac{3x}{6} + \frac{2x}{6} + \frac{60}{6}$$

$$5x = 60$$

$$x = 12$$

$$= \frac{5x + 60}{6}$$

◆ SOLVING EQUATIONS INVOLVING DECIMALS

Since a decimal represents a fraction, we clear decimal coefficients the same way we clear fractions. For instance, to clear a coefficient like 0.035, we think of it as 35/1,000, so we would multiply by 1,000. The multiplication is easier since the LCD is a power of 10. Multiplying by 10 moves a decimal point one place to the right; by 100, two places to the right; by 1,000, three places to the right; and so on. Example 5 illustrates the process.

Example 5
Solving Equations with Decimal Coefficients

Solve $0.1x + 0.5(x - 5) = 3.5$.

Solution

$$0.1x + 0.5(x - 5) = 3.5$$ To clear decimals, multiply each side by 10.

$$10(0.1x) + 10[0.5(x - 5)] = 10(3.5)$$

$$1x + 5(x - 5) = 35$$ The equation is now free of decimals.

$$x + 5x - 25 = 35$$

$$6x = 60$$

$$x = 10$$

Of course, we could retain the decimals and solve the equation as follows:

$$0.1x + 0.5(x - 5) = 3.5$$

$$0.1x + 0.5x - 2.5 = 3.5$$

$$0.6x = 6$$

$$x = \frac{6}{0.6} = 10$$

You may find the first method easier.

Matched Problem 5   Solve $0.25x - 0.05(x - 3) = 1.75$.

Answers to Matched Problems   **1.** 6   **2.** −4   **3.** $\frac{19}{2}$ or 9.5   **4.** 7/6   **5.** 8

EXERCISE 3-5

A *Solve each equation:*

1. $\dfrac{x}{7} = -5$

2. $\dfrac{x}{3} = -2$

3. $\dfrac{x}{4} = \dfrac{3}{2}$

4. $\dfrac{x}{9} = \dfrac{1}{6}$

5. $\dfrac{x}{6} = \dfrac{5}{8}$

6. $\dfrac{x}{8} = \dfrac{5}{12}$

7. $\dfrac{x}{3} = \dfrac{-5}{9}$

8. $\dfrac{x}{5} = \dfrac{-4}{15}$

9. $\dfrac{-x}{4} = \dfrac{7}{12}$

10. $\dfrac{-x}{6} = \dfrac{7}{18}$

11. $\dfrac{x}{7} - 1 = \dfrac{1}{7}$

12. $\dfrac{x}{5} - 2 = \dfrac{3}{5}$

13. $\dfrac{y}{4} + \dfrac{y}{2} = 9$

14. $\dfrac{x}{3} + \dfrac{x}{6} = 4$

15. $\dfrac{x}{2} + \dfrac{x}{3} = 5$

16. $\dfrac{y}{4} + \dfrac{y}{3} = 7$

17. $\dfrac{x}{3} - \dfrac{1}{4} = \dfrac{3x}{8}$

18. $\dfrac{y}{5} - \dfrac{1}{3} = \dfrac{2y}{15}$

19. $\dfrac{n}{5} - \dfrac{n}{6} = \dfrac{6}{5}$

20. $\dfrac{m}{4} - \dfrac{m}{3} = \dfrac{1}{2}$

21. $\dfrac{2}{3} - \dfrac{x}{8} = \dfrac{5}{6}$

22. $\dfrac{5}{12} - \dfrac{m}{3} = \dfrac{4}{9}$

23. $\dfrac{3}{4} - \dfrac{a}{3} = \dfrac{5}{6}$

24. $\dfrac{2}{5} - \dfrac{b}{4} = \dfrac{7}{10}$

25. $\dfrac{4}{5} + \dfrac{x}{2} = -\dfrac{3}{4}$

26. $\dfrac{3}{4} + \dfrac{x}{6} = -\dfrac{2}{3}$

27. $0.8x = 16$

28. $0.5x = 35$

29. $0.3x + 0.5x = 24$

30. $0.7x + 0.9x = 32$

31. $0.04x = -0.8$

32. $0.6x = -1.5$

33. $0.7x - 0.35 = 1.26$

34. $0.8x - 0.2 = 1.5$

B 35. $\dfrac{x+3}{2} - \dfrac{x}{3} = 4$

36. $\dfrac{x-2}{3} + 1 = \dfrac{x}{7}$

37. $\dfrac{2x+1}{4} = \dfrac{3x+2}{3}$

38. $\dfrac{4x+3}{9} = \dfrac{3x+5}{7}$

39. $3 - \dfrac{x-1}{2} = \dfrac{x-3}{3}$

40. $4 - \dfrac{x-3}{4} = \dfrac{x-1}{8}$

41. $3 - \dfrac{2x-3}{3} = \dfrac{5-x}{2}$

42. $1 - \dfrac{3x-1}{6} = \dfrac{2-x}{3}$

43. $\dfrac{2x-5}{6} = \dfrac{3-4x}{4}$

44. $\dfrac{5-x}{2} = \dfrac{3x+4}{5}$

45. $\dfrac{5a+2}{3} = \dfrac{4-a}{5}$

46. $\dfrac{6-2x}{4} = \dfrac{3x+7}{3}$

47. $\dfrac{x+4}{4} = \dfrac{x+3}{3}$

48. $\dfrac{6x+3}{2} = \dfrac{9x-5}{3}$

49. $\dfrac{4x+7}{4} = \dfrac{3x-5}{3}$

50. $\dfrac{2x+5}{3} = \dfrac{10x+2}{15}$

51. $\dfrac{3}{4} + \dfrac{x+1}{2} = \dfrac{2x+5}{4}$

52. $\dfrac{x+4}{5} = \dfrac{1}{5} + \dfrac{2x+6}{10}$

53. $\dfrac{3x+5}{4} + \dfrac{1}{2} = \dfrac{x+9}{8}$

54. $\dfrac{6}{5} + \dfrac{2x+7}{2} = \dfrac{3x+9}{4}$

55. $0.4(x+5) - 0.3x = 17$

56. $0.1(x-7) + 0.05x = 0.8$

57. $0.05x + 0.1(x-5) = 1.15$

58. $0.25x - 0.2(x+1) = 0.35$

59. $0.1 + 0.2(x+3) = 0.4$

60. $0.3 + 0.5(x+7) = 0.9$

61. $0.1x + 3(0.4 + 0.1x) = 1.8$

62. $0.2x + 4(0.3 + 0.2x) = 2.3$

63. $x - 0.4(0.3 - 0.2x) = 3.12$

64. $x - 1.2(0.4 - 0.1x) = 2.88$

65. $0.3(3.4 - 1.6x) - 2(0.8x - 0.1) = 26.18$

66. $-0.6(x - 0.8) - 0.8(0.6 - x) = 2.8$

C 67. $\dfrac{3x-1}{8} - \dfrac{2x+1}{3} = \dfrac{1-x}{12} - 1$

68. $\dfrac{2x-3}{9} - \dfrac{x+5}{6} = \dfrac{3-x}{2} + 1$

69. $\dfrac{3x-4}{2} + \dfrac{5x-3}{10} - \dfrac{8x-1}{5} = 4$

70. $\dfrac{x-5}{3} - \dfrac{6-x}{4} + \dfrac{x-2}{6} = 8$

71. $1 - \dfrac{x-1}{2} - \dfrac{x-2}{3} = \dfrac{x-3}{4}$

72. $5 - \dfrac{4-x}{3} - \dfrac{3-x}{2} = \dfrac{2-x}{4}$

73. $\dfrac{a+1}{2} - \dfrac{2-a}{3} + \dfrac{a+3}{4} = \dfrac{4-a}{5}$

74. $\dfrac{1-b}{8} - \dfrac{2-b}{6} + \dfrac{3-b}{4} = \dfrac{4-b}{2}$

75. $\dfrac{2}{3}\left(\dfrac{x+1}{4}\right) + \dfrac{1}{3}\left(\dfrac{1-x}{2}\right) = \dfrac{3}{4}$

76. $\dfrac{1}{5}\left(\dfrac{2-x}{2}\right) - \dfrac{3}{10}\left(\dfrac{1-x}{2}\right) = \dfrac{17}{20}$

In Problems 77–86, round answers to four decimal places:

77. $0.4312x = 3.1205$ **78.** $2.1038x = 24.6109$

79. $4.3292x = 6.0791x + 38.7415$

80. $23.9308x = 21.0753 - 4.3387x$

81. $2.1463x = -4.3688$

82. $-3.2422x = 8.3346$

83. $3.1415x + 1.0101 = 12.3642$

84. $-3.2614x + 5.1216 = 24.8558$

85. $-4.8536x + 1.3412 = 1.7500x$

86. $7.2735x - 10.8002 = 2.1144x$

87. *Education* A student took eight tests during a term and had an average grade of 79.5. The teacher's records showed grades of 76, 82, 85, 68, 75, 90, and 80, with one grade recorded but illegible. What was the illegible grade?

88. *Education* Repeat Problem 87 for recorded grades of 92, 84, 77, 91, 85, 86, and 80, and an average of 86.25.

89. *Sports* The radio broadcaster for a professional basketball team announces that your favorite player has averaged 34.5 points through the first four playoff games. You have listened to 3 games, in which he scored 38, 29, and 41 points. How many points did he score in the game you missed?

90. *Sports* Repeat Problem 89 for an average of 29 points per game and known scores in three of the four games of 26, 35, and 24.

3-6 Word Problems

♦ A Strategy for Solving Word Problems
♦ Number Problems
♦ Geometric Problems
♦ Percent Problems

We are now ready to consider a variety of word problems and significant applications. This section deals with relatively straightforward number, geometric, and percent problems to give you more practice in translating words into symbolic forms. Additional applications are considered in Sections 3-7 and 3-8. Applications of a slightly more difficult nature are included in Chapter 4.

♦ A STRATEGY FOR SOLVING WORD PROBLEMS

To start our discussion, we will restate the strategy for solving word problems introduced in Section 2-8:

A Strategy for Solving Word Problems

1. Read the problem carefully—several times if necessary—that is, until you understand the problem, know what is to be found, and know what is given.
2. If appropriate, draw figures or diagrams, and label given and unknown parts. Look for formulas connecting the given with the unknown.
3. Let one of the unknown quantities be represented by a variable, say, x, and try to represent all other unknown quantities in terms of x. This is an important step and must be done carefully.
4. Form an equation relating the unknown quantities with known quantities.

> **5.** Solve the equation and write answers for all parts of the problem requested.
> **6.** Check and interpret all solutions in terms of the original problem and not just in the equation found in step 4. A mistake might have been made in setting up the equation in step 4.

Although this is the same strategy introduced in the last chapter, we can, in fact, solve many more problems now. The difference is that we can solve more of the equations that may arise in step 4.

◆ NUMBER PROBLEMS

Recall that if x is a number, then two-thirds x can be written

$$\frac{2}{3}x \text{ or } \frac{2x}{3}$$

The latter form will be more convenient for our purposes.

Example 1
A Number Problem

Find a number such that 10 less than two-thirds the number is one-fourth the number.

Solution   Let $x = $ The number. Symbolize each part of the problem.

$$\left(\begin{array}{c}\text{10 less than} \\ \text{two-thirds} \\ \text{the number}\end{array}\right) \text{ is } \left(\begin{array}{c}\text{one-fourth} \\ \text{the number}\end{array}\right)$$

$$\frac{2x}{3} - 10 = \frac{x}{4}$$

Solve the equation resulting from the symbolic forms:

$$\frac{2x}{3} - 10 = \frac{x}{4} \qquad \text{Multiply by 12, the LCM of 3 and 4.}$$

$$12 \cdot \frac{2x}{3} - 12 \cdot 10 = 12 \cdot \frac{x}{4} \qquad \text{Divide out denominators and simplify.}$$

$$8x - 120 = 3x$$

$$5x = 120$$

$$x = 24$$

Check   Two-thirds of 24 is 16; 10 less than this is 6, which is one-fourth of 24.

Matched Problem 1   Find a number such that 6 more than half the number is two-thirds the number.

♦ GEOMETRIC PROBLEMS

Recall that the **perimeter** of a triangle or rectangle is the distance around the figure. Symbolically:

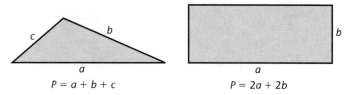

$$P = a + b + c \qquad\qquad P = 2a + 2b$$

Figure 1

Example 2
A Geometric Problem

If one side of a triangle is 7 inches, the second side is two-fifths the perimeter, and the third side is one-fourth the perimeter, what is the perimeter of the triangle?

Solution   Let P = Perimeter. Draw a triangle and label the sides, as shown. Thus,

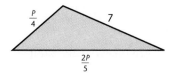

Figure 2

$$P = 7 + \frac{2P}{5} + \frac{P}{4}$$

$$20 \cdot P = 20 \cdot 7 + 20 \cdot \frac{2P}{5} + 20 \cdot \frac{P}{4} \qquad \text{20 is LCM of 4 and 5.}$$

$$20P = 140 + 8P + 5P$$

$$7P = 140$$

$$P = 20 \text{ inches}$$

Check
$$7 + \frac{2 \cdot 20}{5} + \frac{20}{4} = 7 + 8 + 5 = 20 \qquad \text{The perimeter}$$

Matched Problem 2   If one side of a triangle is one-sixth the perimeter, the second side is 14 meters, and the third side is four-ninths the perimeter, what is the perimeter?

Example 3
A Geometric Problem

Find the dimensions of a rectangle with perimeter 224 centimeters if its width is three-fifths its length.

Solution   Let x = length. Draw a rectangle and label the sides, as shown. Thus,

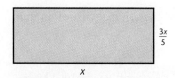

Figure 3

$$2x + 2 \cdot \frac{3x}{5} = 224 \qquad \text{Use the perimeter formula } 2a + 2b = P.$$

$$2x + \frac{6x}{5} = 224$$

$$5 \cdot 2x + 5 \cdot \frac{6x}{5} = 5 \cdot 224$$

$$10x + 6x = 1{,}120$$
$$16x = 1{,}120$$
$$x = 70 \text{ centimeters}$$
$$\frac{3x}{5} = 42 \text{ centimeters}$$

Check $2 \cdot 70 + 2 \cdot 42 = 140 + 84$
$$= 224 \text{ centimeters} \quad \text{The perimeter. Be sure to add all four sides, not just two.}$$

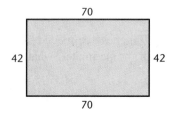

Figure 4

Matched Problem 3   Find the dimensions of a rectangle with perimeter of 108 yards if its width is four-fifths its length.

♦ PERCENT PROBLEMS

Recall that to find a percent of a given quantity, you convert the percent to a decimal and multiply. For example, to find 30% of 180, multiply 0.30 times 180 to obtain 54.

Example 4
A Percent Problem

The sales tax in a given state is $5\frac{1}{2}\%$ of the total sales. Compute the tax on a bill of $114.

Solution $5\frac{1}{2}\%$ of $114 = 5.5\%$ of $114 = 0.055 \times 114 = 6.27$. The tax is $6.27.

Matched Problem 4   The tax on hotel rooms in a certain city is 11%. Find the tax if the room charge is $65.

Example 5
A Percent Problem

A retail store marks up an item 20% above cost. Find the original cost of an item that the store sells for $45.60.

Solution   Let x be the original cost. The markup is 20% of x, that is, $0.2x$. The selling price is cost plus markup:

$$\text{Cost} + \text{Markup} = \text{Selling price}$$
$$x + 0.2x = 45.60$$
$$1.2x = 45.60$$
$$x = \frac{45.60}{1.2} = 38$$

so the cost to the store was $38.

This problem can also be set up by recognizing that 20% above cost is 120% of the cost. Therefore the selling price is 120% of the cost, that is,

$$1.2x = 45.60$$

Matched Problem 5   A worker receives a raise of 6.25%. If the worker's new pay rate is $10.20 per hour, what was her previous rate?

Answers to Matched Problems   **1.** 36   **2.** 36 meters   **3.** 30 yards by 24 yards   **4.** $7.15   **5.** $9.60

EXERCISE 3-6

A *If x represents a number, write an algebraic expression for each of the following numbers:*

1. Half of x
2. One-sixth x
3. Two-thirds x
4. Three-fourths x
5. Two more than one-third x
6. Five more than one-fourth x
7. Eight less than two-thirds x
8. Six less than three-fourths x
9. 80% of x
10. 35% of x
11. Three less than 60% of x
12. 60% of a quantity 3 less than x
13. 20% of a quantity 5 more than x
14. Five more than 10% of x
15. Half the number that is 3 less than twice x
16. One-third the number that is 5 less than four times x
17. 80% of the number two less than 50% of x
18. 95% of the number 5 more than 25% of x

Find numbers meeting each of the indicated conditions:
(A) *Write an equation using x.* **(B)** *Solve the equation.*

19. Two more than one-fourth of a number is $\frac{1}{2}$.
20. Three more than one-sixth of a number is $\frac{2}{3}$.

21. Two less than half a number is one-third the number.
22. Three less than one-third of a number is one-fourth the number.
23. A certain percentage of 58 is 20.3.
24. A certain percentage of 80 is 35.2.
25. Eighty percent of some number is 45.
26. Forty-five percent of some number is 80.
27. Three more than a number is 140% of the number.
28. Five less than a number is 80% of the number.
29. Fourteen less than a number is 30% of the number.
30. Ten more than a number is 120% of the number.
31. Five less than 70% of a number is 16.
32. Ten more than 5% of a number is 56.
33. Seventy less than 60% of a number is 25% of the number.
34. Twenty more than 10% of a number is 35% of the number.
35. Five less than half a number is 3 more than one-third the number.
36. Two less than one-sixth of a number is 1 more than one-fourth the number.
37. Five more than two-thirds of a number is 10 less than one-fourth the number.
38. Four less than three-fifths of a number is 8 more than one-third the number.

B *Solve:*

39. If one side of a triangle is one-fourth the perimeter, the second side is 3 meters, and the third side is one-third the perimeter, what is the perimeter?

40. If one side of a triangle is two-fifths the perimeter, the second side is 70 centimeters, and the third side is one-fourth the perimeter, what is the perimeter?

41. An electrical transmission tower is erected so that it rests on bedrock. If one-twenty-fifth of the tower is in topsoil, 12 meters is in subsoil, and the remaining four-fifths above ground, what is the total height of the tower from the rock foundation to the top (see the figure)?

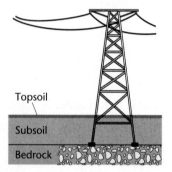

42. On a safari in Africa a group traveled half the distance by Land Rover, 55 kilometers by camel, and the last one-third of the distance by boat. How long was the trip?

43. Find the dimensions of a rectangle with perimeter 72 centimeters if its width is one-third its length.

44. Find the dimensions of a rectangle with perimeter 84 meters if its width is one-sixth its length.

45. Find the dimensions of a rectangle with perimeter 216 meters if its width is two-sevenths its length.

46. Find the dimensions of a rectangle with perimeter 100 centimeters if its width is two-thirds its length.

47. If you paid $168 for a camera after receiving a 20% discount, what was the original price of the camera?

48. The retail price of a record is $9.80. The markup on the cost is 40%. What did the store pay for the record?

49. The retail price of a car stereo is $247. The seller's markup on cost is 25%. How much did the stereo cost the seller?

50. If you paid $27 for a pair of slacks at a "20% off sale," what was the original price of the slacks?

Interest rates are usually expressed as annual rates. To convert an annual rate to a monthly rate, divide by 12. For example, a 9% annual rate is equivalent to 0.75% per month.

51. A checking account pays interest monthly at an annual rate of 5.4% on the average monthly balance. What is the interest for a month when the average balance is $440.00?

52. A money market fund pays interest monthly at an annual rate of 6.6%. What is the monthly return on an investment of $1,500?

53. A credit card company charges 18% annual interest on unpaid balances. What is the monthly charge on a balance of $2,200?

54. A home mortgage has an annual interest rate of 12%. What is the interest for a month with a principal balance of $80,000?

55. If the unemployment rate is 5.6% and the total eligible work force is 140 million people, how many people are unemployed?

56. In 1989 civilian employees at all levels of government numbered 17.7 million and constituted 15.1% of the nation's total work force. What was the number in the total work force? Round your answer to the nearest hundred thousand.

57. It is estimated that by the year 2000, there will be a demand for 1,388,000 secondary schoolteachers in the United States. This would represent an increase of 19% over 1988 levels. What was the demand in 1988? Round your answer to the nearest thousand.

58. The demand for registered nurses in the United States is expected to grow to 2,190,000 by the year 2000. If this represents an increase of 39% from the 1988 level, what was that level? Round your answer to the nearest ten thousand.

59. In 1988 there were 1,141,000 farmers in the country. This level is expected to drop by 23% by the year 2000. What level is projected for that year? Round your answer to the nearest thousand.

60. The attrition rate at a certain university is 14% from freshman to sophomore year, meaning that 14% of a typical freshman class does not return to the school the following year. If the school enrolls 650 freshmen, how many does it expect to return as sophomores?

61. Wage increases are often tied to the Consumer Price Index (CPI), also called the cost-of-living index. If the CPI increase for a given year is 4.6%, how much must an hourly wage of $8.40 be increased to match the cost-of-living increase?

62. Between 1960 and 1990, the CPI (see Problem 61) increased by 435%. What 1960 salary would be equivalent to a 1990 salary of $16,000?

63. A realty office receives a commission of 8% on the sale price of houses listed and sold. If such commissions totaled $60,400 for a month, what was the value of houses listed and sold?

64. Book royalties are often determined as a percentage of gross sales. If a best-selling author gets a royalty of 18%, how much income does she receive from a book that has gross sales of $400,000?

65. A person borrowed a sum of money from a lending group at 12% simple interest. At the end of 2.5 years the loan was cleared by paying $520. How much was originally borrowed? [*Hint:* $A = P + Prt$, where A is the amount repaid, P is the amount borrowed, r is the interest rate expressed as a decimal, and t is time in years.]

66. A money market fund pays interest monthly at an annual rate of 6.6%. If $2,000 is placed in the fund and the interest reinvested, how much is in the fund after 2 months?

67. If the CPI (see Problem 61) increased from 113.6 to 118.3, to what amount must an annual salary of $28,000 increase to keep pace with the cost of living? Round your answer to the nearest hundred.

68. Between 1985 and 1990, the CPI increased from 107.6 to 128.7. If the percentage increase stays the same, what will the index be in 1995? Round your answer to the nearest tenth.

69. Between 1985 and 1990, the population of the Mexico City urban area grew from 17.30 million people to 20.25 million. If the percentage increase stays the same, what will the population be in 1995? Round your answer to the nearest hundredth.

70. One measure of how unusual a record for a sports performance is is the percentage by which it exceeds the next best performance. Ty Cobb's record for the most runs scored in a major league baseball career is 2,245. The next best is 2,174 by Babe Ruth. What number would exceed Cobb's record by the same percentage by which his exceeds Ruth's? Round your answer to the nearest integer.

71. A motel room costs $70 per night. You have a coupon for a 20% discount. The city tax on the amount you pay for the room is 11%. What will the total cost to you be? Does it matter whether you compute the discount before you figure the tax?

72. In a particular year the teachers in a certain public school district received raises of 8%. In the next year all received pay cuts of 5% owing to a state funding crisis. What happened over the 2-year period to a salary of $30,000? Does it matter which came first, the raise or the cut?

73. A retail store with a markup of 30% offers an item on sale for 20% off at a price of $54.08. What was the store's original cost?

74. A retail store has a markup of 20%. An item is subsequently offered for 15% off the list price. What is the sale price if the item cost the store $30?

75. A credit card account charges monthly interest at an annual rate of 19.2% of any unpaid balance and all charges that are made with an unpaid balance. If an account had charges in a month of $135.40 and a total of $325.12 due at the end of the month, what was the beginning unpaid balance?

76. A checking account pays monthly interest at an annual rate of 5.4% on the average monthly balance. If the account had no activity during the month and the balance at the end of the month was $662.97, what was the balance at the start of the month?

77. Find the dimensions of a rectangle with perimeter 112 centimeters if its width is 7 centimeters less than two-fifths its length.

78. Find the dimensions of a rectangle with perimeter 264 centimeters if its width is 11 centimeters less than three-eighths its length.

79. An isosceles triangle has two sides of equal length (see the figure). Find the dimensions if the perimeter is 42 feet and the sides are 3 feet longer than half the base.

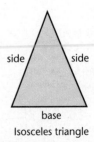

Isosceles triangle

80. In an isosceles triangle (see Problem 79), the sides are 2 centimeters less than twice the base. Find the dimensions if the perimeter is 96 centimeters.

Problems 81–84 refer to a rectangular box as shown in the figure. The girth is defined to be 2w + 2h. The postal size is girth plus length. The surface area is given by 2wh + 2wl + 2hl.

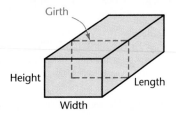

81. Suppose the height and width are the same and equal to one-third the length. Find the dimensions if the postal size is 105 inches.

82. Suppose the height is one-fourth the length and half the width. Find the dimensions if the postal size is 108 inches.

83. Suppose the box has length 44 centimeters, width 36 centimeters, and surface area 5,608 square centimeters. What is the height of the box?

84. Suppose the box has width 27 inches, height 15 inches, and surface area 4,618 square inches. How long is the box?

3-7 Ratio and Proportion

♦ Ratio
♦ Proportion
♦ Direct Variation
♦ Metric Conversion

One of the first applications of algebra in elementary science and technology courses you are likely to encounter involves ratio and proportion. Many problems in these courses can be solved using the methods developed in this section. Ratios and proportions will also provide us with a convenient way to convert metric units into English units and vice versa.

♦ RATIO

The ratio of two quantities is the first divided by the second. That is:

Ratio of *a* to *b*

The ratio of *a* to *b*, assuming $b \neq 0$, is $\dfrac{a}{b}$.

It is also written *a:b* or *a/b* and is read "*a* to *b*."

Thus a ratio of two integers is a rational number. The ratio of two rational numbers is also a rational number.

Example 1 **A Ratio Problem**	If there are 10 men and 20 women in a class, what is the ratio of men to women?
Solution	The ratio is $\dfrac{10}{20}$ or $\dfrac{1}{2}$, which is also written 1:2, and is read ''1 to 2.''

Matched Problem 1   If there are 500 men and 400 women in a school, what is the ratio of:

(A) Men to women   **(B)** Women to men

Ratios are sometimes expressed as an ''average.'' The ratio of distance traveled in miles to the elapsed time in hours is the ''average'' speed in miles per hour. In baseball, the ratio hits/(at bats) gives a player's batting average. The ratio

$$\frac{\text{Earned runs allowed}}{\text{Games (nine innings) pitched}}$$

gives a pitcher's earned run average (ERA). The use of the word ''average'' in Section 3-4 also involves a ratio:

$$\frac{\text{Sum of a set of quantities}}{\text{Number of quantities}}$$

♦ **PROPORTION**

Statements about ratios that include an unknown quantity lead directly to equations that we can solve. It will not be necessary to clear fractions. We can solve these equations directly.

Example 2 **A Ratio Problem**	Suppose you are told that the ratio of women to men in a college is 3:5 and there are 1,450 men. How many women are in the college?
Solution	Let x = Number of women; then the ratio of women to men is $x/1,450$. Thus,

$$\frac{x}{1,450} = \frac{3}{5} \qquad \text{To isolate } x, \text{ multiply both sides by 1,450.}$$

$$x = 1,450 \cdot \frac{3}{5}$$

$$= 870 \text{ women}$$

We could also have solved this problem by cross-multiplying:

$$\frac{x}{1,450} \bowtie \frac{3}{5}$$

$$5x = 4,350$$

$$x = \frac{4,350}{5} = 870$$

Matched Problem 2   If in a college the ratio of men to women is 2:3 and there are 1,200 women, how many men are in the school?

A statement of equality between two ratios is called a proportion. That is:

Proportion

$$\frac{a}{b} = \frac{c}{d} \qquad b, d \neq 0$$

In Example 2, the problem statement lead directly to a proportion. This happens in many situations.

Example 3
A Proportion Problem

If a rental truck can travel 66 miles on 4 gallons of gas, how far will it go on a full tank of 22 gallons?

Solution   Let $x = $ Distance traveled on 22 gallons. Thus,

$$\frac{x}{22} = \frac{66}{4}$$

Both ratios represent miles per gallon. We can isolate x by multiplying both sides by 22—we do not need to use the LCM of 22 and 4.

$$x = 22 \cdot \frac{66}{4}$$

$$= 363 \text{ miles}$$

We could also have isolated x in the solution by cross-multiplying:

$$\frac{x}{22} \diagup\!\!\!\!\diagdown \frac{66}{4}$$

$$4x = 1,452$$

$$x = 363$$

Matched Problem 3   If a car can travel 180 kilometers on 24 liters of gas, how far will it travel on 30 liters? Set up a proportion and solve.

Example 4
A Proportion Problem

If there are 12 centiliters of cola syrup in 64 centiliters of a soft drink, how many centiliters of syrup will be in 48 centiliters of the same drink? Set up a proportion and solve.

Solution   Let $x = $ Number of centiliters of syrup in 48 centiliters of the drink; then

$$\frac{x}{48} = \frac{12}{64}$$

$$x = 48 \cdot \frac{12}{64} = 9 \text{ centiliters}$$

Matched Problem 4 If there are 4 cups of milk in a recipe for 6 people, how many cups of milk should be used in the same recipe for 10 people? $\frac{4}{6} \quad \frac{x}{10} \quad \frac{40}{6} - 6\frac{2}{3}$ or 6.66

♦ **DIRECT VARIATION**

In applied problems, ratios of certain quantities may be constant. In this case the quantities are said to vary directly or to be directly proportional.

> **Direct Variation**
>
> A quantity a is said to **vary directly** with another quantity b if the ratio a/b is a constant. We also say a is **directly proportional** to b.

For example, at a constant speed, distance d is proportional to time t, since

$$\frac{d}{t} = \text{speed (constant)}$$

If a car travels at a constant speed of 20 miles per hour, then the distance is given by $d = 20t$, so $\frac{d}{t} = 20$.

The variation in the two quantities distance and time is "direct" in the sense that as one increases so does the other.

The word "per" is an indicator of a constant ratio, such as in miles *per* gallon, miles *per* hour, dollars *per* hour, cost *per* unit, etc. All these ratios represent rates and are written as

$$\frac{\text{Miles}}{\text{Gallon}} \qquad \frac{\text{Miles}}{\text{Hour}} \qquad \frac{\$}{\text{Hour}} \qquad \frac{\text{Cost}}{\text{Unit}}$$

and so on.

Example 5
A Proportion Problem

A hospital has a supply of morphine for injections that comes in a liquid solution that contains 10 milligrams of morphine per milliliter of solution. If a physician orders a patient to be medicated with 24 milligrams of morphine, how many milliliters should be injected?

Solution The ratio

$$\frac{\text{Milligrams of morphine}}{\text{Milliliters of solution}}$$

is a constant and is equal to 10. The ratio

$$\frac{\text{Milliliters of solution}}{\text{Milligrams of morphine}}$$

is also constant and is equal to 1/10. Let x be the number of milliliters of solution required to provide 24 milligrams of morphine. Then

$$\frac{x}{24} = \frac{1}{10}$$

$$x = \frac{24}{10} = 2.4$$

Another way of looking at this problem is to observe that the amount of morphine required is 2.4 times the amount in 1 milliliter of solution. The amount of solution needed should therefore be 2.4 times 1 milliliter.

Matched Problem 5   The number of traffic accidents on a particular stretch of highway varies directly with the number of vehicles traveling the highway. If 2 accidents are expected per 5,000 vehicles, how many should be expected if 12,500 vehicles use the highway on a busy day?

$$\frac{12500}{5000} = \frac{X}{2}$$

♦   **METRIC CONVERSION**

A summary of metric units is located inside the back cover of the text. Here we show how proportion can be used to convert between the English and metric systems.

Example 6
Metric Conversion

If there is 0.2642 gallon in 1 liter, how many liters are in 18 gallons?

Solution   Let x = Number of liters in 18 gallons. Set up a proportion with x in a numerator; that is, a proportion of the form

$$\frac{\text{Liters}}{\text{Gallons}} = \frac{\text{Liters}}{\text{Gallons}}$$ Each ratio represents liters per gallon.

Thus,

$$\frac{x}{18} = \frac{1}{0.2642}$$

$$x = \frac{18}{0.2642}$$

$$= 68.130\ 2 \approx 68.13 \text{ liters}$$

Matched Problem 6   If there are 3.785 liters in 1 gallon, how many gallons are in 40 liters? Set up a proportion (with the variable in a numerator) and solve.

Example 7
Metric Conversion

If there are 109 yards in 100 meters, how many meters are in 440 yards?

Solution   Let x = Number of meters in 440 yards. Set up a proportion of the form

$$\frac{\text{Meters}}{\text{Yards}} = \frac{\text{Meters}}{\text{Yards}}$$

kilogram weight (see the figure). How much does one whole standard bar of silver weigh?

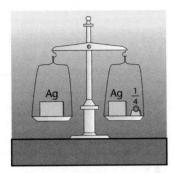

10. If 180 centiliters of a solution contains 10 grams of arsenic, how much arsenic is in 400 centiliters of the same solution?

11. In the study of gases, Boyle's law states that the product of the pressure and volume, as these quantities change and all other variables are held fixed, remains constant. Stated as a formula, $P_1V_1 = P_2V_2$. If 250 cubic centimeters of gas at 60 centimeters of pressure expands to 400 cubic centimeters, what is the resulting pressure?

12. Apply Boyle's law (see Problem 11) to the following situation: 500 cubic centimeters of air at 70 centimeters of pressure is subjected to 100 centimeters of pressure. What is the resulting volume?

LIFE SCIENCE

13. A fairly good approximation for the normal weight of a person over 60 inches (5 feet) tall is given by the formula $w = 5.5h - 220$, where h is height in inches and w is weight in pounds. How tall should a 121-pound person be?

14. Apply the formula from Problem 13 to determine how tall a 209-pound person should be.

15. The weight of a particular variety of snake is proportional to the cube of its length. That is, w/l^3 is constant. If a snake that is 0.6 meter long weighs 100 grams, how much will a snake that is 1 meter long weigh?

16. Refer to Problem 15. If a snake weighs 180 grams when it is 0.8 meter long, how much will it weigh when it reaches 2.2 meters in length?

17. The rate at which a population grows is proportional to the size of the population. If a population is growing at the rate of 40,000 per year when the total population is 1,000,000, at what rate is the population growing when the total population is 1,750,000?

18. A wildlife management team estimated the number of bears in a national forest by the popular capture-mark-recapture technique. Using live traps, they captured and marked 30 bears and then released them. After a period for mixing, they captured another 30 and found 5 marked among them. Assuming that the ratio of the total bear population to the bears marked in the first sample is the same as in the ratio of bears in the second sample to those found marked in the second sample, estimate the bear population in the forest.

EARTH SCIENCE

19. About one-ninth of the height of an iceberg is above water. If 20 meters is observed above water, what is the total height of the iceberg?

20. The mantle of the earth extends from the core of the earth to its crust, varying from 3 to 55 miles below the earth's surface. The mantle accounts for roughly two-thirds of the earth's mass, and silicon dioxide constitutes about 50 percent of the mantle. What percent of the earth's mass is made up of silicon dioxide in the mantle?

21. Pressure in seawater increases by 1 atmosphere (15 pounds per square inch) for each 33 feet of depth. It is 15 pounds per square inch at the surface. Thus, $p = 15 + 15(d/33)$, where p is the pressure in pounds per square inch at a depth of d feet below the surface. How deep is a diver if she observes that the pressure is 200 pounds per square inch?

22. Repeat Problem 21 for a pressure reading of 165 pounds per square inch.

★ 23. As dry air moves upward, it expands and, in so doing, cools. This ascent is known as the "adiabatic process." In Problem 27 of Section 2-8, the drop in temperature was given as 5°F for each 1,000-foot rise. More accurately, the drop is 5.5°F for each 1,000 feet. Write a formula that relates temperature T with altitude h (in feet) and a ground temperature of 80°F. How high is an airplane if the pilot observes that the temperature is 25°F?

★ 24. Repeat Problem 23 for a ground temperature of 50°F and a pilot's observation of a temperature of −43.5°F.

GEOGRAPHY

25. Arid land covers a fifth of the earth's land area. Arid and semiarid land together cover a third. What fraction of the earth's land area is semiarid?

26. Antarctica has a total land area of 5,404,000 square miles and is 98% covered with ice. What is the ice-free land area?

MUSIC

27. Starting with a string tuned to a given note, one can move up and down the scale simply by increasing or decreasing the length of the string according to simple whole-number ratios, while maintaining the same tension. For example, in the figure, the "relative string lengths" show that doubling the string length produces the C below middle C, while halving it produces the C above middle C. By the same reasoning, a string $\frac{8}{9}$ the length of the middle C string produces a D. Chords can also be formed by using two strings whose lengths form the ratios shown in the relative string lengths. Find the length of the shorter string in each of the following cases that will produce the following chords when paired with a 30-inch string:
(A) Octave   1:2   (B) Fifth   2:3
(C) Fourth   3:4   (D) Major third   4:5

a 1-meter wrecking bar and places a fulcrum 10 centimeters from one end, how much can be lifted with a force of 25 kilograms on the long end?

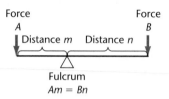

Force A ↓ Distance m Distance n ↓ Force B

Fulcrum

$Am = Bn$

30. How heavy a rock can be moved with an 8-foot steel rod if the fulcrum is placed 1 foot from the rock end and a force of 100 pounds is exerted on the other end? (See Problem 29.)

★ **31.** How far would a fulcrum have to be placed from an end with a 65-kilogram weight to balance 85 kilograms on the other end if the bar is 3 meters long? (See Problem 29.)

32. A child weighing 70 pounds is seated at one end of a teeter-totter, 6 feet from the center. How far from the center on the other side must a child weighing 90

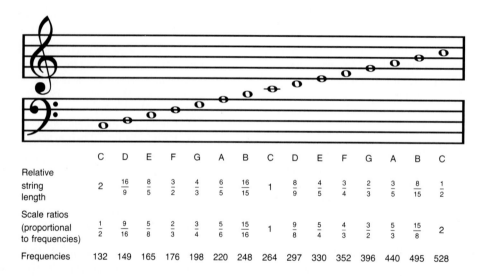

	C	D	E	F	G	A	B	C	D	E	F	G	A	B	C
Relative string length	2	$\frac{16}{9}$	$\frac{8}{5}$	$\frac{3}{2}$	$\frac{4}{3}$	$\frac{6}{5}$	$\frac{16}{15}$	1	$\frac{8}{9}$	$\frac{4}{5}$	$\frac{3}{4}$	$\frac{2}{3}$	$\frac{3}{5}$	$\frac{8}{15}$	$\frac{1}{2}$
Scale ratios (proportional to frequencies)	$\frac{1}{2}$	$\frac{9}{16}$	$\frac{5}{8}$	$\frac{2}{3}$	$\frac{3}{4}$	$\frac{5}{6}$	$\frac{15}{16}$	1	$\frac{9}{8}$	$\frac{5}{4}$	$\frac{4}{3}$	$\frac{3}{2}$	$\frac{5}{3}$	$\frac{15}{8}$	2
Frequencies	132	149	165	176	198	220	248	264	297	330	352	396	440	495	528

28. Continue Problem 27 for these chords:
(A) Minor third   5:6   (B) Major sixth   3:5
(C) Minor sixth   5:8

PHYSICS AND ENGINEERING

29. An important problem in physics and engineering is the lever problem shown in the figure. In order for the system to be balanced (not move), the product of the force and distance on one side must equal the product of the force and distance on the other. If a person has

pounds sit in order to balance the teeter-totter? (See Problem 29.)

PSYCHOLOGY

33. Psychologists experimenting with laboratory animals may use drug dosages proportional to the weight of the animal to attempt to achieve the same effect on different animals. If a rat weighing 300 grams receives a dosage of 2 cubic centimeters (cc), what dosage should be used on a rat weighing 240 grams?

34. Intelligence quotient (IQ) is defined to be 100 times the ratio of mental age to chronological age. If a person with a mental age of 18 years has an IQ of 120, what is the person's chronological (actual) age?

SPORTS

35. Baseball pitcher Sandy Koufax of the Brooklyn and Los Angeles Dodgers won 165 games in a 12-year major league career and had a winning percentage of 65.5%. Winning percentage is the ratio of wins to the sum of wins and losses. How many games did Koufax lose?

36. Hockey players Gordie Howe of the Detroit Red Wings and Phil Esposito of the New York Rangers scored 801 and 717 goals, respectively, in their careers. Howe played 1,767 games, Esposito 1,282. If Esposito had played as many games as Howe and scored at the same rate he did for his career, how many goals would he have scored?

37. From 1980 to 1990, prize money for the Indianapolis 500 race increased from $6,325,803 to $15,322,510. If it experiences the same percentage increase between 1990 and 2000, what will the prize money be in the year 2000?

38. Connie Mack of the old Philadelphia Athletics is baseball's winningest manager with a total of 3,776 wins. However, Mack's teams won only 48.4% of their games. How many games did Mack lose?

39. During the 1950s, baseball's New York Yankees won 8 American League championships, finished second once with 103 wins, and third once. In their championship seasons, the Yankees won 98 games three times and won 99, 97, 96, 95, and 92 games one season each. For the 10-year period they averaged 95.5 victories. How many games did they win in 1959 when they finished third?

40. For the 9 National Football League Super Bowl games played from 1981 to 1989, the scores were 27-10, 26-21, 27-17, 38-9, 28-16, 46-10, 39-20, 42-10, and 20-16. If the 1990 game is included, the average winning score was 34.8 and the average margin of victory was 20.9 points. What was the score in 1990?

UNIT CONVERSION

41. How many miles are in 1 nautical mile if 9,000 nautical miles equals 10,357 miles?

42. A league is 3 nautical miles. How far in miles is "20,000 leagues beneath the sea"? (See Problem 41.)

43. How many furlongs are there in a mile-and-a-quarter horserace if a 7-furlong horserace is seven-eighths of a mile?

44. How many square miles are there in a 200,000-acre ranch if 1 square mile has 640 acres?

45. An acre is 43,560 square feet. How many acres are there in a rectangular suburban lot measuring 100 by 200 feet, and thus having 20,000 square feet?

46. A hectare is the metric unit comparable to acres in the English system. How many hectares are in 1 acre if 1,000 hectares is 2,471 acres?

MISCELLANEOUS

47. A piece of wire 18 inches long is to be cut into 4 pieces. The longest piece is to be twice as long as the next longest. The two shortest pieces are to be the same length, each one-fourth the length of the longest. How long should each piece be?

48. The three angles in a triangle sum to 180°. If the triangle is isosceles and the third angle is 6 times as large as the two equal angles, what are the angles?

49. If you drove from home to a neighboring town 81 miles away, and then returned home, with a total driving time of 3 hours and 36 minutes, what was your average speed?

★ 50. Bicycle races like the Tour de France are run in several stages. Assume that at the beginning of a 90-kilometer stage, the second-place rider trails the leader by 5 minutes. If the leader rides the stage at an average speed of 36 kilometers per hour, what speed must the second-place rider average to finish the stage tied with the leader?

51. A basketball game had a paid attendance of 2,100. General admission tickets were $2.25, and reserved seats $3.50. If the total gate receipts were $5,850, how many reserved seats were sold?

52. A photographer sells two sizes of photographs at an arts and craft show, 9 by 12 and 15 by 21. The smaller works sell for $12.50, the larger for $16.00. If the photographer sells 27 photos and makes $358.50, how many of each size did he sell?

53. A word processing system can print characters in three type sizes: standard, condensed, and enlarged. In condensed print, 22 spaces correspond to 13 standard spaces. How many spaces correspond to 65 standard spaces?

54. Refer to Problem 53. In enlarged print, 8 spaces correspond to 16 standard spaces. How many enlarged characters can be fit in a line that holds 60 standard spaces?

55. In a recent year, the ratio of the number of people included in households in the United States to the number of households was 317:100. How many people would you expect to find in 10,000 households?

★ **56.** In 1990 the Republican candidate for governor of California won the election by a margin of 187,438 votes over his Democratic opponent. This difference represented 2.78% of the votes cast for these two candidates. How many votes did the winner receive?

CHAPTER SUMMARY

3-1 FRACTIONS, DECIMALS, AND PERCENT

A **fraction** represents parts of a whole. The fraction a/b, where a and b are positive integers, represents a parts of a whole divided into b equal parts. Fractions may be represented as **decimals. Percent** means parts per 100.

3-2 RATIONAL NUMBERS AND THE FUNDAMENTAL PRINCIPLE OF FRACTIONS

A **rational number** is a number that can be written in the form a/b, where the **numerator** a and **denominator** b are integers with $b \neq 0$. The rational number a/b is **positive** when a and b have like signs and **negative** when their signs are opposite. The **absolute value** and **opposite** of a rational number are defined as for integers. A rational number a/b can be **raised to higher terms** or the rational number ak/bk **reduced to lower terms** by using the **fundamental principle of fractions:**

$$\frac{a}{b} = \frac{ak}{ab}$$

A fraction is reduced to **lowest terms** if the numerator and denominator have no common factor greater than 1. Rational numbers have the following **sign properties:**

$$\frac{-a}{-b} = \frac{a}{b} \qquad \frac{-a}{b} = \frac{a}{-b} = -\frac{a}{b} = (-1)\frac{a}{b}$$

3-3 MULTIPLICATION AND DIVISION

Multiplication and **division** of rational numbers are defined by

$$\frac{a}{b} \cdot \frac{c}{d} = \frac{ac}{bd} \quad \text{and} \quad \frac{a}{b} \div \frac{c}{d} = \frac{a}{b} \cdot \frac{d}{c} = \frac{ad}{bc}$$

3-4 ADDITION AND SUBTRACTION

Addition and **subtraction** of rational numbers with common denominators are defined by

$$\frac{a}{b} + \frac{c}{b} = \frac{a+c}{b} \quad \text{and} \quad \frac{a}{b} - \frac{c}{b} = \frac{a-c}{b}$$

To add or subtract rational numbers when the denominators are not the same, convert them to equivalent forms with a common denominator by using the fundamental principle of fractions. Computation is generally minimized by using the **least common denominator** (LCD)—that is, the LCM of the denominators.

3-5 SOLVING EQUATIONS INVOLVING FRACTIONS AND DECIMALS

Equations involving fractions or decimals can be solved by first clearing the equation of fractions and decimals by multiplication.

3-6 WORD PROBLEMS

Word problems are solved using the strategy first introduced in Chapter 2:

1. Read the problem very carefully.
2. Write down the important facts and relationships.
3. Identify the unknown quantities in terms of one variable.
4. Find the equation.
5. Solve the equation and write down all solutions asked for.
6. Check the solution.

3-7 RATIO AND PROPORTION

A **ratio** of two quantities is the first divided by the second, written a/b or $a:b$ and read "a to b." A **proportion** is an equality between two ratios. A quantity **varies directly** with another, or is **directly proportional** to another, if their ratio is constant.

CHAPTER REVIEW EXERCISE

Work through all the problems in this chapter review and check the answers in the back of the book. Answers to all review problems are there, and following each answer is a number in italics indicating the section in which that type of problem is discussed. Where weaknesses show up, review appropriate sections in the text.

All variables represent nonzero integers.

A 1. Graph $\{-\frac{7}{4},\ -\frac{3}{4},\ \frac{3}{2}\}$ on a number line.

2. What rational numbers are associated with the points a, b, c, and d on the following number line?

Raise to higher terms or reduce to lower terms as indicated:

3. $\dfrac{3}{8} = \dfrac{?}{40}$ 4. $\dfrac{12}{21} = \dfrac{?}{7}$ 5. $\dfrac{6}{14} = \dfrac{30}{?}$ 6. $\dfrac{72}{30} = \dfrac{?}{5}$

Reduce to lowest terms:

7. $\dfrac{24}{60}$ 8. $\dfrac{108}{63}$ 9. $\dfrac{30xy}{12x}$

Perform the indicated operations and reduce to lowest terms:

10. $\dfrac{3}{2y} \cdot \dfrac{5x}{4}$ 11. $\dfrac{3}{2y} \div \dfrac{5x}{4}$

12. $\dfrac{2x}{7} + \dfrac{3x}{7}$ 13. $\dfrac{3a}{8} - \dfrac{9a}{8}$

14. $\dfrac{y}{2} + \dfrac{y}{3}$ 15. $\dfrac{3}{2y} - \dfrac{5x}{4}$

16. $\dfrac{-3x}{4} + \dfrac{x}{6}$ 17. $\dfrac{2a}{5} \cdot \dfrac{-15}{8}$

18. $\dfrac{-3b}{4} - \dfrac{-b}{10}$ 19. $\dfrac{-2y}{15} \div \dfrac{-5}{12}$

In Problems 20–28, solve the equation:

20. $6x = 5$ 21. $3x - 5 = 5x - 8$

22. $\dfrac{y}{8} = \dfrac{3}{4}$ 23. $0.7x = 4.2$

24. $\dfrac{x}{2} - \dfrac{1}{3} = \dfrac{x}{6}$ 25. $1 - 3x = 5 - 5x$

26. $\dfrac{a}{4} = \dfrac{-10}{7}$ 27. $0.03y = -2.1$

28. $\dfrac{1}{3} - \dfrac{2x}{5} = \dfrac{7}{10}$

29. Find 24% of 70.

30. What percent of 130 is 52?

31. Forty-five percent of what number is 60?
    (A) Write an equation.   **(B)** Solve.

32. Three-tenths of what number is $\frac{2}{5}$?
    (A) Write an equation.   **(B)** Solve.

33. For a particular basketball team the ratio of 3-point shots attempted to 2-point shots attempted is $2:9$.

During the season the team attempted 1,530 2-point shots. How many 3-point shots did it attempt?

34. At a particular fast-food restaurant, the ratio of diet drinks sold to regular drinks sold is 1.5 to 4. If the restaurant sells 20,000 regular drinks in a month, how many diet drinks did it sell?

35. Find a number such that 6 more than two-fifths of the number is seven-tenths of the number.

36. If the width of a rectangle with perimeter 80 centimeters is three-fifths the length, what are the dimensions of the rectangle?

37. The response rate in a mail survey is 46%. If 3,000 surveys were mailed, how many were returned?

B *Reduce to lowest terms:*

38. $\dfrac{210}{110}$ **39.** $\dfrac{105}{252}$ **40.** $\dfrac{84xyz}{66xz}$

Perform the indicated operations and reduce to lowest terms:

41. $\dfrac{3y}{5xz} \cdot \dfrac{10z}{15xy}$ **42.** $\dfrac{3y}{5xy} \div \dfrac{10z}{15xy}$

43. $\dfrac{3y}{5xz} + \dfrac{10z}{15xy}$ **44.** $\dfrac{3y}{5xz} - \dfrac{10z}{15xy}$

45. $\dfrac{3}{4xy^2} \div \dfrac{1}{3x^2y}$ **46.** $\dfrac{3}{4xy^2} - \dfrac{1}{3x^2y}$

47. $\dfrac{3}{x^2} - \dfrac{2}{x} + 1$ **48.** $\dfrac{1}{4y} + \dfrac{3}{2z} - \dfrac{1}{3x} - 2$

49. $\dfrac{-4}{9} - \dfrac{35}{18} - \dfrac{-10}{3}$ **50.** $\dfrac{-2}{5} + \dfrac{-8}{33} \cdot \dfrac{11}{2} - \dfrac{-5}{6}$

51. $\dfrac{x}{3} + \dfrac{5x}{4} - \dfrac{-6x}{5}$ **52.** $\dfrac{3x}{5}\left(\dfrac{10}{27} + \dfrac{15x}{9}\right)$

53. $\left(\dfrac{2x}{3} - \dfrac{3x}{2}\right) \cdot \dfrac{6x}{5}$ **54.** $\dfrac{x}{3}\left(6y - \dfrac{9}{x}\right)$

In Problems 55–60, solve the equation:

55. $\dfrac{x}{4} - \dfrac{x-3}{3} = 2$ **56.** $0.4x - 0.3(x - 3) = 5$

57. $0.4x - \dfrac{x}{0.12} = -4.76$ **58.** $\dfrac{x}{5} + \dfrac{x+1}{6} = 90$

59. $\dfrac{y-4}{3} - \dfrac{3}{4} = \dfrac{y-3}{12}$

60. $\dfrac{3}{5}\left(\dfrac{x-5}{2}\right) = (x + 4) \div \dfrac{25}{3}$

61. If the ratio of all the trout in a lake to the ones that had been captured, marked, and released is $20:3$ and there are 450 marked trout, how many trout are in the lake? Set up an equation and solve.

62. If there are 2.54 centimeters in 1 inch, how many inches are in 40 centimeters? Set up a proportion and solve.

63. In 1989, sales of personal computers in the United States grew 14% to $28 billion. What was the total for 1988 sales?

64. If 1 Dutch guilder is worth 0.45 U.S. dollar, how many guilders is $10.00 worth?

65. The sum of the angles in a quadrilateral (a four-sided figure) is 360°. In a parallelogram (quadrilateral with opposite sides parallel) the opposite angles are equal, as shown in the figure. If the larger angle in a parallelogram is 4 times the size of the smaller, what are the angles?

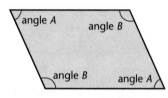

66. An isosceles triangle has perimeter 30 centimeters. The two equal sides are each 3 times as long as the third side. How long are the sides?

67. Find a number such that 1 less than three-quarters of the number is 12 less than the number itself.

68. The federal income tax rate for single taxpayers on net taxable incomes of $16,800 to $27,000 is $2,448 plus 28% of the amount over $16,800. What is the tax on a taxable income of $22,800?

69. The parimutuel odds for a given horse at a racetrack indicate the ratio of profit to wager should the horse win. If your horse wins at odds of $9:2$, what will your profit be on a $5 wager?

70. In a recent year, Japan imported goods worth $150.8 billion. Of this total 23% came from the United States, 23% from southeast Asia, 15% from the middle east, and 14% from western Europe. What was the value of goods imported from the rest of the world?

C *Perform the indicated operations and reduce to lowest terms:*

71. $\dfrac{10x}{9y} \div \left(\dfrac{15xy}{2z} \div \dfrac{3y^2}{-z}\right)$

72. $\dfrac{3}{10x^2} - \dfrac{2}{15xy} + \dfrac{5}{18y^2}$

73. $\dfrac{10x}{9y} + \dfrac{15xy}{2z} + \dfrac{3y^2}{-z}$ **74.** $\dfrac{10x}{9y} - \dfrac{15xy}{2z} \div \dfrac{3y^2}{-z}$

75. $\dfrac{10x}{9y} \cdot \dfrac{15xy}{2z} \cdot \dfrac{3y^2}{-z}$ **76.** $\dfrac{10x}{9y} \cdot \dfrac{15xy}{2z} \div \dfrac{3y^2}{-z}$

77. $\left(\dfrac{10x}{9y} - \dfrac{15xy}{2z}\right) \div \dfrac{3y^2}{-z}$

78. $-\left(\dfrac{10x}{9y} - \dfrac{15xy}{2z}\right) \div \dfrac{3y^2}{-z}$

79. $\dfrac{3}{10x^2} - \left(\dfrac{2}{15xy} + \dfrac{5}{18y^2}\right)$

80. $\left(\dfrac{2}{15xy} - \dfrac{5}{18y^2}\right) \div \dfrac{3}{10x^2}$

Reduce to lowest terms:

81. $\dfrac{105xyz}{140xy^2z^2}$

82. $\dfrac{90a^2b^2c}{168abc^2}$

In Problems 83–90, solve the equation:

83. $\dfrac{x+3}{10} - \dfrac{x-2}{15} = \dfrac{3-x}{6} - 1$

84. $\dfrac{x+1}{3} + \dfrac{x-1}{5} = \dfrac{x-2}{3} + \dfrac{x+2}{5}$

85. $\dfrac{2x+1}{4} + \dfrac{2x-1}{3} = \dfrac{2x-1}{4} + \dfrac{2x+1}{3}$

86. $\dfrac{2x+1}{4} + \dfrac{1}{3} = \dfrac{6x+7}{12}$

87. $\dfrac{2x+1}{4} + \dfrac{2x-1}{3} = \dfrac{6x-1}{4} - \dfrac{x+1}{3}$

88. $\dfrac{x+1}{3} + \dfrac{x-1}{5} = \dfrac{x-2}{3} - \dfrac{x+2}{5}$

89. $\dfrac{2x+1}{4} + \dfrac{2x-1}{3} = \dfrac{2x-1}{4} - \dfrac{2x+1}{3}$

90. $\dfrac{2x+1}{4} - \dfrac{1}{3} = \dfrac{6x+7}{12}$

91. Estimate the total number of squirrels in a forest if a sample of 140 is captured, marked, and released, and after a period for mixing, a second batch of 70 is captured and it is found that 20 of these are marked.

92. A retail store buys an item for \$60 and adds a markup of 30%. A customer pays \$82.68 including sales tax. What is the sales tax rate?

93. Find a number such that three-tenths more than the quantity 29% of the number is 2 less than 75% of the number.

94. In a pentagon (a five-sided figure) the sum of the angles is 540°. If four of the angles are 3, 4, 5, and 5 times the size, respectively, of the remaining angle, what are the angles?

95. In 1985 there were 36,340,000 subscribers to basic cable television in the United States, representing 42.8% of all U.S. households. By 1990 the percentage of households with basic cable television had increased to 56.4%, and the number of such households was 42.8% higher than in 1985. How many total households were there in the United States in 1990? Round your answer to the nearest hundred thousand.

96. Crime statistics in the United States are often reported by cities in rates per 1,000 inhabitants. In a particular year Los Angeles had a reported rate of serious crimes of 88.4 per 1,000. Thus, the ratio of crimes to people was 88.4/1,000 and the ratio of people to crimes was 1,000/88.4. If its total number of reported serious crimes that year was 300,731, what was Los Angeles' population? Round your answer to the nearest hundred thousand.

97. In the game of poker, a flush (a five-card hand with all cards of the same suit) occurs with a probability of 5,148/2,598,560. That is, of every 2,598,560 hands, approximately 5,148 of them will be a flush. Approximately how many flushes should occur in 10,000 poker hands? Round your answer to the nearest whole number.

98. Roger Maris set the major league baseball record for home runs in a season with 61 in 1961. In his 4 seasons before that, he had averaged 24.25 home runs per season. Maris retired after the 1968 season with a career total of 275 home runs. How many home runs per year did he average after his record-setting year?

CHAPTER PRACTICE TEST

The following practice test is provided for you to test your knowledge of the material in this chapter. You should try to complete it in 50 minutes or less. The answers in the back of the book indicate the section in the text that covers the material in the question. Actual tests in your class may vary from this practice test in difficulty, length, or emphasis, depending on the goals of your course or instructor.

1. Find the least common multiple of 10, 24x, and 15x^2.

Reduce to lowest terms:

2. $\dfrac{36}{15}$

3. $\dfrac{24xyz^2}{40xz}$

Rewrite each expression by performing the indicated operations and reducing to lowest terms:

4. $\dfrac{3}{4} + \dfrac{x}{5}$

5. $\dfrac{1}{3} - \dfrac{3}{8}$

6. $\dfrac{12x}{25} \div \dfrac{18x}{60}$

7. $\dfrac{3x}{40} \cdot \dfrac{50}{9}$

8. $\dfrac{1}{7} + \dfrac{x-1}{8}$

9. $\dfrac{4x}{3} + \dfrac{-2x}{5}$

10. $\dfrac{-4}{5} - \dfrac{x-2}{2}$

11. $-\left(\dfrac{12a}{5}\right) \cdot \left(\dfrac{-15b}{16}\right)$

In Problems 12–16, solve the equation:

12. $\dfrac{-x}{3} = -6$

13. $\dfrac{y}{12} = \dfrac{7}{20}$

14. $1.05x + 3.48 = -1.98$

15. $\dfrac{a}{3} + \dfrac{4}{5} = \dfrac{5}{6}$

16. $\dfrac{z-4}{5} - \dfrac{z+2}{4} = \dfrac{3-z}{10}$

17. Find 40% of the average of the numbers 3, 0.6, 5, and 7/5.

18. Enrollment at a small college is currently 1,250, and an increase of 12% is anticipated for next year. What total enrollment is expected?

19. If 1 pound is equivalent to 453.6 grams, how many pounds are equivalent to 1,000 grams (1 kilogram)?

20. Find a number such that 3 less than seven-eighths of the number is 9 less than twice the number.

21. Among the teams in a large youth sports organization, 60% of the teams had losing records during the year, 1/3 had winning records, and the other 12 teams each won the same number as they lost. How many teams are there in the organization?

CUMULATIVE REVIEW EXERCISE—CHAPTERS 1–3

This set of problems reviews the major concepts and techniques of Chapters 1–3. Work through all the problems, and check the answers in the back of the book. Answers to all review problems are there, and following each answer is a number in italics indicating the section in which that type of problem is discussed. Where weaknesses show up, review appropriate sections in the text.

A **1.** Locate the following numbers on a number line: -5, 4.5, $2\frac{3}{4}$, $-\frac{6}{4}$, 3, -2.

2. What rational numbers are associated with the points a, b, c, d, and e on this number line?

3. Which of the numbers listed in Problem 1 are:
(A) Whole numbers?   (B) Integers?
(C) Rational numbers?

4. Which of the numbers plotted in Problem 2 are:
(A) Whole numbers?   (B) Integers?

5. In the expression $3x^2$, identify:
(A) The base   (B) The exponent
(C) The coefficient

6. Find the average of all the prime numbers in the following list: 2, 5, 8, 13, 15, 19, 24.

7. Factor 80 into a product of prime numbers.

Find (A) the opposite and (B) the absolute value of the given number:

8. 12

9. -3.4

10. $\frac{2}{3}$

11. $-\frac{5}{8}$

12. 7.6

13. -5

Find the LCM of the given numbers:

14. 12, 21

15. 6, 10, 14

16. $3x$, $4y$

Evaluate the expressions (A) $x - 3y$ and (B) $3x^2 - y$ for the values of x and y given:

17. $x = 7$, $y = 2$

18. $x = -2$, $y = -5$

19. $x = 1.2$, $y = -2.5$

20. $x = \frac{1}{3}$, $y = -\frac{2}{3}$

Replace each question mark with an appropriate symbol or algebraic expression:

21. $3x + 15y = 3(?)$

22. $5(x - 4y) = 5x + (?)$

23. $a(bc) = (?)c$

24. $x^4 x^7 = x^?$

25. $3.6x + 18x^2 = 3.6(?)$

Remove any grouping symbols present and combine like terms:

26. $3xy + 4x - 5xy + 6x$

27. $3(a - b) - 2(a + 4b)$

28. $\frac{2}{5}x + \frac{1}{4}y + \frac{3}{10}x + \frac{3}{8}y$

Rewrite the given expression by performing the indicated operations and simplifying your answer as much as possible. Like terms should be combined. Fractions should be reduced to lowest terms.

29. $x^2(x + 3x^2)$

30. $\{5 - [x - (5 - x)]\}$

31. $-\dfrac{1}{10}(-3x)(-5x^2)$

32. $\dfrac{a}{4} - \dfrac{3a}{5}$

33. $\dfrac{2x}{3} \div \dfrac{3x}{2}$

34. $\dfrac{-1}{4} \cdot \left(-\dfrac{2b}{3}\right) \cdot \dfrac{-3b}{5}$

In Problems 35–40, solve the equation:

35. $4x + 6 = 13 - 3x$ **36.** $-3y - 6 = -9y - 12$

37. $\dfrac{z}{15} = \dfrac{7}{105}$ **38.** $-12x = 50$

39. $2.3 + 1.8x = 59$ **40.** $\dfrac{1}{2}x - \dfrac{1}{4} = \dfrac{7}{8}$

41. For what number is 90 equal to 45% of the number?

42. Translate the following statement into an algebraic equation using only the variable x, then solve the equation: 27 plus a certain number is 3 less than 3 times the number.

43. The ratio of teenage workers to the total work force in a particular fast-food chain is $11:15$. If the chain employs 8,130 workers, how many of them are teenagers?

44. A rectangle that is 3 times as long as it is wide has a perimeter of 112 centimeters. What are its dimensions?

45. Over a 2-year period a university's budget increased from \$28.4 million to \$36.8 million. If it increases by the same percentage in the next 2-year period, what will its budget be at the end of that time? Round your answer to the nearest hundred thousand.

B **46.** Find the LCM of $15ab^2c$ and $40abc^3$.

47. Factor 525 into a product of primes.

Evaluate the expression $(3 - x)^2 + 3 - x^2$ for the given value of x:

48. $x = 1$ **49.** $x = -1$ **50.** $x = \frac{2}{3}$ **51.** $x = -2.3$

Evaluate the expression $-x + 2|x|$ for the given value of x:

52. $x = 5$ **53.** $x = -5$ **54.** $x = \frac{3}{4}$ **55.** $x = -3.4$

Evaluate the expression $3xy^2 - 2xy - x$ for the given values of x and y:

56. $x = 3,\ y = -2$ **57.** $x = -1.1,\ y = -2.3$

58. $x = \dfrac{1}{3},\ y = \dfrac{2}{3}$ **59.** $x = -\dfrac{1}{4},\ y = \dfrac{-3}{4}$

Rewrite the given expression by performing the indicated operations and simplifying your answer as much as possible. Like terms should be combined. Fractions should be reduced to lowest terms.

60. $(3x^3y)(2xy^2)(xy)$ **61.** $\dfrac{3x^3y}{5} \cdot \dfrac{2xy^2}{3} \div \dfrac{-4xy}{5}$

62. $(2x + 1)(x - 1)$ **63.** $\dfrac{x^2}{2} + \dfrac{x}{3} - \dfrac{1}{4}$

64. $3ab^2(4a^2b - 3ab^2) - 10a^3b^3 + 10a^2b^4$

65. $(3x^2 - 2y^2) - (3y^2 - 2x^2)$

Factor out all common factors:

66. $6xy + 15xy^2 - 9x^2y$ **67.** $12a^3 - 18a^2 + 27a^4$

In Problems 68–73, solve the equation:

68. $5(z - 3) + 2(4 + z) = 7(z - 1)$

69. $5(z - 3) + 2(4 + z) = 7(z - 2)$

70. $5(z - 3) + 2(4 + z) = -7(z - 1)$

71. $\dfrac{y - 5}{4} + \dfrac{3}{5} = \dfrac{y + 1}{2}$

72. $\dfrac{2y - 5}{4} + \dfrac{3}{5} = \dfrac{y + 1}{2}$

73. $\dfrac{2y + 5}{4} - \dfrac{3}{4} = \dfrac{y + 1}{2}$

74. A newspaper coin rack accepts only quarters and dimes. The paper costs 25 cents, and if a person deposits 3 dimes, no change is given. At the end of the day, 33 papers have been sold, and the coin box contains \$8.45. How many quarters were in the box?

75. The sum of 4 consecutive integers is 150. Find the numbers.

76. A long-distance runner runs at a pace that covers a mile every 7 minutes. How fast is she running in miles per hour?

77. A credit card customer at a restaurant figures the tip at exactly 15% of the bill, excluding tax. If the total paid, including tip and the 7.5% tax on the bill, is \$33.81, how much of a tip was left?

C *Solve the equation:*

78. $\dfrac{x - 1}{3} + \dfrac{x + 2}{4} = \dfrac{x + 6}{4} + \dfrac{x - 4}{3}$

79. $\dfrac{x - 1}{3} + \dfrac{x + 2}{4} = \dfrac{x + 3}{4} + \dfrac{x - 4}{3}$

80. $\dfrac{x - 1}{3} + \dfrac{x + 2}{4} = \dfrac{x + 3}{4} - \dfrac{x - 4}{3}$

81. $\dfrac{x}{5} + \dfrac{3x}{10} - \dfrac{x + 2}{15} = \dfrac{2x - 1}{5}$

82. $\dfrac{x}{5} + \dfrac{3x}{10} - \dfrac{x + 2}{15} = \dfrac{2x - 1}{5} + \dfrac{x + 2}{30}$

Solve the equation for the indicated variable:

83. $w = 5.5h - 220;\ h$

84. $p = 15 + 15(d/33);\ d$

85. $\dfrac{P_1}{V_2} = \dfrac{P_2}{V_1}; \; P_1$

Rewrite the given expression by performing the indicated operations and simplifying your answer as much as possible. Like terms should be combined. Fractions should be reduced to lowest terms.

86. $\dfrac{x}{5} + \dfrac{3x}{10} - \dfrac{x+2}{15}$

87. $\dfrac{x}{5} + \dfrac{3x}{10} - \dfrac{x+2}{15} + \dfrac{x+2}{30}$

88. $\dfrac{x}{6} + 3x^2 + \dfrac{3x}{2} - 4x^2 + 2x + \dfrac{x^2}{2}$

*In Problems 89–96, match the statement to the property in the list **a–n** that it illustrates:*

a. Associative property for addition
b. Associative property for multiplication
c. Commutative property for addition
d. Commutative property for multiplication
e. Distributive property of multiplication over addition
f. Reflexive property of equality
g. Symmetric property of equality
h. Transitive property of equality
i. Addition property of equality
j. Subtraction property of equality
k. Multiplication property of equality
l. Division property of equality
m. First law of exponents
n. Fundamental principle of fractions

89. $3(x^2 + 2) = 3x^2 + 6$

90. If $3x = 5$, then $x = 5/3$

91. If $c = a + b$ and $a + b = 5$, then $c = 5$

92. $4(5z) = 20z$

93. $10(x + 3) = (x + 3) \cdot 10$

94. If $y^2 + 6 = 11$, then $y^2 = 5$

95. $3 + (4 + x) = 7 + x$

96. If $\dfrac{z}{4} = 5$, then $z = 20$

97. Raffle tickets are sold for a quarter apiece or 5 for a dollar. If 830 raffle tickets are sold and $189 taken in, how many tickets were purchased at the 5-for-a-dollar rate?

98. In a relay race between two teams, the last runner for team A can run a 400-meter lap in 48 seconds; the last runner for team B can run it in 45 seconds. Team B starts its last 400-meter lap 24 meters behind. Can its last runner make up the difference?

99. If 1 U.S. dollar is worth 1.195 Canadian dollars and worth 2,500 Mexican pesos, how many pesos can be purchased with 100 Canadian dollars?

100. The volume V of a sphere is directly proportional to the cube of the diameter d. Thus, the ratio V/d^3 is constant. If a sphere of diameter 4 inches has volume 33.51 cubic inches, what is this constant ratio?

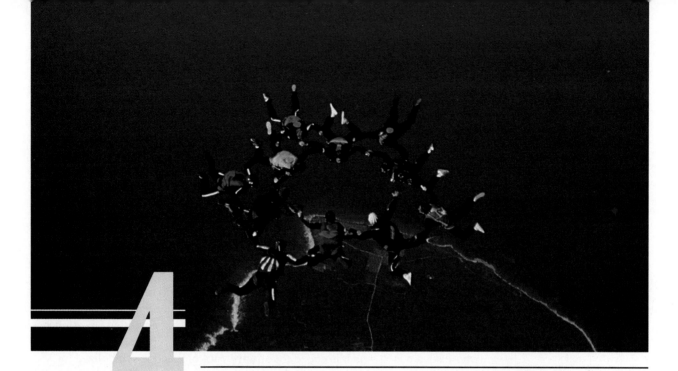

4

Graphing and Linear Equations

By extending the integers to the rational numbers in Chapter 3, we significantly increased our ability to perform certain operations, to solve equations, and to attack practical problems. It appears that the rational numbers are capable of satisfying all our number needs. Can we stop here, or do we need to go further?

In Section 2-8, we mentioned that almost everything you will do in algebra will fall into one of six categories: evaluate, translate, rewrite, solve, graph, use. We have discussed five of these processes already. The sixth, graphing, is introduced in this chapter. To use graphs, we will need a number associated with each point on the number line. We will see that there are lengths, and therefore points, on the number line that do not correspond to rational numbers. An expanded number system, the *real numbers,* is needed.

The real-number system provides an exact correspondence between numbers and points on the number line. This correspondence is utilized in Section 4-1 to identify points in the plane with pairs of numbers. This provides a link between algebra and geometry (graphing) that is explored in the rest of the chapter.

Photo reference: see Exercise 4-7, Problem 33.

4-1 Real Numbers and the Cartesian Coordinate System

♦ Square Roots and Irrational Numbers
♦ The Set of Real Numbers
♦ Rectangular Coordinates
♦ Graphing Ordered Pairs (a, b)

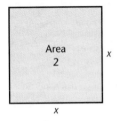

Area
2

x

x

Figure 1

Suppose we wish to find the length of the side of a square with area 2. Referring to Figure 1, we see that $x^2 = 2$. We ask: Are there any rational numbers whose square is 2? It turns out that one can prove that there is no rational number whose square is 2. If a square of area 2 is to have a number that represents the length of a side, then we must invent a new kind of number. This new kind of number is called an **irrational number.** In this case, the length is represented by a number whose square is 2, which is called the **square root** of 2 and is symbolized by $\sqrt{2}$.

The rational and irrational numbers together provide a number associated with each point on the number line. In this section we will also see how pairs of such numbers are associated with every point in a plane.

♦ SQUARE ROOTS AND IRRATIONAL NUMBERS

We define the square root of a number as follows:

> **Definition of Square Root**
>
> x is a square root of y if $x^2 = y$.

Example 1
Square Roots

Find two square roots of 4.

Solution Since $2^2 = 4$, 2 is a square root of 4. In addition, -2 is a square root of 4, since $(-2)^2 = 4$.

Matched Problem 1

Find two square roots of 9.

Except for the perfect squares 1, 4, 9, 16, 25, etc., natural numbers have square roots that are not rational numbers. Such nonrational numbers were "discovered" by the Pythagoreans, mathematical followers of the Greek mathematician Pythagoras, in the sixth century B.C. Pythagoras is known for, among other things, the Pythagorean theorem in geometry that states that the square of the length of the hypotenuse of a right triangle is equal to the sum of the squares of the lengths of the two sides. See Figure 2.

Hypotenuse
c

Side
a

Side b

$a^2 + b^2 = c^2$

Figure 2

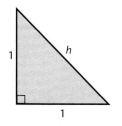

Figure 3

If we apply the Pythagorean theorem to the right triangle in Figure 3, we obtain

$$h^2 = 1^2 + 1^2 = 2$$

Thus, h is a square root of 2. The Pythagoreans could prove that this number was not the ratio of two integers, so that nonrational, *irrational,* numbers exist. An **irrational number** can be thought of as a length, or its negative, that is not the ratio of two integers.

Square roots provide some examples of irrational numbers, but there are many more. In fact, in a certain sense there are more irrational numbers than rational numbers.

Square roots will be discussed in more detail in Chapter 8. It will be useful, however, to have available the standard square root notation. Every positive number will have two square roots, just as we found for the number 4 in Example 1. We will use the notation $\sqrt{}$ to denote the *positive* square root of a number and $-\sqrt{}$ to denote the negative square root of a number. Thus,

$$\sqrt{2} \text{ represents the positive square root of } 2$$

$$-\sqrt{2} \text{ represents the negative square root of } 2$$

$$\sqrt{9} = 3 \qquad \text{and} \qquad -\sqrt{9} = -3$$

♦ THE SET OF REAL NUMBERS

Both rational and irrational numbers can be represented in decimal form. Every rational number has a **repeating decimal** representation. For example:

$$5 = 5.000\cdots = 5.\overline{0} \qquad 3.14 = 3.1400\cdots = 3.14\overline{0}$$

$$\frac{4}{3} = 1.33\cdots = 1.\overline{3} \qquad \frac{5}{7} = 0.\overline{714285}$$

$$\frac{71}{330} = 0.21515\cdots = 0.2\overline{15}$$

where the overbar indicates the block of numbers that continues to repeat indefinitely. When the block of repeating numbers is all zeros, as in $5 = 5.\overline{0}$, the decimal representations are often called **terminating decimals.**

Every irrational number, when represented in decimal form, has an infinite **nonrepeating decimal** representation. For example:

$$\sqrt{2} = 1.4142135\cdots \qquad \pi = 3.1415926\cdots \qquad -\sqrt{6} = -2.4494897\cdots$$

and no block of numbers will continue to repeat.

The set of rational and irrational numbers form the **real-number system.** Table 1 compares the various sets of numbers we have discussed.

Table 1   **The Set of Real Numbers**

Symbol	Number Set	Description	Examples
N	Natural numbers	Counting numbers (positive integers)	1; 3; 3,525
Z	Integers	Set of counting numbers, their opposites, and 0	-31, -1, 0, 4, 702
Q	Rationals	Any number that can be represented in the form a/b, $b \neq 0$, where a and b are integers	-4, $-\frac{3}{8}$, 0, $\frac{4}{3}$, 3.57
R	Reals	Set of all rational and irrational numbers	$-\sqrt{2}$, $-\frac{4}{7}$, 0, 62.48, $\sqrt{5}$, π, 3,407

The following diagram shows the relationships among the number sets. Each set is contained in the sets above it.

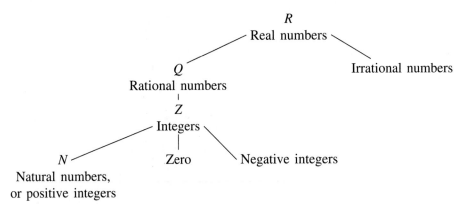

Figure 4 shows some real numbers on a **real number line.**

Figure 4

We can summarize our assumptions about the real numbers as follows:

Real Numbers

Each point on the number line corresponds to exactly one real number. Conversely, each real number corresponds to exactly one point on the number line.

If we write

$$\sqrt{2} \approx 1.414 \quad \text{and} \quad \pi \approx 3.1415$$

where $\approx$ means **approximately equal to,** then we are using rational number approximations for the irrational numbers $\sqrt{2}$ and π.

Example 2
Approximating
Irrational Numbers

Use a calculator to approximate the following irrational numbers by rounding the decimal representation to three decimal places:

(A) $\sqrt{5}$ **(B)** $\sqrt{6}$

Solution   **(A)**   Your calculator should show $\sqrt{5} = 2.236067977 \cdots$. The number of decimal places displayed will vary depending on the particular calculator used. Thus $\sqrt{5} \approx 2.236$.

(B) $\sqrt{6} = 2.449489743 \cdots \approx 2.449$

Matched Problem 2   Use a calculator to approximate the following irrational numbers by rounding the decimal representation to three decimal places:

(A) $\sqrt{3} \approx 1.732$ **(B)** $\sqrt{8} \approx 2.8284$

It is possible to define addition, subtraction, multiplication, and division on the set of real numbers. We will not do so in this text. In Chapter 8, some exact computations with irrational numbers will be considered, but, otherwise, irrational numbers will be approximated by rational numbers as necessary. However, all the common properties of the number systems we have developed thus far extend to the real numbers. These include the equality properties and the associative, commutative, and distributive properties. The properties are summarized on the inside front cover of the text. Thus, we can treat variables representing real numbers just as we have treated variables restricted to rational numbers.

Now that we have each point on a number line labeled with a real number, our next goal is to label points in the plane. We accomplish this by introducing a *rectangular coordinate system.*

◆ **RECTANGULAR COORDINATES**

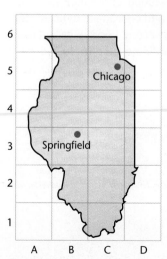

Figure 4 above illustrates the real number *line.* We now move to a *plane* and develop a system, called a **rectangular coordinate system,** that will enable us to graph equations and inequalities with two variables instead of just one.

The idea behind rectangular coordinates can be illustrated by the familiar way of locating places on maps. Consider the map of the state of Illinois shown here. Each square region of the map is designated by a letter-number pair. An index for the map would locate the city of Chicago in region C5 and the state capital, Springfield, in region B3. The pairs B5 and C3 are examples of rectangular coordinates. We want to use pairs to assign such a location designator not just to regions in a map but to every point in the plane. We will use the idea that every point on a number line has a real number corresponding to it. We begin with a horizontal number line:

Figure 5

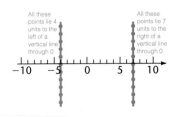

Figure 6

The real numbers attached to each point on the line indicate how far to the right or left of 0 the point lies. This is also true for points above or below the line as shown in Figure 6.

We will use the number that indicates horizontal distance to 0 as the first number in our coordinate pair. To completely locate a point, we need indicate only how far above or below the horizontal line the point lies. To do this we use a vertical number line as shown in Figure 7.

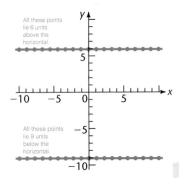

Figure 7

The number that indicates vertical distance to the horizontal line will be the second number in our coordinate pair. Two numbers completely identify a point in the plane. The pair (6, 3) locates a point 6 units to the right of 0 and 3 units above the horizontal, as may be seen in Figure 8. Conversely, the point *P* shown in the figure must have coordinates (−7, 5).

The coordinate system developed above is also called a **cartesian coordinate system,** after the seventeenth-century French mathematician-philosopher René Descartes, who first made significant use of it. The horizontal and vertical number lines are called the **horizontal axis** and **vertical axis,** respectively. Together they are called the **coordinate axes.** The point where the axes cross is called the **origin.** The axes divide the plane into four regions called **quadrants,** numbered as shown in Figure 9.

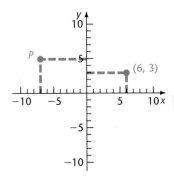

Figure 8

In a coordinate pair (*a*, *b*) that locates a point, the numbers *a* and *b* are called the **coordinates** of the point. The first number in the pair is called the **first coordinate,** or **abscissa.** The second number is called the **second coordinate,** or **ordinate.** The first coordinate gives the horizontal distance to the origin. This distance is a **directed distance,** such that a negative number indicates that the point is to the left of the origin; a positive number, to the right. The second coordinate gives the directed vertical distance to the origin. A negative second coordinate indicates that the point is below the horizontal axis, and a positive second coordinate indicates that the point is above the axis. Thus for the coordinate pair (*a*, *b*) we have

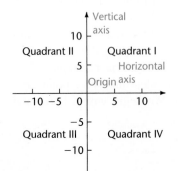

Figure 9

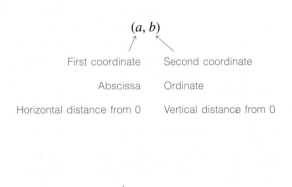

Example 3
Coordinates of Points

Find the coordinates of each of the points *A, B, C, D, E,* and *F*:

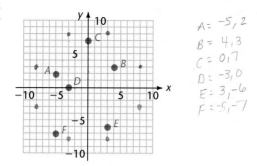

A = -5, 2
B = 4, 3
C = 0, 7
D = -3, 0
E = 3, -6
F = -5, -7

Solution $A(-5, 2)$ $B(4, 3)$ $C(0, 7)$ $D(-3, 0)$ $E(3, -6)$ $F(-5, -7)$

Matched Problem 3   Find the coordinates, using the figure in Example 1, for each of the following points:

(A) 1 unit to the right and 2 units up from *A*   *-4, 4*
(B) 3 units to the left and 9 units down from *C*   *-3, -2*
(C) 1 unit up and 1 unit to the right of *E*   *4, -5*
(D) 2 units to the right of *D*   *-1, 0*

♦ **GRAPHING ORDERED PAIRS** (*a, b*)

To **graph** an ordered pair of numbers (*a, b*), we start at the origin. We move left or right depending on whether the first coordinate *a* is negative or positive, then move down or up depending on whether the second coordinate *b* is negative or positive. Consider the four cases shown in Figure 10:

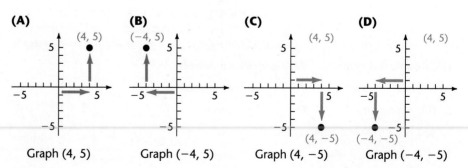

Figure 10

Example 4
Graphing Ordered Pairs

Graph the following coordinates in the same rectangular coordinate system; that is, associate each ordered pair of numbers with a point: (3, 8), (8, 3), (−8, 3), (3, −8), (−8, −3), (−3, −8), (−3, 8), (8, −3).

Solution

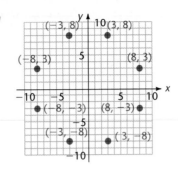

 CAUTION It is very important to note that the ordered pair (3, 8) and the set {3, 8} are not the same thing: {3, 8} = {8, 3} but (3, 8) ≠ (8, 3).

Matched Problem 4 Graph the following coordinates in the same rectangular coordinate system: (2, 5), (−2, 5), (−2, −5), (2, −5), (0, 5), (−2, 0).

Answers to Matched Problems
1. +3, −3   2. (A) 1.732   (B) 2.828
3. (A) (−4, 4)   (B) (−3, −2)   (C) (4, −5)   (D) (−1, 0)
4.

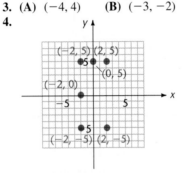

EXERCISE 4-1

A *Indicate whether the following are true (T) or false (F):*

1. 5 is a natural number.   2. $\frac{2}{3}$ is a rational number.

3. −3 is an integer.   4. $\frac{3}{4}$ is a natural number.

5. 0 is an integer.   6. $-\frac{2}{3}$ is an integer.

7. $\sqrt{2}$ is an irrational number.

8. π is an irrational number.

9. $\sqrt{5}$ is a real number.   10. 7 is a real number.

11. −5 is a real number.   12. $\frac{3}{7}$ is a real number.

Use a calculator to approximate the following irrational numbers by decimals rounded to three decimal places:

13. $\sqrt{10}$ 14. $\sqrt{11}$ 15. $\sqrt{112}$

16. $\sqrt{113}$ 17. $\sqrt{2,114}$ 18. $\sqrt{2,115}$

Give the coordinates for each labeled point in the graph:

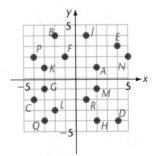

19. *A, B, C, D*   20. *E, F, G, H*

21. *J, K, L, M*   22. *N, P, Q, R*

Give the coordinates for each labeled point in the graph:

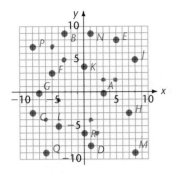

23. *A, B, C, D* **24.** *E, F, G, H*

25. *J, K, L, M* **26.** *N, P, Q, R*

Graph each set of ordered pairs of numbers on the same coordinate system:

27. $A(-3, 5)$, $B(2, -6)$, $C(-4, -1)$, $D(5, 2)$

28. $A(-1, 4)$, $B(3, -5)$, $C(-2, -7)$, $D(6, 1)$

29. $A(3, 2)$, $B(1, -4)$, $C(-6, -4)$, $D(-5, 7)$

30. $A(4, 8)$, $B(5, -8)$, $C(-8, -2)$, $D(-2, 6)$

B *Give the coordinates to the nearest half unit for each labeled point in the graph:*

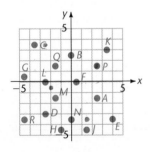

31. *A, B, C, D* **32.** *E, F, G, H*

33. *J, K, L, M* **34.** *N, P, Q, R*

Graph each set of ordered pairs on the same coordinate system:

35. $A(-2.5, 3.5)$, $B(1.5, -3.5)$,
  $C(-2, -0.5)$, $D(2.5, 1.5)$

36. $A(1.5, 2.5)$, $B(0.5, -4)$,
  $C(-3.5, -4.5)$, $D(-2, 1.5)$

37. $A(4.25, 2.75)$, $B(3.5, -1.25)$,
  $C(-1.75, 0.5)$, $D(-2, -1.25)$

38. $A(1.75, 3.25)$, $B(2.5, 4.25)$,
  $C(-0.75, -0.5)$, $D(-1, 3.25)$

Tell, without graphing, which of the quadrants contain each point:

39. **(A)** $(-30, 22)$ **(B)** $(21, -40)$ **(C)** $(-30, -100)$

40. **(A)** $(35, -25)$ **(B)** $(25, 45)$ **(C)** $(-35, 15)$

41. **(A)** $(-300, 200)$ **(B)** $(-200, -400)$
  (C) $(300, -500)$

42. **(A)** $(350, -250)$ **(B)** $(-250, 450)$ **(C)** $(350, 150)$

C *Express each rational number as a repeating decimal by dividing each denominator into the numerator, and continuing until a repeating pattern is observed:*

43. $\frac{1}{4}$ **44.** $\frac{7}{4}$ **45.** $\frac{23}{9}$ **46.** $\frac{10}{6}$

47. $\frac{1}{7}$ **48.** $\frac{3}{8}$ **49.** $\frac{5}{9}$ **50.** $\frac{1}{6}$

In Problems 51–58, express each rational number as a repeating decimal with the aid of a calculator:

51. $\frac{32}{99}$ **52.** $\frac{10}{33}$ **53.** $\frac{4}{7}$ **54.** $\frac{7}{13}$

55. $\frac{7}{11}$ **56.** $\frac{3}{32}$ **57.** $\frac{1}{64}$ **58.** $\frac{1}{27}$

59. In which quadrants is the product of the two coordinates of a point positive?

60. In which quadrants is the product of the two coordinates of a point negative?

The following technique can be used to find a fraction corresponding to some repeating decimal. For example, given a number x represented as $0.35353535 \cdots = 0.\overline{35}$, we have

$$x = 0.3535 \cdots \quad \text{and} \quad 100x = 35.3535 \cdots$$

Subtracting the first equation from the second, left side from left side and right side from right side, yields

$$99x = 35$$

so, $x = 35/99$. Note how the decimals subtracted out. In this case, since the block of repeating decimals has 2 digits in each repeat, multiplying by 10^2 removed the decimals. Use this technique to find a fraction, in lowest terms, corresponding to the repeating decimals given.

61. $0.\overline{42}$ **62.** $0.\overline{51}$ **63.** $0.\overline{27}$ **64.** $0.\overline{38}$

65. $0.\overline{123}$ **66.** $0.\overline{456}$ **67.** $0.\overline{246}$ **68.** $0.\overline{135}$

4-2   Equations and Straight Lines

- ♦ Graphing Straight Lines
- ♦ Vertical and Horizontal Lines
- ♦ Graphing with Different Scales and Restricted Values
- ♦ Information from Graphs

The development of the rectangular coordinate system represented a very important advance in mathematics. It was through the use of this system that Descartes was able to transform geometric problems requiring long, tedious reasoning into algebraic problems that could be solved almost mechanically. This joining of algebra and geometry has now become known as **analytic geometry.**

When an equation involves only one variable, a solution is a number that can replace the variable and that makes the equation true (see Section 2-7). The set of all solutions is called the **solution set.** When an equation involves two variables, a **solution** must consist of a pair of numbers, one for each variable. The set of solution pairs is again called the **solution set,** and the graph of all these pairs in a coordinate system is called the **graph of the equation.**

The graph of an equation can provide insight and intuition about the relationship an equation describes. The geometric picture is often "worth a thousand words" in helping to understand such a relationship. A graph can provide information that may be difficult to see in an equation or in a table of numbers.

There are two fundamental problems in relating the geometry of graphs and the algebra of equations in analytic geometry:

1. Given an equation, find its graph; that is, find the graph of its solution set.
2. Given a geometric figure, such as a straight line, circle, or ellipse, find an equation that has this figure as the graph of its solution set.

In this course we will be interested mainly in the first problem, with particular emphasis on equations whose graphs are straight lines. We explore these equations in this and the next section.

♦ GRAPHING STRAIGHT LINES

We begin with the equation

$$y = 2x + 3$$

and ask what is its solution set and what does the graph of its solution set look like?

The solutions will be pairs of numbers. For example, $x = 0$ and $y = 3$ is a solution, since

$$3 = 2 \cdot 0 + 3$$

We agree to write such a solution as $(0, 3)$ with the value of x first and y second. We can find other solutions by assigning any value to x and solving for y. If $x = 1$, then

$$y = 2 \cdot 1 + 3 = 5$$

and if $x = 2$, $\qquad\qquad\qquad y = 2 \cdot 2 + 3 = 7$

Therefore $(1, 5)$ and $(2, 7)$ are also solutions to the equation. Since we can assign any value to x, it should be clear that the solution set will be infinite. We can list some of the solutions in a table:

Table 1

Assigned Value of x	Computation of $2x + 3 = y$	Solution (x, y)
-3	$2(-3) + 3 = -3$	$(-3, -3)$
-2	$2(-2) + 3 = -1$	$(-2, -1)$
-1	$2(-1) + 3 = 1$	$(-1, 1)$
0	$2 \cdot 0 + 3 = 3$	$(0, 3)$
1	$2 \cdot 1 + 3 = 5$	$(1, 5)$
2	$2 \cdot 2 + 3 = 7$	$(2, 7)$
3	$2 \cdot 3 + 3 = 9$	$(3, 9)$

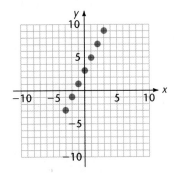

Figure 1

We can graph these points in a rectangular coordinate system. Since we have identified the first coordinate with a value of x, we label the horizontal axis as the **x axis.** The vertical axis then becomes the **y axis.** See Figure 1. It appears that these solutions lie in a straight line, and this suggests that perhaps all the solutions lie on this line. This, in fact, is true, not only for this example but also for equations like this in general. We will use, but not prove, the following result:

Equations and Straight Lines

The graph of any equation of the form

$$Ax + By = C$$

where A, B, and C are constants, A and B are not both 0, and x and y are variables, is a straight line. Every straight line in a rectangular coordinate system is the graph of an equation of this form.

Our equation $y = 2x + 3$ can be rewritten as $-2x + y = 3$, which shows it is a special case of $Ax + By = C$. Thus, the graph of $y = 2x + 3$ is a straight line. In general, the graph of

$$y = mx + b$$

where m and b are constants, is a straight line, since it can be written in the form $-mx + y = b$, which is a special case of $Ax + By = C$.

An equation of the form $Ax + By = C$ or $y = mx + b$ is called a **linear equation** because its graph is a straight line. As a special case, this applies to the equations in one variable such as were solved in Section 2-7, where the term "linear" equation was first used.

Since graphing a straight line requires only two points, we have a simple procedure for graphing equations of the form

$$Ax + By = C \qquad \text{or} \qquad y = mx + b$$

Mechanics of Graphing Equations of the Form $Ax + By = C$ or $y = mx + b$

1. Find any two solutions to the equation. (A third solution is useful as a check point.)
2. Draw coordinate axes and label them.
3. Indicate a scale on each axis by numbering appropriate points on each axis.
4. Plot (graph) the solutions found in step 1.
5. Draw a line through points plotted in step 4, using a straightedge or ruler.

Our graphs, of course, are only part of the complete graph, since the line extends infinitely far in either direction. We usually are most interested in the portion of the graph near the origin, but will indicate the infinite extension of the graph by including arrows at the ends of the graph.

Example 1
Graphing a Linear Equation

Graph:

(A) $y = 2x + 3$ **(B)** $x + 2y = 6$

Solution **(A)** Since this is the same example with which we started this section, choose two points from the table already constructed, say, $(-3, -3)$ and $(2, 7)$. Plot these two points and graph the line through them with a straightedge or ruler. See Figure 2. We can use a third point, say, $(0, 3)$, as a check. If this point had not been on the line, we would check for an error in finding our solutions.

(B) Find two solutions by assigning a value to one variable and solving for the other. If we let $x = 0$, then

$$0 + 2y = 6$$
$$2y = 6$$
$$y = 3$$

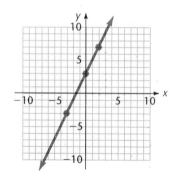

Figure 2

so $(0, 3)$ is a solution. If we let $y = 0$, then

$$x + 2 \cdot 0 = 6$$
$$x = 6$$

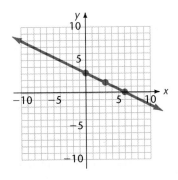

Figure 3

so $(6, 0)$ is a solution. Plot the two points and the line they determine. See Figure 3. Find a checkpoint by assigning another value to one variable, say, $x = 3$, and solving for the other:

$$3 + 2y = 6$$

$$2y = 3$$

$$y = \frac{3}{2}$$

Thus, $(3, \frac{3}{2})$ should lie on the line, and it does.

The choices of $x = 0$ and $y = 0$ used in part (B) above are often convenient ones, since they make for easy calculations in solving for the other variable. Geometrically, the points are where the line crosses the two coordinate axes, and they are therefore called the **x and y intercepts.**

Finding the x and y Intercepts

To find the x intercept, let $y = 0$ and solve for x. To find the y intercept, let $x = 0$ and solve for y.

Matched Problem 1   Graph:

    (A) $y = 2x - 1$ **(B)** $2x + y = 4$

♦ **VERTICAL AND HORIZONTAL LINES**

Vertical and horizontal lines are graphed easily and have simple equations. The x coordinate at any point on a vertical line is constant. Thus, to graph a vertical line, we need to know only the x coordinate common to all points on the line. The equation will be x equal to a constant. Similarly, a horizontal line will have equation y equal to a constant.

Example 2
Graphing Horizontal and Vertical Lines

Graph the equations $x = 3$ and $y = -2$.

Solution   To graph $x = 3$, recognize that this is the same as $x + 0 \cdot y = 3$, so the graph is a line. No matter what value is assigned to y, $x = 3$, so the points $(3, 0)$ and $(3, 1)$ are on the line and the line is vertical. See Figure 4.

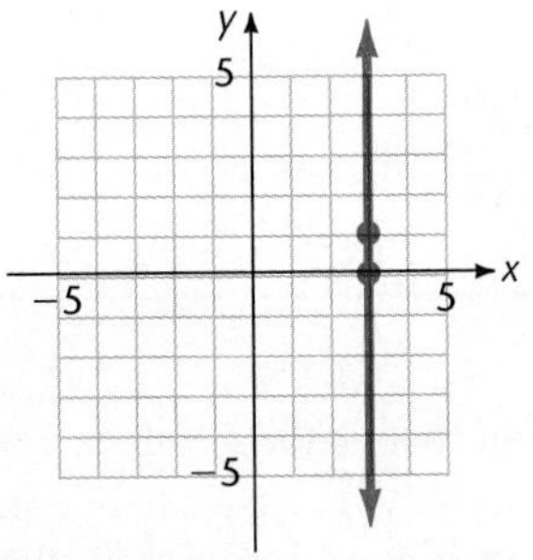

Figure 4

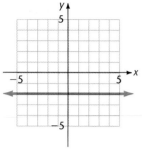

Figure 5

Similarly, $y = -2$ is the same as $0 \cdot x + y = -2$, so this equation represents a horizontal line with x assuming any value and $y = -2$. See Figure 5.

In general:

Vertical and Horizontal Lines

$x = a$ represents a vertical line through $(a, 0)$

$y = b$ represents a horizontal line through $(0, b)$

Matched Problem 2 Graph the equations $x = -3$ and $y = 2$.

♦ GRAPHING WITH DIFFERENT SCALES AND RESTRICTED VALUES

We may be interested in graphing an equation with one or both of the variables restricted to a particular set of values. Such restrictions often arise in applied problems because of real-world considerations. Frequently variables must be restricted to nonnegative values. It may also occur that the values of interest for one variable are significantly larger than those for the other. In this instance, it is helpful to use a different scale on the two axes.

Example 3
Graphing with Different Scales

Graph $y = 55x$ for x between 0 and 5.

Solution Find three solutions (two points and a checkpoint). Since x is restricted to be between 0 and 5, we plot the extreme values of x, 0 and 5, and an intermediate value $x = 3$ as a check. We note the large values of y corresponding to the smaller values for x, so we adjust the scale on the y axis accordingly. We also note that the whole graph will be drawn in the first quadrant. If we think of x as representing time in hours traveled, and 55 as representing a speed in miles per hour, then y represents the distance traveled in miles. The final graph is shown in Figure 6.

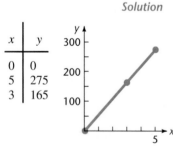

x	y
0	0
5	275
3	165

Figure 6

Matched Problem 3 Graph $y = 30x + 10$ for x between 0 and 6.

♦ INFORMATION FROM GRAPHS

If we look at the graph in Example 3, it should be apparent that as the x values get larger, so do the values of y. This is the kind of insight that may be easier to see in the graph than in a table of values. Consider the graph in Figure 7 depicting the

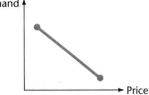

Demand ↑

Price →

Figure 7

relationship between price and demand as it might be shown in an introductory economics text. Even with no units shown on the axes, it is clear that as price increases, demand gets smaller. This confirms our intuition.

Answer to Matched Problems

1. (A)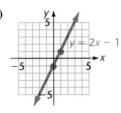

$y = 2x - 1$

(B)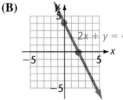

$2x + y = 4$

2.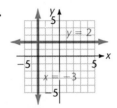

$y = 2$

$x = -3$

3.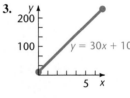

$y = 30x + 10$

EXERCISE 4-2

Graph each equation in a rectangular coordinate system:

A
1. $y = x + 3$
2. $y = x - 4$
3. $y = -x + 1$
4. $y = -x + 2$
5. $y = x - 1$
6. $y = -x - 1$
7. $y = -x - 3$
8. $y = x + 1$
9. $y = -x$
10. $y = x$
11. $y = 3x$
12. $y = -2x$
13. $y = -3x$
14. $y = 2x$
15. $x + 2y = 4$
16. $2x - y = 4$
17. $-x + 2y = 4$
18. $-2x - y = 4$
19. $3x - 2y = 6$
20. $2x + 3y = 6$
21. $-3x - 2y = 6$
22. $-2x + 3y = 6$
23. $x = 4$
24. $x = -2$
25. $y = -3$
26. $y = 1$
27. $x = -4$
28. $x = 2$
29. $y = 3$
30. $y = -1$

B
31. $y = \frac{1}{3}x$
32. $y = -\frac{1}{3}x$
33. $y = -0.5x$
34. $y = 0.5x$

35. $y = -\frac{1}{2}x + 1$
36. $y = \frac{1}{2}x + 1$
37. $y = \frac{1}{2}x - 1$
38. $y = -\frac{1}{2}x - 1$
39. $x + 2y = 3$
40. $x - 2y = 3$
41. $3x - 2y = 4$
42. $3x + 2y = 4$
43. $-3x + 2y = 4$
44. $-3x - 2y = 4$
45. $-x - 2y = 3$
46. $-x + 2y = 3$
47. $y = \frac{1}{2}x + \frac{1}{5}$
48. $y = -\frac{1}{2}x + \frac{2}{3}$
49. $y = -\frac{1}{2}x + \frac{1}{5}$
50. $y = \frac{1}{2}x - \frac{2}{3}$
51. $x = 0$
52. $y = 0$

Write each equation in the form $y = mx + b$ and graph. For example, to write $-3x + y = -2$ in the form $y = mx + b$, we proceed as follows:

$$-3x + y = -2$$

$$\boxed{+3x - 3x + y = -2 + 3x}$$

$$y = 3x - 2 \qquad \text{Form } y = mx + b; \\ m = 3, b = -2$$

53. $3x + 4y = 12$
54. $4x - 3y = 12$
55. $5x - 2y = 10$
56. $-5x + 2y = 10$

57. $-4x - 2y = 8$

58. $-6x - 3y = 9$

59. $-4x + 2y = 8$

60. $6x + 3y = 9$

Write in the form $Ax + By = C$, $A > 0$, and graph. For example, to write $y = 3x - 2$ in the form $Ax + By = C$, $A > 0$, we proceed as follows:

$$y = 3x - 2$$

$$\boxed{-3x + y = 3x - 2 - 3x}$$

$$-3x + y = -2$$

$$3x - y = 2 \qquad \text{Form } Ax + By = C,$$
$$A > 0;\ A = 3,\ B = -1,$$
$$C = 2$$

61. $y = -0.2x + 1$

62. $y = -0.2x - 1$

63. $y = -0.5x - 1$

64. $y = -0.5x + 1$

65. $y = \frac{1}{2}x - 1$

66. $y = \frac{2}{3}x + 4$

67. $x = \frac{3}{4}y + 2$

68. $x = \frac{5}{3}y - 1$

C *In Problems 69–76, graph the equation for the given values of x. Use a different scale on the vertical axis to keep the size of the graph within reason.*

69. $y = 65x$ for x between 0 and 5

70. $y = 20x$ for x between 0 and 15

71. $y = 30x + 100$ for x between 0 and 6

72. $y = 40x + 50$ for x between 0 and 8

73. $y = 100 - 20x$ for x between 0 and 5

74. $y = 60 - 4x$ for x between 0 and 15

75. $y = -25x + 150$ for x between 0 and 6

76. $y = -3x + 60$ for x between 0 and 10

77. Graph $y = 2x + 3$ and $x + y = 0$ on the same coordinate system. Estimate the coordinates of the point where the lines intersect. Substitute these coordinates into both equations and see if the equations are satisfied.

78. Repeat Exercise 77 with $2x + y = 4$ and $y = 3x - 1$.

79. Graph $y = \frac{1}{2}x + b$ for $b = -4$, $b = 0$, and $b = 4$, all on the same coordinate system.

80. Graph $y = mx + 1$ for $m = -2$, $m = -\frac{1}{2}$, $m = 0$, $m = \frac{1}{2}$, and $m = 2$, all on the same coordinate system.

81. Graph $y = mx + 3$ for $m = -3$, $m = -\frac{1}{3}$, $m = 0$, $m = \frac{1}{3}$, $m = 3$, all on the same coordinate system.

82. Graph $y = 2x + b$ for $b = -5$, $b = 0$, $b = 5$, all on the same coordinate system.

83. Graph $y = |x|$. [*Hint:* Graph $y = x$ for $x \geq 0$ and $y = -x$ for $x < 0$.]

84. Graph $y = -|x|$.

85. Graph $y = -2x$ and $y = \frac{1}{2}x$ on the same coordinate system. Do the lines appear perpendicular?

86. Graph $y = 2x$ and $y = -\frac{1}{2}x$ on the same coordinate system. Do the lines appear perpendicular?

APPLICATIONS

Problems 87–94 refer to the following graphs:

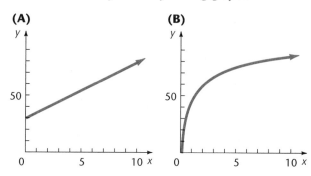

(A)

(B)

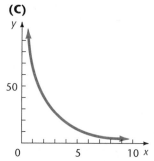

(C)

87. For which of the graphs (A), (B), (C) do the values of y get larger as x increases?

88. For which of the graphs (A), (B), (C) do the values of y get smaller as x increases?

89. Estimate the value of y corresponding to $x = 3$ in graph (C).

90. Estimate the value of y corresponding to $x = 2$ in graph (B).

91. *Business* Graph (A) might come from a problem involving production costs. Suppose x represents the number of units produced in thousands and y the cost in thousands of dollars.
  (A) Estimate the cost associated with producing 5,000 units.
  (B) Estimate how many units are produced if the cost is $70,000.

(C) The graph shows a cost of $30,000 associated with production of 0 units. What might account for this?

92. Business Graph (A) might also represent the costs associated with renting a small bulldozer for a day. Let x represent the number of hours for which the bulldozer is rented and y the cost in dollars.
   (A) Estimate the cost associated with renting the bulldozer for 6 hours.
   (B) Estimate how many hours of rental time are used if the cost is $60.
   (C) There is a cost of $30 associated with 0 hours rental. What might account for this?

93. Education Graph (B) might be used to represent measurements of learning over time. Suppose x represents weeks of intense foreign language study and y the percent achieved on a test of proficiency.
   (A) When does the proficiency increase the fastest, in the first weeks or later?
   (B) Estimate how long it takes to achieve a proficiency score of 70%.

94. Economics Graph (C) might portray a relationship between price and demand. Let x represent the number of units of a commodity in hundreds of tons and y the price per ton in hundreds of dollars.
   (A) What happens to the price as the demand increases?
   (B) What price corresponds to a demand of 5,000 tons?

Choose horizontal and vertical scales to produce maximum clarity in the graphs:

95. Earth science As dry air moves upward, it expands and in doing so cools at the rate of about 5.5°F for each 1,000 feet in rise (see Problem 23, Section 3-8). If the ground temperature is 80°F, the equation

$$T = 80 - 5.5h$$

relates temperature T with altitude h measured in 1,000-foot units. Graph this equation for h between 0 and 20.

96. Earth science Pressure in seawater increases by 1 atmosphere (15 pounds per square inch) for each 33 feet of depth; it is 15 pounds per square inch at the surface. Thus, $p = 15 + 15(d/33)$, where p is pressure in pounds per square inch at a depth of d feet below the surface (see Problem 21, Section 3-8). Graph this equation for d between 0 and 330.

97. Finance Simple interest is calculated by the formula $A = P + Prt$, where P represents the principal, or amount borrowed; A represents the amount due,

including both principal and interest; r represents the annual interest rate; and t represents the time in years over which interest accumulates. For $P = $1,000$ and $r = 9\% = 0.09$, the formula becomes $A = 1000 + 90t$. Graph this equation for values of t between 0 and 6.

98. Finance Repeat Problem 97 for $P = $2,400$ and $r = 10\%$. Graph the resulting equation for values of t between 0 and 5.

99. Finance If an asset initially worth P dollars is allowed to depreciate at an annual rate of $100r$ percent, its value V after t years is given by $V = P - Prt$. If a $100,000 building is depreciated at an annual rate of 12%, the formula becomes $V = 100,000 - 12,000t$. Graph this equation for values of t between 0 and 8.

100. Finance Repeat Problem 99 for a $20,000 automobile depreciated at 15%. Graph the resulting equation for values of t between 0 and 6.

101. Physics Fahrenheit temperature F and Celsius temperature C are related by the formula $F = 1.8C + 32$. Graph this equation for values of C between 0 and 100.

102. Physics Fahrenheit temperature F and Celsius temperature C are also related by the formula

$$C = \frac{5}{9}\left(F - 32\right)$$

Graph this equation for values of F between 32 and 212.

103. Life science Springtime phenomena, such as flowers blooming and certain species of insects arriving, occur later at higher altitudes. How much later is given approximately by the formula $d = h/125$, where d is in days and the difference in altitude h is in feet. Graph this equation for values of h between 0 and 8,000.

104. Life science The formula $w = 5.5h - 220$ gives an approximate normal weight w in pounds for a person of height h in inches, provided the person is at least 5 feet tall. (See Problem 13, Section 3-8.) Graph this equation for heights between 60 and 80 inches.

105. Psychology IQ (intelligence quotient) is related to chronological age CA and mental age MA by the formula $MA = 0.01IQ \cdot CA$. (See Problem 46, Section 3-8.) Graph this equation for 18-year-old people ($CA = 18$) and values of IQ from 80 to 150.

106. *Psychology* In experiments a psychologist discovered this relationship between the pull p in grams that a rat exerts on a spring harness and the distance d in centimeters the rat is placed from a reward:

$$p = -0.2d + 70$$

Graph this equation for distances between 10 and 30 centimeters.

  # 4-3 Slope of a Line

♦ Slope of a Line
♦ Slope-Intercept Form
♦ Parallel and Perpendicular Lines
♦ Application

From Section 4-2 we know that every linear equation has a graph that is a straight line, and given a particular linear equation, we can graph it. In this section we will consider the reverse situation: given a straight line, or at least certain information about it, find an equation to describe it.

Two points are sufficient to determine a line. It is also enough to know one point and the "direction" or "steepness" of the line. The notion of direction or steepness is measured as the *slope* of the line and is considered in this section.

♦ SLOPE OF A LINE

Given two points on a line, say, the points $(-3, -3)$ and $(2, 7)$ on the line $y = 2x + 3$, we define the **slope** of the line as the ratio of the change in y to the change in x as we move from one point to the other. For the given points, moving from $(-3, -3)$ to $(2, 7)$, the slope would be

$$\frac{\text{Change in } y}{\text{Change in } x} = \frac{7 - (-3)}{2 - (-3)} = \frac{10}{5} = 2$$

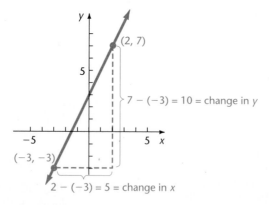

Figure 1

If we reverse the order of the two points, we calculate

$$\frac{-3 - 7}{-3 - 2} = \frac{-10}{-5} = 2$$

If we use the point $(0, 3)$ on the line instead of $(-3, -3)$, we calculate

$$\frac{7-3}{2-0} = \frac{4}{2} = 2$$

Thus it appears that no matter what two points on the line we choose, and no matter in what order we choose them, we obtain the same ratio, namely, 2. The slope of this line is therefore 2. The same thing happens in general, so we can define slope:

Slope Formula

If a line passes through $P_1(x_1, y_1)$ and $P_2(x_2, y_2)$, then its slope is given by the formula

$$m = \frac{y_2 - y_1}{x_2 - x_1} \qquad x_1 \neq x_2$$

$$= \frac{\text{Vertical change (rise)}}{\text{Horizontal change (run)}}$$

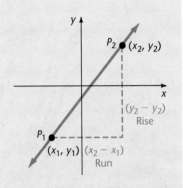

CAUTION

Note that we must be consistent: whichever y component we use first in the numerator, we must use the x coordinate of the same point first in the denominator. That is,

$$m \neq \frac{y_2 - y_1}{x_1 - x_2}$$

The reason that any two points on a line will give the same value for the slope is that all triangles like that drawn for the line in the definition are similar. The ratio of the vertical to the horizontal side is therefore constant.

Example 1
Slope

Find the slope of the line passing through $(2, 1)$ and $(4, -3)$.

Solution Let $(x_1, y_1) = (2, 1)$ and $(x_2, y_2) = (4, -3)$. Then

$$\text{Slope} = m = \frac{y_2 - y_1}{x_2 - x_1} = \frac{-3 - 1}{4 - 2} = \frac{-4}{2} = -2$$

Matched Problem 1 Find the slope of the line passing through $(-3, 2)$ and $(1, 0)$

$$\frac{0-2}{1-3} = \frac{-2}{4} = \frac{1}{2}$$

Example 2
Slope

Find the slope of the line passing through $(-2, -5)$ and $(3, -1)$.

Solution

$$m = \frac{-1 - (-5)}{3 - (-2)} = \frac{4}{5}$$

Matched Problem 2   Find the slope of the line passing through $(-1, 4)$ and $(-7, -4)$.

For a vertical line, the slope is not defined. Such a line has equation $x = a$, so $x_1 = x_2 = a$. Therefore, $x_2 - x_1 = 0$, and we cannot divide by zero in

$$\frac{y_2 - y_1}{x_2 - x_1}$$

On the other hand, for a horizontal line $y = b$, $y_1 = y_2 = b$, $y_2 - y_1 = 0$, and the slope is 0.

In general, the slope of a line may be positive, negative, 0, or not defined. Each of these cases is interpreted geometrically as shown in Table 1.

Table 1   **Going from Left to Right**

Line	Slope	Example
Rising	Positive	
Falling	Negative	
Horizontal	0	
Vertical	Not defined	

Thus, the sign of the slope indicates whether the line is falling or rising from left to right. The magnitude of the slope indicates how steep the line is, as shown in the following figure:

As a horizontal line is rotated counterclockwise, the resulting lines rise from left to right and have positive slope. The magnitude of the slopes and the steepness of the lines both increase.

As a horizontal line is rotated clockwise, the lines fall from left to right. The slopes are negative but the magnitude (absolute value) of the slopes gets larger. The lines fall more steeply.

Figure 2

◆ SLOPE-INTERCEPT FORM

In our discussion at the beginning of this section, we found that the slope of the line $y = 2x + 3$ is 2. The fact that the slope and the coefficient of x are the same is not coincidence: the constant m in the form $y = mx + b$ is always equal to the slope of the line.

The constant b in the form $y = mx + b$ also has geometric significance. If we let $x = 0$, then

$$y = m \cdot 0 + b = 0 + b = b$$

Thus, b is the y coordinate when $x = 0$; that is, b is the y intercept. Therefore, in the example $y = 2x + 3$, the y intercept is 3, as we already knew from Example 1 in Section 4-2.

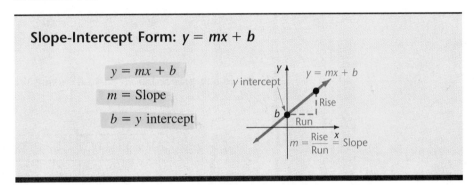

Slope-Intercept Form: $y = mx + b$

$$y = mx + b$$
$$m = \text{Slope}$$
$$b = y \text{ intercept}$$

Example 3
Slope-Intercept Form

(A) Find the slope and y intercept for the line $y = 3x - 1$.

(B) Find an equation for the line with slope $-\frac{2}{3}$ and y intercept 1.

Solution

(A) $y = 3x - 1$
 $y = 3x + (-1)$
 $\qquad\;\; \uparrow \qquad\quad \uparrow$
 $\qquad\text{Slope} \qquad y \text{ intercept}$

(B) Since $m = -\frac{2}{3}$ and $b = 1$, then $y = mx + b$ becomes $y = -\frac{2}{3}x + 1$.

Matched Problem 3

(A) Find the slope and y intercept for the line $y = \frac{2}{3}x + 2$.

(B) Find an equation for the line with slope $\frac{1}{2}$ and y intercept -1.

In Example 3(B), we found the equation of the line given the slope and y intercept. In this case the line was specified by one point, the y intercept, and its direction (slope). Frequently, however, the given point will not be the y intercept. Also, the line may be specified by two points instead of a point and direction. These situations are only slightly more difficult to handle. The following example shows how to proceed:

Example 4
Equations for a Line

(A) Find the equation of the line that passes through the point $(1, 3)$ and has slope $-\frac{1}{3}$.

(B) Find the equation of the line that passes through $(1, 3)$ and $(4, 1)$.

Solution **(A)** Since the slope is $-\frac{1}{3}$, the equation must have the form

$$y = -\tfrac{1}{3}x + b$$

To find b, use the fact that $(1, 3)$ lies on the line. Thus

$$3 = -\tfrac{1}{3} \cdot 1 + b$$
$$3 = -\tfrac{1}{3} + b$$
$$\tfrac{10}{3} = b$$

Therefore, the equation is $y = -\tfrac{1}{3}x + \tfrac{10}{3}$.

(B) First find the slope m:

$$m = \frac{y_2 - y_1}{x_2 - x_1} = \frac{1 - 3}{4 - 1} = \frac{-2}{3}$$

Now we can proceed as in part (A), using the coordinates of either point to find b:

Using $(1, 3)$	Using $(4, 1)$
$y = -\tfrac{2}{3}x + b$	$y = -\tfrac{2}{3}x + b$
$3 = -\tfrac{2}{3} \cdot 1 + b$	$1 = -\tfrac{2}{3} \cdot 4 + b$
$\tfrac{11}{3} = b$	$\tfrac{11}{3} = b$

The equation is therefore $y = -\tfrac{2}{3}x + \tfrac{11}{3}$.

Matched Problem 4 **(A)** Find the equation of the line that passes through the point $(1, 2)$ and has slope $\tfrac{3}{4}$.

(B) Find the equation of the line that passes through $(-2, 1)$ and $(3, 4)$.

The above methods for finding an equation from various given information can be summarized as follows:

Finding Equations of Lines

1. If given the slope m and the y intercept b, the equation is $y = mx + b$.
2. If given the slope m and one point, substitute the point into $y = mx + b$ and solve for b. The equation is $y = mx + b$.
3. If given two points, use them to find m and then proceed as in 2.

An equation in slope-intercept form allows us to graph its line very efficiently, as the following example shows.

Example 5
Graphing from Slope-Intercept Form

Graph $y = \frac{3}{4}x - 2$.

Solution

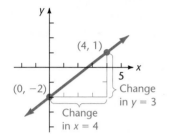

Figure 3

The y intercept is -2 so the point $(0, -2)$ is on the graph. Since the slope is

$$\frac{\text{Change in } y}{\text{Change in } x} = \frac{3}{4}$$

we know that if we increase x by 4, y must increase by 3. Starting from $(0, -2)$, this yields the point $(4, 1)$, and the line can be graphed as shown in Figure 3.

Matched Problem 5

Graph $y = -\frac{1}{2}x + 1$.

Any linear equation with coefficient of y not zero can be rewritten in the form $y = mx + b$ by dividing by the coefficient of y and shifting all terms except y to the other side. For example, if the equation is $3x + 4y = 5$, then

$$\frac{3}{4}x + y = \frac{5}{4}$$
$$y = -\frac{3}{4}x + \frac{5}{4}$$

If the coefficient of y is zero, the line must be vertical and is also easily graphed.

There are a number of other standard forms for the equations of straight lines. One follows directly from the definition of the slope. If a line with slope m passes through the point (x_1, y_1), then the line is represented by the equation

$$\frac{y - y_1}{x - x_1} = m$$

or

$$y - y_1 = m(x - x_1)$$

This last form is called the **point-slope form** for the line. It will be studied further in subsequent courses.

♦ **PARALLEL AND PERPENDICULAR LINES**

Since the slope of a line gives an indication of direction, it should not be surprising that the slopes can provide information about parallel or perpendicular lines.

1. For two lines to be parallel, they must have the same slope.
2. Lines having different slopes will intersect.
3. A vertical line is perpendicular only to a horizontal line. For two nonvertical lines to be perpendicular, it can be shown that the product of their slopes must be -1.

♦ APPLICATION

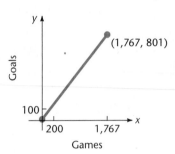

Figure 4

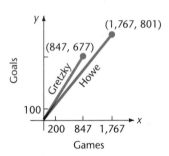

Figure 5

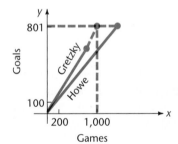

Figure 6

An example will illustrate how the slope of a line can convey useful information. Hockey player Gordie Howe holds the National Hockey League goal scoring record. In a 26-year career, Howe scored 801 goals in 1,767 games. If we plot the pair (games, goals) for the start and end of Howe's career, we get the points $(0, 0)$ and $(1,767, 801)$. The line segment connecting these points is shown in Figure 4. The slope of this segment is $801/1767 \approx 0.45$. This slope represents the ratio goals/games or *goals per game*. That is, the slope gives a rate at which Howe scored goals.

Now also consider the record of a current player, Wayne Gretzky, who scored 677 goals in his first 847 games. Plotting the segment for Gretzky on the same graph with Howe's we obtain Figure 5. From the figure, it is apparent that Howe scored more goals since his segment ends at a higher y value. However, looking at the slopes, it is clear that Gretzky is scoring at a higher rate, since his slope is larger, and he will overtake Howe if he plays long enough and continues scoring at this rate. The picture also suggests how long this might take. Extending Gretzky's line until it reaches a y value of 801, we see that the corresponding x value is approximately 1,000. See Figure 6. The exact value is obtained by solving the equation

$$801 = \frac{677}{847}x$$

This gives $x = 1,002.137 \cdots$. Thus, at this rate, Gretzky should surpass Howe in his 1,003d game.

The applications at the end of Exercise 4-3 include questions where you are asked to consider what the slope represents in various contexts.

Answers to Matched Problems

1. $-\frac{1}{2}$ 2. $\frac{4}{3}$ 3. **(A)** Slope $= \frac{2}{3}$, y intercept $= 2$ **(B)** $y = \frac{1}{2}x - 1$
4. **(A)** $y = \frac{3}{4}x + \frac{5}{4}$ **(B)** $y = \frac{3}{5}x + \frac{11}{5}$
5.

EXERCISE 4-3

A *Find the slope of the line determined by the given points:*

1. $(2, 3)$ and $(4, 5)$ 2. $(3, 2)$ and $(6, 5)$

3. $(1, 5)$ and $(3, 9)$ 4. $(0, 2)$ and $(3, 4)$

5. $(\frac{1}{3}, \frac{1}{2})$ and $(\frac{1}{2}, \frac{5}{6})$ 6. $(\frac{1}{4}, \frac{2}{5})$ and $(\frac{1}{3}, \frac{4}{5})$

7. $(\frac{1}{5}, \frac{1}{5})$ and $(\frac{1}{2}, \frac{1}{3})$ 8. $(\frac{1}{2}, \frac{1}{2})$ and $(\frac{2}{3}, \frac{3}{5})$

Find the slope and y intercept of the line with the given equation:

9. $y = 3x + 5$ 10. $y = 2x - 7$

11. $y = -2x + 4$

12. $y = -2x - 1$

13. $y = \frac{1}{5}x - \frac{2}{5}$

14. $y = -\frac{1}{10}x + \frac{7}{10}$

15. $y = -0.35x + 0.8$

16. $y = 0.20x - 1.5$

Find the slope-intercept form of the equation for the line with slope and y intercept as given:

17. slope $= 3$, y intercept $= -1$

18. slope $= 2$, y intercept $= 5$

19. slope $= -4$, y intercept $= 1$

20. slope $= -1$, y intercept $= -4$

21. slope $= \frac{2}{3}$, y intercept $= \frac{1}{2}$

22. slope $= -\frac{1}{2}$, y intercept $= \frac{2}{3}$

23. slope $= -3.5$, y intercept $= 6.25$

24. slope $= 7.5$, y intercept $= -0.75$

Find the slope-intercept form of the equation for the line that passes through the given point and has the indicated slope:

25. point $(1, 2)$; $m = 2$

26. point $(4, 3)$; $m = 3$

27. point $(1, 2)$; $m = -2$

28. point $(4, 3)$; $m = -3$

29. point $(\frac{1}{4}, \frac{1}{5})$; $m = \frac{1}{2}$

30. point $(\frac{2}{3}, \frac{1}{4})$; $m = \frac{1}{5}$

31. point $(1.2, 3.4)$; $m = 2.5$

32. point $(-1.5, 3.0)$; $m = -4.5$

Find the slope-intercept form of the equation of the line that passes through the given two points:

33. $(2, 3)$ and $(4, 5)$

34. $(3, 2)$ and $(6, 5)$

35. $(1, 5)$ and $(3, 9)$

36. $(0, 2)$ and $(3, 4)$

37. $(\frac{1}{3}, \frac{1}{2})$ and $(\frac{1}{2}, \frac{5}{6})$

38. $(\frac{1}{4}, \frac{2}{5})$ and $(\frac{1}{3}, \frac{4}{5})$

39. $(\frac{1}{5}, \frac{1}{5})$ and $(\frac{1}{2}, \frac{1}{3})$

40. $(\frac{1}{2}, \frac{1}{2})$ and $(\frac{3}{5}, \frac{3}{5})$

B *Find the slope of the line determined by the given points:*

41. $(1, 2)$ and $(5, -1)$

42. $(4, 1)$ and $(-1, 7)$

43. $(-1, 1)$ and $(4, -4)$

44. $(3, -1)$ and $(-2, 3)$

45. $(\frac{1}{4}, 1)$ and $(\frac{1}{2}, -1)$

46. $(-\frac{2}{5}, 3)$ and $(\frac{1}{5}, 1)$

47. $(-2.6, 1.8)$ and $(3.9, -2.1)$

48. $(3.1, -6.2)$ and $(-2.1, 4.2)$

Find the slope and y intercept of the line with the given equation:

49. $2x - 3y = 4$

50. $-2x + 5y = -7$

51. $-3x + 5y = 4$

52. $4x - 3y = 2$

53. $\frac{1}{2}x - \frac{1}{3}y = 1$

54. $\frac{1}{4}x + \frac{1}{5}y = 1$

55. $3.2x + 4.8y = 2.4$

56. $-2.5x + 5y = 7.5$

Find the slope-intercept form of the equation for the line that passes through the given point and has the indicated slope:

57. $(1, 3)$; $m = \frac{1}{5}$

58. $(2, 3)$; $m = -\frac{3}{4}$

59. $(\frac{1}{3}, 1)$; $m = -\frac{1}{2}$

60. $(1, \frac{2}{3})$; $m = \frac{1}{2}$

61. $(3, 1.4)$; $m = 0.8$

62. $(-2.6, 1.4)$; $m = 4.4$

63. $(-3.2, -8.2)$; $m = -4$

64. $(2, -3.6)$; $m = -1.6$

Find the slope-intercept form of the equation for the line that passes through the given two points:

65. $(1, 2)$ and $(5, -1)$

66. $(4, 1)$ and $(-1, 7)$

67. $(-1, 1)$ and $(4, -4)$

68. $(3, -1)$ and $(-2, 3)$

69. $(\frac{1}{4}, 1)$ and $(\frac{1}{2}, -1)$

70. $(-\frac{2}{5}, 3)$ and $(\frac{1}{5}, 1)$

71. $(-2.6, 1.8)$ and $(3.9, -2.1)$

72. $(3.1, -6.2)$ and $(-2.1, 4.2)$

Graph:

73. $y = 3x + 5$

74. $y = 2x - 7$

75. $y = -2x + 4$

76. $y = -2x - 1$

77. $2x - 3y = 4$

78. $-2x + 5y = -7$

79. $-3x + 5y = 4$

80. $4x - 3y = 2$

81. $3.6x - 9y = 7.2$

82. $4.5x + 7.5y = -6$

C **83.** $y = \frac{1}{3}x + 1$

84. $y = \frac{1}{2}x - 1$

85. $y = -\frac{1}{2}x + 2$

86. $y = -\frac{1}{3}x + 1$

87. $y - 3x = 1$

88. $y + x = -1$

89. $x - 2y = 6$

90. $-x + 3y = 6$

91. $x + \frac{1}{2}y = 3$

92. $2x - \frac{1}{3}y = 1$

93. $3x - \frac{1}{2}y = 2$

94. $x + \frac{2}{3}y = 2$

APPLICATIONS

95. *Earth science* Pressure in seawater increases by 1 atmosphere (15 pounds per square inch) for each 33 feet of depth; it is 15 pounds per square inch at the surface. Thus,

$$p = 15 + 15\left(\frac{d}{33}\right) = 15 + \frac{15}{33}d$$

where p is pressure in pounds per square inch at a depth of d feet below the surface (see Problem 21, Section 3-8). What is the slope of this line, and how is it related to the physical problem?

96. *Earth science* As dry air moves upward, it expands and in doing so cools at the rate of about 5.5°F for each 1,000 feet in rise (see Problem 23, Section 3-8). Let the ground temperature be 80°F.

(A) If the altitude h is expressed in feet, the equation

$$T = -\frac{5.5}{1,000}h + 80$$

relates temperature and altitude. What is the slope of this line, and how is it related to the physical problem?

(B) If the altitude is expressed in 1,000-foot units, the equation becomes

$$T = -5.5h + 80$$

What is the slope of this line, and how is it related to the physical problem?

97. *Business* A music store sells compact discs that cost $11.00 for $17.50, and tape cassettes that cost $9.80 for $15.70.

(A) If the markup policy of the store for items is assumed to be linear and is reflected in the pricing of these two items, write an equation that relates retail price R with cost C.

(B) Find the slope and y intercept for the line determined by this equation.

(C) The store's markup policy depends on the cost of an item, a fixed charge that is the same for all items, and a percentage of the cost. Describe the exact policy in words. How are the slope and y intercept related to this policy?

98. *Business* The management of a small copy center estimates the cost of business at $400 per day at zero output and $1,000 per day at an output of 10,000 copies.

(A) Assuming that total cost per day C is linearly related to total output per day x, write an equation relating these two quantities.

(B) What is the slope of the line determined by this equation, and how is it related to the problem?

99. *Physics* Water freezes at 32°F and 0°C, and boils at 212°F and 100°C. The relationship between Fahrenheit degrees F and Celsius degrees C is linear.

(A) Use two points to write an equation in the form $F = mC + b$ for this relationship.

(B) Use two points to write a linear equation in the form $C = mF + b$ for this relationship.

(C) Solve the equation you obtained in part (A) for C in terms of F. You should obtain the same equation you obtained in part (B).

100. *Physics* See Problem 99. In the Kelvin, or absolute, scale, water freezes at 273.15 K and boils at 373.15 K. The relationship between Kelvin degrees K and Celsius degrees C is linear.

(A) Use two points to write an equation in the form $K = mC + b$ for this relationship.

(B) Use two points to write a linear equation in the form $C = mK + b$ for this relationship.

(C) Solve the equation you obtained in part (A) for C in terms of K. You should obtain the same equation you obtained in part (B).

101. *Sports* Baseball player Pete Rose holds the major league record for most hits, 4,256 in 3,562 games. Robin Yount, still playing, had 2,878 hits in his first 2,579 games. Graph this information as was done in the Howe-Gretzky example in the text. What does the slope represent in this case? How many games would Yount have to play, continuing at the same rate, to surpass Rose's record?

102. *Sports* Basketball player Kareem Abdul-Jabbar holds the National Basketball Association record for most points in a career, 38,387 in 1,560 games. Michael Jordan, still playing, had 14,016 points in his first 427 games. Graph this information as was done in the Howe-Gretzky example in the text. What does the slope represent in this case? How many games would Jordan have to play, continuing at the same rate, to surpass Abdul-Jabbar's record?

4-4 Systems of Equations and Solving by Graphing

♦ Systems of Equations
♦ Solving by Graphing

In Section 2-8, we considered the following problem: find the dimensions of a rectangle with perimeter 52 centimeters if its length is 5 more than twice its width. We solved the problem by letting x be the width, $2x + 5$ the length, and then solving the equation $2(2x + 5) + 2x = 52$. The width turned out to be 7 and the length 19. We set the problem up in terms of one variable x, but we could also have set it up using two variables as follows:

Let x = width, y = length.
Then

$$2x + 2y = 52 \qquad \text{Perimeter is 52}$$

$$y = 2x + 5 \qquad \text{Length is 5 more than twice width}$$

In this way we obtain a **system of equations**—two equations in two variables. We consider such systems in the next three sections.

◆ **SYSTEMS OF EQUATIONS**

A **solution of a system of equations** such as

$$2x + 2y = 52$$

$$y = 2x + 5$$

is an ordered pair of real numbers that satisfies both equations. Thus, $(7, 19)$ is a solution to this system, since

$$2 \cdot 7 + 2 \cdot 19 = 52$$

$$19 = 2 \cdot 7 + 5$$

To solve the system of equations means to find all such pairs. We will see below that $(7, 19)$ is the only solution to this system. More generally, we will consider systems of two linear equations in two variables, that is, systems of the type

$$ax + by = m$$

$$cx + dy = n$$

where a, b, c, d, m, and n are real-number constants and x and y are variables.

◆ **SOLVING BY GRAPHING**

Since each equation in our system represents a straight line, the system represents two straight lines. Geometrically, two different lines are parallel or they intersect at one point. In some cases the two lines in the system might not be different; that is, the two equations might represent the *same* line. We can determine the nature of the solution of a system of equations and get an approximation to the solution itself by graphing the lines.

Example 1
Solving by Graphing

Solve by graphing:

$$2x + 2y = 52$$

$$y = 2x + 5$$

Solution   Graph the two equations on the same coordinate system. It is clear that the lines intersect, so there is exactly one solution. The intersection point appears to be (7, 19). We already know this is the solution, as we have checked by substituting $x = 7$ and $y = 19$ into both equations.

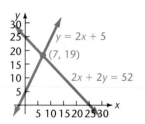

In general, we cannot hope to read the exact solution from the graph. If the solution in Example 1 had been (7.01, 18.93), we would not have found it exactly by graphing but we could have approximated it. A good graph can, however, suggest integer solutions.

Figure 1

Matched Problem 1   Solve by graphing and check. The solution has integer values.

$$2x + 3y = 12 \qquad 3y = 2x + 12$$
$$3x - 4y = 1 \qquad\qquad y = \frac{2}{3}$$

The three systems in Figure 2 show all three solution possibilities for a system of two equations in two unknowns:

(A) $2x - 3y = 6$ **(B)** $4x - 6y = -12$ **(C)** $2x - 3y = -6$
     $-3x + \;y = 5$ $2x - 3y = 6$ $-\frac{1}{3}x + \frac{1}{2}y = 1$

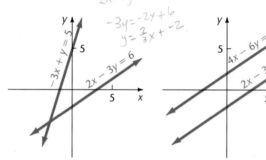

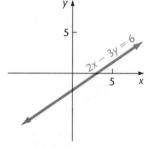

Lines intersect at one point only: *exactly one solution* $x = -3$, $y = -4$

Lines are parallel: *no solution*

Lines coincide: *infinite number of solutions*

Figure 2

The nature of the solution can be determined by considering the slopes of the lines involved. Rewrite each equation in the form $y = mx + b$ and recall that m represents the slope:

(A) $\begin{cases} 2x - 3y = 6 & y = \frac{2}{3}x - 2 & \text{Slope} = \frac{2}{3} \\ -3x + \;y = 5 & y = 3x + 5 & \text{Slope} = 3 \end{cases}$

Since the slopes are different, the lines intersect at one point.

(B) $\begin{cases} 4x - 6y = -12 & y = \frac{2}{3}x + 2 & \text{Slope} = \frac{2}{3} \\ 2x - 3y = 6 & y = \frac{2}{3}x - 2 & \text{Slope} = \frac{2}{3} \end{cases}$

Since the slopes are equal, the lines are parallel. Since the lines have different y intercepts, they will never meet.

$$\text{(C)} \begin{cases} 2x - 3y = -6 & y = \tfrac{2}{3}x + 2 \\ -\tfrac{1}{3}x + \tfrac{1}{2}y = 1 & y = \tfrac{2}{3}x + 2 \end{cases}$$

Both equations can be transformed into the same one, so both represent the same line.

Now we know exactly what to expect when solving a system of two linear equations in two unknowns:

(A) Exactly one pair of numbers as a solution (lines intersecting at exactly one point)
(B) No solutions (parallel lines)
(C) An infinite number of solutions (two equations representing the same line)

In most applications the first case prevails. If we find a pair of numbers that satisfies the system of equations and the graphs of the equations meet at only one point, then that pair of numbers is the only solution of the system, and we need not look further for others.

The graphing method of solving systems of equations yields considerable information about the solution to a system of two linear equations in two variables. Moreover, graphs frequently reveal relationships in problems that would otherwise be hidden. On the other hand, if one is interested in precise solutions, approximation by the graphing method is not practical. The methods of elimination, to be considered in the next two sections, will take care of this deficiency.

Answers to Matched Problem   **1.** $x = 3,\ y = 2$

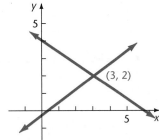

(3, 2)

EXERCISE 4-4

Solve by graphing and check your answers. In all systems with unique solutions, the solution will have integer values.

A
1. $x + y = 5$
$\quad x - y = 1$

2. $x + y = 6$
$\quad x - y = 2$

3. $x + y = 5$
$\quad 2x - y = 4$

4. $2x + y = 6$
$\quad x - y = -3$

5. $x - 3y = 3$
$\quad x - y = 7$

6. $x - 2y = -4$
$\quad 2x - y = 10$

7. $-x + y = 6$
$\quad x - y = 8$

8. $-x + y = 4$
$\quad x - y = 2$

9. $x - y = 7$
$\quad -x + y = -7$

10. $-x + y = 3$
$\quad x - y = -6$

11. $3x + y = 6$
$\quad x - y = -2$

12. $4x - y = 8$
$\quad -x + 2y = 5$

13. $-x + 2y = 6$
$\quad 5x - 5y = -15$

14. $4x + y = 24$
$\quad x - y = 6$

15. $3x - y = 9$
$\quad -4x + y = -12$

16. $-x + 5y = 5$
$\quad x - y = -5$

17. $4x - 8y = 40$
$\quad -2x + 4y = 12$

18. $-x + 3y = 12$
$\quad 2x + 6y = -12$

B 19. $3x - y = 2$
$\quad x + 2y = 10$

20. $x - 2y = 2$
$\quad 2x + y = 9$

21. $-2x + 3y = 12$
$\quad 2x - y = 4$

22. $3x - 2y = 12$
$\quad 7x + 2y = 8$

C 39. $\frac{1}{3}x + \frac{1}{2}y = \frac{7}{2}$
$\quad \frac{1}{6}x + \frac{1}{5}y = \frac{5}{2}$

40. $\frac{2}{3}x + \frac{1}{4}y = \frac{1}{2}$
$\quad \frac{1}{2}x + \frac{1}{5}y = 0$

23. $-3x + y = 9$
$\quad 3x + 4y = -24$

24. $x + 5y = -10$
$\quad -x + 2y = 3$

41. $\frac{3}{4}x + \frac{1}{2}y = 0$
$\quad \frac{1}{2}x + \frac{1}{3}y = 0$

42. $\frac{1}{5}x + \frac{1}{4}y = 1$
$\quad \frac{1}{10}x + \frac{1}{4}y = \frac{7}{2}$

25. $x + 2y = 4$
$\quad 2x + 4y = -8$

26. $3x + 5y = 15$
$\quad 6x + 10y = -30$

43. $\frac{1}{6}x + \frac{1}{4}y = 1$
$\quad \frac{1}{3}x + \frac{1}{4}y = \frac{5}{2}$

44. $\frac{1}{3}x + \frac{1}{2}y = \frac{2}{3}$
$\quad \frac{1}{4}x + \frac{3}{4}y = \frac{1}{2}$

27. $\frac{1}{2}x - y = -3$
$\quad -x + 2y = 6$

28. $3x - 5y = 15$
$\quad x - \frac{5}{3}y = 5$

45. $\frac{1}{6}x - \frac{1}{2}y = \frac{11}{2}$
$\quad \frac{1}{2}x + \frac{1}{5}y = -\frac{1}{2}$

46. $\frac{1}{4}x - \frac{1}{2}y = \frac{9}{4}$
$\quad -\frac{1}{3}x + \frac{1}{6}y = -\frac{5}{2}$

29. $-x + 2y = 7$
$\quad 2x - y = -2$

30. $-4x + y = -6$
$\quad 3x - y = 3$

47. $-\frac{2}{3}x - \frac{1}{2}y = 0$
$\quad \frac{1}{5}x - \frac{1}{2}y = \frac{13}{5}$

48. $-\frac{3}{5}x - \frac{1}{5}y = \frac{6}{5}$
$\quad x + \frac{1}{3}y = -2$

31. $4x - 10y = -10$
$\quad -x + 2y = -3$

32. $3x + 2y = 8$
$\quad 2x + y = 6$

Use slopes and intercepts to determine whether each pair of equations represents intersecting lines, parallel lines, or the same line. Do not solve the systems.

33. $-2x + 5y = z$
$\quad 3x - y = -11$

34. $5x - 2y = -2$
$\quad -x + 2y = 10$

49. $y = 3x + 5$
$\quad x - \frac{1}{3}y = 8$

50. $y = 6x - 1$
$\quad 3x - \frac{1}{2}y = \frac{1}{2}$

35. $6x + 2y = 4$
$\quad -3x - y = -2$

36. $-4x + y = 6$
$\quad 2x - \frac{1}{2}y = 3$

51. $y = -4x + 12$
$\quad \frac{1}{2}x + \frac{1}{8}y = \frac{3}{2}$

52. $y = 6x - 3$
$\quad 2x - \frac{1}{3}y = 1$

37. $5x - 10y = -10$
$\quad -\frac{1}{2}x + y = 1$

38. $2x + \frac{1}{2}y = -5$
$\quad -4x - y = 10$

53. $y = 2x + 3$
$\quad x + \frac{1}{2}y = -3$

54. $y = -3x - 5$
$\quad \frac{2}{5}x - \frac{1}{5}y = 1$

4-5 Solving Systems Using Elimination by Substitution

We now consider a method—called **elimination by substitution** or, more simply, just **substitution**—for solving systems of linear equations exactly. This method will produce solutions, if they exist, to any decimal accuracy desired. The method is most useful when one of the equations in the system can easily be solved for one variable in terms of the other. Examples should make the process clear.

Example 1
Solving by Substitution

Solve by substitution and check:

$$y = 2x - 3$$
$$2x - 3y = 1$$

Solution Since in the first equation y is already expressed in terms of x, we substitute the right side of this equation for y in the second equation and then solve for x. After finding x, we substitute it into the first equation to find y.

$$y = 2x - 3 \quad \text{Substitute } 2x - 3 \text{ for } y \text{ in the second equation.}$$

$$2x - 3y = 1$$

$$2x - 3(2x - 3) = 1 \quad \text{Now solve this equation for } x.$$

$$2x - 6x + 9 = 1$$

$$-4x + 9 = 1$$

$$-4x = -8$$

$$\boxed{x = 2}$$

Replace x with 2 in the first equation, $y = 2x - 3$, to find y:

$$y = 2x - 3$$
$$y = 2(2) - 3$$
$$\boxed{y = 1}$$

The solution is $x = 2$, $y = 1$.

Check

$$y = 2x - 3 \qquad\qquad 2x - 3y = 1$$
$$(1) \overset{?}{=} 2(2) - 3 \qquad 2(2) - 3(1) \overset{?}{=} 1$$
$$1 \overset{\checkmark}{=} 1 \qquad\qquad\qquad 1 \overset{\checkmark}{=} 1$$

It is not sufficient to check just one equation. The solution must satisfy *both* equations. You can also get a visual check of your solution by graphing the two lines. They should cross at the point (2, 1).

Matched Problem 1 Solve by substitution and check:

$$y = 3x + 1$$
$$3x - 2y = 1$$

$$y = 3x + 1$$
$$3x - 2(3x+1) = 1$$
$$3x - 6x - 2 = 1$$
$$y = -2 \qquad -3x = 3$$
$$x = -1$$

Example 2
Solving by Substitution

Solve by substitution and check:

$$5x + y = 4$$
$$2x - 3y = 5$$

Solution Solve either equation for one variable in terms of the other; then proceed as in Example 1. In this example we can avoid fractions by choosing the first equation and solving for y in terms of x:

$$5x + y = 4 \qquad\qquad \text{Solve the first equation for } y \text{ in terms of } x.$$
$$y = \underline{4 - 5x} \qquad\qquad \text{Substitute into the second equation.}$$
$$2x - 3y = 5$$
$$2x - 3(4 - 5x) = 5 \qquad\qquad \text{Solve for } x.$$
$$2x - 12 + 15x = 5$$
$$17x = 17$$
$$\boxed{x = 1}$$

Now replace x with 1 in $y = 4 - 5x$ to find y:

$$y = 4 - 5x$$
$$y = 4 - 5(1)$$
$$\boxed{y = -1}$$

The solution is $x = 1$, $y = -1$.

Check
$$5x + y = 4 \qquad\qquad 2x - 3y = 5$$
$$5(1) + (-1) \overset{?}{=} 4 \qquad 2(1) - 3(-1) \overset{?}{=} 5$$
$$4 \overset{\checkmark}{=} 4 \qquad\qquad\qquad 5 \overset{\checkmark}{=} 5$$

Matched Problem 2   Solve by substitution and check:

$$3x + 2y = -2$$
$$2x - y = -6$$

2y = -3x - 2

y = -\tfrac{3}{2}x - 1

Answers to   **1.** $x = -1$, $y = -2$ **2.** $x = -2$, $y = 2$
Matched Problems

2x - (-\tfrac{3}{2}x - 1) = -6

7x = -49

2x + \tfrac{3}{2}x + 1 = -6   *x =*

\tfrac{4}{2}x + \tfrac{3}{2}x = -7   *\tfrac{7}{2}x = -7 \cdot \tfrac{2}{1}*

\tfrac{7}{2}x = -7

EXERCISE 4-5

Solve by substitution and check your answers. If you cannot obtain a unique answer, graph the system to see what kind of solution to expect.

A

1. $y = x + 1$
$x + y = 5$

2. $y = x - 1$
$x + y = 7$

3. $y = 4x + 11$
$x + 2y = 4$

4. $y = 3x + 5$
$x - 3y = -7$

5. $y = 3x - 5$
$2x + y = 20$

6. $y = 5x - 3$
$x - 2y = 15$

7. $y = 2x - 3$
$2x - y = 3$

8. $y = 4x - 7$
$3x - y = 4$

9. $y = -2x - 1$
$2x + y = -1$

10. $y = -3x - 4$
$4x + y = 2$

11. $y = 2x - 7$
$y = x - 4$

12. $y = 3x + 2$
$y = 2x + 1$

13. $y = 3x - 2$
$y = 2x + 5$

14. $y = 4x - 3$
$y = 11x - 10$

15. $y = -2x - 4$
$y = -x + 5$

16. $y = -x - 2$
$y = x + 10$

17. $y = -x - 7$
$y = -2x - 8$

18. $y = -3x - 6$
$y = -x + 6$

B

19. $2x + y = 6$
$x - y = -3$

20. $2x - y = 3$
$x + 2y = 14$

21. $3x - y = 2$
$x + 2y = 10$

22. $x - y = 4$
$x + 3y = 12$

23. $2m - n = 10$
$m - 2n = -4$

24. $3m - n = 7$
$2m + 3n = 1$

25. $3u - v = -3$
$5u + 3v = -19$

26. $2u - 3v = 9$
$u + 2v = -13$

27. $x + 5y = 12$
$2x - 4y = 10$

28. $3x - 4y = 1$
$-x + 5y = 10$

29. $x + 5y = 12$
$-2x - 10y = -6$

30. $2x - 4y = 1$
$-x + 2y = 2$

31. $4x + 5y = 20$
$x - 2y = 18$

32. $6x - 2y = -4$
$2x - y = -3$

33. $-x + 3y = -4$
$2x + 10y = -16$

34. $x - 6y = -1$
$-2x + 2y = 12$

C

35. $y = 0.4x$
$y = 50 + 0.2x$

36. $y = 0.6x$
$y = 30 + 0.3x$

37. $y = 0.07x$
$y = 80 + 0.05x$

38. $y = 0.08x$
$y = 100 + 0.04x$

39. $y = \tfrac{1}{2}x - 5$
$y = \tfrac{7}{8}x - 8$

40. $y = \tfrac{1}{3}x + 9$
$y = \tfrac{1}{2}x + 8$

41. $y = \tfrac{2}{5}x - 4$
$x - 5y = 5$

42. $y = \tfrac{2}{3}x - 8$
$x - 3y = 15$

43. $y = \tfrac{3}{8}x + 1$
$2x - 2y = \tfrac{1}{2}$

44. $y = \tfrac{1}{4}x + 4$
$2x - 2y = 7$

45. $3x + y = 5$
$12x - 10y = 20$

46. $-4x + 8y = 12$
$10x - 2y = -3$

47. $5x - 15y = 10$
$3x + 7y = 22$

48. $8x - 3y = 14$
$12x + y = 14$

49. $2x - 12y = 6$
$-4x + 24y = -12$

50. $2x - 12y = -12$
$-4x + 24y = 6$

51. $2x - 12y = -6$
$-4x + 24y = -12$

52. $2x - 12y = -6$
$-4x + 24y = 12$

4-6 Solving Systems Using Elimination by Addition

♦ Equivalent Systems
♦ Solution Using Elimination by Addition
♦ Inconsistent and Dependent Systems

In this section we consider a third method for solving systems of equations, **elimination by addition.** This method involves the replacement of systems of equations with simpler equivalent systems, by performing appropriate operations, until we get a system whose solution is obvious.

♦ EQUIVALENT SYSTEMS

Equivalent systems are systems with the same solution set. What operations on a system produce equivalent systems?

Operations Producing Equivalent Systems

The following operations on a system of equations will produce an equivalent system:

1. Multiplying an equation by a nonzero constant
2. Adding one of the equations to another

These are direct results of the properties of equality first listed in Section 2-7.

♦ SOLUTION USING ELIMINATION BY ADDITION

Through the appropriate use of the above operations, we can eliminate one of the variables in one of the equations and obtain a system that has an obvious solution. The method is called **elimination by addition** or sometimes, more briefly, just **elimination.** It is a very important method in that it generalizes to large-scale systems involving many equations and many variables. Solving linear systems by this method is illustrated in the following examples.

Example 1
Solving by Elimination

Solve, using elimination by addition, and check:

$$3x + 2y = 13$$
$$2x - \ y = 4$$

Solution   We use the operations presented above to eliminate one of the variables and thus obtain a system whose solution is obvious:

$$3x + 2y = 13$$
$$2x - y = 4$$

If we multiply the bottom equation by 2 and add this to the top equation, we can eliminate y.

$$3x + 2y = 13$$
$$\underline{4x - 2y = 8}$$
$$7x \qquad = 21$$

Now solve for x.

$$x = 3$$

$$2 \cdot 3 - y = 4$$
$$-y = -2$$
$$y = 2$$

Substitute $x = 3$ back into either of the two original equations, preferably the simpler of the two, and solve for y. We choose the second equation.

The solution is $x = 3$, $y = 2$.

Check

$$3x + 2y = 13 \qquad 2x - y = 4$$
$$3(3) + 2(2) \overset{?}{=} 13 \qquad 2(3) - 2 \overset{?}{=} 4$$
$$9 + 4 \overset{\checkmark}{=} 13 \qquad 6 - 2 \overset{\checkmark}{=} 4$$

Matched Problem 1   Solve the system:

$$2x + 3y = 7$$
$$3x - y = 5$$

Example 2
Solving by Elimination

Solve the system:

$$2x + 3y = 1$$
$$5x - 2y = 12$$

Solution

$$2x + 3y = 1$$
$$5x - 2y = 12$$

If we multiply the top equation by 2 and the bottom equation by 3 and add, we can eliminate y.

$$4x + 6y = 2$$
$$\underline{15x - 6y = 36}$$
$$19x \qquad = 38$$
$$x = 2$$

$$2 \cdot 2 + 3y = 1$$
$$3y = -3$$
$$y = -1$$

Substitute $x = 2$ back into either of the two original equations, then solve for y.

The solution is $x = 2$, $y = -1$.

Check

$$2x + 3y = 1 \qquad 5x - 2y = 12$$
$$2(2) + 3(-1) \overset{?}{=} 1 \qquad 5(2) - 2(-1) \overset{?}{=} 12$$
$$4 - 3 \overset{\checkmark}{=} 1 \qquad 10 + 2 \overset{\checkmark}{=} 12$$

Matched Problem 2   Solve the system:

$$3x - 2y = 8$$
$$2x + 5y = -1$$

Example 3
An Application

College freshman athletes must have their high school grades converted to a grade point average in order to determine their eligibility at some schools. Quality points are awarded for grades in high school courses as follows: A = 4 points, B = 3, C = 2, and D = 1. If all courses are 1-unit courses, the grade point average is computed as the total quality points divided by the number of courses. A particular student had a grade point average of 2.59 resulting from 57 quality points earned in 22 courses. If the student earned only B's and C's, how many of each did he earn?

Solution   Let x represent the number of B's and y the number of C's. Then

$$x + y = 22$$ The number of grades is 22.

$$3x + 2y = 57$$ The total number of quality points is 57. Each of the x B's earns 3 quality points; each of the y C's earns 2.

To solve the system by elimination, begin by multiplying the first equation by 2:

$$2x + 2y = 44$$
$$\underline{3x + 2y = 57}$$ Subtract the first equation from the second.
$$x \qquad = 13$$ Next substitute $x = 13$ into $x + y = 22$ and solve for y.

$$13 + y = 22$$
$$y = 9$$

Thus, the student earned 13 B's and 9 C's.

Matched Problem 3

If a student had all A's and B's and had a total of 74 quality points for 22 classes, how many of each grade did she have?

♦ **INCONSISTENT AND DEPENDENT SYSTEMS**

Systems of equations do not always have a single solution. As we saw in Section 4-4, a system may not have a solution at all or it may have infinitely many. Recall that a system with no solution corresponds to parallel lines. A system with infinitely many solutions corresponds to the situation in which both equations represent the same line. How do we recognize these cases when solving a system using elimination by addition? The next two examples show how.

Example 4
A System with
No Solution

Solve the system:

$$x + 3y = 2$$
$$2x + 6y = -3$$

Solution

$$x + 3y = 2$$ Multiply the top equation by -2 and add.
$$2x + 6y = -3$$

$$-2x - 6y = -4$$
$$\underline{2x + 6y = -3}$$
$$0 = -7$$ A contradiction

Hence, there is no solution.

Our assumption that there are values for x and y that satisfy both equations simultaneously must be false, because otherwise we have proved that $0 = -7$. Thus, the system has no solutions. Systems of this type are said to be **inconsistent,** since conditions have been placed on the variables x and y that are impossible to meet. Geometrically, the graphs of the two equations must be parallel lines.

Matched Problem 4   Solve the system:

$$2x - y = 2$$
$$-4x + 2y = 1$$

Example 5
A System with Infinitely Many Solutions

Solve the system:

$$-2x + y = -8$$
$$x - \tfrac{1}{2}y = 4$$

Solution

$$-2x + y = -8$$
$$x - \tfrac{1}{2}y = 4$$

Multiply the bottom equation by 2 and add.

$$-2x + y = -8$$
$$\underline{2x - y = 8}$$
$$0 = 0$$

Both variables have been eliminated. Actually, if we had multiplied the bottom equation by -2, we would have obtained the top equation. When one equation is a constant multiple of the other, the system is said to be **dependent** and their graphs will coincide. There are infinitely many solutions to the system, and any solution of one equation is a solution of the other.

Matched Problem 5   Solve the system:

$$4x - 2y = 3$$
$$-2x + y = -\tfrac{3}{2}$$

$$4x - 2y = 3$$
$$-4x + 2y = -\tfrac{6}{2}$$
$$0 = 0$$

Answers to Matched Problems
1. $x = 2, y = 1$ 2. $x = 2, y = -1$ 3. 8 A's, 14 B's   4. No solution
5. Infinitely many solutions. Any solution of one equation is a solution of the other.

EXERCISE 4-6

Solve by the elimination method and check:

A 1. $x + y = 5$
$\quad x - y = 1$

2. $x - y = 6$
$\quad x + y = 10$

3. $x + 3y = 13$
$\quad -x + y = 3$

4. $-x + y = 1$
$\quad x - 2y = -5$

5. $2x + y = 0$
$\quad 3x + y = 2$

6. $x + 5y = 16$
$\quad x - 2y = 2$

7. $2x + 3y = 1$
$\quad 3x - y = 7$

8. $3x - y = -3$
$\quad 5x + 3y = -19$

9. $3x - 4y = 1$
$\quad -x + 3y = 3$

10. $-x + 5y = -3$
$\quad 2x - 3y = -1$

11. $2x + 4y = 6$
$\quad -3x + y = 5$

12. $3x + y = -8$
$\quad -5x + 3y = 4$

B 13. $3x + 2y = -2$
$\quad 4x + 5y = 2$

14. $5x + 7y = 8$
$\quad 3x + 2y = 7$

15. $6x + 5y = 4$
$\quad 7x + 2y = -3$

16. $9x + 4y = 1$
$\quad 4x + 3y = -2$

17. $11x + 2y = 1$
$9x - 3y = 24$

18. $3x - 11y = -7$
$4x + 3y = 26$

19. $3p + 8q = 4$
$15p + 10q = -10$

20. $5m - 3n = 7$
$7m + 12n = -1$

21. $4m + 6n = 2$
$6m - 9n = 15$

22. $5a - 4b = 1$
$3a - 6b = 6$

23. $3x + 5y = 15$
$6x + 10y = -5$

24. $x + 2y = 4$
$2x + 4y = -9$

25. $3x - 5y = 15$
$x - \frac{5}{3}y = 5$

26. $\frac{1}{2}x - y = -3$
$-x + 2y = 6$

Write in standard form

$$ax + by = c$$
$$dx + ey = f$$

and solve:

27. $y = 3x - 3$
$6x = 8 + 3y$

28. $3x = 2y$
$y = -7 - 2x$

29. $3y = 5x + 7$
$9x = 1 + 2y$

30. $3x = 4y + 5$
$5y = 8x - 1$

31. $3m + 2n = 2m + 2$
$2m + 3n = 2n - 2$

32. $2x - 3y = 1 - 3x$
$4y = 7x - 2$

33. $2x + 3y = 4x + 5y$
$5x - 4y = 3x - 2y$

34. $x + 2y = 3x + 4y$
$5x - 3y = x + y$

35. $x + y = 3x + 3y$
$2x - 2y = 6x + 2y$

36. $5x - 2y = -4x + 7y$
$2x + 5y = x + 6y$

37. $1 + 3x - 4y = 4x - 3y + 1$
$1 + 5x - 6y = 6x - 5y - 1$

38. $2 - x + 3y = 3x - y + 2$
$2 + x - 3y = -3y + x - 2$

C *Solve using elimination by addition. [Hint: Multiply both sides of each equation first by a constant that will eliminate decimals or fractions.]*

39. $0.3x - 0.6y = 0.18$
$0.5x + 0.2y = 0.54$

40. $0.8x - 0.3y = 0.79$
$0.2x - 0.5y = 0.07$

41. $0.15x + 0.20y = 0.25$
$0.45x - 0.35y = 1.70$

42. $0.25x - 0.60y = 2.60$
$0.10x + 0.20y = -1.60$

43. $\dfrac{x}{3} + \dfrac{y}{2} = 4$
$\dfrac{x}{3} - \dfrac{y}{2} = 0$

44. $\dfrac{x}{4} + \dfrac{y}{3} = 0$
$-\dfrac{x}{4} + \dfrac{y}{3} = -4$

45. $\dfrac{x}{2} + \dfrac{y}{3} = 1$
$\dfrac{2x}{3} + \dfrac{y}{2} = 2$

46. $\dfrac{a}{4} - \dfrac{2b}{3} = -2$
$\dfrac{a}{2} - b = -2$

47. $\frac{3}{4}x + \frac{5}{6}y = \frac{1}{3}$
$\frac{1}{2}x + \frac{5}{9}y = \frac{1}{12}$

48. $\frac{1}{3}x - \frac{3}{5}y = \frac{4}{5}$
$-\frac{1}{4}x + \frac{9}{20}y = \frac{3}{5}$

Determine whether the given system is equivalent to the system

$$2x + 3y = 4$$
$$-x - 5y = 6$$

To show that the systems are equivalent, you can simply give the operation on this system that changes it to the one in the problem. To show the systems are not equivalent, show they have different solutions.

49. $2x + 3y = 4$
$x + 5y = -6$

50. $-2x - 3y = -4$
$-x - 5y = 6$

51. $2x + 3y = 4$
$x - 2y = 10$

52. $3x + 8y = -10$
$-x - 5y = 6$

53. $2x + 3y = 4$
$-x - 5y = 10$

54. $3x + 8y = 10$
$-x - 5y = 6$

APPLICATIONS

55. *Domestic*   If 3 limes and 12 lemons cost 81 cents, and 2 limes and 5 lemons cost 42 cents, what is the cost of 1 lime and of 1 lemon?

56. *Business*   Find the capacity of each of two trucks if 3 trips of the larger and 4 trips of the smaller result in a total haul of 41 tons, and if 4 trips of the larger and 3 trips of the smaller result in a total haul of 43 tons.

57. *Sports*   A basketball team makes 34 field goals, accounting for 75 points. Some of the field goals were 3-point shots, the rest 2-point shots. How many of each did the team score?

58. *Domestic*   You buy a combination of 19-cent and 29-cent stamps, spending $18.30, and getting a total of 70 stamps. How many of each did you buy?

4-7 Rate-Time Problems

♦ Rate-Time Formulas
♦ Rate-Time Examples
♦ Checking Units

Many problems that can be solved using the one-equation–one-variable methods discussed in Chapter 3 can also be solved using two-equation–two-variable methods. In fact, many practical problems are more naturally set up using two variables rather than one. It is important to remember that if you introduce two variables into a problem, you will need two linear equations involving those two variables.

In this and the following two sections, we will work a variety of problems using both one- and two-variable methods.

♦ RATE-TIME FORMULAS

If a car travels 400 kilometers in 5 hours, then the ratio

$$\frac{400 \text{ kilometers}}{5 \text{ hours}} \qquad \text{or} \qquad 80 \text{ kilometers per hour}$$

is called the rate of motion, or speed. It is the number of kilometers traveled in each unit of time. Similarly, if a person types 420 words in 6 minutes, the ratio

$$\frac{420 \text{ words}}{6 \text{ minutes}} \qquad \text{or} \qquad 70 \text{ words per minute}$$

is the rate of typing. It is the number of words produced in each unit of time. In general, if q is the quantity produced in t units of time, then the ratio

$$\frac{\text{Quantity}}{\text{Time}} = \text{Rate}$$

or
$$\frac{q}{t} = r \tag{1}$$

Thus, the **rate** r is the amount of q produced in each unit of time. If both sides of Equation (1) are multiplied by t, we obtain the more commonly encountered form

$$q = rt \tag{2}$$

If q is distance d, then
$$d = rt \tag{3}$$

a special form of Equation (2) with which you may be more familiar. Formulas (2) and (3) enter into the solutions of many rate-time problems.

Important Rate-Time Formulas

$$\text{Quantity} = (\text{Rate})(\text{Time})$$
$$q = rt \qquad\qquad\qquad (2)$$

$$\text{Distance} = (\text{Rate})(\text{Time})$$
$$d = rt \qquad\qquad\qquad (3)$$

♦ **RATE-TIME EXAMPLES**

We now consider a variety of examples involving the rate-time concept.

Example 1
A Rate-Time Example

If a woman jogs 9 miles in 2 hours, what is her speed in miles per hour?

4.5

Solution This problem has only one variable. We write Equation (3) in the form

$$rt = d$$

and let $t = 2$ and $d = 9$, then solve for r:

$$r \cdot 2 = 9$$

$$r = \frac{9}{2} = 4.5 \text{ miles per hour}$$

Matched Problem 1   If a gas station pump can dispense 10 gallons in 4 minutes, what is its rate of pumping in gallons per minute?

2.5

Example 2
A Rate-Time Example

If a postal service bar code sorter can sort 28,000 pieces of mail per hour, how long will it take the machine to sort 19,600 pieces?

Solution We write Equation (2) in the form

$$rt = q$$

and let $r = 28{,}000$ and $q = 19{,}600$, then solve for t:

$$28{,}000t = 19{,}600$$

$$t = \frac{19{,}600}{28{,}000} = 0.7 \text{ hour}$$

You can convert 0.7 hour to 42 minutes using proportions, if an answer in minutes is preferred.

Matched Problem 2 If an airplane flies at 800 kilometers per hour, how long will it take to fly 2,600 kilometers?

Example 3
A Rate-Time Example

An airplane flew out to an island from the mainland and back in 5 hours. How long did each portion of the trip take if the pilot averaged 600 miles per hour going to the island and 400 returning? How far is the island from the mainland?

Solution We will use two variables. Let x be the time in hours spent flying to the island, y the time spent flying back. Thus, since the total time is 5 hours,

$$x + y = 5$$

The distance is the same flying out as flying back, and each is equal to rate times time, so

$$\left(\begin{array}{c}\text{Distance} \\ \text{flying out}\end{array}\right) = \left(\begin{array}{c}\text{Distance} \\ \text{flying back}\end{array}\right)$$

$$\left(\begin{array}{c}\text{Rate of} \\ \text{flight out}\end{array}\right)\left(\begin{array}{c}\text{Time for} \\ \text{flight out}\end{array}\right) = \left(\begin{array}{c}\text{Rate of} \\ \text{flight back}\end{array}\right)\left(\begin{array}{c}\text{Time for} \\ \text{flight back}\end{array}\right)$$

$$600 \qquad x \qquad = \qquad 400 \qquad y$$

The system

$$x + y = 5$$
$$600x = 400y$$

can be solved by substitution, rewriting $x + y = 5$ as $y = 5 - x$, and substituting into the other equation:

$$600x = 400(5 - x)$$

Solve:
$$600x = 2,000 - 400x$$
$$1,000x = 2,000$$
$$x = 2 \text{ hours}$$
$$y = 5 - x = 3 \text{ hours}$$

The distance to the island is then $600x = 1,200$ miles. (This problem could also have been solved using only one variable by letting $5 - x$ represent the time of the return trip right from the start.)

Matched Problem 3 An airplane flew from San Francisco to a distressed ship at sea and back in 7 hours. How far was the ship from San Francisco if the pilot averaged 400 miles per hour going and 300 returning?

Example 4
A Rate-Time Example

A car leaves a town traveling at 90 kilometers per hour. How long will it take a second car traveling at 105 kilometers per hour to catch up to the first car if it leaves 2 hours later?

Solution   Here we will demonstrate both the two- and one-variable method.
Using Two Variables: Let x be the time the first car travels, and y be the time the second car travels. Draw a diagram and label known and unknown parts.

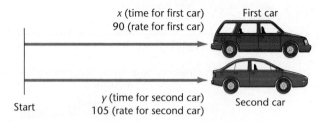

The distance traveled is then $90x$ for the first car and $105y$ for the second. When the second car catches up to first car, both cars will have traveled the same distance; hence

$$\begin{pmatrix} \text{Distance first} \\ \text{car travels} \end{pmatrix} = \begin{pmatrix} \text{Distance second} \\ \text{car travels} \end{pmatrix}$$

$$\begin{pmatrix} \text{Rate of} \\ \text{first car} \end{pmatrix}\begin{pmatrix} \text{Time for} \\ \text{first car} \end{pmatrix} = \begin{pmatrix} \text{Rate of} \\ \text{second car} \end{pmatrix}\begin{pmatrix} \text{Time for} \\ \text{second car} \end{pmatrix}$$

$$90 \qquad x \quad = \quad 105 \qquad y$$

We thus obtain this system of equations:

$$90x = 105y \qquad \text{The distances traveled must be the same.}$$

$$x = y + 2 \qquad \text{The first car leaves 2 hours earlier so travels 2 hours more.}$$

We can solve this system by substitution, substituting $y + 2$ for x:

$$90(y + 2) = 105y$$

$$90y + 180 = 105y$$

$$-15y = -180$$

$$y = \frac{-180}{-15}$$

$$y = 12$$

Thus, it takes 12 hours for the second car to catch up to the first. Since $x = y + 2$, $x = 14$; that is, the first car has traveled for 14 hours.
Using One Variable: Let $t =$ Number of hours for the second car to catch up. Then the time the first car has traveled is $t + 2$. The cars will have traveled the same distance so

$$\begin{pmatrix} \text{Distance first} \\ \text{car travels} \end{pmatrix} = \begin{pmatrix} \text{Distance second} \\ \text{car travels} \end{pmatrix}$$

$$\begin{pmatrix} \text{Rate of} \\ \text{first car} \end{pmatrix}\begin{pmatrix} \text{Time for} \\ \text{first car} \end{pmatrix} = \begin{pmatrix} \text{Rate of} \\ \text{second car} \end{pmatrix}\begin{pmatrix} \text{Time for} \\ \text{second car} \end{pmatrix}$$

$$90 \qquad (t + 2) \quad = \quad 105 \qquad t$$
$$90t + 180 = 105t$$

This is the same equation solved above, so $t = 12$ hours.

Matched Problem 4 An older copying machine can print 90 copies per minute, and a newer machine can print 120 per minute. If the newer machine is brought on the job 5 minutes after the first machine starts and both continue until the job is done, how long will it take to print 2,900 handbills?

♦ **CHECKING UNITS**

The rate-time formulas come from the proportion

$$\frac{\text{Quantity}}{\text{Time}} = \text{Rate}$$

where the rate is a constant or average for the given time. The units for the rate must be units of quantity per unit of time, for example, miles per hour or gallons per minute. When written in the $q = rt$ form, the units look like

$$\text{Units of quantity} = \frac{\text{Units of quantity}}{\text{Units of time}} \times \text{Units of time}$$

For example:

$$\text{Miles} = \frac{\text{Miles}}{\text{Hours}} \times \text{Hours}$$

Note how the units of time "cancel" to leave just the units of quantity on both sides of the equation. Checking that units balance this way can help you set up the equations correctly, not just in rate-time problems but in almost any problem.

The units of time used in the above examples were the usual ones of minutes, hours, etc. Other units can also be associated with time. For example, in professional basketball, a quarter is 12 minutes and a game is 4 quarters, or 48 minutes. Quarters and games can be thought of as units of time.

Example 5
A Rate-Time Example

In his first 427 games, Michael Jordan of the Chicago Bulls scored 14,016 points. At what rate did he score per game?

Solution The rate-time equation here takes the form

$$\text{Points} = \frac{\text{Points}}{\text{Games}} \times \text{Games} \qquad \text{Notice how the units balance.}$$

or $\quad$ Points = (Points per game) × Games

Thus, with r representing the unknown rate

$$14,016 = r \cdot 427$$

$$\frac{14,016}{427} = r$$

so $r = 32.824 \approx 32.8$ points per game.

Matched Problem 5 In his 1,560-game career, Kareem Abdul-Jabbar scored 38,387 points. At what rate did he score per game?

Answers to Matched Problems

1. $r \cdot 4 = 10$, $r = 2.5$ gallons per minute   2. $800t = 2,600$, $t = 3.25$ hours

3. $\dfrac{x}{400} + \dfrac{x}{300} = 7$, $x = 1,200$ miles

4. $90t + 120(t - 5) = 4,100$, $t = 16\frac{2}{3}$ minutes   5. 24.6 points per game

EXERCISE 4-7

Solve using either a one-variable or a two-variable method:

A 1. A car is traveling at an average rate of 48 miles per hour. How long will it take to go 156 miles?

2. How long will it take a car traveling at 88 feet per second to travel 300 feet?

3. If a data entry specialist can enter 20 pieces of numerical data in 1 minute, how long will it take to enter 12,000 pieces?

4. If a typist can type 76 words per minute, how long will it take to type 1,520 words?

5. If a water pump on a boat can pump at the rate of 12 gallons per minute, how long will it take to pump 30 gallons?

6. How long will it take to make 180 copies of a page if your copy machine can make 120 per minute?

7. If an automobile dealer's service department charges $517.50 for $7\frac{1}{2}$ hours of work on your car, what is the hourly rate?

8. If the labor costs for a painter are $480, how long did the painter spend on the job if she receives $16 per hour?

9. If a person earns $220 for a 40-hour week, what is the rate per hour?

10. If a teacher earns $28,800 in a year, what is her salary per month?

11. If a keyboard operator can enter 10,500 items of statistical data in 50 hours, what is the rate of entry per hour?

12. If a typist can type 2,400 words in 50 minutes, what is the rate per minute?

13. If a car travels 550 kilometers in 5.5 hours, what is its rate?

14. If a plane flies 3,300 miles in 5.5 hours, what is its speed?

15. If an English instructor can grade 15 essays in $2\frac{1}{2}$ hours, at what rate is he working per hour?

16. If a printing press can print the evening run of 12,000 papers in 2.5 hours, what is its rate?

17. Two cars leave New York at the same time and travel in opposite directions. If one travels at 55 miles per hour and the other at 50 miles per hour, how long will it take them to be 630 miles apart?

18. Two airplanes leave Atlanta at the same time and fly in opposite directions. If one travels at 600 kilometers per hour and the other at 500 kilometers per hour, how long will it take them to be 3,850 kilometers apart?

19. A baseball player gets 180 hits in 600 at bats. At what rate, Hits/At bats, is he hitting? This ratio is his *batting average*. Here an "at bat" is treated as a unit of time.

20. Over the course of several seasons a football team wins 50 out of 80 games. At what rate is the team winning? The ratio Wins/Games is the team's *winning percentage*. One game is treated as a unit of time here.

B 21. If one machine can fill and cap 20 bottles per minute, and another machine can do 30, how long will it take both together to complete a 30,000-bottle order?

22. Two people volunteer to fold and stuff envelopes for a political campaign. If one person can produce 6 per minute and the other 8, how long will it take both together to prepare 1,267 envelopes?

23. A car leaves town traveling at 45 miles per hour. How long will it take a second car traveling at 50 miles per hour to catch up to the first car if it leaves 1 hour later?

24. Repeat Problem 23 if the second car leaves 2 hours later and travels at 55 miles per hour.

25. Find the total amount of time to complete the job in Problem 21 if the second (faster) machine is brought on the job 1 hour after the first machine starts and both continue until the job is finished.

26. Find the total amount of time to complete the job in Problem 22 if the second (faster) person is brought on the job 28 minutes after the first person starts and both continue until the job is finished.

27. A research chemist charges $35 per hour for her services and $21 per hour for her assistant. On a given job a customer received a bill for $1,505. If the chemist worked 5 hours less on the job than her assistant, how much time did each spend?

28. At a particular garage, the hourly labor rate in the mechanical shop is $39. The rate in the body shop is $42 per hour. Following an accident, your insurance company is billed a total of $1,041 for labor. How many hours of the labor was in the body shop if it was 8 hours less than that in the mechanical shop?

29. The city lets you fill your swimming pool from a fire hydrant, which releases water into the pool at a rate of 60 gallons per minute. Two hours after you begin, you get impatient and also turn on your garden hose, which lets water into the pool at a rate of 12 gallons per minute. If the pool holds 30,960 gallons, what will be the total time required to fill it?

30. Repeat Problem 29 for a pool holding 30,000 gallons, a fire hydrant releasing 60 gallons per minute, and a garden hose started 2 hours after the fire hydrant and releasing 15 gallons per minute.

C 31. An earthquake emits a primary wave and a secondary wave. Near the surface of the earth, the primary wave travels at about 5 miles per second and the secondary wave at about 3 miles per second. From the time lag between the two waves arriving at a given seismic station, it is possible to estimate the distance to the quake. (The "epicenter" can be located by getting distance bearings at three or more stations.) Suppose a station measured a time difference of 12 seconds between the arrival of the two waves. How far would the earthquake be from the station?

32. Repeat Problem 31 for a time difference of 15 seconds.

33. A sky diver free-falls (because of air resistance) at about 176 feet per second, or 120 miles per hour; with parachute open the rate of fall is about 22 feet per second or 15 miles per hour. If the sky diver opened the chute halfway down and the total time for the descent was 9 minutes, how high was the plane when the sky diver jumped?

34. Repeat Problem 33 for a total time of 6 minutes.

35. A baseball player has a batting average (see Problem 19) of .320 with 192 hits in 600 at bats. His average against left-handed pitching is .250. Against right-handed pitching it is .350. How many times did he bat against left-handers?

36. Repeat Problem 35 with the player having a .275 average against right-handed pitching and a .3875 average against left-handers.

37. Over a decade, a baseball team won 891 out of 1,620 regular season games for a winning percentage of 55%, or 0.55. It won 50% of the games it played on natural grass fields and 58.1% of the games it played on artificial turf. How many games did it play on each?

38. Repeat Problem 37 for a team that won 729 out of 1,620 games and had winning percentages of 40% and 51.2% on natural grass and artificial turf, respectively.

39. A company representative knows that on the highway she averages 58 miles per hour, while in the city she averages only 24 miles per hour. If she drives 616 miles in a week and spends 16.6 hours driving, how many of the miles were city driving?

40. Repeat Problem 39 for a week in which the representative drove 639 miles in 13.5 hours.

41. A tile layer earns $24 per hour and can lay tile at the rate of 12 square feet per hour. His apprentice earns $14 per hour and can lay tile at a rate of 8 square feet per hour. If having 444 square feet of tile laid costs a company $852, how many hours did the apprentice work?

42. Repeat Problem 41 for 452 square feet and $860.

  # 4-8 **Mixture Problems**

♦ Money Problems
♦ Solution Problems

A variety of applications can be classified as mixture problems, including some of the time-rate problems considered in the previous section. Even though the problems come from different areas, their mathematical treatment is essentially the same.

◆ **MONEY PROBLEMS**

In working coin problems, be careful not to confuse the *value* of the coins with the *number* of coins.

**Example 1
A Coin Problem**

Suppose you have 45 coins consisting only of nickels and quarters that are worth $4.25. How many of each type of coin do you have?

Solution   We solve the problem using two variables. Let

$$x = \text{Number of nickels}$$
$$y = \text{Number of quarters}$$

Then
$$x + y = 45 \qquad \text{Number of coins}$$
$$5x + 25y = 425 \qquad \text{Value of coins in cents}$$

To solve, multiply the top equation by -5 and add:

$$
\begin{aligned}
-5x - 5y &= -225 \\
\underline{5x + 25y} &= \underline{425} \\
20y &= 200 \\
y &= 10 \qquad \text{Quarters}
\end{aligned}
$$

$$x + 10 = 45 \qquad \text{Now solve for } x \text{ using the top equation.}$$
$$x = 35 \qquad \text{Nickels}$$

Check $35 + 10 = 45$ coins; $35 \cdot 5 + 10 \cdot 25 = 175 + 250 = 425$ cents $= \$4.25$.

We could also have solved the problem using only one variable. If we let

$$x = \text{Number of nickels}$$

then
$$45 - x = \text{Number of quarters}$$

and the equation to solve is

$$5x + 25(45 - x) = 425 \qquad \text{Not 4.25}$$

Matched Problem 1   A noon concert brought in $2,000 on the sale of 1,500 tickets. If tickets sold for $1 and $2, how many of each were sold?

**Example 2
A Mix Value Problem**

A candy shop wishes to mix candy that sells for $1.10 per pound with candy that sells for $1.50 per pound. How much of each should be used to produce 40 pounds of the new mix selling at $1.34 per pound?

Solution   We will solve the problem using two variables. Let

$$x = \text{Amount of \$1.10-per-pound candy used}$$
$$y = \text{Amount of \$1.50-per-pound candy used}$$

Then
$$x + y = 40 \qquad \text{Pounds in mix}$$
$$1.1x + 1.5y = 1.34(40) = 53.6 \qquad \text{Value of mix}$$

To solve, multiply the top equation by -11, multiply the bottom equation by 10, and then add:

$$
\begin{array}{r}
-11x - 11y = -440 \\
\underline{11x + 15y = 536} \\
24y = 96 \\
y = \tfrac{96}{4} = 24 \text{ pounds of \$1.50-per-pound candy}
\end{array}
$$

$$x + 24 = 40 \qquad \text{Now solve for } x \text{ using the top equation.}$$
$$x = 16 \text{ pounds of \$1.10-per-pound candy}$$

Check $24 + 16 = 40$ pounds; $1.5(24) + 1.1(16) = 36 + 17.6 = \$53.60 = $ Value of mix.
  The problem could also be solved by one-variable methods, replacing y by $40 - x$ in the second equation of the system.

Matched Problem 2   Repeat Example 2 but suppose that the two candies to be used in the mix cost \$0.80 and \$1.44 per pound, respectively, and the final mix is to sell for \$1.20 per pound.

♦ **SOLUTION PROBLEMS**

We now consider mixture problems involving percent. Recall that 12% in decimal form is 0.12, 3.5% is 0.035, and so on. Also recall that 30% of 50 means 0.30×50 and 2.5% of 50 means 0.025×50.

Example 3
A Solution Problem

How many centiliters of pure alcohol must be added to 50 centiliters of a 20% solution to obtain a 40% solution?

Solution   There is only one variable in this problem. We let $x = $ Number of centiliters (cl) of pure alcohol to be added. Let us illustrate the situation before and after mixing. The amount of alcohol present in the two solutions to be mixed must equal the amount of alcohol present after mixing. Note that pure alcohol is 100% alcohol.

$$
\begin{array}{ccccc}
\begin{pmatrix}\text{Amount of}\\ \text{pure alcohol}\\ \text{in first solution}\end{pmatrix} & + & \begin{pmatrix}\text{Amount of}\\ \text{pure alcohol}\\ \text{in second solution}\end{pmatrix} & = & \begin{pmatrix}\text{Amount of}\\ \text{pure alcohol}\\ \text{in mixture}\end{pmatrix} \\
0.2(50) & + & x & = & 0.4(x + 50) \\
& & 10 + x & = & 0.4x + 20 \\
& & 0.6x & = & 10 \\
& & x & = & \dfrac{50}{3} \\
& & & = & 16\tfrac{2}{3} \text{ centiliters}
\end{array}
$$

Decimals could have been cleared first, if desired.

The information provided by the illustration also can be given in tabular form:

	Volume, in Centiliters	Percent Alcohol	Amount of Alcohol
Component 1: 20% solution	50	20%	0.2(50)
Component 2: pure alcohol	x	100%	x
Mixture: 40% solution	$50 + x$	40%	$0.4(50 + x)$

The equation is then obtained from the right-hand column of the table. The amount of alcohol is to be 40% of the volume, that is,

$$0.2(50) + x = 0.4(50 + x)$$

Matched Problem 3 How many centiliters of distilled water must be added to 80 centiliters of a 60% acid solution to obtain a 50% acid solution? Set up an equation and solve. Use the fact that distilled water has 0% acid.

Example 4
A Solution Problem

A chemical storeroom has a 40% acid solution and an 80% solution. How many deciliters (dl) must be used from each to obtain 12 deciliters of a 50% solution?

Solution Let

$$x = \text{Amount of 40\% solution used}$$
$$y = \text{Amount of 80\% solution used}$$

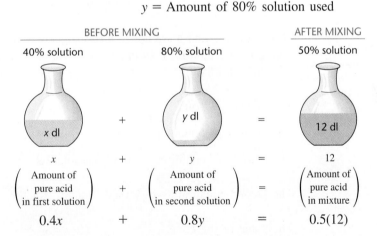

BEFORE MIXING		AFTER MIXING
40% solution	80% solution	50% solution

$$
\begin{array}{ccccc}
x & + & y & = & 12 \\
\left(\begin{array}{c}\text{Amount of}\\ \text{pure acid}\\ \text{in first solution}\end{array}\right) & + & \left(\begin{array}{c}\text{Amount of}\\ \text{pure acid}\\ \text{in second solution}\end{array}\right) & = & \left(\begin{array}{c}\text{Amount of}\\ \text{pure acid}\\ \text{in mixture}\end{array}\right) \\
0.4x & + & 0.8y & = & 0.5(12)
\end{array}
$$

Thus, we obtain two equations and two variables:

$$
\begin{aligned}
x + y &= 12 \\
0.4x + 0.8y &= 6
\end{aligned}
$$
Multiply by 10 to clear decimals.

$$
\begin{aligned}
x + y &= 12 \\
4x + 8y &= 60
\end{aligned}
$$
Divide by −4 to simplify.

$$
\begin{aligned}
x + y &= 12 \\
\underline{-x - 2y} &= \underline{-15} \\
-y &= -3
\end{aligned}
$$
Add to eliminate x.

$$y = 3 \text{ deciliters of 80\% solution}$$

$$
\begin{aligned}
x + y &= 12 \\
x + 3 &= 12 \\
x &= 9 \text{ deciliters of 40\% solution}
\end{aligned}
$$
Now solve for x using the top equation.

The information provided by the illustration also can be given in tabular form:

	Volume, in Deciliters	Percent Acid	Amount of Acid
Component 1: 40% solution	x	40%	$0.4x$
Component 2: 80% solution	y	80%	$0.8y$
Mixture: 50% solution	12	50%	$0.5(12)$

The volumes and amounts of acid in the components must add to that in the mixture, that is,

$$x + y = 12$$

$$0.4x + 0.8y = 0.5(12)$$

The problem also can be solved using one variable, replacing y with $12 - x$ in the second equation of the system.

Matched Problem 4 In Example 4 how many deciliters of each stockroom solution must be used to obtain 20 deciliters of a 70% solution?

Examples 1, 2, and 4 are not so different as they might at first appear. It is worth noting their similarities. In each case there is an equation representing the total amount in a mixture:

The number of coins

The number of pounds in the mix

The number of centiliters in the final solution

The second equation gives a total "value" for the mixture:

The value of the coins

The value of the mix

The percentage value of acid in the solution

The same pattern will be true in many of the exercises.

Answers to Matched Problems

1. $x + y = 1,500$; $x + 2y = 2,000$; $x = 1,000$ ($1 tickets); $y = 500$ ($2 tickets)
2. $x + y = 40$; $0.8x + 1.44y = 1.2(40)$; $x = 15$ pounds ($0.80 candy); $y = 25$ pounds ($1.44 candy)
3. $0.6(80) + 0.0x = 0.5(x + 80)$; $x = 16$ centiliters
4. $x + y = 20$; $0.4x + 0.8y = 0.5(20)$; $x = 5$ deciliters (40% solution); $y = 15$ deciliters (80% solution)

EXERCISE 4-8

A **1.** A newspaper vending machine takes only dimes and quarters. If the machine has 137 coins worth a total of $26, how many of each kind of coin are in the machine?

2. A candy vending machine takes only nickels and quarters. If the machine contains 174 coins that have a total value of $26.70, how many of each kind of coin are in the machine?

3. A high school basketball game had a total attendance of 734 and gate receipts of $939. Student tickets were $1 each, general admission tickets $1.50. How many of each kind of ticket were sold?

4. A student production brought in $7,000 on the sale of 3,000 tickets. If tickets sold for $2 and $3 each, how many of each type were sold?

5. A basketball player made 280 baskets and scored 648 points during a season on a combination of 2- and 3-point shots. How many of each type shot did she make?

6. In baseball, total bases are computed by the formula

Number of singles + 2(Number of doubles)
  + 3(Number of triples) + 4(Number of home runs)

If a baseball player had 138 hits during a season, all singles and doubles, and had 157 total bases, how many of each kind of hit did he have?

7. A university mathematics department hires full-time and part-time faculty. Full-time faculty teach four classes per term, part-time faculty two. In a term with 45 faculty employed and 146 classes being taught, how many full-time faculty are there?

8. Repeat Problem 7 for a term with 39 faculty employed and 130 classes being taught.

9. A park district purchases birdseed in 20- and 50-pound bags for winter feeding. If it has 60 bags weighing a total of 1,680 pounds, how many of each kind does it have?

10. Repeat Problem 9 for 40 bags weighing 1,610 pounds.

11. In hockey, standings are kept by points, with 2 points awarded for a win, 1 for a tie, and 0 for a loss. In a recent National Hockey League season, a team lost 33 games out of 80 and finished with 88 points. How many games did it tie?

12. Repeat Problem 11 for a team that lost 41 games out of 80 and finished with 64 points.

13. A farm family planted 540 acres in soybeans and corn. Each acre of corn yielded a profit of $62.20, each acre of soybeans $58.70. If their total profit was $32,783, how many acres of each crop did they plant?

14. Answer the question in Problem 13 if the total profit was $32,835.50.

B **15.** How many deciliters of pure alcohol must be added to 60 deciliters of a 35% solution to obtain a 60% solution?

16. How many centiliters of pure acid must be added to 80 milliliters of 30% solution to obtain a 50% solution?

17. How many deciliters of distilled water must be added to 400 deciliters of a 35% solution to obtain a 20% solution?

18. How many centiliters of distilled water must be added to 800 deciliters of a 70% solution to obtain a 40% solution?

19. A chemical storeroom has two solutions in stock. One is 20% acid, the other 60% acid. How many centiliters of each solution should be mixed to obtain 200 centiliters of a 44% solution?

20. A chemical storeroom has in stock a 40% acid solution and a 75% acid solution. How many milliliters of each solution should be mixed to obtain 100 milliliters of a 54% solution?

21. A coffee and tea shop wishes to blend a $3.50-per-pound coffee with a $4.75-per-pound coffee in order to produce a blend selling for $4 per pound. How much of each would have to be used to produce 100 pounds of the new blend?

22. A coffee and tea shop wishes to blend a $2.50-per-pound tea with a $3.25-per-pound tea to produce a blend selling for $3 per pound. How much of each should be used to produce 75 pounds of the new blend?

23. An investor has $40,000 to invest and is going to buy two mutual funds. The first fund yields an 8% return; the second yields 13%. If the investor seeks a total return of $4,000, how much should she invest in each fund?

24. If an investor has $50,000 to invest and is going to buy two securities that yield 7 and 12 percent returns, how much should he invest in each to get a total yield of $4,400?

25. The octane rating for a gasoline is a percentage. It measures the percentage of a particular chemical in mixture with heptane that has the same antiknock properties as the gasoline in question. Thus, a gasoline with an octane rating of 89 can be thought of as an 89% solution. Suppose you want to fill your car's 22-gallon gas tank with gasoline having an octane rating of 90. If the only two gasolines available have octane ratings of 87 and 92, how much of each should you use?

26. Repeat Problem 25 for the situation in which the available gasolines have octane ratings of 86 and 93.

27. A 10-liter radiator contains a 60% solution of antifreeze in distilled water. How much should be drained and replaced with pure antifreeze to obtain an 80% solution?

28. A 3-gallon radiator contains a 50% solution of antifreeze in distilled water. How much should be drained and replaced with pure antifreeze to obtain a 70% solution?

29. A park district purchases birdseed in 20- and 50-pound bags for winter feeding as in Problem 9. The 50-pound bags cost $6.70 per bag, and the smaller bags $3.10 per bag. If the district has a total bill of $315.80 for 2,200 pounds of seed, how many of each kind did it purchase?

30. Repeat Problem 29 for a bill of $454 for 3,200 pounds of seed?

31. A carton contains 100 packages. Some of the packages weigh $\frac{1}{2}$ pound each and the rest weigh $\frac{1}{3}$ pound each. If the whole carton weighs 45 pounds, how many of each kind of package does it contain?

32. Repeat Problem 31 for a carton weighing 37.5 pounds.

33. A gardener needs a fertilizer containing 10 pounds of nitrogen and 15 pounds of potassium. Two commercial products are available. One is 10% nitrogen and 10% potassium. The other is 8% nitrogen and 18% potassium. How much of each product should the gardener buy to obtain the required mix?

34. Repeat Problem 33 if the two commercial products have amounts of nitrogen and potassium as given in this table:

	Nitrogen	Potassium
Product 1	5%	20%
Product 2	12%	8%

35. A *karat* is a measure of gold in an article by weight. A 24-karat-gold item is all gold. An item that is 14-karat gold is $\frac{14}{24}$ gold. A jeweler wants to make a 14-karat-gold ring weighing 1 ounce. Two alloys are available: one is 8-karat gold, and the other 18-karat gold. How much of each should be used?

36. The jeweler in Problem 35 wishes to make an 18-karat-gold cup weighing 250 grams. A 14-karat alloy and pure gold are available. How much of each should the jeweler use?

  # 4-9   Supplemental Applications

This section provides another supplemental exercise set with a variety of applications. Some of the problems may be solved by one-variable methods; some may be better approached using two variables.

EXERCISE 4-9

The problems in this exercise are grouped according to subject area. The most difficult problems are marked with two stars (**), and the moderately difficult problems with one star (*). The easier problems are not marked.

GEOMETRY

1. An 18-foot board is cut into two pieces so that one piece is 4 feet longer than the other piece. How long is each piece?

2. If the sum of two angles in a right triangle is 90° and their difference is 14°, find the two angles.

3. A rectangle with length equal to twice the width has perimeter 600 meters. What are the dimensions?

★ **4.** Find the dimensions of a rectangle with perimeter 168 centimeters if its length is 1.8 times its width.

★ **5.** In a triangle, the sum of the first and third angles is twice the second angle. The sum of the first and second angle is 10° more than the third. What are the angles? Recall that the sum of the three angles is 180°.

6. In an isosceles triangle each of the two equal angles is 30° more than the third angle. What are the angles?

7. In a parallelogram the opposite angles are equal, and the sum of all the angles is 360° (see the figure). If one angle of a parallelogram is 3 times as large as the one adjacent to it, what are the angles?

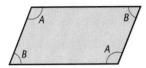

★ 8. A rectangle has area equal to another rectangle that is 4 feet longer and 2 feet narrower. It also has the same area as a third rectangle that is 3 feet shorter and 5 feet wider. What are its dimensions?

NUMBER

9. Two numbers that add to 81 differ by 15. What are the numbers?

★ 10. The product of two numbers is the same as the product of 5 more than the first times 2 less than the second. It is also the same as the product of 2 more than the first times 1 less than the second. What are the numbers?

BUSINESS

11. A jazz concert brought in $75,000 on the sale of 8,200 tickets. If the tickets sold for $7 and $11 each, how many of each type were sold?

12. Attendance at a minor league baseball game was 3,450, and gate receipts totaled $8,950. General admission tickets sold for $2.25, and reserved seats for $3.50. How many of each were sold?

★ 13. A person has $8,000 to invest. If part is invested at 6% and the rest at 8%, how much should be invested at each rate to have a total return of $520 per year?

★ 14. How should $25,000 be split between two investments to yield $2,580 if one investment returns 7% and the other 12%?

★ 15. A coffee shop wishes to blend a $3.70-per-pound coffee with a $5.20-per-pound coffee to produce a blend selling for $4.60 per pound. How much of each should be used to produce 100 pounds of the new blend?

★ 16. A nut and candy shop wants to combine two mixes, one plain and one fancy, to obtain 20 pounds of a sale mix. The plain mix sells for $2.00 a pound, and the fancy mix for $3.20 a pound. How much of each should be used if the sale mix is to sell for $2.33 a pound?

CHEMISTRY

17. Alcohol and distilled water were mixed to produce 120 centiliters of solution. If 20 centiliters more water was used than alcohol, how many centiliters of each was mixed?

18. Hydrochloric acid and distilled water were mixed to produce 96 milliliters of solution. If 36 milliliters more water was used than acid, how much of each was mixed?

★ 19. A chemist has two solutions in a stockroom. One is a 30% acid solution, and the other is a 70% acid solution. How many milliliters of each should be mixed to obtain 100 milliliters of a 60% solution?

★ 20. A chemical storeroom has a 20% acid solution and a 50% acid solution. How many centiliters must be taken from each to obtain 90 centiliters of a 40% solution?

LIFE SCIENCE

21. If laboratory animals in a diet experiment have a combined weight of 800 grams and one weighs 200 grams more than the other, how much does each weigh?

22. If two monkeys have a combined weight of 10 kilograms and one weighs 2 kilograms more than the other, how much does each weigh?

★★ 23. Animals in an experiment are to be kept on a strict diet. Each animal is to receive, among other things, 20 grams of protein and 6 grams of fat. The laboratory technician is able to purchase two food mixes of the following compositions:

	Protein	Fat
Mix 1	10%	6%
Mix 2	20%	2%

How many grams of each mix should be used to obtain the right diet for a single animal?

★★ 24. A farmer placed an order with a chemical company for a fertilizer that would contain, among other things, 120 pounds of nitrogen and 90 pounds of phosphoric acid. The company had two mixtures on hand with the following compositions:

	Nitrogen	Phosphoric Acid
Mixture A	20%	10%
Mixture B	6%	6%

How many pounds of each mixture should the company mix to fill the order?

PHYSICS AND ENGINEERING

25. Where should the fulcrum be placed on a 12-foot bar if it is to balance with a 14-pound weight on one end and a 42-pound weight on the other?

Hint: 14 ____ x ____ y 42 ____ $14x = 42y$

26. Where should the fulcrum be placed on a 240-centimeter bar to balance 30 kilograms on one end and 50 kilograms on the other?

MUSIC

27. If a string on a stringed instrument is divided in the ratio of 4:5, a major third chord will result. What will be the length of each part if a 36-inch string is used?

[*Hint:* Use the proportion $\dfrac{x}{y} = \dfrac{4}{5}$ for one of the equations.]

28. If a string on a stringed instrument is divided in the ratio of 5:8, a minor sixth chord will result. How would you divide a 39-inch string to produce a minor sixth?

MISCELLANEOUS

★★ **29.** If 1 flask and 4 mixing dishes balance 16 test tubes and 2 mixing dishes, and if 2 flasks balance 2 test tubes and 6 mixing dishes, how many test tubes will balance 1 flask, and how many test tubes will balance 1 mixing dish?

★★ **30.** Two boys trading baseball cards agree that 6 Will Clark cards and 3 José Canseco cards are worth the same to them as 2 Nolan Ryan cards. They also agree that 9 Canseco cards are worth the same as 12 Clark cards plus 2 Ryan cards. How many Canseco cards have the same value to them as 2 Ryan cards? How many Clark cards have the same value to them as 2 Ryan cards?

31. A packing carton contains 150 small packages, some weighing $\frac{1}{4}$ pound each and the others $\frac{1}{2}$ pound each. How many of each type are in the carton if the total contents weigh 52 pounds?

32. Repeat Problem 31 if the small packages weigh $\frac{1}{3}$ pound and $\frac{3}{4}$ pound, respectively, and the total contents weigh 90 pounds?

★ **33.** A canoeist can paddle 12 miles downstream in $1\frac{1}{2}$ hours and return in 6 hours. Find the rate at which the canoeist paddles in still water and the rate of the current in the stream.

★ **34.** A driver leaves Chicago, Illinois, by car at 9 A.M. driving west. At 10 A.M. a driver in another car is 290 miles west of Chicago heading east. The drivers meet at noon. If the eastbound driver drove 5 miles per hour slower than the driver from Chicago, at what speed did each driver travel?

35. The formula $s = a + (n - 1)d$ relates terms in a sequence of numbers. When $n = 4$, $s = 26$. When $n = 10$, $s = 68$. Find a and d.

36. Repeat Problem 35 under these conditions: when $n = 5$, $s = 35$, and when $n = 12$, $s = 14$.

37. A man is 24 years older than his son. Eighteen years ago he was 4 times as old as his son. What are their current ages?

★ **38.** A number with tens digit t and units digit u is equal to $10t + u$. For example, $35 = 10 \cdot 3 + 5$. A particular two-digit number is 9 more than the sum of its digits. However, if the digits are reversed, the resulting number is 6 more than 7 times the sum of its digits. What is the original number?

39. A company representative knows that her car averages 22.4 miles per gallon on the highway and 16 miles per gallon in the city. In a week, she drove 508 miles and used 25.5 gallons of gasoline. How many of the miles were highway miles? It may be easier first to compute how many gallons were used on the highway. There is no obvious unit of time in this problem, but it is solved in the same way as rate-time problems, with gallons treated analogously to hours.

40. Repeat Problem 39 for a week in which the representative traveled 552 miles and used 27 gallons of gasoline.

41. A basement is flooded with 7,200 gallons of water. A pump that can remove 20 gallons per minute is used to pump out the water. An hour after the first pump is started, a second pump is added that works at the same rate. How long does it take to empty the basement?

42. Repeat Problem 41 with the second pump working at a rate of 30 gallons per minute.

★ **43.** A basement is flooded with 7,200 gallons of water as in Problem 41. A pump that can remove 20 gallons per minute is used to pump out the water. Sometime later, a second pump is added that works at the same

rate. If it takes 3 hours and 45 minutes to pump out the basement, how long did the second pump run?

★ **44.** Repeat Problem 43 with the second pump working at a rate of 30 gallons per minute and a total time of 3 hours and 12 minutes.

CHAPTER SUMMARY

4-1 REAL NUMBERS AND THE CARTESIAN COORDINATE SYSTEM

When numbers are associated with points on a number line, those that cannot be represented in the form a/b, where a and b are integers, are called **irrational numbers.** Rational numbers have decimal representations that repeat; irrational numbers have nonrepeating decimal representations. The rational and irrational numbers together make up **real numbers.** A **rectangular coordinate system** is formed with two perpendicular real number lines—one as a **horizontal axis,** the other as a **vertical axis**—intersecting at their origins. These axes divide the plane into four **quadrants.** Every point in the plane corresponds to its **coordinates,** a pair (a, b) where a is the coordinate of the point projected to the horizontal axis and b is the coordinate of the point projected to the vertical axis. The point $(0, 0)$ is the **origin.**

4-2 EQUATIONS AND STRAIGHT LINES

The **solution to an equation in two variables** is an ordered pair of numbers that satisfy the equation. The **graph of an equation** is the graph of its solution set. The graph of an equation of the form $Ax + By = C$ is a straight line; conversely, every straight line is the graph of such an equation, called a **linear equation.**

4-3 SLOPE OF A LINE

The **slope** of a nonvertical line is given by

$$\frac{y_2 - y_1}{x_2 - x_1} \qquad x_1 \neq x_2$$

where (x_1, y_1) and (x_2, y_2) are any two distinct points on the line. The equation $y = mx + b$ represents a line in **slope-intercept form;** m is the slope of the line, and b is the **y intercept,** the y coordinate of the point where the line crosses the y axis. A **vertical line** has an equation of the form $x = c$ and no slope. A **horizontal line** has an equation of the form $y = c$ and slope 0. Two nonvertical lines are parallel when their slopes are the same; they are perpendicular when the product of their slopes is -1.

4-4 SYSTEMS OF EQUATIONS AND SOLVING BY GRAPHING

To **solve a system of two linear equations in two variables** is to find all ordered pairs of real numbers that satisfy both equations. An approximate solution can be found by **graphing** the lines in the system. The system will have one solution when the lines intersect, no solution when the lines are parallel, and an infinite number of solutions when the equations represent the same line.

4-5 SOLVING SYSTEMS USING ELIMINATION BY SUBSTITUTION

The system also can be solved using **elimination by substitution:** solve one equation for one variable in terms of the other and then substitute into the other equation.

4-6 SOLVING SYSTEMS USING ELIMINATION BY ADDITION

Equivalent systems are systems with the same solution set. These operations produce equivalent systems:

1. Multiplying an equation by a nonzero constant
2. Adding one equation to another

The method of **elimination by addition** uses these operations to produce an equivalent system with an obvious solution. A system with no solution is called **inconsistent;** a system with an infinite number of solutions is called **dependent.**

4-7, 4-8 RATE-TIME AND MIXTURE PROBLEMS

A variety of word problems can be solved by two-variable methods that include solving a system of equations. These include rate-time problems that utilize the rate-time formula **Quantity = Rate × Time** ($q = rt$).

CHAPTER REVIEW EXERCISE

Work through all the problems in this chapter review and check the answers in the back of the book. Answers to all review problems are there, and following each answer is a number in italics indicating the section in which that type of problem is discussed. Where weaknesses show up, review appropriate sections in the text.

A *Problems 1 and 2 refer to this figure:*

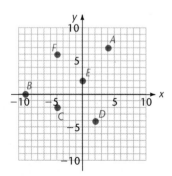

1. Give the coordinates for the points labeled *A*, *B*, and *C*.

2. Give the coordinates for the points labeled *D*, *E*, and *F*.

Graph in a rectangular coordinate system:

3. The points $(3, -4)$, $(-3, 4)$, $(3, 4)$, and $(-3, -4)$

4. The points $(2, 0)$, $(0, 3)$, $(-4, 0)$, and $(0, -5)$

5. $y = 2x - 3$ **6.** $y = \dfrac{x}{2} + 2$

7. $2x + y = 6$ **8.** $4x - 3y = 12$

9. $y = -4x - 2$ **10.** $x - 2y = 8$

11. $-x + 5y = 10$ **12.** $\frac{1}{2}x + \frac{2}{3}y = 1$

Find the slope of the given line:

13. The line through $(2, 3)$ and $(5, 15)$

14. The line having equation $y = -2x + 15$

15. The line having equation $y = 7$

16. The line having equation $x = 7$

17. The line having equation $3x + 5y = 8$

Find the y intercept of the given line:

18. The line through $(5, 22)$ and $(0, 19)$

19. The line having equation $y = -11x + 13$

Find the equation of the given line. Put your answer in slope–intercept form $y = mx + b$ if possible.

20. The line through $(0, 5)$, having slope -3

21. The line through $(1, 3)$, having slope -2

22. The line having slope 3 and y intercept -1

23. The horizontal line through $(5, -14)$

24. The vertical line through $(5, -14)$

25. The line through $(3, 0)$ and $(5, 6)$

26. The line through $(-1, 3)$ and $(2, -3)$

Solve graphically. Any unique solutions have integer values.

27. $x - y = 5$
$x + y = 7$

28. $3x + 5y = 15$
$4x + 10y = 20$

29. $x + y = 5$
$-x - y = 1$

30. $y = 3x + 5$
$6x - 2y = -10$

In Problems 31–36, solve the system by an elimination method:

31. $y = 2x - 7$
$x = 2y - 1$

32. $y = 4x + 3$
$5x - y = 4$

33. $2x + 3y = 7$
$3x - y = 5$

34. $3x + 2y = 1$
$5x + 6y = 7$

35. $y = 5x - 2$
$x = \frac{1}{5}y + \frac{2}{5}$

36. $2x + 5y = 8$
$y = -\frac{2}{5}x - 8$

37. If an automobile can travel 546 miles using 19.5 gallons of gasoline, what is its fuel economy rate in miles per gallon?

38. If the tropical rain forest on earth is disappearing at the rate of 50 acres per minute, how much will be lost in 1 day? A day consists of 24×60 minutes.

39. The perimeter of a rectangle is 480 yards. What are the dimensions if the length is $1\frac{1}{2}$ times as great as the width? Solve using two equations in two variables.

40. Solve using two equations and two unknowns: if you have 30 nickels and dimes in your pocket worth $2.30, how many of each do you have?

B **41.** Indicate true (T) or false (F):
(A) $\sqrt{2}$ is a real number.
(B) -5 is a real number.
(C) 3.47 is a real number.
(D) $-\frac{3}{5}$ is a real number.

42. Indicate true (T) or false (F):
(A) $-\frac{3}{4}$ is a rational number.
(B) 5 is an integer.
(C) -4 is a real number.
(D) $\sqrt{3}$ is an irrational number.

43. Approximate the irrational number $\sqrt{20}$ to three decimal places.

Find the slope of the given line:

44. The line through $(1, 6)$ and $(3, -7)$

45. The line having equation $\frac{2}{5}x + \frac{3}{8}y = 2$

46. The line through $(0, 0)$ parallel to the line having equation $y = 5x - 17$

Find the equation of the given line. Put your answer in slope-intercept form.

47. The line through $(-1, -4)$ and $(2, 1)$

48. The line through $(1, -3)$ and parallel to the line having equation $y = 4x - 2$

49. The line with slope $\frac{2}{7}$, passing through $(28, 5)$

50. The line with y intercept -3, passing through $(5, -4)$

Graph in a rectangular coordinate system:

51. The line through $(0, 0)$ having slope 3.

52. The line through $(10, 0)$ with slope $-\frac{1}{5}$

53. $y = \frac{1}{3}x - 2$

54. $4x - 3y = 10$

55. $y = \frac{1}{3}x - \frac{1}{2}$

56. $0.2x + 0.5y = 1$

57. $\frac{2}{5}x + \frac{1}{3}y = 2$

58. $-5x = 8y$

59. $y = 50x - 20$ for values of x between 0 and 5. Choose appropriate scales.

60. $y = -40x + 200$ for values of x between 0 and 5. Choose appropriate scales.

Solve graphically. Any unique solutions have integer values.

61. $2x - 3y = -3$
$3x + y = 12$

62. $x = 5y + 2$
$x = 6y + 6$

63. $y = \frac{1}{2}x + \frac{13}{2}$
$y = \frac{1}{3}x + \frac{20}{3}$

64. $x - y = 0$
$\frac{1}{3}x + \frac{1}{4}y = \frac{7}{6}$

65. $2x - 6y = -3$
$-\frac{2}{3}x + 2y = 1$

66. $0.75x = 6$
$0.5x + 0.25y = 5.5$

Solve by an elimination method:

67. $6u + 4v = -2$
$5u + 3v = -1$

68. $5m - 3n = 4$
$-2m + 4n = -10$

69. $3x + 5y = 10$
$-x - 4y = -8$

70. $-2x + 3.5y = 4$
$1.5x - 0.4y = 5.9$

71. $0.2x = 3.6y$
$1.8x = 2.4y + 2.1$

72. $3.5x - 1.6y = 4.1$
$-2.5x + 1.9y = 0.1$

Rewrite each linear equation in the form $y = mx + b$:

73. $3x - 5y = 7$

74. $-2x + 9y = 15$

75. $\frac{1}{2}x + y = \frac{2}{5}$

76. $0.24x - 1.44y = 3$

In Problems 77–80, rewrite each equation in the form $Ax + By = C$:

77. $y = 5x - 11$

78. $x = 4y + 6$

79. $1.5y - 3.2x = 3.3$

80. $\frac{2}{3}y = \frac{1}{6}x$

81. If 100 U.S. dollars can be exchanged for 666 French francs, what is the exchange rate in francs per dollar?

82. A batting average is a baseball player's rate of hits per at bat. If a player has a batting average of .375, how many hits did he get in 200 at bats?

83. In a bake sale, brownies were sold for 25 cents apiece and cupcakes for 30 cents apiece. If 310 of these items were sold and brought in $84.50, how many of each were sold?

84. Two boats leave from opposite ports along the same shipping route, which is 2,800 miles long. If one boat travels at 22 miles per hour and the other at 13 miles per hour, how long will it take them to meet? Set up an equation and solve.

85. If one car leaves town traveling 48 miles per hour, how long will it take a second car traveling at 54 miles per hour to catch up to the first car, if the second car leaves 1 hour later? Set up an equation and solve.

86. Part of $6,000 is to be invested at 10% and the rest at 6%. How much should be invested at each rate if the total annual return from both investments is to be $440? Set up two equations with two unknowns and solve.

87. A chemical storeroom contains a 50% alcohol solution and a 70% solution. How much of each should be used to obtain 100 milliliters of a 66% solution? Set up two equations with two unknowns and solve.

C *Solve by an elimination method:*

88. $x - 4y = 12$
$-\dfrac{x}{4} + y = 4$

89. $\frac{1}{2}x + \frac{1}{4}y = 1$
$x - y = \frac{1}{3}$

In Problems 90–93, determine the nature of the solution without graphing the equations. That is, determine if there is no solution, a unique solution, or an infinite number of solutions.

90. $x + 2y = 3$
$2x - 3y = 4$

91. $0.3x - 1.2y = 5.6$
$-1.2x + 4.8y = -1.4$

92. $x + 5y = \frac{3}{4}$
$8x + 40y = 6$

93. $y = 0.35x - 1.25$
$-20y + 7x = 25$

94. Two numbers whose sum is 90 have the property that two-thirds of one is 4 less than twice the other. What are the numbers?

95. In a triangle with perimeter 130 centimeters, the longest side is twice as long as the shortest side and $1\frac{1}{2}$ times as long as the remaining side. How long is each side?

96. A baseball team that won 90 out of 160 games in a season won more games in the second half of the season than in the first. The difference between its winning percentages in the two halves was $12\frac{1}{2}$ percent, that is, 0.125. How many games did it win in each half of the season?

97. A car leaves a town at 9 A.M. At 9:30 A.M., another car leaves from the same location on the same route and passes the first car at 12 noon. If the second car had not left until 10:00 A.M., it would have taken until 3:00 P.M. to overtake the first car. How fast is each car going?

98. If one printing press can print 90 leaflets per minute and a newer press can print 110, how long will it take to print 6,000 leaflets if the newer press is brought on the job 20 minutes after the first press starts and both continue until finished? Set up an equation and solve.

99. A radiator with a capacity of 12 liters contains a 40% solution of antifreeze in distilled water. How much should be drained and replaced with pure antifreeze to bring the level up to 50%? Set up an equation and solve.

100. Wishing to log some flying time, you have rented an airplane for 2 hours. You decide to fly due east until you have to turn around in order to be back at the airport at the end of the 2 hours. The cruising speed of the plane is 120 miles per hour in still air. Solve (A) and (B) using the two-equation–two-unknown method.
   (A) If there is a wind blowing from the east at 30 miles per hour, how long should you head east before you turn around, and how long will it take you to get back?
   (B) How far from the airport were you when you turned back?

CHAPTER PRACTICE TEST

The following practice test is provided for you to test your knowledge of the material in this chapter. You should try to complete it in 50 minutes or less. The answers in the back of the book indicate the section in the text that covers the material in the question. Actual tests in your class may vary from the practice test in difficulty, length, or emphasis, depending on the goals of your course or instructor.

Graph in a rectangular coordinate system:

1. The points $(2, -5)$, $(-4, 0)$, $(-1, 4)$

2. The line having equation $3x + 4y = 6$

3. The rectangle enclosed by the following four lines: $x = 2$, $y = 5$, the vertical line through $(8, 3)$, and the horizontal line having y intercept 2.

In Problems 4–5, find the slope-intercept form of the equation of the given line:

4. The line with slope 4 and y intercept -1

5. The line having equation $3x + 5y = 15$

6. Find the slope of the line through $(5, 12)$ and $(-4, -6)$.

7. Find the slope and y intercept of the line $-x + 2y = -7$.

8. Solve graphically. Any unique solution has integer values.

$$2x + 5y = 10$$
$$-x + 2y = 4$$

9. Solve using elimination by substitution:

$$5x - y = 8$$
$$-3x + 2y = 5$$

10. Solve using elimination by addition:

$$6x + 7y = 8$$
$$3x + 5y = 1$$

11. If a television network charged \$70,000 for a 20-second commercial, at what rate in dollars per minute are they charging?

12. A rectangle with perimeter 1,050 meters is $2\frac{1}{2}$ times as long as it is wide. What are the dimensions?

13. One trail mix consists of 30% dried fruit by weight. Another consists of 55% dried fruit by weight. How much of each should be used to obtain 10 pounds of a mix that is 40% dried fruit by weight?

14. Two members of a family begin an innertube float trip down a river, floating at $3\frac{1}{2}$ miles per hour. Three hours later, a third family member sets out after them in a canoe that she can paddle at 7 miles per hour with the assistance of the current. How long will it take her to catch up?

5

Inequalities in One and Two Variables

We have already observed that there is a natural order among the real numbers: one number is greater than another if it is to the right of the other on the number line. In this chapter we will explore this inequality relationship further. In particular, we will take three concepts we have considered for equations, namely,

1. Solving an equation in one variable
2. Graphing an equation in two variables
3. Solving a system of equations in two variables and develop the same ideas for inequalities.

Photo reference: see Exercise 5-2, Problem 78.

5-1 **Inequality Statements and Line Graphs**

♦ Inequality and Notation
♦ Inequality Statements and Line Graphs

To begin our treatment of inequalities, we will review what we already know about the inequality relation $a < b$ in the various number systems and consider the solution of inequalities in terms of points on the number line.

♦ INEQUALITY AND NOTATION

We have assumed since Chapter 1 that for any two natural numbers the larger one is easily recognized. We routinely extended this to the integers in Chapter 2 so that one number is larger than another if it is to the right of the other on the number line. The same idea is extended to the rational numbers (Chapter 3), but in practice it may be harder to recognize when one rational number is to the right of another. For example, which of $\frac{4}{7}$ and $\frac{6}{11}$ is larger? For this reason, we developed a more formal definition of "<" and ">" that applied to all real numbers:

Definition of $<$ and $>$

If a and b are real numbers, then we write

$$a < b$$

if there exists a positive real number p such that $a + p = b$. Equivalently,

$$b > a$$

if the difference $b - a$ is positive.

Thus to compare $\frac{4}{7}$ and $\frac{6}{11}$, we subtract:

$$\frac{4}{7} - \frac{6}{11} = \frac{44}{77} - \frac{42}{77} = \frac{2}{77}$$

Since $\frac{2}{77}$ is positive, $\frac{4}{7} > \frac{6}{11}$. If the difference had been negative, we would have concluded the opposite. We could also have compared $\frac{4}{7}$ and $\frac{6}{11}$ by considering their decimal approximations:

$$\frac{4}{7} = 0.57 \cdots \qquad \frac{6}{11} = 0.54 \cdots$$

However, understanding the formal definition of "<" and ">" given above is necessary for developing further properties of inequalities.

We can also use decimal representations to compare irrational numbers or to compare a rational and an irrational number. For example, to compare π and $\sqrt{10}$, we find

$$\pi = 3.14\cdots \qquad \sqrt{10} = 3.16\cdots$$

so $\sqrt{10} > \pi$.

Example 1
Real-Number
Inequalities

Which of the two numbers is larger?

(A) $\dfrac{4}{5}, \dfrac{14}{17}$ **(B)** $\dfrac{13}{19}, \dfrac{15}{23}$ **(C)** $\dfrac{\pi}{2}, \sqrt{2.5}$ **(D)** $2\pi, \sqrt{40}$

Solution

(A) $\dfrac{4}{5} - \dfrac{14}{17} = \dfrac{4 \cdot 17}{5 \cdot 17} - \dfrac{5 \cdot 14}{5 \cdot 17} = \dfrac{68}{85} - \dfrac{70}{85} = -\dfrac{2}{85}$

Therefore, $\dfrac{14}{17} - \dfrac{4}{5} = \dfrac{2}{85}$ is positive, so $\dfrac{14}{17} > \dfrac{4}{5}$.

(B) $\dfrac{13}{19} - \dfrac{15}{23} = \dfrac{13 \cdot 23}{19 \cdot 23} - \dfrac{19 \cdot 15}{19 \cdot 23} = \dfrac{299}{437} - \dfrac{285}{437} = \dfrac{14}{437}$ is positive so $\dfrac{13}{19} > \dfrac{15}{23}$.

(C) $\dfrac{\pi}{2} = 1.57\cdots \qquad \sqrt{2.5} = 1.58\cdots$

Therefore $\sqrt{2.5} > \pi/2$.

(D) $2\pi = 6.28\cdots \qquad \sqrt{40} = 6.32\cdots$

Therefore $\sqrt{40} > 2\pi$.

Matched Problem 1

Which of the two numbers is larger?

(A) $\dfrac{3}{8}, \dfrac{7}{19}$ **(B)** $\dfrac{2}{7}, \dfrac{3}{11}$ **(C)** $3\pi, \sqrt{89}$ **(D)** $\dfrac{\pi}{4}, \dfrac{15}{19}$

The inequality symbols $<$ and $>$ may be combined with the equality symbol $=$ to form statements such as $a \le b$ and $a \ge b$, so we have four basic inequality statements:

$a < b$	a is less than b
$a > b$	a is greater than b
$a \le b$	a is less than or equal to b
$a \ge b$	a is greater than or equal to b

♦ **INEQUALITY STATEMENTS AND LINE GRAPHS**

The inequality symbols have a very clear geometric interpretation on the real number line. If $a < b$, then a is to the left of b; if $c > d$, then c is to the right of d.

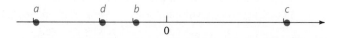

Now let us turn to inequality statements of the form

$$x \geq -3 \qquad -2 < x \leq 3$$

To solve such inequality statements is to find the set of all replacements of the variable x from some specified set of numbers that makes the inequality true. This set is called the **solution set** for the inequality. To **graph** an inequality statement on a real number line is to graph its solution set.

Example 2
Graphing Inequalities

Graph each inequality statement on a real number line:

(A) $x < 2$ **(B)** $x \geq -3$

Solution   **(A)** The solution set for

$$x < 2$$

is the set of all real numbers less than 2. Graphically, the solution can be represented as follows:

The parenthesis at the point 2 is used to indicate that 2 is not included in the solution set.

(B) The solution set for

$$x \geq -3$$

is the set of all real numbers greater than or equal to -3. Graphically, the solution can be represented as follows:

The bracket at the point -3 is used to indicate that -3 is included in the solution set.

Matched Problem 2   Graph each inequality statement on a real number line:

(A) $x > -3$ **(B)** $x \leq 2$

Example 3
Graphing Inequalities

Graph each inequality statement on a real number line:

(A) $-2 < x \leq 3$ **(B)** $-2 \leq x < 3$

Solution   **(A)** The **double inequality**

$$-2 < x \leq 3$$

is a short way of writing

$$-2 < x \qquad \text{and} \qquad x \leq 3$$

which means that x is greater than -2 and at the same time x is less than or equal to 3. In other words, x can be any real number between -2 and 3, excluding -2 but including 3. This set is graphed as follows:

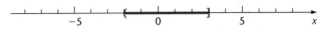

(B) The solution set for

$$-2 \le x < 3$$

is again the set of all real numbers between -2 and 3, but now including -2 and excluding 3. This set is graphed as follows:

Matched Problem 3   Graph each inequality statement on a real number line:

(A) $-4 < x < 2$ **(B)** $-4 \le x \le 2$

Solution sets like those encountered in Examples 2 and 3 are called **intervals.** In some situations you may find intervals graphed using open and closed dots rather than parentheses and brackets, respectively. That is,

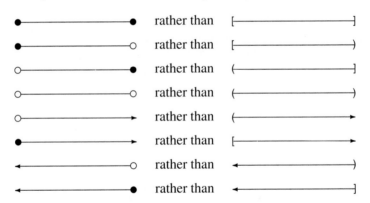

However, the parenthesis-and-bracket notation has advantages that will be apparent in subsequent algebra courses. For this reason parenthesis-and-bracket notation is used throughout this text.

Answers to Matched Problems

1. (A) $\dfrac{3}{8}$ **(B)** $\dfrac{2}{7}$ **(C)** $\sqrt{89}$ **(D)** $\dfrac{15}{19}$

2. (A) ⟶ **(B)** ⟶

3. (A) ⟶ **(B)** ⟶

EXERCISE 5-1

A *Which of the two numbers is larger?*

1. $-8, -10$ **2.** $-11, -4$ **3.** $-9, -6$

4. $-5, -7$ **5.** $\dfrac{4}{9}, \dfrac{11}{26}$ **6.** $\dfrac{3}{7}, \dfrac{8}{19}$

7. $\dfrac{2}{3}, \dfrac{6}{11}$ **8.** $\dfrac{5}{8}, \dfrac{8}{13}$ **9.** $\dfrac{2}{13}, \dfrac{3}{20}$

10. $\dfrac{11}{15}, \dfrac{23}{32}$ **11.** $\pi, \dfrac{22}{7}$ **12.** $2\pi, \dfrac{43}{7}$

13. $2\pi, \dfrac{439}{70}$ **14.** $\pi, \dfrac{217}{69}$ **15.** $\pi, \sqrt{9.9}$

16. $\dfrac{\pi}{2}, \sqrt{\dfrac{17}{7}}$ **17.** $\dfrac{\pi}{2}, \sqrt{\dfrac{23}{9}}$ **18.** $\pi, \sqrt{\dfrac{69}{7}}$

Referring to the number line below, replace each question mark in Problems 19–24 with either $<$ or $>$:

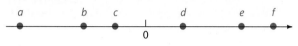

19. $a \, ? \, d$ **20.** $e \, ? \, a$ **21.** $b \, ? \, a$

22. $0 \, ? \, d$ **23.** $e \, ? \, f$ **24.** $d \, ? \, e$

Graph each inequality statement on a real number line.

25. $x > -4$ **26.** $x < 1$ **27.** $x \le -1$

28. $x \ge -1$ **29.** $x < -3$ **30.** $x > 5$

31. $x \ge 2$ **32.** $x \le 4$

B *Which of the two numbers is larger?*

33. $-\dfrac{6}{11}, -\dfrac{8}{15}$ **34.** $-\dfrac{5}{7}, -\dfrac{14}{19}$

35. $-\pi, -3.14$ **36.** $-\dfrac{\pi}{2}, -1.58$

37. $5\sqrt{3}, 6\sqrt{2}$ **38.** $\sqrt{3} \div 6, \sqrt{2} \div 5$

39. $\sqrt{3} \div 2, \sqrt{10} \div 4$ **40.** $5\sqrt{7}, \sqrt{11}$

Graph each inequality statement on a real number line for $x = $ a real number:

41. $-3 < x \le 2$ **42.** $-5 \le x < 0$

43. $-3 < x < -1$ **44.** $-5 < x < -3$

45. $-3 \le x \le 3$ **46.** $-1 \le x \le 4$

47. $2 < x \le 5$ **48.** $3 \le x < 6$

49. $1 < x < 4$ **50.** $3 \le x < 5$

Rewrite the pair of inequalities as one double inequality:

51. $x \le 6$ and $x > -1$ **52.** $x \ge 0$ and $x < 6$

53. $x > 2$ and $x \le 6$ **54.** $x \le 5$ and $x > -2$

C *In Problems 55–60, determine which of the two numbers is larger?*

55. $-\sqrt{2}, 8 - 3\pi$ **56.** $2\pi - 6, -\dfrac{2}{7}$

57. $2\pi - 7, -\sqrt{2} \div 2$ **58.** $5 - 2\pi, \sqrt{7} \div 2$

59. $\pi + \sqrt{2}, \sqrt{21}$ **60.** $\sqrt{2} + \sqrt{3}, \pi$

61. If we add a positive number to any real number, will the sum be to the right or left of the original number on a real number line?

62. If we subtract a positive number from any real number, will the difference be to the right or left of the original number on a real number line?

APPLICATIONS

63. *Meteorology* A tropical rainstorm is officially designated as a *tropical storm* if its winds are at least 39 miles per hour (mph) and no more than 73 mph. It is designated a *hurricane* if its winds exceed 73 mph. Let w represent wind speed, and translate these two statements into inequalities.

64. *Astronomy* The temperature on the planet Mercury ranges between a low of $-346°F$ to a high of $950°F$. Use t as a variable and express this statement as a double inequality.

65. *Astronomy* Temperatures on the earth's moon range from $-170°C$ on the dark side to $134°C$ on the bright side. Use t as a variable and express this statement as a double inequality.

66. *Meteorology* The stratosphere is the second gaseous layer in the earth's atmosphere. It is found between 7 and 50 miles, inclusive, out from the earth's surface. Write a double inequality expressing this statement with the distance from the earth's surface denoted by d.

Problems 67–74 define some terms used by climatologists to describe climates of the world depending on average rainfalls, in centimeters, and average temperatures, in degrees Celsius. Use the following variables to translate each definition into inequalities:

$R = $ *Average annual rainfall*

$R_d = $ *Average rainfall in the driest month*

$R_w = $ *Average rainfall in the wettest month*

$T = $ *Average annual temperature*

$T_h = $ *Average temperature in the hottest month*

$T_c = $ *Average temperature in the coldest month*

67. In a *subtropical climate* the average temperature in the coldest month is between $-3°C$ and $18°C$, inclusive.

68. A *dry summer subtropical climate* is a subtropical climate (see Problem 67) also having at least 3 times as much rain in the wettest month as in the driest, and having less than 4 centimeters of rain in the driest.

69. In a *continental climate* the average temperature in the warmest month is 10°C or more, and the average temperature in the coldest month is −3°C or below.

70. In a *polar climate* the average temperature in the warmest month is less than 10°C.

71. In a *humid tropical climate* the average temperature of every month is 18°C or higher. Note that this is

equivalent to saying something about the coldest month.

72. A *rain forest climate* is a humid tropical climate (see Problem 71) where in addition the driest month has at least 6 centimeters of rain.

73. A *monsoon climate* is a humid tropical climate (see Problem 71) where in addition the amount of rainfall in the driest month is less than 6 centimeters but greater than or equal to $10 - (R/25)$.

74. A *savannah climate* is a humid tropical climate (see Problem 71) where the amount of rainfall in the driest month is less than $10 - (R/25)$.

  ## 5-2 Solving Linear Inequalities in One Variable

♦ Inequality Properties
♦ Solving Inequalities
♦ Applications

In Section 5-1 we introduced inequality statements of the form

$$x < 5 \qquad -3 \le x < 4 \qquad x \ge -4$$

with obvious solutions. In this section we will consider inequality statements that do not have obvious solutions. For example, can you guess the real-number solutions for

$$3x - 4 > x - 2$$

The methods of this section will make solving this type of inequality almost as easy as solving *equations* of the corresponding type.

♦ INEQUALITY PROPERTIES

When solving equations, we made considerable use of the addition, subtraction, multiplication, and division properties of equality. We can use similar properties to help us solve inequalities. We will start with several numerical examples and generalize from these.

1. Add the same quantity to each side of an inequality:

Add a positive quantity to each side.	Add a negative quantity to each side.
$-2 < 4$	$-2 < 4$
$-2 + 3 \quad 4 + 3$	$-2 + (-3) \quad 4 + (-3)$
$1 < 7$	$-5 < 1$
The sense of inequality remains the same.	The sense of inequality remains the same.

2. Subtract the same quantity from each side of an inequality:

Subtract a positive quantity from each side.	Subtract a negative quantity from each side.
$-2 < 4$	$-2 < 4$
$-2 - 5 \quad 4 - 5$	$-2 - (-7) \quad 4 - (-7)$
$-7 < -1$	$5 < 11$
The sense of inequality remains the same.	The sense of inequality remains the same.

3. Multiply each side of an inequality by the same nonzero quantity:

Multiply both sides by a positive quantity.	Multiply both sides by a negative quantity.
$-2 < 4$	$-2 < 4$
$2(-2) \quad 2(4)$	$(-2)(-2) \quad (-2)(4)$
$-4 < 8$	$4 > -8$
The sense of inequality remains the same.	The sense of inequality reverses.

└──── Note difference ────┘

4. Divide each side of an inequality by the same nonzero quantity:

Divide both sides by a positive quantity.	Divide both sides by a negative quantity.
$-2 < 4$	$-2 < 4$
$\dfrac{-2}{2} \quad \dfrac{4}{2}$	$\dfrac{-2}{-2} \quad \dfrac{4}{-2}$
$-1 < 2$	$1 > -2$
The sense of inequality remains the same.	The sense of inequality reverses.

└──── Note difference ────┘

These descriptions generalize completely and are summarized as follows without proof:

Inequality Properties

For any real numbers a, b, and c:

1. If $a < b$, then $a + c < b + c$.　　　　Addition Property

$\qquad -2 < 4 \qquad -2 + 3 < 4 + 3$

2. If $a < b$, then $a - c < b - c$.　　　　Subtraction Property

$\qquad -2 < 4 \qquad -2 - 3 < 4 - 3$

3. If $a < b$ and c is positive,   Multiplication Property
then $ca < cb$.

$$-2 < 4 \qquad 3(-2) < 3(4)$$ Note the difference between 3 and 4.

4. If $a < b$ and c is negative,   Multiplication Property
then $ca > cb$.

$$-2 < 4 \qquad (-3)(-2) > (-3)(4)$$

5. If $a < b$ and c is positive,   Division Property
then $\dfrac{a}{c} < \dfrac{b}{c}$.

$$-2 < 4 \qquad \dfrac{-2}{2} < \dfrac{4}{2}$$ Note the difference between 5 and 6.

6. If $a < b$ and c is negative,   Division Property
then $\dfrac{a}{c} > \dfrac{b}{c}$.

$$-2 < 4 \qquad \dfrac{-2}{-2} > \dfrac{4}{-2}$$

Similar properties hold if each inequality sign is reversed or if $>$ is replaced with $\geq$ and $<$ is replaced with $\leq$. Thus, we find that we can perform essentially the same operations on inequality statements to produce equivalent statements that we perform on equations to produce equivalent equations with this exception:

The inequality sign reverses if we multiply or divide both sides of an inequality by a negative number.

♦ SOLVING INEQUALITIES

We will now solve some inequalities using the properties presented above. Unless otherwise stated, in **solving an inequality** we find *all* real-number solutions.

Example 1
Solving Inequalities

Solve each inequality:

(A) $x + 3 < -2$ **(B)** $5x \geq 10$ **(C)** $-3x - 2 < 7$ **(D)** $-\dfrac{x}{2} > -3$

$x < -5$ $x \geq 2$ $-3x < 9$ $x > 6$
                                       $x < -3$

Solution   **(A)** $x + 3 < -2$ Subtract 3 from each side.

$$x + 3 - 3 < -2 - 3$$ The inequality sign does not reverse.

$$x < -5$$ The solution set is the set of all real numbers less than -5.

(B) $5x \geq 10$ Divide each side by 5.

$$\boxed{\frac{5x}{5} \geq \frac{10}{5}}$$ The inequality sign does not reverse, since we divided by a positive number.

$x \geq 2$ The solution set is the set of all real numbers greater than or equal to 2.

(C) $-3x - 2 < 7$ Add 2 to each side.

$$\boxed{-3x - 2 + 2 < 7 + 2}$$ The inequality sign does not reverse.

$-3x < 9$ Divide each side by -3.

$$\boxed{\frac{-3x}{-3} > \frac{9}{-3}}$$ The inequality sign reverses, since we divided by a negative number.

$x > -3$ The solution set is the set of all real numbers greater than -3.

(D) $-\dfrac{x}{2} > -3$ Multiply each side by -2.

$$\boxed{(-2)\left(-\frac{x}{2}\right) < (-2)(-3)}$$ The inequality sign reverses, since we multiplied by a negative number.

$x < 6$ The solution set is the set of all real numbers less than 6.

Matched Problem 1   Solve each inequality:

(A) $x - 3 \geq 5$ **(B)** $-6x < 18$ **(C)** $3x + 5 \leq -1$ **(D)** $-\dfrac{x}{3} > -4$

$x \geq 8$ $x < -3$ $x \leq -2$ $x > 12$

Example 2
Solving Inequalities

Solve and graph:

(A) $3(x - 1) + 5 \leq 5(x + 2)$ **(B)** $\dfrac{3x - 2}{2} - 5 > 1 - \dfrac{x}{4}$

$3x+2 \leq 5x+10$ $-2x \leq 8$ $6x-4-20 > 4-x$
$x=-4$

Solution   **(A)** $3(x - 1) + 5 \leq 5(x + 2)$ Simplify left and right sides. $7x=28$

$3x - 3 + 5 \leq 5x + 10$ Isolate x on the left side. $x > 4$

$3x + 2 \leq 5x + 10$

$3x \leq 5x + 8$

$-2x \leq 8$

$x \geq -4$ The inequality sign reverses.

(B) $\dfrac{3x - 2}{2} - 5 > 1 - \dfrac{x}{4}$ Multiply both sides by 4, the LCM of the denominators.

$4\left(\dfrac{3x - 2}{2} - 5\right) > 4\left(1 - \dfrac{x}{4}\right)$ The inequality sign does not reverse.

$2(3x - 2) - 20 > 4 - x$ Simplify the left side.

$6x - 4 - 20 > 4 - x$

$6x - 24 > 4 - x$ Isolate x on the left side.

$6x > 28 - x$

$7x > 28$

$x > 4$ The inequality sign does not reverse.

$$\begin{array}{c} \text{4} \\ \underset{-5 \quad\quad 0 \quad\quad 5}{\longmapsto\!\longmapsto} \quad x \end{array}$$

Matched Problem 2   Solve and graph:

(A) $2(2x + 3) \geq 6(x - 2) + 10$ **(B)** $\dfrac{2x - 3}{3} - 2 > \dfrac{x}{6} - 1$

$4x + 6 \geq 6x - 8 + 10$
$-2x \geq -4 \quad x \geq 2 \quad 2$

$4x - 6 - 12 > x - 6$
$3x > 12 \quad x > 4$

Example 3
Solving Double Inequalities

Solve and graph:

(A) $-8 \leq 3x - 5 < 7$ **(B)** $-1 < 1 - 2x < 5$

$-3 \leq 3x < 12$
$-1 \leq x < 4$

Solution   **(A)** We proceed as above, except that we try to isolate x in the middle:

$-8 \leq 3x - 5 < 7$ Add 5 to each member.

$\boxed{-8 + 5 \leq 3x - 5 + 5 < 7 + 5}$

$-3 \leq 3x < 12$ Divide each member by 3.

$\boxed{\dfrac{-3}{3} \leq \dfrac{3x}{3} < \dfrac{12}{3}}$ The inequality signs do not reverse.

$-1 \leq x < 4$

$$\begin{array}{c} \text{-1} \quad\quad \text{4} \\ \underset{-5 \quad\quad 0 \quad\quad 5}{\longmapsto\!\longmapsto} \quad x \end{array}$$

We could have split the double inequality into two simpler inequalities

$-3 \leq 3x - 5$ and $3x - 5 < 7$,

solved each separately, and then combined the solutions. However, the same steps would apply to each inequality. The first solution process solves the two at the same time.

(B) $-1 < 1 - 2x < 5$ Subtract 1 from each member.

$$\boxed{-1 - 1 < 1 - 2x - 1 < 5 - 1}$$

$$-2 < -2x < 4$$ Divide each member by -2.

$$\boxed{\frac{-2}{-2} > \frac{-2x}{-2} > \frac{4}{-2}}$$ The inequality signs reverse.

$$1 > x > -2$$

or

$$-2 < x < 1$$

Matched Problem 3   Solve and graph:

(A) $-3 < 2x + 3 \leq 9$ **(B)** $-2 \leq 4 - 3x < 1$

$-6 < 2x \leq 6$ $-3 < x < 3$ $-6 \leq -3x < -3$
$2 > x > -1$

◆ APPLICATIONS

We conclude this section with a word problem and an application.

Example 4
A Number Problem

What numbers satisfy the condition "4 more than twice a number is less than or equal to that number"?

Solution   Let $x =$ The number. Then

$$2x + 4 \leq x$$

$$x \leq -4 \quad ?$$

Thus, all numbers less than or equal to -4 satisfy the condition.

Matched Problem 4   What numbers satisfy the condition "6 less than 3 times a number is greater than or equal to 9"?

$3x - 6 \geq 9$ $3x \geq 15$
$x \geq 5$

Example 5
Inequality Application

If the temperature for a 24-hour period in Antarctica ranged between $-49°F$ and $14°F$, that is, $-49 \leq F \leq 14$, what was the temperature range in Celsius degrees? Recall that $F = \frac{9}{5}C + 32$.

Solution   Since $F = \frac{9}{5}C + 32$, we replace F in $-49 \leq F \leq 14$ with $\frac{9}{5}C + 32$ and solve the double inequality:

$$-49 \le \frac{9}{5}C + 32 \le 14$$

$$-49 - 32 \le \frac{9}{5}C + 32 - 32 \le 14 - 32$$

$$-81 \le \frac{9}{5}C \le -18$$

$$\left(\frac{5}{9}\right)(-81) \le \left(\frac{5}{9}\right)\left(\frac{9}{5}C\right) \le \left(\frac{5}{9}\right)(-18)$$

$$-45 \le C \le -10$$

Matched Problem 5   Repeat Example 5 for $-31 \le F \le 5$.

Answers to   **1. (A)** $x \ge 8$ **(B)** $x > -3$ **(C)** $x \le -2$ **(D)** $x < 12$
Matched Problems   **2. (A)** $x \le 4$

(B) $x > 4$

3. (A) $-3 < x \le 3$

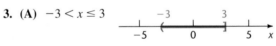

(B) $1 < x \le 2$

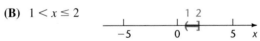

4. $x \ge 5$
5. $-35 \le C \le -15$

EXERCISE 5-2

A *Graph on a real number line:*

1. $x < 2$ **2.** $x > 3$ **3.** $x > -2$

4. $x < -3$ **5.** $x \le 3$ **6.** $x \ge 2$

7. $x \ge -2$ **8.** $x \le -1$

Write in inequality notation:

9.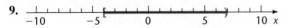

10.

11.

12.

13.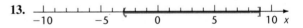

14.

15.

16.

Solve and graph:

17. $x - 2 > 5$ **18.** $x - 4 < -1$

19. $x + 5 < -2$ **20.** $x + 3 > -4$

21. $2x > 8$

22. $3x < 6$

23. $-2x \geq 8$

24. $-3x \leq 6$

25. $\dfrac{x}{3} < -7$

26. $\dfrac{x}{5} > -2$

27. $\dfrac{x}{-3} \leq -7$

28. $\dfrac{x}{-5} \geq -2$

29. $3x + 7 < 13$

30. $2x - 3 > 5$

31. $-2x + 8 < 4$

32. $-4x - 7 > 5$

33. $7x - 8 \leq 4x + 7$

34. $6m + 2 \leq 4m + 6$

35. $4y - 7 \geq 9y + 3$

36. $x - 1 \leq 4x + 8$

B *Solve and graph:*

37. $3 - (2 + x) > -9$

38. $2(1 - x) \geq 5x$

39. $3 - x \geq 5(3 - x)$

40. $2(x - 3) + 5 < 5 - x$

41. $3(u - 5) - 2(u + 1) \geq 2(u - 3)$

42. $4(2u - 3) < 2(3u + 1) - (5 - 3u)$

43. $\dfrac{m}{6} - \dfrac{1}{2} > \dfrac{2}{3} + m$

44. $\dfrac{x}{5} - 3 < \dfrac{3}{5} - x$

45. $-2 - \dfrac{1 + x}{3} < \dfrac{x}{4}$

46. $\dfrac{x}{4} - \dfrac{2x + 1}{6} < -1$

47. $2 < x + 3 < 5$

48. $-3 \leq x - 5 \leq 8$

49. $-4 \leq 5x + 6 \leq 21$

50. $2 < 3x - 7 < 14$

51. $-4 \leq \dfrac{9}{5}C + 32 \leq 68$

52. $-1 \leq \dfrac{2}{3}m + 5 \leq 11$

C **53.** $-10 \leq \dfrac{5}{9}(F - 32) \leq 25$

54. $-5 \leq \dfrac{5}{9}(F - 32) \leq 10$

55. $-3 \leq 3 - 2x < 7$

56. $-5 < 7 - 4x \leq 15$

57. $4 < 5 - 6x < 7$

58. $-4 \leq -3 - 2x \leq -1$

59. $-2 < 4 - 10x \leq 16$

60. $5 \leq 10 - 15x < 20$

61. $0.25 \leq 1.5x - 3 \leq 1.75$

62. $1.5 < 3x - 2.5 \leq 5.5$

63. $0.75 \leq 5 - 2.5x \leq 4.75$

64. $3.5 < 4.5 - 3x \leq 5.5$

65. $\frac{1}{6} < \frac{1}{3} - \frac{1}{4}x \leq \frac{1}{2}$

66. $\frac{1}{20} < \frac{1}{4} - \frac{1}{2}x \leq \frac{1}{5}$

67. $\frac{3}{4} \leq 2 - \frac{1}{8}x < \frac{9}{4}$

68. $-\frac{4}{5} < \frac{1}{2} - \frac{7}{10}x < 2$

APPLICATIONS

In Problems 69–92, set up appropriate inequality statements and solve. The moderately difficult problems are marked with a star (*). The easier problems are not marked.

69. *Number* What numbers satisfy the condition ''3 less than twice the number is greater than or equal to -6''?

70. *Number* What numbers satisfy the condition ''5 less than 3 times the number is less than or equal to 4 times the number''?

71. *Geometry* If the perimeter of a rectangle with a length of 10 centimeters must be smaller than 30 centimeters, how small must the width be?

72. *Geometry* If the area of a rectangle of length 10 inches must be greater than 65 square inches, how large must the width be?

73. *Weather* Recorded temperatures in the United States have ranged from $-62°C$ in Prospect Creek, Alaska, to $57°C$ in Death Valley, California. What is this range in degrees Fahrenheit? Recall that $C = \frac{5}{9}(F - 32)$.

74. *Psychology* IQ is defined by the formula

$$IQ = 100\left(\frac{\text{Mental age}}{\text{Chronological age}}\right)$$

(See Problem 32, Section 3-8.) What is the mental age range for a group of 18-year-old students with IQs in the range $90 \leq IQ \leq 130$?

75. *Earth science* If the temperature on the ground is $70°F$, the temperature T in degrees Fahrenheit at an altitude of A thousand feet is given by

$$T = 70 - 5.5A$$

(See Problem 23, Section 3-8.) If the temperature varies between $50°F$ and $20°F$ during a portion of a balloon ascent, what is the range of altitudes for this period?

76. *Earth science* The formula in Problem 75 can be rewritten as

$$A = \frac{140}{11} - \frac{2}{11}T$$

What is the range of temperatures encountered as a balloon descends from 20,000 feet to 500 feet?

77. **Earth science** Seawater pressure p in pounds per square inch at a depth d feet is given by

$$p = 15 + \frac{15}{33}d$$

(See Problem 21, Section 3-8.) What range of depths corresponds to $45 \le p \le 90$?

78. **Earth science** The formula in Problem 77 can be rewritten as

$$d = \frac{33}{15}(p - 1)$$

What range of pressures is encountered as a diver descends from 20 feet below the surface to 60 feet below the surface?

★ 79. **Physics** The velocity at time t of an object shot upward with an initial velocity of 160 feet per second is given by

$$v = 160 - 32t$$

(See Problem 25, Section 2-8.) What is the range of velocities corresponding to $2 \le t \le 5$?

★ 80. **Physics** The formula in Problem 79 can be rewritten as

$$t = 5 - \frac{1}{32}v$$

What range of time corresponds to $144 \ge v \ge 16$?

★ 81. **Life science** The body-mass index (BMI) is defined by

$$\text{BMI} = \frac{\text{Weight in kilograms}}{(\text{Height in meters})^2}$$

and is used to determine satisfactory weights for a given height. An index between 20 and 25, inclusive, is considered satisfactory. What are the corresponding satisfactory weights for a person 178 centimeters tall?

★ 82. **Life science** Repeat Problem 81 for a person 200 centimeters tall.

83. **Business** A building initially valued at $240,000 is depreciated over 20 years by the formula

$$V = 240{,}000 - 12{,}000t$$

where V gives the value of the building at time t in years. Over what range in time is the building valued between $100,000 and $150,000?

84. **Business** The formula in Problem 83 can be rewritten as

$$t = 20 - \frac{1}{12{,}000}V$$

What range in values correspond to $3.5 \le t \le 8.5$?

85. **Finance** The value of f francs is equivalent to d U.S. dollars, as given by the formulas

$$f = \frac{20}{3}d \qquad \text{or} \qquad d = \frac{3}{20}f$$

If a restaurant in France has prices ranging from 40 francs to 200 francs, what is the equivalent range in dollars?

86. **Finance** A restaurant in New York City has a price range in dollars d of $12 \le d \le 60$. Use the formulas from Problem 85 to find the equivalent range in French francs.

★ 87. **Sports** The international rules for soccer specify a playing field of between 100 and 130 yards long. The rules convert this to a range in meters of 90 to 120. What is the actual range in meters? Use the conversion table inside the back cover of the text.

★ 88. **Sports** The international rules for soccer specify a playing field of between 50 and 100 yards wide. The rules convert this to a range in meters of 45 to 90. What is the actual range in meters? Use the conversion table inside the back cover of the text.

89. **Chemistry** How many deciliters of pure alcohol must be added to 60 deciliters of a 35% solution to obtain a solution that is at least 60% alcohol?

90. **Chemistry** How many centiliters of pure acid must be added to 80 milliliters of 30% solution to obtain a solution that is at least 50% acid?

91. **Chemistry** How many deciliters of distilled water must be added to 400 deciliters of a 35% solution to obtain a solution that is at most a 20% solution?

92. **Chemistry** How many centiliters of distilled water must be added to 800 deciliters of a 70% solution to obtain a solution that is at most a 40% solution?

5-3 Graphing Linear Inequalities in Two Variables

♦ Half-Planes and Boundary Lines
♦ Graphing Linear Inequalities

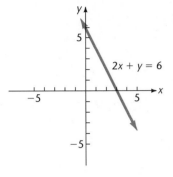

Figure 1

Linear equations in one variable, such as $3x - 2 = 7$, were introduced in Chapters 2 and 3. The solution of such an equation is generally a single point on the real number line, in this case, $x = 3$. Linear inequalities in one variable, such as $3x - 2 < 7$, were introduced in Sections 5-1 and 5-2. The solution for these inequalities is generally an interval on the real number line, in this case, the interval $x < 3$.

Linear equations in two variables were introduced in Section 4-2. An equation such as $2x + y = 6$ has as its solution the set of all points on a straight line in the plane. See Figure 1.

In this section, we will consider linear inequalities in two variables, such as $2x + y < 6$.

♦ **HALF-PLANES AND BOUNDARY LINES**

A line divides a plane into two halves called **half-planes.** A vertical line divides it into right and left half-planes; a nonvertical line divides it into upper and lower half-planes. See Figure 2.

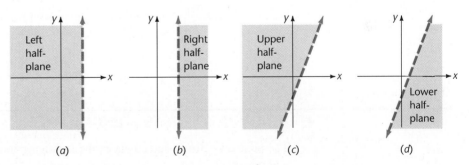

(a) (b) (c) (d)

Figure 2

To investigate the half-planes determined by $2x + y = 6$, rewrite the equation as $y = -2x + 6$. For any given value of x, there is exactly one value for y such that (x, y) lies on the line. For example, for $x = 1$, $y = (-2) \cdot 1 + 6 = 4$. For the same value of x and smaller values of y, the point (x, y) will lie below the line, since $y < -2x + 6$. For example, if x still equals 1 and $y < 4$, then $y < -2x + 6$. Thus the lower half-plane corresponds to the solution to the inequality $y < -2x + 6$. Similarly, the upper half-plane corresponds to $y > -2x + 6$. See Figure 3.

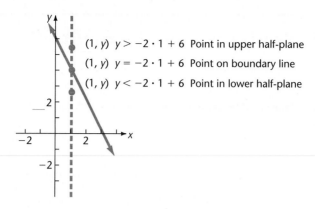

Figure 3

The four inequalities formed from $y = -2x + 6$ by replacing the $=$ sign by $<$, $>$, $\leq$, or $\geq$ are

$$y < -2x + 6 \qquad y \leq -2x + 6$$
$$y > -2x + 6 \qquad y \geq -2x + 6$$

The graph of each is a half-plane, excluding the boundary line for $<$ and $>$, and including it for $\leq$ and $\geq$. The half-planes are graphed as shaded regions in Figure 4. Excluded boundary lines are shown as broken lines, included boundaries as solid lines.

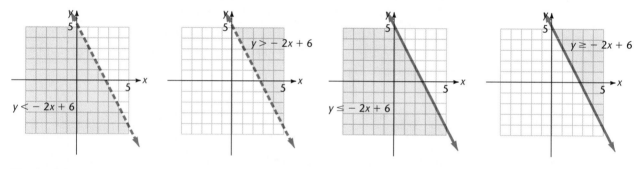

Figure 4

♦ GRAPHING LINEAR INEQUALITIES

The preceding discussion suggests the following important result, which we state without proof.

Graphing Linear Inequalities

The graph of a linear inequality

$$Ax + By < C \qquad \text{or} \qquad Ax + By > C$$

with $B \neq 0$ is either the upper half-plane or the lower half-plane, but not both, determined by the line $Ax + By = C$. If $B = 0$, the graph of

---
Example 3
A Solution Problem

A chemical storeroom has a 40% acid solution and an 80% solution. How many deciliters of each can be used to obtain a solution that is at least 50% acid? Set up an appropriate inequality statement, simplify it if necessary, and graph the solution in the first quadrant.

Solution   Let

$$x = \text{Amount of 40\% solution used}$$

$$y = \text{Amount of 80\% solution used}$$

	Volume, in Deciliters	Percent Acid	Amount of Acid
Component 1: 40% solution	x	40%	$0.4x$
Component 2: 80% solution	y	80%	$0.8y$
Mixture: 50% solution	$x + y$	$\geq 50\%$	$0.4x + 0.8y$

The amounts of acid in the components must add to *at least* 50% of the volume of the mixture, that is,

$$0.4x + 0.8y \geq 0.5(x + y)$$

We simplify this inequality using the inequality properties introduced in the last section. We would like to rewrite it in the form $Ax + By \geq C$ so that we can graph the solution.

$$0.4x + 0.8y \geq 0.5x + 0.5y$$

$$-0.1x + 0.3y \geq 0 \qquad \text{Multiply by 10 to clear the decimal fractions.}$$

$$-x + 3y \geq 0$$

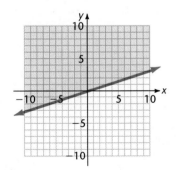

Figure 9

We graph this inequality in Figure 9.

Since x and y represent amounts of chemical solutions, neither can be negative. Thus, we are interested only in the portion of the graph that lies in the first quadrant, the quadrant where $x \geq 0$ and $y \geq 0$. The graph of the possible values of x and y that yield a strong enough mixture is therefore the portion of the above graph that lies in the first quadrant as shown in Figure 10.

The inequality $-x + 3y \geq 0$ is equivalent to $3y \geq x$, and this allows us to say something about the proportions in the mixture. The volume in the mixture is $x + y$, so

$$\text{Volume} = x + y \leq 3y + y = 4y \qquad \text{Since } x \leq 3y, \ x + y \leq 3y + y.$$

$$\frac{\text{Volume}}{4} \leq y$$

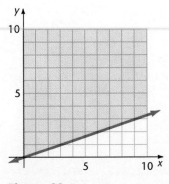

Figure 10

That is, to obtain a solution that is at least 50% acid, at least one-fourth of the mixture must be the 80% solution.

Matched Problem 3 In Example 3 how many deciliters of each stockroom solution can be used to obtain a solution that is at least 70% acid? Set up an appropriate inequality statement, simplify it if necessary, and graph the solution in the first quadrant.

Answers to Matched Problems

1. $3x - 4y \geq 12$

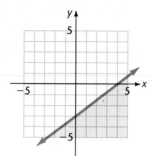

2. (A)

(B)

(C)

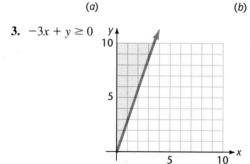

(a)   (b)   (c)

3. $-3x + y \geq 0$

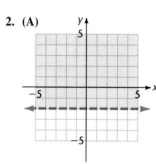

EXERCISE 5-3

Graph each inequality in a rectangular coordinate system:

A **1.** $x \leq -1$ **2.** $y \geq 3$ **3.** $y \leq -2$

4. $x \geq 4$ **5.** $x + y \leq 2$ **6.** $x + y \leq 3$

7. $x - y > 1$ **8.** $x - y < 3$ **9.** $x + y > -3$

10. $x + y > -2$ **11.** $x - y \leq -4$

12. $x - y \geq -1$ **13.** $y \leq x + 3$

14. $y \geq x - 2$ **15.** $y > -x + 3$

16. $y < -x + 1$ **17.** $x < 0$

18. $x \geq 0$ **19.** $y \geq 0$ **20.** $y < 0$

B **21.** $3x - 4y \leq 12$ **22.** $3x + y \geq 6$

23. $x + 2y \leq 4$ **24.** $2x - 5y < 10$

25. $-4x + 3y > 12$ **26.** $x - 5y > 10$

27. $3x + 6y \geq 12$ **28.** $4x - y > 8$

29. $y \leq -\frac{1}{3}x + 2$ **30.** $y \leq \frac{1}{2}x + 3$

31. $y < \frac{1}{2}x - 1$ **32.** $y > \frac{5}{3}x - 2$

33. $y \geq \frac{3}{5}x + 1$ **34.** $y < -\frac{3}{4}x + 6$

35. $y > -\frac{3}{2}x + 2$ **36.** $y \geq -\frac{3}{5}x + 3$

37. $y \geq -x$ **38.** $y \leq -x$

39. $y > x$ **40.** $y < x$

C **41.** $\frac{1}{2}x > \frac{2}{3}y$ **42.** $\frac{1}{3}x \geq \frac{3}{4}y$

43. $\frac{1}{3}x \leq \frac{2}{5}y$ **44.** $\frac{2}{3}x < \frac{5}{4}y$

45. $\frac{1}{2}x + \frac{2}{3}y \leq 1$ **46.** $\frac{1}{3}x - \frac{1}{2}y \geq 3$

47. $\frac{1}{3}x - \frac{1}{5}y > 1$ **48.** $3x - \frac{1}{2}y < 4$

49. $\frac{1}{2}x - \frac{1}{2}y < 0$ **50.** $-2x + y \leq \frac{1}{3}$

51. $x + \frac{2}{3}y > 0$ **52.** $3x - \frac{1}{2}y \geq 0$

53. $2x + \frac{1}{3}y \leq 0$ **54.** $4x - \frac{1}{2}y > -3$

APPLICATIONS

In Problems 55–68, set up an appropriate inequality statement, simplify if necessary, and graph the statement in the first quadrant. All variables in these problems will represent quantities that cannot be negative, so the first quadrant graph will represent the solution to the applied problem.

55. *Finance* An investor is going to buy a combination of two securities. The first yields an 8% return, the second a 13% return. What combinations will yield a total return of at least 10%?

56. *Finance* Repeat Problem 55 when the securities yield 10% and 15% and the investor wants a total return of at least 13%.

57. *Domestic* Two kinds of birdseed are available. The first costs 15 cents per pound and the second 20 cents per pound. What combinations of the two kinds can be purchased without spending more than $48?

58. *Domestic* Repeat Problem 57 if the birdseeds cost 18 and 25 cents per pound and at most $45 is to be spent.

59. *Chemistry* A chemical storeroom has two solutions in stock. One is 20% acid, the other 60% acid. How many centiliters of each solution can be mixed to obtain a solution that is at least 45% acid?

60. *Chemistry* A chemical storeroom has in stock a 40% acid solution and a 75% acid solution. How many milliliters of each solution can be mixed to obtain a solution that is at least 60% acid?

61. *Agriculture* Two kinds of fertilizer are available to a farmer. The percentages of nitrogen and phosphoric acid in each are given in the following table:

	Nitrogen	Phosphoric Acid
Mixture 1	20%	10%
Mixture 2	6%	6%

What combinations will give a mixture that is at least 15% nitrogen?

62. *Agriculture* What combinations of the mixtures given in Problem 61 will give a mixture that is at most 9% phosphoric acid?

63. *Agriculture* If corn yields a profit of $64 per acre and soybeans yield a profit of $56 per acre, what combinations of acres planted in corn and soybeans will yield a total profit of at least $4,464?

64. *Agriculture* Repeat Problem 63 with the profits for corn and soybeans falling to $52.80 and $48.40 per acre, respectively, and a total profit of at least $5,808 being sought.

65. *Business* A nut and candy shop wants to combine a mix worth $2.00 per pound with a mix worth $3.20 per pound to obtain a sale mix worth at least $2.20 per pound. What combinations will work?

66. *Business* Repeat Problem 65 with the value of the sale mix to be at most $2.50 per pound.

67. *Life science* Animals in an experiment are to be kept on a balanced diet. Two kinds of feed are being used, with protein and fat composition as follows:

	Protein	Fat
Feed 1	10%	6%
Feed 2	20%	2%

What combinations of the feeds will result in a diet that is at least 16% protein?

68. *Life science* What combinations of the feeds in Problem 67 will result in a diet that is at most 5% fat?

5-4 Systems of Linear Inequalities in Two Variables

Systems of linear equations were introduced in Chapter 4. Graphically, the solution is the intersection of the straight lines the equations represent. We now consider systems of linear inequalities. Each inequality represents a half-plane as we saw in Section 5-3. The graphical solution of the system is the intersection of these half-planes, that is, the area where the half-planes overlap.

Example 1 **Graphing a System** **of Inequalities**	Solve the following system graphically: $$x \geq 0$$ $$y \geq 0$$ $$2x + 5y \leq 10$$

Solution   The half-planes for $x \geq 0$ and $y \geq 0$ are shown in Figures 1*a* and *b*. Together these two inequalities determine the first quadrant (Figure 1*c*).

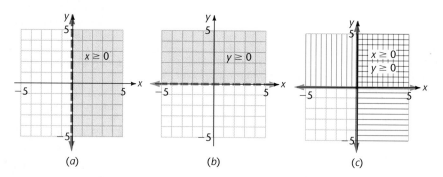

Figure 1

We graphed the third inequality in Example 1 of Section 5-3 as shown here in Figure 2*a*. For the solution to the system we need the points in this last half-plane that also lie in the first quadrant, as shown in Figure 2*b*.

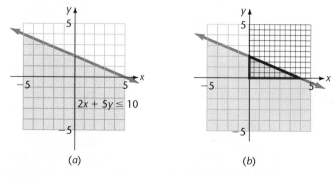

Figure 2

Matched Problem 1   Solve the following system graphically:

$$x \leq 0$$
$$y \leq 0$$
$$2x + y > -4$$

Example 2
**Graphing a System
of Inequalities**

Solve the following system graphically:

$$-2 \leq x \leq 0$$
$$0 \leq y < 3$$

Solution   Here there are actually four inequalities, each of which is easily graphed individually:

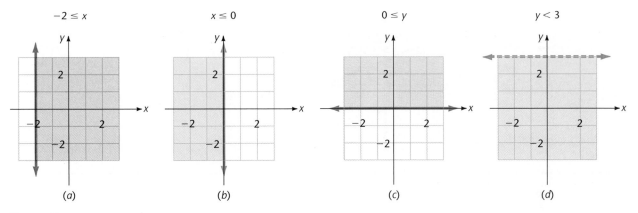

$-2 \leq x$ $x \leq 0$ $0 \leq y$ $y < 3$

(a) (b) (c) (d)

Figure 3

To see the intersection, it is helpful to draw all four boundary lines, broken or solid as needed, on the same graph and to mark the appropriate side of the line with arrows to indicate the half-plane each inequality determines as shown in Figure 4a. The solution can then be graphed as shown in Figure 4b.

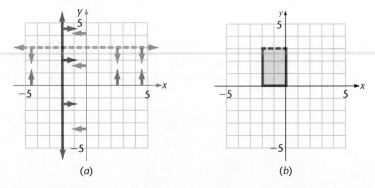

(a) (b)

Figure 4

Matched Problem 2   Solve the following system graphically:

$$1 < x < 4$$
$$0 \le y \le 3$$

Example 3
**Graphing a System
of Inequalities**

Solve the following system graphically:

$$x \ge 0$$
$$y \ge 0$$
$$2x + 5y \le 10$$
$$3x + 4y \le 12$$

Solution   Although the boundary lines are not all vertical or horizontal, we can still apply the same procedure used in Example 2 to produce the graphical solution (Figure 5).

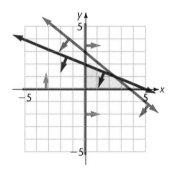

Figure 5

Matched Problem 3   Solve the following system graphically:

$$x \le 0$$
$$y \ge 0$$
$$2x + 5y \le 10$$
$$-3x + 4y \le 12$$

In Section 5-3, we considered an example of a mixture problem that led to the graph of the inequality $-x + 3y \ge 0$ in the first quadrant. Restricting the solution to the first quadrant is the same as solving the system

$$-x + 3y \ge 0$$
$$x \ge 0$$
$$y \ge 0$$

The following variation of the same example includes an additional inequality.

Example 4
A Solution Problem

A chemical storeroom has a 40% acid solution and an 80% solution. How many deciliters of each can be used to obtain a solution that is at least 50% acid and at most 60% acid? Set up an appropriate system of inequalities and graph the solution.

Solution Let

$$x = \text{Amount of 40\% solution used}$$
$$y = \text{Amount of 80\% solution used}$$

	Volume, in Deciliters	Percent Acid	Amount of Acid
Component 1: 40% solution	x	40%	$0.4x$
Component 2: 80% solution	y	80%	$0.8y$
Mixture: 50% solution	$x + y$		$0.4x + 0.8y$

The amounts of acid in the components must add to *at least* 50% and *at most* 60% of the volume, in deciliters, of the mixture, that is,

$$0.4x + 0.8y \geq 0.5(x + y)$$
$$0.4x + 0.8y \leq 0.6(x + y)$$

These inequalities can be simplified to

$$-x + 3y \geq 0 \quad \text{and} \quad -x + y \geq 0$$

Since neither x nor y can be negative, we also include the inequalities

$$x \geq 0 \quad \text{and} \quad y \geq 0$$

to obtain the system and its graph:

$$-x + 3y \geq 0$$
$$-x + y \geq 0$$
$$x \geq 0$$
$$y \geq 0$$

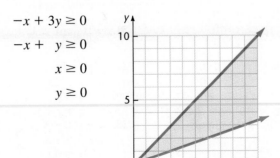

Matched Problem 4

In Example 4 how many deciliters of each stockroom solution must be used to obtain a solution that is at least 70% acid and at most 75% acid? Set up an appropriate system of inequalities and graph the solution.

Answers to Matched Problems

1.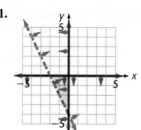

2.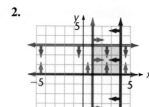

3.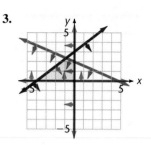

4. $-3x + y \geq 0$

$-7x + y \leq 0$

$x \geq 0$

$y \geq 0$

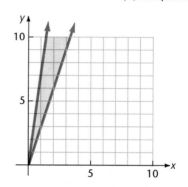

EXERCISE 5-4

Solve each system graphically:

A **1.** $x \geq 0$
$y \geq 0$
$x + y \leq 1$

2. $x \leq 0$
$y \geq 0$
$y - x \leq 2$

3. $x \leq 0$
$y \leq 0$
$y + x \geq -3$

4. $x \geq 0$
$y \leq 0$
$x - y \leq 4$

5. $x \leq 0$
$y \geq 0$
$x - y \leq -5$

6. $x \leq 0$
$y \leq 0$
$y + x \leq -6$

7. $x \geq 0$
$y \leq 0$
$x - y \leq 7$

8. $x \geq 0$
$y \geq 0$
$x + y \geq 8$

9. $x \leq 0$
$y \leq 0$
$2x + 3y \geq -6$

10. $x \geq 0$
$y \geq 0$
$3x - 2y \leq 6$

11. $x \geq 0$
$y \geq 0$
$2x + 3y < 6$

12. $x \leq 0$
$y \geq 0$
$x - 2y \geq 4$

13. $x \geq 0$
$1 \leq y \leq 3$

14. $y \leq 3$
$0 \leq x \leq 2$

15. $x \leq 0$
$-1 \leq y \leq 3$

16. $y \geq 2$
$-3 \leq x \leq 3$

17. $-2 \leq y \leq 3$
$y > 2x + 1$

18. $-4 \leq y \leq -1$
$y < -3x + 2$

19. $-5 \leq x \leq -2$
$y \leq -x + 3$

20. $-1 \leq x \leq 2$
$y < 2x - 3$

B **21.** $1 \leq x \leq 4$
$y \leq 2x + 3$

22. $0 \leq x \leq 3$
$y > x - 1$

23. $0 \leq y \leq 5$
$x \leq y + 2$

24. $-2 \leq y \leq 2$
$x < y$

25. $x \leq y + 1$
$x \geq y$

26. $-x \leq y$
$x \leq y + 2$

27. $y \geq x$
$y \leq -2x + 3$

28. $y \geq 4x - 1$
$y \leq -x$

29. $2x + 3y \geq 6$
$-x + 2y \leq 4$

30. $3x - 2y \leq 6$
$5x + y \leq 10$

31. $-2x + 6y \geq 12$
$3x - 4y \geq 12$

32. $-2x + 5y \geq 10$
$5x + 2y \geq 10$

33. $x + y \leq 5$
$x - y \geq -5$
$y \geq 0$

34. $y + x \leq 6$
$y - x \geq -6$
$x \geq 0$

35. $x - y \geq -3$
$x + y \geq -3$
$x \leq 0$

36. $x + y \geq -4$
$x - y \leq 4$
$y \leq 0$

C **37.** $x + y \leq 4$
$y \geq 2x - 4$
$x \geq 0$

38. $x + y > -3$
$x - y > -2$
$x \leq 0$

39. $x + y \geq -4$
$x \leq y + 4$
$y \leq 0$

40. $x - y \leq 8$
$y \geq -3x - 8$
$y \leq 0$

41. $y \le x + 1$
$\quad y \ge -x + 1$
$\quad y \le x - 1$

42. $y \le x + 1$
$\quad y \le -x + 1$
$\quad y \ge x - 1$

43. $y \ge x + 1$
$\quad y \ge -x + 1$
$\quad y \ge x - 1$

44. $y \le x + 1$
$\quad y \ge -x + 1$
$\quad y \ge x - 1$

45. $y \le 2x + 2$
$\quad y \le -2x + 2$
$\quad y \ge x - 3$

46. $y \ge 2x - 2$
$\quad y \le -2x + 2$
$\quad y \ge -x - 2$

47. $x - y \le 4$
$\quad x + y \le 4$
$\quad\quad y \ge -4x - 4$

48. $2x + 3y \le 6$
$\quad 3x + 2y \le 6$
$\quad\quad x + y \ge 2$

APPLICATIONS

In Problems 49–62, set up an appropriate system of inequalities and graph the solution. All variables in these problems will represent quantities that cannot be negative, so $x \ge 0$ and $y \ge 0$ should be included in the system, and the graph should therefore lie in the first quadrant.

49. *Finance*   An investor is going to buy a combination of two securities. The first yields an 8% return, the second a 13% return. What combinations will yield a total return of between 10% and 12%, inclusive?

50. *Finance*   Repeat Problem 49 for when the securities yield 10% and 15%.

51. *Domestic*   Two kinds of birdseed are available. The first costs 15 cents per pound and the second 20 cents per pound. What combinations of the two kinds can be purchased spending at least $40 but not spending more than $48?

52. *Domestic*   Repeat Problem 51 if the birdseeds cost 18 and 25 cents per pound.

53. *Chemistry*   A chemical storeroom has two solutions in stock. One is 20% acid, the other 60% acid. How many centiliters of each solution can be mixed to obtain a solution that is at least 45% acid and at most 50% acid?

54. *Chemistry*   Repeat Problem 53, with the two solutions 30% and 70% acid.

55. *Agriculture*   Two kinds of fertilizer are available to a farmer. The percentages of nitrogen and phosphoric acid in each are given in the following table:

	Nitrogen	Phosphoric Acid
Mixture 1	20%	10%
Mixture 2	6%	6%

What combinations will give a mixture that is at least 15% nitrogen and at most 9% phosphoric acid?

56. *Agriculture*   What combinations of the mixtures given in Problem 55 will give a mixture that is at most 15% nitrogen and at least 8% phosphoric acid?

57. *Agriculture*   If corn yields a profit of $64 per acre and soybeans yield a profit of $56 per acre, what combinations of acres planted in corn and soybeans will yield a total profit between $4,464 and $5,320, inclusive?

58. *Agriculture*   Repeat Problem 57 with the profits for corn and soybeans falling to $52.80 and $48.40 per acre, respectively, and a total profit of between $5,808 and $7,634, inclusive, being sought.

59. *Business*   A nut and candy shop wants to combine a mix worth $2.00 per pound together with a mix worth $3.20 per pound to obtain a sale mix worth between $2.20 and $2.50 per pound. What combinations will work?

60. *Business*   Repeat Problem 59 with the value of the sale mix being between $2.50 and $2.80 per pound.

61. *Life science*   Animals in an experiment are to be kept on a balanced diet. Two kinds of feed are being used, with protein and fat composition as follows:

	Protein	Fat
Feed 1	10%	6%
Feed 2	20%	2%

What combinations of the feeds will result in a diet that is at most 16% protein and at most 5% fat?

62. *Life science*   What combinations of the feeds in Problem 61 will result in a diet that is at least 18% protein and at least 3% fat?

CHAPTER SUMMARY

5-1 INEQUALITY STATEMENTS AND LINE GRAPHS

The inequality $a < b$ means there exists a positive real number p such that $a + p = b$ or, equivalently, that a is to the left of b on the number line. When graphing an inequality statement in one variable on a number line, a bracket, ''['' or ''],'' is used to indicate an endpoint that is included, a parentheses, "(" or ")," indicates an endpoint that is not included.

5-2 SOLVING LINEAR INEQUALITIES IN ONE VARIABLE

Inequality relations satisfy these **inequality properties:**

1. If $a < b$, then $a + c < b + c$.
2. If $a < b$, then $a - c < b - c$.
3. If $a < b$ and c is positive, then $ac < bc$.
4. If $a < b$ and c is negative, then $ac > bc$.
5. If $a < b$ and c is positive, then $a/c < b/c$.
6. If $a < b$ and c is negative, then $a/c > b/c$.

Solving an inequality means finding all real-number solutions.

5-3 GRAPHING LINEAR INEQUALITIES IN TWO VARIABLES

A line divides the plane into two halves called **half-planes.** The graph of a linear inequality in two variables is a half-plane. The boundary line is included in the graph if equality is included in the original statement; otherwise, it is not.

5-4 SYSTEMS OF LINEAR INEQUALITIES IN TWO VARIABLES

The **solution of a system of linear inequalities** in two variables is the intersection of the half-planes representing each inequality.

CHAPTER REVIEW EXERCISE

Work through all the problems in this chapter review and check the answers in the back of the book. Answers to all review problems are there, and following each answer is a number in italics indicating the section in which that type of problem is discussed. Where weaknesses show up, review appropriate sections in the text.

A *Select the larger of the two numbers:*

1. $\dfrac{3}{11}, \dfrac{5}{18}$ **2.** $\dfrac{\pi}{4}, \dfrac{19}{24}$ **3.** $\sqrt{7}, \dfrac{13}{5}$ **4.** $2\pi, \sqrt{39}$

Graph on a real number line:

5. $-3 < x \le 2$ **6.** $-3 \le x < 2$

7. $-2 < x$ **8.** $x \le 3$

9. $-1 \le x \le 1$ **10.** $3 < x < 6$

In Problems 11–21, solve the inequality or inequalities:

11. $5x \ge 20$ **12.** $\dfrac{x}{8} < 2$

13. $x - 12 \le 13$ **14.** $4 < x + 14$

15. $2 - x > 3$ **16.** $8 \le x + 8$

17. $\dfrac{x}{-3} > -2$ **18.** $-4x \le 12$

19. $3x + 9 \le -3 - x$ **20.** $-14 \le 3x - 2 \le 7$

21. $2 \le 3x + 4 \le 5$

22. Temperatures on the planet Mars vary from $-189°$F to $39°$F. Express this range in double inequality notation.

23. What numbers satisfy the condition "5 less than 5 times the number is less than or equal to 10"? Write an inequality and solve.

24. A rectangle with length equal to one more than twice the width is to have a perimeter of at most 164 meters. What are the possible widths?

B *Solve and graph on a real number line:*

25. $3x - 9 < 7x - 5$

26. $2x - (3x + 2) > 5 - 2(3 - 2x)$

27. $\dfrac{x}{2} - \dfrac{x-1}{3} \ge -1$ **28.** $-9 \le \dfrac{2}{3}x - 5 < 7$

29. $2 < 4 - 0.5x \le 8$ **30.** $\dfrac{1}{3} < \dfrac{3}{4}x - \dfrac{1}{6} < \dfrac{1}{2}$

Graph in a rectangular coordinate system:

31. $x < y$ **32.** $y \le x + 1$

33. $x \geq 3$

34. $y > -2$

35. $x - y < 3$

36. $x + y \leq 5$

37. $2x + 3y > 6$

38. $y \geq \dfrac{1}{2}x + 2$

In Problems 39–40, use $F = \frac{9}{5}C + 32$ to rewrite the temperature range in degrees Celsius. Set up a double inequality and solve.

39. A roast is to be cooked to a temperature of between 140°F for rare and 170°F for well-done.

40. A chemical is to be kept between 59°F and 86°F.

C *In Problems 41–47, find the solution set of the system of inequalities graphically:*

41. $x > 3$
$y \leq 2$

42. $x + y \leq 3$
$x + y \geq -3$

43. $x + y \leq 6$
$x - y \geq 6$
$-2 \leq x \leq 4$

44. $-1 \leq x \leq 3$
$0 \leq y \leq 2$

45. $3x + y \leq 6$
$x \geq 0$
$y \geq 3$

46. $2x - y \geq -1$
$y + 3x \leq 1$
$y \geq 0$

47. $x + y \leq 8$
$3x - 2y \leq 6$
$x \geq 1$
$1 \leq y \leq 5$

48. A company manufactures two metal products A and B, both of which involve stamping and finishing work. The amount of time in minutes that each product requires is given in this table:

	Stamping	Finishing
Product A	2 minutes	3 minutes
Product B	3 minutes	5 minutes

On a given day the company has 360 minutes of stamping time and 420 minutes of finishing time available. Set up a system of inequalities that describes the combinations of the two products that can be produced that day.

49. A farmer's chickens are to be fed a combination of corn and a commercial chicken feed. The chickens require at least 150 units of protein and at least 120 units of vitamins a day. The units of each of these nutritional ingredients included in each pound of the two food sources are given in the following table:

	Protein	Vitamins
Corn	15	20
Commercial feed	40	50

What combinations of corn and commercial feed will meet the nutritional requirements? Set up a system of inequalities and graph the solution.

50. Which, if any, of the following statements are true? For each statement that is false give an example that illustrates this.
(A) If $x < y$, then $x - 3 < y - 3$
(B) If $x \leq y$, then $xz \leq yz$
(C) If $x \geq y$, then $3 + x \geq 3 + y$
(D) If $x > y$, then $x/z > y/z$

CHAPTER PRACTICE TEST

The following practice test is provided for you to test your knowledge of the material in this chapter. You should try to complete it in 50 minutes or less. The answers in the back of the book indicate the section in the text that covers the material in the question. Actual tests in your class may vary from this practice test in difficulty, length, or emphasis, depending on the goals of your course or instructor.

1. Select the larger of the two numbers $2\sqrt{3} + 3$ and $\sqrt{11} + \pi$.

2. Graph $-5 < x \leq 3$ on a real number line.

Solve and graph on a real number line:

3. $x - 5 \geq 3 - 4x$

4. $-\dfrac{1}{3}x \leq \dfrac{1}{6}$

5. $-4 \leq 2 - 3x < 14$

Graph in a rectangular coordinate system:

6. $y \geq -3x + 2$ **7.** $x > 3$ **8.** $y \leq 0$

In Problems 9–10, solve graphically:

9. $x \leq 4$
$y > -2$

10. $3x - 5y \geq 15$
$5x + 4y \leq 20$
$x \geq 0$
$y \leq 0$

11. A rectangle with width 12 centimeters is to have a perimeter between 30 and 60 centimeters, inclusive. What is the range of possible lengths?

12. Two variables u and v are related by the formula

$$u = 3v - 2$$

What values of v will result in values of u that are at least 4 but less than 10?

13. Each time a landscaper places an order with a supplier of a certain mulch, it costs a fixed charge of $28 plus a per unit cost of $34.50 per ton. Past orders have ranged from $235 to $580. What is the range for the number of tons ordered?

CUMULATIVE REVIEW EXERCISE—CHAPTERS 1–5

This set of problems reviews the major concepts and techniques of Chapters 1–5. Work through all the problems, and check the answers in the back of the book. Answers to all review problems are there, and following each answer is a number in italics indicating the section in which that type of problem is discussed. Where weaknesses show up, review appropriate sections in the text.

A *Find, evaluate, or calculate the quantity requested:*

1. The least common multiple of 4, 15, and 24

2. The least common denominator of $\frac{5}{9}$, $\frac{7}{12}$, and $\frac{11}{15}$

3. The least common multiple of $3x$, $2x^2$, and $4xy$

4. $(-4)(-5) - (-6)$

5. $(-12) \div 3 + (-5)$

6. x^3 for $x = -2$

7. $-4x^2$ for $x = -3$

8. The absolute value of $-3 + (-4)$

9. The opposite of $-5 - (-7)$

10. The smaller of the two numbers $\frac{3}{14}$ and $\frac{7}{32}$

11. The decimal approximation, to three decimal places, of $\sqrt{14}$

12. The slope of the line containing the points $(2, -5)$ and $(4, 7)$

13. The y intercept of the line $x - 2y = 8$

14. The equation of the line having slope -4 and y intercept equal to 9

15. The equation of the vertical line through $(-5, 41)$

16. The average of the numbers $-8, -3, 5, 12$

17. A number such that 80% of the number is 96

18. The numbers associated with the points labeled a, b, and c on the number line shown

19. The coordinates of the points labeled A, B, and C in the coordinate system shown

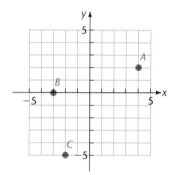

Rewrite the given expression or quantity in the form requested:

20. 288 as a product of prime factors

21. $3x(4 - 5y)$ multiplied out

22. $2xy + 5x - 4xy + 2x$ with like terms combined

23. $\frac{3}{5}$ in higher terms, with denominator 30

24. $\frac{42}{70}$ in lowest terms

Solve the equation, inequality, or system:

25. $3 + 4x = 8x - 15$

26. $x + 6 \geq 9 - 4x$

27. $2x + 7y = 3$
    $y = x - 6$

28. $x - 3y = 15$
    $x + 4y = -6$

29. $-2 \leq x + 3 \leq 5$

30. $\dfrac{2x}{5} = \dfrac{1}{2}(x - 3)$

31. $y = -3x + 8$; solve for x

Solve graphically:

32. $y = 2x - 3$
    $x + y = 6$

33. $3 < x < 5$
    $-2 \leq y \leq 1$

Graph:

34. The points -3, 0.5, and 4 on a number line

35. The points $(-3, 2)$, $(2, 0.5)$, and $(0, -2)$ in a rectangular coordinate system

36. $x \geq 2$ on a number line

37. $x \geq 2$ in a rectangular coordinate system

38. The line having equation $y = 3x + 1$

39. The line through $(2, 4)$ having slope -1

40. $y \leq x - 5$

Set up the appropriate equation, or equations, and solve:

41. A mixture of 80 coins consisting of dimes and quarters has a total value of $14.30. How many of each type of coin are in the mix?

42. At a particular travel agency the ratio of domestic to foreign trips booked in a given year is $9:4$. If the agency booked 432 domestic trips in the year, how many foreign trips did it book?

43. The average of three numbers is 36. The largest is 4 more than the next largest and 3 times the smallest. What are the numbers?

44. The area of a regular pentagon is proportional to the square of the length of a side. If a regular pentagon with a side 10 centimeters long has area 172 square centimeters, what is the area of a regular pentagon with sides of length 20 centimeters?

B *Find, evaluate, or calculate the quantity requested:*

45. The value of $ab^2 - ba^2$ for $a = 2$ and $b = 3$

46. The value of $-a \div b - a \div b^2$ for $a = 3$ and $b = -2$

47. $\dfrac{x}{3} + \dfrac{x}{4}$ **48.** $\dfrac{2y}{5} \div \dfrac{5y}{2}$

49. The least common denominator for $\dfrac{13}{x}$, $\dfrac{10}{xy}$, and $\dfrac{xy + 1}{y^2}$

50. The average of $\dfrac{x}{3}$ and x

51. The equation of the line through the points $(1, 5)$ and $(4, -8)$

Translate the expression or statement into an algebraic expression, equality, or inequality with x as the variable:

52. Two more than 3 times a number

53. Three more than a number is 2 less than 4 times the number.

54. Six less than the square of a number is 1 less than twice the number.

55. Four more than 3 times a number is at least 40.

56. The sum of four consecutive integers beginning with x

57. Five is less than the square of the quantity 1 less than a number.

58. Four more than 7 times a number

59. Eleven less than twice a number is less than twice the number.

60. The average of three consecutive integers beginning with x

61. The product of a number and 3 more than the number is 154.

Rewrite the given expression or quantity in the form requested:

62. $\dfrac{3xy^2}{8y}$ in higher terms with numerator $15x^2y^3$

63. $\{x - [1 - (x - 1)]\}$ without grouping symbols and with like terms combined

64. $a - 3b - 4c$ in the form $a - (?)$

65. $12xy + 4x^2 - 36xy^2$ with all common factors factored out

66. $y = 7x - 11$ in the form $Ax + By = C$

67. $3x + 4y = 12$ in the form $y = mx + b$

68. $-3a(b - 2c) + 2c(3a - 4b)$ without grouping symbols and with like terms combined

69. $4 + 2x - 3x^2 - 8 + 2x^2 - 5x$ with like terms combined

Solve the equation, inequality, or system:

70. $3(x - 2) = x - (6 - 2x)$

71. $\dfrac{x}{3} \leq 1 + x$

72. $2x + 4y = 15$ **73.** $3x - 4y = 12$
$2x - 6y = -10$ $y = \dfrac{3}{4}x - 12$

74. $4 - 2(x - 4) = x - 3(x - 4)$

75. $5x + 2y = 10$
$y = 2.5x + 5$

76. $A = \dfrac{h(a + b)}{2}$; solve for h

Solve graphically:

77. $3x + 2y \leq 12$ **78.** $3x - 2y \leq 12$
$x \geq 0$ $-2x + 3y \leq 12$
$y \geq 0$ $y \leq 0$

Graph:

79. The line having equation $5x + 4y = 10$

80. The line through the point $(1, 2)$ and having the same slope as the line having equation $3x - 4y = 6$

Set up the appropriate equation, or equations, and solve:

81. In a rectangle with perimeter 180 yards, the ratio of length to width is $11:7$. What are the dimensions?

82. In air at temperature t degrees Celsius the speed of sound V in feet per second is given by the formula

$$V = \frac{1{,}087\sqrt{273t}}{16.52}$$

What is the temperature, to the nearest whole number, when the speed of sound is 43,487 feet per second?

83. A long ton is a measure of weight slightly larger than a ton. Fifty-six tons is equal in weight to fifty long tons. How many long tons are in 80 tons?

84. A chemical company has two production processes for a product. The older process has an output of 1,500 gallons per hour, and the newer process has an output of 2,100 gallons per hour. In a particular week the output from the two processes totaled 88,200 gallons, and the older process ran 6 hours more than the newer. How many hours did each run?

85. A car leaves a city at noon traveling south at 60 miles per hour. A second car leaves the same city an hour later traveling north at 65 miles per hour. At what time will the cars be 610 miles apart?

86. An employee pays 7.65% of his earnings into the Social Security system. If the employee pays in $140 more this year than last, how much of a salary increase, to the nearest dollar, did the employee receive?

87. A manufacturer of motorcycle racing helmets is willing to supply 2,000 helmets for $40 each and to supply 3,000 if the price is increased to $50 each. If there is a linear relationship between the number the manufacturer will supply and the price, write the linear equation and find the number that would be supplied at a price of $60 per helmet.

C *Rewrite the given expression without grouping symbols and with like terms combined:*

88. $1 - \{1 - [1 - (1 - x)]\}$

89. $2x[1 - 3(x - y)] + 4y[2 - 5(y - x)]$

Solve the equation, inequality, or system:

90. $2x - 7y = 31$
$5x + 3y = 16$

91. $4x - \quad 3y = 7$
$5x - 3.75y = 9$

92. $2x - 7y \le 14$
$y \le -x$
$x \ge 0$

93. $x + y \ge 4$
$x \ge y$
$y \ge 0$

Graph using appropriate scales:

94. $y = -60x + 900$ for $0 \le x \le 6$

95. $100x - 3y = -1200$ for $0 \le x \le 6$

Set up the appropriate equation, or equations, and solve:

96. How much water must be added to 3 liters of a 60% acid solution to obtain a 45% solution?

97. A driver who has driven 7 hours at 55 miles per hour increases her speed for the next 9 hours so that she will average 59.5 miles per hour for the entire trip. To what does she increase her speed?

98. A coffee and tea shop creates a blend of two coffees selling for $2.45 and $3.60 per pound. How much of each should be used to make 23 pounds of the blend if the blend is to sell for $3.15 per pound?

99. A 10-quart radiator contains a 75% solution of antifreeze in distilled water. How much should be drained and replaced by distilled water to obtain a 60% solution?

100. A copy center charges 2 cents per page for one-sided copying and 2.5 cents per page when two original document sides are printed front and back onto one page. If your bill for having 600 original sides copied is $9.30, how many sides were copied to single-sided pages and how many to two-sided pages?

6

Operations on Polynomials and Factoring

The equations we have solved and graphed thus far have been linear, involving no powers of the variables higher than 1. The algebraic expressions we have used have, for the most part, been similarly restricted. To be able to deal with more complicated equations, graphs, or word problems, we will have to be able to manipulate more complicated algebraic expressions. In this chapter, we consider in more detail the same kinds of expressions we have worked with thus far but now allowing higher powers of the variables. This leads us to define a particular kind of algebraic expression, a *polynomial,* which is an expression such as

$$4x^3 - 6x^2 + 5x + 2$$

We will introduce addition, subtraction, multiplication, and division of polynomials. We will also consider the problem of *factoring* a polynomial into the product of polynomials that are in some sense simpler.

Photo reference: see Exercise 6-8, Problem 91.

6-1   Polynomial Addition and Subtraction

♦ Terminology
♦ Addition and Subtraction

Recall that an **algebraic expression** is an expression built up from constants, variables, mathematical operations, and grouping symbols. The mathematical operations are addition, subtraction, multiplication, division, and raising to natural number powers. We will later include other powers as well as the taking of roots. Algebraic expressions can be classified in a variety of ways for more efficient study. We will start with the important class of expressions called polynomials and introduce addition and subtraction for these expressions in this section. Multiplication of polynomials will be considered in Section 6-2 and division in Section 6-9.

♦ TERMINOLOGY

A **polynomial** is an algebraic expression built up from constants and variables using only addition, subtraction, or multiplication. If one or two variables are involved, then, in addition to constants, the polynomial will consist of terms of the form ax^n or $bx^m y^n$, where a and b are real-number coefficients and m and n are natural number exponents. In a polynomial, a variable cannot appear (1) in a denominator, (2) as an exponent, or (3) within a radical sign. Here are some examples of polynomials and nonpolynomials:

Polynomials	Nonpolynomials
$2x^2 - 5x + 8$	$\dfrac{3x - 1}{2x^2 + 3x - 5}$
$2x - 1$	2^x
x	$x^3 - \dfrac{2\sqrt{x}}{y} + 3y^4$
5^2	$\dfrac{1}{x}$
0	
$x^2 - \dfrac{1}{3}xy + \sqrt{2}y^2$	
$2x^3 - 3x^2y - 4xy^2 + y^3$	

Note that the constants in the polynomials may occur in a denominator, as radicals, or as exponents, but the variables never do. You should be sure you recognize why each of the nonpolynomials listed above is not a polynomial.

Polynomials, particularly in one and two variables, are encountered with great frequency at all levels in mathematics and science. As a consequence, they receive a great deal of attention in beginning and intermediate algebra.

It is helpful to be able to classify and discuss certain types of polynomials. The concept of degree is used for this purpose. If a term in a polynomial has only one variable, the **degree of the term** is the power of the variable. If more than one variable is present in a term, then the sum of the powers of the variables in the term is the degree of the term. Thus, the degree of the term $2x^5$ is 5, and the degree of the term $4x^3y^4$ is 7. The **degree of a nonzero constant** is defined to be 0, and the degree of 0 is left undefined. The **degree of a polynomial** is the degree of the term with the highest degree in the polynomial.

Example 1
Degree

What is the degree of

(A) $3x^5$?   (B) $7x^2y^3$?   (C) $4x^7 - 3x^5 + 2x^3 - 1$?

Solution

(A) $3x^5$ is of degree 5.
(B) $7x^2y^3$ is of degree $2 + 3 = 5$.
(C) In $4x^7 - 3x^5 + 2x^3 - 1$, the highest-degree term is the first, with degree 7. Thus, the degree of the polynomial is 7.

Matched Problem 1

What is the degree of

(A) $5x^3$?   (B) $2x^3y^4$?   (C) $6x^3 - 2x^2 + x - 1$?
(D) $2x^2 - 3xy + y^2 + x - y + 1$?

A polynomial with only one term is called a **monomial,** one with two terms is called a **binomial,** and one with three terms is called a **trinomial.**

Three terms	Two terms	One term	One term
$3x^2 - 2x + 1$	$2x - 3y$	$3x^4y^2$	8
Trinomial	Binomial	Monomial	Monomial
Degree 2	Degree 1	Degree 6	Degree 0

◆ ADDITION AND SUBTRACTION

As algebraic expressions, polynomials represent real numbers, so we should be able to add, subtract, multiply, and divide them. We have already spent some time on these operations with simpler polynomials. In this chapter we will review and extend these processes to more complex forms. We start with addition and subtraction. We add polynomials by combining like terms.

Example 2
Adding Polynomials

Add $x^2 + 1$, $-x^3 + 2x - 3$, and $x^3 - 2x^2 + x$ horizontally and vertically.

Solution

Method 1 Add horizontally:

$$(x^2 + 1) + (-x^3 + 2x - 3) + (x^3 - 2x^2 + x)$$ Clear parentheses— be careful of signs.

$$= x^2 + 1 - x^3 + 2x - 3 + x^3 - 2x^2 + x$$ Combine like terms.

$$= -x^2 + 3x - 2$$

Method 2   Add vertically. Line up like terms and add coefficients. This method is generally preferred when you have several polynomials to add:

$$
\begin{array}{r}
x^2 + 1 \\
-x^3 + 2x - 3 \\
x^3 - 2x^2 + x \\
\hline
- x^2 + 3x - 2
\end{array}
$$

Note the spaces for missing powers; for example, think of the first polynomial as $0 \cdot x^3 + x^2 + 0 \cdot x + 1$.

Matched Problem 2   Add $2x^3 - x^2 + 4$, $x^3 - 3x^2 - x + 1$, and $x^2 - 6$ horizontally and vertically.

We subtract a second polynomial from a first by taking the negative of the second and adding this result to the first.

Example 3
Subtracting
Polynomials

Subtract $2x^2 - x + 2$ from $x^2 + 3x + 5$.

Solution   *Method 1*   Work horizontally. This method is often preferred for subtraction, since only two polynomials are involved.

$$(x^2 + 3x + 5) - (2x^2 - x + 2)$$

Notice which polynomial goes on the right. We are subtracting $2x^2 - x + 2$ from $x^2 + 3x + 5$. Clear parentheses—be careful of signs.

$$= x^2 + 3x + 5 - 2x^2 + x - 2 \qquad \text{Combine like terms.}$$
$$= -x^2 + 4x + 3$$

Method 2   Work vertically. Notice which polynomial goes on the bottom.

$$
\begin{array}{r}
x^2 + 3x + 5 \\
2x^2 - x + 2 \\
\hline
\end{array}
\longleftarrow \text{Change signs and add.} \longrightarrow
\begin{array}{r}
x^2 + 3x + 5 \\
-2x^2 + x - 2 \\
\hline
-x^2 + 4x + 3
\end{array}
$$

 Extra care should be taken when working subtraction problems vertically, since sign errors here are common. This is why the horizontal method is usually preferred.

Check   Since $A - B = C$ means $A = B + C$, subtraction can be checked by adding the difference to the quantity subtracted to see if the result is the quantity we subtracted from:

            Difference   Quantity Subtracted

$$(-x^2 + 4x + 3) + (2x^2 - x + 2) = -x^2 + 4x + 3 + 2x^2 - x + 2$$
$$= x^2 + 3x + 5$$

                          Quantity subtracted from

Matched Problem 3   Subtract $3x^2 + 2x - 4$ from $2x^2 + 3x + 2$.

Answers to
Matched Problems
   1. (A) 3   **(B)** 7   **(C)** 3   **(D)** 2   **2.** $3x^3 - 3x^2 - x - 1$
   3. $-x^2 + x + 6$

EXERCISE 6-1

A *State the degree of each monomial:*

1. x^3 **2.** y^5 **3.** 4 **4.** -5

5. $-11x^7$ **6.** $7x^2$ **7.** xy **8.** $-xy$

9. x^3y^2 **10.** x^4y^3 **11.** $8x^5y^2$ **12.** $4x^6y^3$

Add:

13. $2x + 1$ and $5x - 3$ **14.** $4x - 3$ and $x + 4$

15. $2x + 3$, $x - 4$, and $5x + 1$

16. $6x - 4$, $3x + 5$, and $2x - 1$

17. $-3x + 4$, $2x - 5$, and $2x$

18. $5x - 3$, $-4x + 5$, and $3x$

19. $2x - 1$, $-3x + 4$, $-x - 2$, and $5x + 7$

20. $-4x + 2$, $x + 5$, $-3x - 6$, and $6x + 1$

21. $x^2 + 3x + 2$, $x + 4$, and $3x - 5$

22. $3x^2 - x + 4$, $2x - 1$, and $x - 6$

23. $2x^2 - 3x + 4$, $5x + 2$, and $-3x - 5$

24. $x^2 + 5x - 11$, $-4x + 2$, and $3x - 9$

Subtract:

25. $3x + 1$ from $6x + 5$ **26.** $4x + 3$ from $5x + 8$

27. $2x - 5$ from $3x + 1$ **28.** $3x - 2$ from $5x + 3$

29. $x^2 - 2x + 3$ from $3x^2 + x + 4$

30. $x^2 + 3x - 2$ from $2x^2 + x - 4$

31. $2x^2 + x - 1$ from $5x^2 - 2x + 4$

32. $3x^2 + 2x + 1$ from $8x^2 + 2x + 5$

B *State the degree of each polynomial:*

33. $2x - 3$ **34.** $4x^2 - 2x + 3$

35. $3x^3 - x + 7$ **36.** $2x - y$

37. $x^2 - 3xy + y^2$ **38.** $x^3 - 2x^2y + xy^2 - 3y^3$

39. $2x^6 - 3x^5 + x^2 - x + 1$

40. $x^5 - 2x^2 + 5$

41. $-3x^7 + 5x^3 - 6$ **42.** $-11x^8 - 2x^4 + x - 10$

Add:

43. $4x^3 + 3x^2 + 2$, $-x^2 + 2x + 3$, and $x - 5$

44. $2x^3 + x - 1$, $x^3 + x^2 + 1$, and $x^2 + x$

45. $3x + 4y$ and $x - 5y$

46. $x^2 + xy + y^2$ and $2x^2 - xy - 2y^2$

47. $2x - 2y + z$ and $-x + y - 2z$

48. $a^2 - b^2$ and $a^2 + 2ab + b^2$

49. $x + xy + xy^2$ and $3y^2 - 2xy - 4x$

50. $4x^2 - 2xy + y^2$ and $x^2 + 4xy + 4y^2$

Subtract:

51. $2x^3 - x^2 + 4$ from $x^3 + 3x^2 + 2x + 1$

52. $4x^3 + x^2 - x + 2$ from $5x^3 - 2x^2 + x - 3$

53. $2a - 3b$ from $a - b$

54. $x^2 + xy - y^2$ from $x^2 - 2xy - y^2$

55. $3a - 2b + c$ from $a - b + c$

56. $2x^2 + y^2$ from $x^2 - xy - y^2$

57. $x^2 - y^2$ from $x^2 - 4xy + 4y^2$

58. $2x^2 + 9y^2$ from $x^2 - 3xy + y^2$

C *In Problems 59–72, let P, Q, R, and S be these polynomials:*

$$P = x^2 + 2x + 3 \qquad Q = -2x^2 + 3x - 4$$
$$R = -x^2 - 2x + 1 \qquad S = 2x^2 - x + 2$$

Find the polynomial indicated:

59. $P + Q$ **60.** $P - Q$ **61.** $R - S$ **62.** $R + S$

63. $P + Q + R$ **64.** $Q + R + S$

65. $(P - R) + Q$ **66.** $(R - Q) + S$

67. $(S - Q) - P$ **68.** $(P - R) - Q$

69. $(P - Q) + (R - S)$ **70.** $(P - S) + (R - Q)$

71. $(P + Q) - (R + S)$ **72.** $(P + S) - (R + Q)$

73. What polynomial should be added to $4x^3 + 2x^2 - 5x - 7$ to produce the sum $2x^3 + x^2 - 6x + 4$?

74. What polynomial should be added to $6x^3 + x^2 + 3x - 1$ to produce the sum $3x^3 - x^2 + 2x - 3$?

75. If the polynomials $x^3 - 2x^2 + bx - 6$ and $2x^3 + 4x^2 + 5x - c$ have the sum $3x^3 + 2x^2 + x$, what are b and c?

76. If the polynomials $ax^3 - 2x^2 + x - 6$ and $2x^3 + bx^2 + 5x - 2$ have the sum $x^3 - 2x^2 + 6x - 8$, what are a and b?

77. What polynomial subtracted from $2x^3 + 5x^2 - 3x - 7$ yields a difference of $3x^3 - 4x^2 + 5x + 6$?

78. What polynomial subtracted from $-x^3 + 2x^2 - 3x + 4$ yields a difference of $2x^3 - 3x^2 + 4x + 5$?

79. If the polynomials $x^3 - 2x^2 + bx - 6$ and $2x^3 + 4x^2 + 5x - c$ have the difference $-2x^3 - 6x^2 + x$, what are b and c?

80. If the polynomials $ax^3 - 2x^2 + x - 6$ and $2x^3 + bx^2 + 5x - 2$ have the difference $x^3 - 2x^2 - 5x - 4$, what are a and b?

  ## 6-2 **Polynomial Multiplication**

♦ Multiplication
♦ Special Products of the Form $(ax + b)(cx + d)$
♦ Squaring a Binomial

In this section we introduce polynomial multiplication. The multiplication of certain binomials can be done very quickly. We consider two such situations: first the product of two binomials of degree 1, $(ax + b)(cx + d)$, and then the square of any binomial $(A + B)^2$.

♦ MULTIPLICATION

The distributive property is the important principle behind multiplying polynomials. This property leads directly to the following rule:

> **Multiplying Polynomials**
>
> To multiply two polynomials, multiply each term of the first one by each term of the second one; then add like terms.

Example 1
Multiplying Polynomials

Multiply $(x - 3)(x^2 - 2x + 3)$.

Solution   *Method 1*   Horizontal arrangement:

$$(x - 3)(x^2 - 2x + 3)$$ Use the distributive property from right to left.

$$= x(x^2 - 2x + 3) - 3(x^2 - 2x + 3)$$ Use the distributive property from left to right.

$$= x^3 - 2x^2 + 3x - 3x^2 + 6x - 9$$ Combine like terms.
$$= x^3 - 5x^2 + 9x - 9$$

Method 2   Vertical arrangement. The same multiplication can be performed vertically, in much the same way we multiply large positive integers:

$$
\begin{array}{r}
x^2 - 2x + 3 \\
x - 3 \\
\hline
x^3 - 2x^2 + 3x \\
- 3x^2 + 6x - 9 \\
\hline
x^3 - 5x^2 + 9x - 9
\end{array}
$$

Notice that we start multiplying from the left first, that is, by x. Then multiply by -3, line up like terms, and add.

Notice that the product of a first-degree polynomial and second-degree polynomial produces a third-degree polynomial. The fact that the degree of the product is the sum of the degrees of the factors is true in general. Thus, the product of two second-degree polynomials is a fourth-degree polynomial, the product of two first-degree polynomials is a second-degree polynomial, and so forth.

Matched Problem 1   Multiply $(3x^2 - 2x + 1)(2x^2 + 3x - 2)$ horizontally or vertically.

♦ **SPECIAL PRODUCTS OF THE FORM $(ax + b)(cx + d)$**

Let us multiply $(2x - 1)$ and $(3x + 2)$, using a horizontal arrangement. We apply the distributive law:

$$(2x - 1)(3x + 2) = 2x(3x + 2) + (-1)(3x + 2)$$
$$= (2x)(3x) + (2x)(2) + (-1)(3x) + (-1)(2)$$

Each term in the first factor has been multiplied by each term in the second. The order of the four terms in the product leads to a way to perform multiplications such as this quickly and efficiently. The product $(2x)(3x)$ is the product of the *first* terms in the factors. The product $(2x)(2)$ is the product of the *outer* terms of the factors, $(-1)(3x)$ is the product of the *inner* terms, and $(-1)(2)$ is the product of the *last* terms:

First terms   Outer terms

$(2x - 1)(3x + 2)$ $(2x - 1)(3x + 2)$

Inner terms   Last terms

The sequence of products *First*, *Outer*, *Inner*, *Last*, or FOIL for short, written on one line yields

F	O	I	L
First terms' product	Outer terms' product	Inner terms' product	Last terms' product

$$(2x - 1)(3x + 2) = \ 6x^2 \ + \ 4x \ - \ 3x \ - \ 2$$

The outer and inner terms' products are like terms and hence combine into one term. Thus,

$$(2x - 1)(3x + 2) = 6x^2 + x - 2$$

With practice we can speed up the process and combine the outer and inner products mentally. The procedure just described is called the **FOIL method.**

This method provides a simple way to multiply two binomials mentally. However, it is only applicable to this particular kind of product. Term-by-term multiplication must still be used in general.

A simple three-step process for carrying out the FOIL method is illustrated in Example 2.

Example 2
Multiplying Binomials

Multiply by the FOIL method:

(A) $(2x - 1)(3x + 2)$ (B) $(2a - b)(a + 3b)$ (C) $(2x - 3y)(2x + 3y)$

Solution

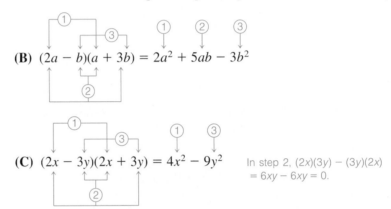

(A) $(2x - 1)(3x + 2) = 6x^2 + x - 2$

The like terms are picked up in step 2 and combined mentally.

(B) $(2a - b)(a + 3b) = 2a^2 + 5ab - 3b^2$

(C) $(2x - 3y)(2x + 3y) = 4x^2 - 9y^2$ In step 2, $(2x)(3y) - (3y)(2x)$
$= 6xy - 6xy = 0$.

Notice that the middle term disappeared, since its coefficient is 0.

Matched Problem 2

Multiply by the FOIL method:

(A) $(2x + 3)(x - 1)$ (B) $(a - 2b)(2a + 3b)$ (C) $(x - 2y)(x + 2y)$

In Section 6-4, we will consider the reverse problem: Given a second-degree polynomial, such as $6x^2 + x - 2$ or $2a^2 + 5ab - 3b^2$, find first-degree factors with integer coefficients that will produce these second-degree polynomials as products. The problem is analogous to factoring in the integers. It is easy to show that $37 \times 59 = 2,183$, but much harder to find these factors from 2,183. To be able to factor second-degree polynomial forms with any degree of efficiency, it is important that you know how to mentally multiply first-degree factors of the types illustrated in this section quickly and accurately.

♦ **SQUARING A BINOMIAL**

Since by multiplying out

$$(A + B)^2 = (A + B)(A + B) = A^2 + 2AB + B^2$$

we can formulate a simple mechanical rule for squaring a binomial directly from $(A + B)^2$ without having to write $(A + B)(A + B)$ first:

Squaring a Binomial

Step 1   Square the first term.

Step 2   Take twice the product of the first and second terms.

Step 3   Square the second term.

Schematically:

Square of first term

Square of second term

$$(A + B)^2 = A^2 + 2AB + B^2 \qquad (A - B)^2 = A^2 - 2AB + B^2$$

Twice the product of
first and second terms

$2A(-B) = -2AB$

$A^2 + 2AB + B^2$ and $A^2 - 2AB + B^2$ are called **perfect square** trinomials because they are the squares of $(A + B)$ and $(A - B)$, respectively.

Example 3
Squaring Binomials

Square each binomial, using the mechanical rule:

(A) $(2x + 3)^2$ **(B)** $(3x - 2)^2$ **(C)** $(3x - 2y)^2$

Solution   **(A)** $(2x + 3)^2 = (2x)^2 + 2(2x)(3) + 3^2 = 4x^2 + 12x + 9$

(B) $(3x - 2)^2 = (3x)^2 + 2(3x)(-2) + (-2)^2 = 9x^2 - 12x + 4$

(C) $(3x - 2y)^2 = (3x)^2 + 2(3x)(-2y) + (-2y)^2$
$$= 9x^2 - 12xy + 4y^2$$

Matched Problem 3   Square each binomial, using the mechanical rule:

(A) $(2x + 1)^2$ **(B)** $(3x - 4)^2$ **(C)** $(5x - 2y)^2$

Answers to
Matched Problems

1. $6x^4 + 5x^3 - 10x^2 + 7x - 2$
2. (A) $2x^2 + x - 3$ **(B)** $2a^2 - ab - 6b^2$ **(C)** $x^2 - 4y^2$
3. (A) $4x^2 + 4x + 1$ **(B)** $9x^2 - 24x + 16$ **(C)** $25x^2 - 20xy + 4y^2$

EXERCISE 6-2

A *Multiply:*

1. $(2x - 3)(x + 2)$ **2.** $(3x - 5)(2x + 1)$

3. $(2x - 1)(x^2 - 3x + 5)$

4. $(3y + 2)(2y^2 + 5y - 3)$

5. $(x - 3y)(x^2 - 3xy + y^2)$

6. $(m + 2n)(m^2 - 4mn - n^2)$

7. $(a + b)(a^2 - ab + b^2)$

8. $(a - b)(a^2 + ab + b^2)$

Square each binomial, using the mechanical rule:

9. $(y - 5)^2$ **10.** $(y + 7)^2$

11. $(x + 3)^2$ **12.** $(x - 4)^2$

Multiply mentally:

13. $(x + 1)(x + 2)$ **14.** $(y + 3)(y + 1)$

15. $(y + 3)(y + 4)$ **16.** $(x + 3)(x + 2)$

17. $(x - 5)(x - 4)$ **18.** $(m - 2)(m - 3)$

19. $(n - 4)(n - 3)$ **20.** $(u - 5)(u - 3)$

21. $(s + 7)(s - 2)$ **22.** $(t - 6)(t + 4)$

23. $(m - 12)(m + 5)$ **24.** $(a + 8)(a - 4)$

25. $(u - 3)(u + 3)$ **26.** $(t + 4)(t - 4)$

27. $(x + 8)(x - 8)$ **28.** $(m - 7)(m + 7)$

29. $(y + 7)(y + 9)$ **30.** $(x + 8)(x + 11)$

31. $(c - 9)(c - 6)$ **32.** $(u - 8)(u - 7)$

33. $(x - 12)(x + 4)$ **34.** $(y - 11)(y + 7)$

35. $(a + b)(a - b)$ **36.** $(m - n)(m + n)$

37. $(x + y)(x + 3y)$ **38.** $(m + 2n)(m + n)$

B   **39.** $(x + 2)(3x + 1)$ **40.** $(x + 3)(2x + 3)$

41. $(4t - 3)(t - 2)$ **42.** $(2x - 1)(x - 4)$

43. $(3y + 7)(y - 3)$ **44.** $(t + 4)(2t - 3)$

45. $(2x - 3y)(x + 2y)$ **46.** $(3x + 2y)(x - 3y)$

47. $(2x - 1)(3x + 2)$ **48.** $(3y - 2)(3y - 1)$

49. $(3y + 2)(3y - 2)$ **50.** $(2m - 7)(2m + 7)$

51. $(5s - 1)(s + 7)$ **52.** $(a - 6)(5a + 6)$

53. $(3m + 7n)(2m - 5n)$ **54.** $(6x - 4y)(5x + 3y)$

55. $(4n - 7)(3n + 2)$ **56.** $(5x - 6)(2x + 7)$

57. $(2x - 3y)(3x - 2y)$ **58.** $(2s - 3t)(3s - t)$

Multiply:

59. $(x + 2y)^3$ **60.** $(2m - n)^3$

61. $(x^2 - 3x + 5)(2x^2 + x - 2)$

62. $(2m^2 + 2m - 1)(3m^2 - 2m + 1)$

63. $(2x^2 - 3xy + y^2)(x^2 + 2xy - y^2)$

64. $(a^2 - 2ab + b^2)(a^2 + 2ab + b^2)$

Square each binomial, using the mechanical rule:

65. $(2x - 3)^2$ **66.** $(3x + 2)^2$ **67.** $(2x - 5y)^2$

68. $(3x + 4y)^2$ **69.** $(4a + 3b)^2$ **70.** $(5m - 3n)^2$

71. $(3u + 2v)^2$ **72.** $(5a - 2b)^2$

Simplify:

73. $(3x - 1)(x + 2) - (2x - 3)^2$

74. $(2x + 3)(x - 5) - (3x - 1)^2$

75. $2(x - 2)^3 - (x - 2)^2 - 3(x - 2) - 4$

76. $(2x - 1)^3 - 2(2x - 1)^2 + 3(2x - 1) + 7$

C *Multiply:*

77. $(x - 1)(x^2 + 2x - 1)$

78. $(x + 2)(x^2 - 3x + 2)$

79. $(2x - 3)(3x^2 - 2x + 4)$

80. $(3x + 2)(2x^2 - 3x - 2)$

81. $(x^2 - 3x + 2)(x^2 + x - 3)$

82. $(x^2 + 2x - 3)(x^2 - 3x - 2)$

Square each binomial, using the mechanical rule:

83. $(x^2 + 1)^2$ **84.** $(a^2 - 3)^2$

85. $(2a^3 - b^2)^2$ **86.** $(x^3 + 4y^2)^2$

87. $(3x^3 + 4y^2)^2$ **88.** $(5a^4 - 2b^3)^2$

In Problems 89–98, let P, Q, R, and S be these polynomials:

$$P = x^2 + 2x + 3 \qquad Q = -2x^2 + 3x - 4$$
$$R = -x^2 - 2x + 1 \qquad S = 2x^2 - x + 2$$

Find the polynomial indicated:

89. PQ **90.** RS **91.** P^2 **92.** Q^2

93. $PS + R$ **94.** $QR + P$ **95.** PQR **96.** QRS

97. $P(Q + S)$ **98.** $S(P + R)$

Multiply:

99. $(A - B)(A + B)(A^2 + B^2)$

100. $(A - B)(A + B)(A^2 - AB + B^2)(A^2 + AB + B^2)$

6-3   Factoring Out Common Factors

- ♦ Factoring Out Common Monomial Factors
- ♦ Factoring Out Common Binomial Factors
- ♦ Factoring by Grouping

You have already had some experience in Chapter 1 in factoring out common factors. The distributive property of real numbers in the form

$$ab + ac = a(b + c)$$

$$\underset{\substack{\text{Sum of}\\\text{two terms}}}{ab+ac} = \underset{\substack{\text{Product of}\\\text{two factors}}}{a(b+c)}$$

is the important property behind the process. In rewriting $ab + ac$ as $a(b + c)$, we have converted a sum of two terms to a product with two factors, a and $(b + c)$, by factoring out the common factor a from $ab + ac$. The process of rewriting polynomials as products is called **factoring.** The word is used here in the same sense as it is with regard to factoring integers. In this section, we begin to look at factoring polynomials by considering the most direct application of the distributive law, factoring out common factors. Factoring is further developed in the next four sections.

♦ FACTORING OUT COMMON MONOMIAL FACTORS

We begin our discussion of factoring by factoring out common monomial factors.

Example 1
Factoring Out Common Factors

Factor out factors common to all terms:

(A) $6x^2 + 15x$ **(B)** $2u^3v - 6u^2v^2 + 8uv^3$

Solution

(A) $6x^2 + 15x \boxed{= 3x \cdot 2x + 3x \cdot 5} = 3x(2x + 5)$

(B) $2u^3v - 6u^2v^2 + 8uv^3 \boxed{= 2uv \cdot u^2 - 2uv \cdot 3uv + 2uv \cdot 4v^2}$
$$= 2uv(u^2 - 3uv + 4v^2)$$

Matched Problem 1

Factor out factors common to all terms:

(A) $12y^2 - 28y$ **(B)** $6m^4 - 15m^3n + 9m^2n^2$

♦ FACTORING OUT COMMON BINOMIAL FACTORS

We also can factor out common factors that are not monomials. Look closely at the following four examples and try to see what they all have in common:

$$2xy + 3y = y(2x + 3)$$

$$2xA + 3A = A(2x + 3)$$

Because of the commutative property of multiplication, a common factor may be taken out on either the left or the right.

$$2x(x - 4) + 3(x - 4)$$
$$= (x - 4)(2x + 3)$$

$$2x(3x + 1) + 3(3x + 1)$$
$$= (3x + 1)(2x + 3)$$

The factoring involved in each example is essentially the same. The only difference is the nature of the common factors being taken out. In the first two examples common monomial factors are taken out. In the second two examples common binomial factors are taken out. In these last two examples think of $(x - 4)$ and $(3x + 1)$ as single numbers, just as A represents a single number in the second example.

Example 2 **Factoring Out** **Common Factors**	Remove factors common to all terms: **(A)** $5x(x - 1) - (x - 1)$ **(B)** $3x(2x - y) - 2y(2x - y)$

Solution   **(A)** $5x(x - 1) - (x - 1) = 5x(x - 1) - 1(x - 1)$ Note: $x - 1 = 1(x - 1)$
$$= (x - 1)(5x - 1)$$
(B) $3x(2x - y) - 2y(2x - y) = (2x - y)(3x - 2y)$

Matched Problem 2   Remove factors common to all terms:

(A) $3m(m + 2) - (m + 2)$ **(B)** $2u(u + 3v) - 3v(u + 3v)$

♦ FACTORING BY GROUPING

Some polynomials can be factored by grouping terms in such a way that we obtain results similar to the expressions in Example 2. We can then complete the factoring by removing common factors. This process will prove useful in Section 6-6, where an efficient method is developed for factoring a second-degree polynomial into the product of two first-degree polynomials.

Example 3 **Factoring by Grouping**	Factor by grouping: **(A)** $2x^2 - 8x + 3x - 12$ **(B)** $5x^2 - 5x - x + 1$ **(C)** $6x^2 - 3xy - 4xy + 2y^2$

Solution   **(A)** $2x^2 - 8x + 3x - 12$ Group the first two and last two terms.
$$= (2x^2 - 8x) + (3x - 12)$$ Remove common factors from each group.
$$= 2x(x - 4) + 3(x - 4)$$ The common factor $(x - 4)$ can be taken out.

$$= (x - 4)(2x + 3)$$ The factoring is complete.

(B) $5x^2 - 5x - x + 1$

Group the first two and last two terms. Notice what happens to the signs in the second grouping. If we clear parentheses, we must get back to where we started.

$$= (5x^2 - 5x) - (x - 1)$$

Remove common factors from each group.

$$= 5x(x - 1) - 1(x - 1)$$

The common factor $(x - 1)$ can be taken out of both terms.

$$= (x - 1)(5x - 1)$$

The factoring is complete.

(C) $6x^2 - 3xy - 4xy + 2y^2$

Group the first two and last two terms.

$$= (6x^2 - 3xy) - (4xy - 2y^2)$$

Note that the signs change inside the second parentheses. Factor within each parenthesis.

$$= 3x(2x - y) - 2y(2x - y)$$

The common factor $(2x - y)$ can be taken out.

$$= (2x - y)(3x - 2y)$$

The factoring is complete.

Matched Problem 3   Factor by grouping:

(A) $6x^2 + 2x + 9x + 3$ **(B)** $3m^2 + 6m - m - 2$
(C) $2u^2 + 6uv - 3uv - 9v^2$

In Example 3, the polynomials were arranged in such a way that grouping the first two terms and the last two terms led to common factors. The process is not always this neat, however, and you will sometimes have to rearrange terms in order to group them profitably for factoring.

Example 4
Factoring by Rearranging and Grouping

Factor $y^2 + xz + xy + yz$ by grouping.

Solution   If we proceed as in Example 3, no common factor can be factored out to complete the factoring:

$$y^2 + xz + xy + yz = (y^2 + xz) + (xy + yz)$$
$$= (y^2 + xz) + y(x + z)$$

Rearrange the terms and proceed again as in Example 3:

$$y^2 + xz + xy + yz = y^2 + xy + xz + yz$$
$$= y(y + x) + (x + y)z$$
$$= y(x + y) + z(x + y)$$
$$= (y + z)(x + y)$$

Matched Problem 4   Factor $ac + bd + bc + ad$ by grouping.

It is important to note that many polynomials of the forms considered in Examples 3 and 4 do not have polynomial factors with smaller degrees and integer coefficients. Consider the following two polynomials:

$$2x^2 + 2x + x - 4 \qquad y^2 - xz + xy + yz$$

No matter how we group the terms, we will not find a factoring of either polynomial using integer coefficients. Try factoring these polynomials by grouping to see what happens.

Answers to   **1. (A)** $4y(3y - 7)$ **(B)** $3m^2(2m^2 - 5mn + 3n^2)$
Matched Problems   **2. (A)** $(m + 2)(3m - 1)$ **(B)** $(u + 3v)(2u - 3v)$
3. (A) $(3x + 1)(2x + 3)$ **(B)** $(m + 2)(3m - 1)$ **(C)** $(u + 3v)(2u - 3v)$
4. $(a + b)(c + d)$

EXERCISE 6-3

Write each of the following expressions in factored form by removing factors common to all terms:

A 1. $2xA + 3A$ **2.** $xM - 4M$ **3.** $10x^2 + 15x$

4. $9y^2 - 6y$ **5.** $14u^2 - 6u$ **6.** $20m^2 + 12m$

7. $6u^2 - 10uv$ **8.** $14x^2 - 21xy$

9. $10m^2n - 15mn^2$ **10.** $9u^2v + 6uv^2$

11. $2x^3y - 6x^2y^2$ **12.** $6x^2y^2 - 3xy^3$

13. $3x(x + 2) + 5(x + 2)$ **14.** $4y(y + 3) + 7(y + 3)$

15. $3m(m - 4) - 2(m - 4)$ **16.** $x(x - 1) - 4(x - 1)$

17. $x(x + y) - y(x + y)$ **18.** $m(m - n) + n(m - n)$

B 19. $6x^4 - 9x^3 + 3x^2$ **20.** $6m^4 - 8m^3 - 2m^2$

21. $8x^3y - 6x^2y^2 + 4xy^3$

22. $10u^3v + 20u^2v^2 - 15uv^3$

23. $8x^4 - 12x^3y + 4x^2y^2$

24. $9m^4 - 6m^3n - 6m^2n^2$

25. $4x^2yz + 8xy^2z + 12xyz^2$

26. $9abc - 3ab^2c^3 + 6a^3b^2c$

27. $3x(2x + 3) - 5(2x + 3)$

28. $2u(3u - 8) - 3(3u - 8)$

29. $x(x + 1) - (x + 1)$ **30.** $3u(u - 1) - (u - 1)$

31. $4x(2x - 3) - (2x - 3)$

32. $3y(4y - 5) - (4y - 5)$

33. $2x^2(y^2 + 2z) - 3x(y^2 + 2z)$

34. $3ab^2(c^2 - 1) + 2(c^2 - 1)$

Replace question marks with algebraic expressions that will make both sides equal:

35. $3x^2 - 3x + 2x - 2 = (3x^2 - 3x) + (?)$

36. $2x^2 + 4x + 3x + 6 = (2x^2 + 4x) + (?)$

37. $3x^2 - 12x - 2x + 8 = (3x^2 - 12x) - (?)$

38. $2y^2 - 10y - 3y + 15 = (2y^2 - 10y) - (?)$

39. $8u^2 + 4u - 2u - 1 = (8u^2 + 4u) - (?)$

40. $6x^2 + 10x - 3x - 5 = (6x^2 + 10x) - (?)$

41. $2x^2 + 3x^3 - 2y - 3xy = (2x^2 + 3x^3) - (?)$

42. $10a - 2b^2 + 15a^2 - 3ab^2 = (10a - 2b^2) + (?)$

43. $2x^2 + 3x^3 - 2y - 3xy = (2x^2 - 2y) + (?)$

44. $10a - 2b^2 + 15a^2 - 3ab^2 = (10a + 15a^2) - (?)$

Factor out common factors from each group, then complete the factoring if possible:

45. $(3x^2 - 3x) + (2x - 2)$

46. $(2x^2 + 4x) + (3x + 6)$

47. $(3x^2 - 12x) - (2x - 8)$

48. $(2y^2 - 10y) - (3y - 15)$

49. $(8u^2 + 4u) - (2u + 1)$

50. $(6x^2 + 10x) - (3x + 5)$

51. $(2x^2 + 3x^3) - (2y + 3xy)$

52. $(10a - 2b^2) + (15a^2 - 3ab^2)$

53. $(2x^2 - 2y) + (3x^3 - 3xy)$

54. $(10a + 15a^2) - (2b^2 + 3ab^2)$

Factor as the product of two first-degree factors using grouping. (These problems are related to Problems 45–54.)

55. $3x^2 - 3x + 2x - 2$ **56.** $2x^2 + 4x + 3x + 6$

57. $3x^2 - 12x - 2x + 8$ **58.** $2y^2 - 10y - 3y + 15$

59. $8u^2 + 4u - 2u - 1$ **60.** $6x^2 + 10x - 3x - 5$

61. $2x^2 + 3x^3 - 2y - 3xy$

62. $10a - 2b^2 + 15a^2 - 3ab^2$

63. $2x^2 - 2y + 3x^3 - 3xy$

64. $10a + 15a^2 - 2b^2 - 3ab^2$

Factor as the product of two first-degree factors using grouping:

65. $2m^2 - 8m + 5m - 20$ **66.** $5x^2 - 10x + 2x - 4$

67. $6x^2 - 9x - 4x + 6$ **68.** $12x^2 + 8x - 9x - 6$

C **69.** $3u^2 - 12u - u + 4$ **70.** $6m^2 + 4m - 3m - 2$

71. $6u^2 + 3uv - 4uv - 2v^2$

72. $2x^2 - 4xy - xy + 2y^2$

73. $6x^2 + 3xy - 10xy - 5y^2$

74. $4u^2 - 16uv - 3uv + 12v^2$

75. $3u^2 + 4 - 12u - u$ **76.** $6m^2 - 2 + 4m - 3m$

77. $6u^2 - 2v^2 + 3uv - 4uv$

78. $2x^2 + 2y^2 - 4xy - xy$

79. $6x^2 - 5y^2 + 3xy - 10xy$

80. $4u^2 + 12v^2 - 3uv - 16uv$

81. $3a^2 + 3b^2 + 9ab + ab$ **82.** $a^2 + b^2 + ab + ab$

83. $uw + vx - vw - ux$ **84.** $2ab + 12 + 6b + 4a$

6-4 Factoring Some Basic Forms of Second-Degree Polynomials

♦ Recognizing Perfect Squares
♦ Sum and Difference of Two Squares

The products

$$(A + B)^2 = A^2 + 2AB + B^2$$

$$(A + B)(A - B) = A^2 - B^2$$

also provide factoring formulas for the expressions on the right. If we can recognize that an expression has either of these two forms, the factoring is straightforward. The process is developed in this section.

♦ RECOGNIZING PERFECT SQUARES

Multiplying out $(x + b)^2$ yields

$$(x + b)^2 = x^2 + 2bx + b^2$$

For example, $(x + 3)^2 = x^2 + 6x + 9$. Notice that the last term 9 is the square of 3, which is half the coefficient of x. In the general expression, the last term b^2 is the square of half of $2b$, the coefficient of x. Whenever this is the case, we can factor the trinomial as a perfect square.

Example 1
Recognizing
Perfect Squares

Which of the following trinomials are perfect squares? Factor those that are.

(A) $x^2 + 4x + 2$ **(B)** $x^2 + 4x + 4$

(C) $x^2 - 6x + 9$ **(D)** $x^2 + 9x + 6$

Solution **(A)** The last term 2 is not the square of half the coefficient of x. The coefficient of x is 4; the square of $\frac{1}{2}$ of 4 is $2^2 = 4$. Thus, the polynomial is not a perfect square.

(B) The last term 4 is the square of half the coefficient of x, that is, 4 is 2^2. Thus, the polynomial is a perfect square, $(x + b)^2$ with $b = 2$:

$$x^2 + 4x + 4 = (x + 2)^2$$

(C) $x^2 - 6x + 9 = (x - 3)^2$

(D) This is not a perfect square.

Matched Problem 1   Which of the following trinomials are perfect squares? Factor those that are.

(A) $x^2 + 14x + 49$ **(B)** $x^2 - 5x + 25$
(C) $x^2 - 8x + 16$ **(D)** $x^2 + 16x + 16$

The polynomial $4x^2 - 12x + 9$ is also a perfect square, since

$$(2x - 3)^2 = 4x^2 - 12x + 9$$

but it is not so easily recognizable, because the coefficient of x^2 is not 1. However, the more general factoring methods considered in the following two sections will handle this sort of problem.

◆ SUM AND DIFFERENCE OF TWO SQUARES

Look at the following products and try to determine what they all have in common:

$$(x - 3)(x + 3) = x^2 - 9 \qquad (x - 2y)(x + 2y) = x^2 - 4y^2$$
$$(2x + 4)(2x - 4) = 4x^2 - 16 \qquad (A - B)(A + B) = A^2 - B^2$$

We note that the binomial factors on the left side of each equation are the same except for the signs. When each pair is multiplied, the middle term disappears. Looking at the expressions on the right, we see that each is the difference of two squares. Writing the last equation in reverse order, we obtain a **factoring formula for the difference of two squares:** $A^2 - B^2 = (A - B)(A + B)$. On the other hand, the sum of two squares, $A^2 + B^2$, cannot be factored using integer coefficients unless A and B have common factors. We will see why in Section 6-6.

Sum and Difference of Two Squares

1. The sum of two squares $A^2 + B^2$ cannot be factored using integer coefficients, unless A and B have common factors.
2. The difference of two squares

$$A^2 - B^2 = (A - B)(A + B)$$ Learn this factoring formula, and use it for difference-of-two-squares forms.

In words, a first expression squared minus a second expression squared is the first expression minus the second expression, times the sum of the two expressions.

<table>
<tr><td>**Example 2**
Factoring Sums and Differences of Squares</td><td>Factor, if possible, using integer coefficients:

(A) $x^2 - y^2$ **(B)** $4x^2 - 9$ **(C)** $U^2 + V^2$ **(D)** $9m^2 - 25n^2$</td></tr>
</table>

Solution   **(A)** $x^2 - y^2 = (x - y)(x + y)$
(B) $4x^2 - 9 = (2x)^2 - (3)^2 = (2x - 3)(2x + 3)$
(C) $U^2 + V^2$ is not factorable using integer coefficients.
(D) $9m^2 - 25n^2 = (3m)^2 - (5n)^2 = (3m - 5n)(3m + 5n)$

Matched Problem 2   Factor, if possible, using integer coefficients:

(A) $x^2 - 4$ **(B)** $4x^2 - 9y^2$ **(C)** $4m^2 + n^2$ **(D)** $16x^2 - 5$

Example 3
Factoring   Factor, if possible, using integer coefficients:

(A) $5x^2 - 180$ **(B)** $3x^2 + 12x + 12$

Solution   **(A)** $5x^2 - 180 = 5(x^2 - 36)$ Take out common factors. Factor the resulting term in parentheses as the difference of two squares.

$$= 5(x - 6)(x + 6)$$

(B) $3x^2 + 12x + 12$ Take out common factors.

$$= 3(x^2 + 4x + 4)$$ The factor in parentheses is a perfect square.

$$= 3(x + 2)^2$$

Matched Problem 3   Factor, if possible, using integer coefficients:

(A) $5x^2 + 30x + 45$ **(B)** $3x^2 + 12$ **(C)** $6x^2 - 150$

Answers to   **1. (A)** $(x + 7)(x + 7) = (x + 7)^2$ **(B)** Not a perfect square
Matched Problems   **(C)** $(x - 4)(x - 4) = (x - 4)^2$ **(D)** Not a perfect square
2. (A) $(x - 2)(x + 2)$ **(B)** $(2x - 3y)(2x + 3y)$
(C) Not factorable using integers   **(D)** Not factorable using integers
3. (A) $5(x + 3)^2$ **(B)** $3(x^2 + 4)$ **(C)** $6(x - 5)(x + 5)$

EXERCISE 6-4

A *Decide whether the trinomial is a perfect square. If it is, factor it.*

1. $x^2 + 10x + 100$ **2.** $x^2 + 14x + 49$

3. $x^2 + 18x + 81$ **4.** $x^2 + 10x + 25$

5. $x^2 + 4x + 4$ **6.** $x^2 + 8x + 64$

7. $x^2 - 6x - 9$ **8.** $x^2 - 14x + 7$

9. $x^2 - 18x + 9$ **10.** $x^2 - 16x + 16$

11. $x^2 - 7x + 49$ **12.** $x^2 - 6x + 144$

13. $x^2 + 12x + 144$ **14.** $x^2 - 10x - 20$

15. $x^2 + 24x + 36$ **16.** $x^2 + 20x + 100$

17. $x^2 + 22x + 121$ **18.** $x^2 - 24x + 144$

19. $x^2 + 16x + 256$ **20.** $x^2 - 14x + 196$

Factor using integer coefficients, if possible. If not factorable, say so.

21. $x^2 - 16$ **22.** $x^2 - 49$ **23.** $x^2 + 25$

24. $x^2 + 9$ **25.** $x^2 - 100$ **26.** $x^2 - 121$

27. $x^2 + 144$ **28.** $x^2 + 81$ **29.** $x^2 - 40$

30. $x^2 - 160$ **31.** $x^2 + 121$ **32.** $x^2 - 200$

B *Factor as far as possible in the integers. Take out any common factors and then factor further if possible.*

33. $3x^2 + 30x + 75$

34. $5x^2 + 20x + 20$

35. $4x^2 + 24x + 36$

36. $8x^2 + 48x + 72$

37. $6x^2 + 36x + 60$

38. $9x^2 + 36x + 18$

39. $4x^2 - 24x + 36$

40. $2x^2 - 16x + 16$

41. $4x^2 - 16x + 8$

42. $3x^2 - 18x + 27$

43. $6x^2 - 60x + 150$

44. $8x^2 - 64x + 32$

45. $4x^2 + 16x + 64$

46. $6x^2 + 24x + 144$

47. $2x^2 - 12x + 36$

48. $6x^2 - 12x + 24$

49. $5x^2 + 10x + 20$

50. $4x^2 + 8x + 16$

51. $6x^2 - 12$

52. $4x^2 - 24$

53. $8x^2 - 72$

54. $3x^2 - 75$

55. $3x^2 + 48$

56. $4x^2 + 64$

57. $2x^2 - 5x$

58. $3x^2 - 15x$

59. $6x^2 + 36x$

60. $8x^2 + 64x$

61. $3x^2 - x$

62. $5x^2 - x$

Determine the value of c that makes the expression a perfect square:

63. $x^2 + 14x + c$

64. $x^2 + 22x + c$

65. $x^2 + 18x + c$

66. $x^2 + 12x + c$

67. $x^2 - 20x + c$

68. $x^2 - 30x + c$

69. $x^2 - 28x + c$

70. $x^2 - 34x + c$

Determine the value of B that makes the expression a perfect square:

71. $x^2 + Bx + 64$

72. $x^2 + Bx + 81$

73. $x^2 + Bx + 25$

74. $x^2 + Bx + 49$

75. $x^2 + Bx + 169$

76. $x^2 + Bx + 121$

77. $x^2 - Bx + 16$

78. $x^2 - Bx + 144$

79. $x^2 - Bx + 196$

80. $x^2 - Bx + 1$

81. $x^2 + Bx$

82. $x^2 - Bx$

6-5 Factoring Second-Degree Polynomials by Trial and Error

♦ Factoring Polynomials of the Form $x^2 + bx + c$
♦ Factoring Polynomials of the Form $ax^2 + bx + c$
♦ Factoring Polynomials of the Form $ax^2 + bxy + cy^2$

We now turn our attention to factoring arbitrary second-degree polynomials such as

$$x^2 - 9x - 36 \qquad 2x^2 - 5x - 3 \qquad 2x^2 + 3xy - 2y^2$$

into the product of two first-degree polynomials with integer coefficients. Using the techniques of Section 6-2, we can see that

$$(x + 5)(x + 7) = x^2 + 12x + 35$$

But can you reverse the process for arbitrary second-degree polynomials? Can you, for example, find integer coefficients to enter into the boxes to complete this factoring

$$2x^2 - 5x - 3 = (\square x + \square)(\square x + \square)$$

Factoring is usually much harder than simply multiplying. In the integers, factoring 1,517 into the prime factors 37 and 41 is not as easy as multiplying 37 times 41.

Similarly, factoring second-degree polynomials with integer coefficients as the product of two first-degree polynomials with integer coefficients is not as easy as multiplying first-degree polynomials. In this section, we develop a trial and error method of factoring the general second-degree polynomials

$$ax^2 + bx + c \quad \text{and} \quad ax^2 + bxy + cy^2$$

that works fairly quickly so long as a and c do not have too many factors or are not too large. In the next section we will provide an approach that builds on the method of factoring by grouping discussed in Section 6-3. This second approach may take a little more effort to learn and apply, but it will work more efficiently on many second-degree polynomials.

◆ FACTORING POLYNOMIALS OF THE FORM $x^2 + bx + c$

Let us start with a polynomial in factored form:

$$x^2 + 12x + 35 = (x + 5)(x + 7) \tag{1}$$

There are three related ways in which this polynomial can also be factored:

$$(x + 7)(x + 5) \tag{2}$$
$$(-x - 5)(-x - 7) \tag{3}$$
$$(-x - 7)(-x - 5) \tag{4}$$

The factors in Equation (2) result from applying the commutative property of multiplication to (1). The factors in (3) and (4) result from multiplying both factors in (1) and (2) by -1. We will consider all four of these ways of factoring $x^2 + 12x + 35$ to be essentially the same, just as we would view writing the integer 15 as

$$3 \times 5 \quad 5 \times 3 \quad (-3) \times (-5) \quad \text{or} \quad (-5) \times (-3)$$

as really the same factorization.

Usually we are interested in *any* factorization that works. We generally pick the simplest form, and this is most often the form that involves the least number of negative signs. In the example above this is either (1) or (2). Thus, in trying to factor a polynomial like $x^2 + 12x + 35$, we need consider only factors of the form

$$(x + r)(x + s)$$

Now consider a polynomial without having its factors given:

$$x^2 + 6x + 8$$

Our problem is to write this polynomial as the product of two first-degree factors with integer coefficients, if they exist. We start by writing

$$x^2 + 6x + 8 = (x + r)(x + s) = x^2 + (r + s)x + rs$$

Thus we need a pair of integers r and s such that

$$r + s = 6 \quad \text{and} \quad rs = 8$$

The possibilities for the product rs are limited. These are the only products of integers that yield 8:

$$
\begin{array}{cccc}
1 \cdot 8 & 2 \cdot 4 & 4 \cdot 2 & 8 \cdot 1 \\
(-1)(-8) & (-2)(-4) & (-4)(-2) & (-8)(-1)
\end{array}
$$

For which of these products is the sum of the integers equal to 6? The pair 2, 4 works. Thus, $x^2 + 6x + 8$ can be factored as $(x + 2)(x + 4)$.

Since reversing the order of r and s, will lead to essentially the same factorization, we did not actually need to consider all eight possible products for rs in this case. We could have restricted our search to the products

$$1 \cdot 8 \quad 2 \cdot 4 \quad (-1)(-8) \quad (-2)(-4)$$

Moreover, since the sum $r + s$ had to be positive, we also did not need to consider $(-1)(-8)$ or $(-2)(-4)$. This kind of analysis can simplify our work.

The same approach will work to factor any polynomial in the form $x^2 + bx + c$. We set

$$x^2 + bx + c = (x + r)(x + s) = x^2 + (r + s)x + rs$$

Thus, we seek a pair of integers r and s such that

$$r + s = b \quad \text{the coefficient of } x$$
$$rs = c \quad \text{the constant term}$$

Because we factor the polynomial by *trying* possible factors of the constant term until we find a pair that add to the coefficient of x, the process is referred to as **trial and error factoring.** You may also see this method referred to as the *product-sum rule,* since we are looking for a pair of integers whose product is the constant term and whose sum is the coefficient of x.

Trial and Error Factoring of $x^2 + bx + c$

1. Consider all pairs of integers r and s that yield a product equal to c. Find one, if it exists, with sum $r + s$ equal to b. This gives the factors $(x + r)(x + s)$. In checking all pairs, only one order need be checked. That is, it is not necessary to check both r, s and s, r.
2. If no such pair exists, the polynomial cannot be factored using integer coefficients.

Example 1
Trial and Error Factoring of $x^2 + bx + c$

Factor using integer coefficients:

(A) $x^2 + 9x + 8$ **(B)** $x^2 - 6x + 8$ **(C)** $x^2 + 7x - 8$
(D) $x^2 - 2x - 8$ **(E)** $x^2 + 4x - 8$

when trying to find the factors by trial and error. We would eventually find that the pair 1, −4 works with the first form.

These observations lead to the following procedure:

Trial and Error Factoring of $ax^2 + bx + c$

1. Consider pairs of integers with a product equal to a. Only one order for each pair needs to be considered. If $a > 0$, only pairs of positive integers need be considered. If $a < 0$, the integers will have opposite signs but we need consider only one choice as to which integer in the pair is negative. These pairs determine the possible forms of the factors. List all these forms.
2. List all pairs of integers r and s with a product equal to c. Both orders for each pair need to be considered. Analyzing the signs in the polynomial may eliminate some of the pairs from consideration.
3. Test every combination of a form from step 1 and a pair from step 2, until a combination, if it exists, yields the necessary middle term b.
4. If no combination yields the middle term, the polynomial cannot be factored using integer coefficients.

Notice that step 1 limits the pairs of factors of a that need to be considered. For example, if $a = 2$, we need consider only one of the pairs

$$2, 1 \qquad 1, 2 \qquad -2, -1 \qquad -1, -2$$

If $a = -3$, we need consider only one of the pairs

$$-3, 1 \qquad 1, -3 \qquad 3, -1 \qquad -1, 3$$

However, if $a = 4$, we need to consider one of the pairs

$$4, 1 \qquad 1, 4 \qquad -4, -1 \qquad -1, -4$$

and one of the pairs

$$2, 2 \qquad -2, -2$$

Again, we try to make choices that reduce the number of negative signs in the final result to a minimum.

Example 2
Trial and Error Factoring for $ax^2 + bx + c$

Factor each polynomial, if possible, using integer coefficients:

(A) $2x^2 + 3x - 2$ **(B)** $2x^2 - 3x + 4$ **(C)** $6x^2 + 5x - 4$

Solution

(A) $2x^2 + 3x - 2$

For $a = 2$, we need consider only the pair 2, 1 with product 2. We then write the form for the factors:

$$2x^2 + 3x - 2 = (2x + r)(x + s)$$
$$= 2x^2 + (2s + r)x + rs$$

Thus we need $2s + r = 3$ and $rs = -2$. Since their product is negative, r and s must have opposite signs. These are the possibilities:

r	s
1	-2
2	-1
-1	2
-2	1

Testing until we find the correct coefficient for x, that is, a pair that gives $2s + r = 3$, we find that $r = -1$ and $s = 2$ works:

$$2x^2 + 3x - 2 = (2x - 1)(x + 2)$$

(B) $2x^2 - 3x + 4$
We set

$$2x^2 - 3x + 4 = (2x + r)(x + s)$$
$$= 2x^2 + (2s + r)x + rs$$

so $2s + r = -3$ and $rs = 4$. Since their product is positive, r and s must have the same sign. These are the possibilities:

r	s
1	4
2	2
4	1
-1	-4
-2	-2
-4	-1

No choice produces the middle term. That is, no choice has $2s + r = -3$. Thus

$$2x^2 - 3x + 4$$

is not factorable using integer coefficients.

(C) $6x^2 + 5x - 4$
For $6x^2 + 5x - 4$, we have two different choices for the product that yields the coefficient 6, 1×6 and 2×3, so we have to consider two possible forms for the factors

$$(x + r)(6x + s) = 6x^2 + (6r + s)x + rs$$

and

$$(2x + r)(3x + s) = 6x^2 + (3r + 2s)x + rs$$

Since $rs = -4$, r and s must have opposite signs. The possibilities are

r	s
1	-4
2	-2
4	-1
-1	4
-2	2
-4	1

Will any combination of choices for the form of the factors and the r, s pair give us the correct middle term? After some checking we would find that choosing the second form and the pair -1, 4 for r, s will work. Thus,

$$6x^2 + 5x - 4 = (2x - 1)(3x + 4)$$

Since we had two choices for the form and six for the r, s pair, we might have had to check $2 \times 6 = 12$ possible combinations until we found one that works.

Matched Problem 2   Factor each polynomial, if possible, using integer coefficients:

(A) $x^2 - 8x + 12$ **(B)** $x^2 + 2x + 5$

♦ **FACTORING POLYNOMIALS OF THE FORM**
$ax^2 + bxy + cy^2$

The same trial and error rules developed above apply to polynomials of the form $ax^2 + bxy + cy^2$. The presence of the second variable y does not make the problem more complicated.

Example 3
Trial and Error
Factoring of
$ax^2 + bxy + cy^2$

Factor each polynomial, if possible, using integer coefficients:

(A) $x^2 + 9xy + 8y^2$ **(B)** $2x^2 + 3xy - 2y^2$

Solution   **(A)** We set

$$x^2 + 9xy + 8y^2 = (x + ry)(x + sy)$$
$$= x^2 + (s + r)xy + rsy^2$$

We need $s + r = 9$, and $rs = 8$. These are the same conditions encountered in Example 1(A), so the solution is the same: $r = 1$, $s = 8$, and

$$x^2 + 9xy + 8y^2 = (x + y)(x + 8y)$$

(B) We set

$$2x^2 + 3xy - 2y^2 = (2x + ry)(x + sy)$$
$$= 2x^2 + (2s + r)xy + rsy^2$$

We need $2s + r = 3$ and $rs = -2$. These are the same conditions encountered in Example 2(A), so the solution is the same: $r = -1$, $s = 2$, and

$$2x^2 + 3xy - 2y^2 = (2x - y)(x + 2y)$$

Matched Problem 3   Factor each polynomial, if possible, using integer coefficients:

(A) $x^2 + xy - 12y^2$ **(B)** $2x^2 - 11xy - 6y^2$
(C) $2x^2 + 7xy - 4y^2$ **(D)** $4x^2 - 15xy - 4y^2$

The difficulty with factoring by trial and error becomes apparent with problems like that found in Example 2(C) involving the factoring of $6x^2 + 5x - 4$. As the integers a and c in $ax^2 + bx + c$ get larger and larger with more and more factors, the number of combinations that need to be checked in step 3 of the trial and error process increases very rapidly. And it is quite possible in most practical situations that none of the combinations will work. It is important, however, that you understand the approach presented above, since it is effective for many of the simpler factoring problems you will encounter. In the next section we will present a systematic approach to factoring that will substantially reduce the amount of trial and error and even tell you whether the polynomial can be factored before you proceed too far.

In conclusion, we point out that if a, b, and c are selected at random out of the integers, it is much more likely that

$$ax^2 + bx + c$$

is not factorable using integer coefficients than that it is. But even being able to factor some second-degree polynomials leads to marked simplification of algebraic expressions and a quick way to solve certain second-degree equations, as we will see later.

Answers to
Matched Problems

1. (A) $(x - 3)(x - 5)$ **(B)** $(x + 1)(x + 15)$ **(C)** $(x + 5)(x - 3)$
   (D) Not factorable using integer coefficients   **(E)** $(x + 1)(x - 15)$
2. (A) $(x - 2)(x - 6)$ **(B)** Not factorable using integer coefficients
3. (A) $(x + 4y)(x - 3y)$ **(B)** $(2x + y)(x - 6y)$
   (C) $(2x - y)(x + 4y)$ **(D)** $(4x + y)(x - 4y)$

EXERCISE 6-5

Factor in the integers, if possible. If not factorable, say so.

A 1. $x^2 + 5x + 4$ 2. $x^2 + 4x + 3$

3. $x^2 + 5x + 6$ 4. $x^2 + 7x + 10$

5. $x^2 - 4x + 3$ 6. $x^2 - 5x + 4$

7. $x^2 - 7x + 10$ 8. $x^2 - 5x + 6$

9. $y^2 + 3y + 3$ 10. $y^2 + 2y + 2$

11. $y^2 - 2y + 6$ 12. $x^2 - 3x + 5$

13. $x^2 - 7x + 12$ 14. $x^2 + 7x + 12$

15. $x^2 + 8x + 12$ 16. $x^2 - 8x + 12$

17. $x^2 + 13x + 12$ 18. $x^2 - 13x + 12$

19. $x^2 + 8xy + 15y^2$ 20. $x^2 + 9xy + 20y^2$

21. $x^2 - 8x + 16$ 22. $x^2 - 8x - 16$

23. $x^2 - 10x - 25$ 24. $x^2 - 10x + 25$

25. $x^2 - x - 12$ 26. $x^2 + x - 12$

27. $x^2 + 4x - 12$ 28. $x^2 - 4x - 12$

29. $x^2 - 11x - 12$ 30. $x^2 + 11x - 12$

31. $x^2 - 10xy + 21y^2$ 32. $x^2 - 10xy + 16y^2$

33. $u^2 + 4uv + v^2$ 34. $u^2 + 5uv + 3v^2$

B 35. $3x^2 + 7x + 2$ 36. $2x^2 + 7x + 3$

37. $3x^2 - 7x + 4$ 38. $2x^2 - 7x + 6$

39. $x^2 + 2x - 1$ 40. $x^2 - 2x + 1$

41. $x^2 + 2x + 1$ 42. $x^2 - 2x - 1$

43. $3x^2 - 14x + 8$ 44. $2y^2 - 13y + 15$

45. $3x^2 - 11xy + 6y^2$ 46. $2x^2 - 7xy + 6y^2$

47. $n^2 - 2n - 8$ 48. $n^2 + 2n - 8$

49. $x^2 - 4x - 6$ 50. $x^2 - 3x - 8$

51. $3x^2 - x - 2$ 52. $6m^2 + m - 2$

53. $x^2 + 4xy - 12y^2$ 54. $2x^2 - 3xy - 2y^2$

55. $3u^2 - 11u - 4$ 56. $8u^2 + 2u - 1$

57. $6x^2 + 7x - 5$ 58. $2m^2 - 3m - 20$

59. $3s^2 - 5s - 2$ 60. $2s^2 + 5s - 3$

61. $3x^2 + 2xy - 3y^2$ 62. $2x^2 - 3xy - 4y^2$

63. $5x^2 - 8x - 4$ 64. $12x^2 + 16x - 3$

65. $6u^2 - uv - 2v^2$ 66. $6x^2 - 7xy - 5y^2$

67. $8x^2 + 6x - 9$ 68. $6x^2 - 13x + 6$

69. $3u^2 + 7uv - 6v^2$ 70. $4m^2 + 10mn - 6n^2$

71. $4u^2 - 19uv + 12v^2$ 72. $12x^2 - xy - 6y^2$

C 73. $12x^2 - 40xy - 7y^2$ 74. $15x^2 + 17xy - 4y^2$

75. $12x^2 + 19xy - 10y^2$ 76. $24x^2 - 31xy - 15y^2$

77. $x^2 - y^2$ 78. $x^2 - 9y^2$ 79. $4x^2 - y^2$

80. $9x^2 - 4y^2$ 81. $x^2 + y^2$ 82. $x^2 + 9y^2$

83. $4x^2 + y^2$ 84. $9x^2 + 4y^2$

85. $9x^2 - 12xy + 4y^2$ 86. $9x^2 + 12xy + 4y^2$

87. $25x^2 + 10xy + y^2$ 88. $25x^2 - 10xy + y^2$

89. $x^2 + xy + y^2$ 90. $x^2 - xy + y^2$

6-6   Factoring Second-Degree Polynomials Systematically

♦   The *ac* Test

♦   Which Factoring Method—*ac* Test or Trial and Error?

We continue our discussion of factoring second-degree polynomials of the type

$$ax^2 + bx + c \qquad ax^2 + bxy + cy^2 \qquad (1)$$

with integer coefficients into the product of two first-degree factors with integer coefficients. In the last section we found that the number of cases that had to be tested tended to increase very rapidly as the coefficients *a* and *c* increased in size and had more possible factors. And then, in realistic situations, it turns out that it is quite unlikely that any of the combinations tested will work. It would be useful to know ahead of time if the polynomials shown above are, in fact, factorable before we start looking for the factors. In this section we provide a procedure to determine this.

♦ **THE ac TEST**

We now provide a test, called the ac test for factorability, that not only tells us if the polynomials in (1) can be factored using integer coefficients but, in addition, leads to a direct way of factoring those that are factorable.

---

ac Test for Factorability

To test whether a polynomial like those in (1) is factorable:

1. From all possible integer factors of the product ac, look for a pair of factors p and q such that

$$pq = ac \quad \text{and} \quad p + q = b$$

2. If such a pair of factors exists, then the polynomial can be factored into two first-degree factors with integer coefficients. If no such factors exist, then the polynomial does not have first-degree factors with integer coefficients.

If a polynomial like those in (1) is found to be factorable by the ac test, the numbers p and q found in the test will help us factor the polynomial.

Using the ac Test to Factor

1. Use the integers p and q found in the ac test to satisfy $pq = ac$ and $p + q = b$.
2. Split the coefficient b into $b = p + q$ and rewrite the polynomial as

$$ax^2 + px + qx + c \quad \text{or} \quad ax^2 + pxy + qxy + cy^2$$

3. Factor the resulting polynomial by grouping.

To justify the ac test, we would procede as we did in Section 6-5 to set up the trial and error procedure. We would write $ax^2 + bx + c = (mx + r)(nx + s)$ and look for conditions on $m, n, r,$ and s. The details are left to subsequent courses. We now make the technique concrete through detailed examples.

Example 1
Using the ac Test

Factor, if possible, using integer coefficients:

(A) $2x^2 + 11x - 6$ **(B)** $4x^2 - 7x + 4$ **(C)** $6x^2 + 5xy - 4y^2$

Solution   **(A)** $2x^2 + 11x - 6$

Step 1   Test for factorability, using the *ac* test:

$$ax^2 + bx + c \left.\begin{array}{l} a = 2 \\ b = 11 \\ c = -6 \end{array}\right\}$$
$$2x^2 + 11x - 6$$

Multiply *a* and *c* to obtain —————— Don't forget the negative sign.

$$ac = (2)(-6) = -12$$

Now try to find two integer factors of -12 with a sum of $b = 11$. We write, or think, of all the ways of factoring -12 into the product of two integers p and q. For each product pq, we check the sum of the two factors $p + q$:

pq	$p + q$
$(1)(-12)$	-11
$(-1)(12)$	11
$(2)(-6)$	-4
$(-2)(6)$	4
$(3)(-4)$	-1
$(-3)(4)$	1

Each pair produces -12 as a product, and the second pair adds up to 11, the coefficient of the middle term in $2x^2 + 11x - 6$. Thus the conditions of the *ac* test are met:

$$\overset{p\quad q}{(-1)(12)} = \overset{ac}{-12} \quad \text{and} \quad \overset{p}{(-1)} + \overset{q}{(12)} = \overset{b}{11}$$

We conclude that $2x^2 + 11x - 6$ can be factored using integer coefficients. Notice that we could also use $p = 12$ and $q = -1$, but we will get the same result and factorization either way. Thus, in looking for the factors of -12, we did not enter both pairs $(-1)(12)$ and $(12)(-1)$.

Step 2   Since the *ac* test is satisfied, we can factor the polynomial by grouping. Split the middle term in $2x^2 + 11x - 6$, using $p = -1$ and $q = 12$ found in step 1. This is possible since $p + q = (-1) + (12) = 11 = b$.

Use p and q found in step 1.

$$2x^2 + \overset{b}{11x} - 6 = 2x^2 \overset{p}{- 1x} \overset{q}{+ 12x} - 6$$

Step 3   We can now complete the factoring by grouping. This will always work if the *ac* test is satisfied. Moreover, it doesn't matter if we reverse the values for p and q.

$$\begin{aligned} 2x^2 - x + 12x - 6 & \qquad \text{Group the first two and last two terms.} \\ = (2x^2 - x) + (12x - 6) & \qquad \text{Factor out common factors.} \\ = x(2x - 1) + 6(2x - 1) & \qquad \text{Factor out the common factor } (2x - 1). \\ = (2x - 1)(x + 6) & \qquad \text{The factoring is complete.} \end{aligned}$$

Thus, $2x^2 + 11x - 6 = (2x - 1)(x + 6)$

Without the commentary, this process can be reduced to a few key operational steps. Some of the steps are done mentally. The only trial and error occurs in step 1, and with practice this will go fairly quickly.

(B) $4x^2 - 7x + 4$
Compute *ac:*

$$ac = (4)(4) = 16$$

Write, or think, all two-integer factors of 16, and try to find a pair whose sum is -7, the coefficient of the middle term:

pq	$p + q$
$(4)(4)$	8
$(-4)(-4)$	-8
$(2)(8)$	10
$(-2)(-8)$	-10
$(1)(16)$	17
$(-1)(-16)$	-17

None of the sums is $-7 = b$; thus, according to the *ac* test,

$$4x^2 - 7x + 4$$

is not factorable using integer coefficients.

(C) $6x^2 + 5xy - 4y^2$
Compute *ac:*

———— Don't forget the negative sign.

$$ac = (6)(-4) = -24$$

Does -24 have two integer factors whose sum is $5 = b$? A little trial and error, either mentally or by listing, gives us

$$\underset{p \quad q}{\underset{}{}} \quad \underset{ac}{} \qquad\qquad \underset{p \quad q \qquad b}{}$$
$$(8)(-3) = -24 \quad \text{and} \quad (8) + (-3) = 5$$

Now we split the middle term $5xy$ into $8xy - 3xy$, using the p and q just found, and write

$$\underset{b}{} \qquad\qquad \underset{p \qquad q}{}$$
$$6x^2 + 5xy - 4y^2 = 6x^2 + 8xy - 3xy - 4y^2$$

We then complete the factoring by grouping:

$$6x^2 + 8xy - 3xy - 4y^2 = (6x^2 + 8xy) - (3xy + 4y^2)$$
$$= 2x(3x + 4y) - y(3x + 4y)$$
$$= (3x + 4y)(2x - y)$$

Thus, $6x^2 + 5xy - 4y^2 = (3x + 4y)(2x - y)$

Matched Problem 1   Factor, if possible, using integer coefficients:

   (A) $4x^2 + 4x - 3$ **(B)** $6x^2 - 3x - 4$ **(C)** $6x^2 - 25xy + 4y^2$

If we apply the *ac* test to $x^2 + y^2$, we find $ac = 1$ and $b = 0$. No p and q exist with $pq = 1$ and $p + q = 0$. This is why the sum of two squares cannot be factored using integer coefficients, unless the terms have a common factor.

◆ **WHICH FACTORING METHOD—*ac* TEST OR TRIAL AND ERROR?**

When do we use the *ac* test and when do we use the trial and error method described in the preceding section? The two methods both involve testing products and sums. When *a* in $ax^2 + bx + c$ is equal to 1, the two methods are essentially the same but the trial and error approach may be quicker. Otherwise, the *ac* test will usually prove more effective.

Answers to Matched Problem   **1.** **(A)** $(2x - 1)(2x + 3)$ **(B)** Not factorable   **(C)** $(6x - y)(x - 4y)$

EXERCISE 6-6

Factor, if possible, using integer coefficients. Use the ac test and proceed as in Example 1.

A **1.** $3x^2 - 7x + 4$

3. $x^2 + 4x - 6$

5. $2x^2 + 5x - 3$

7. $3x^2 - 5x + 4$

9. $6x^2 + 5x + 1$

11. $4x^2 + 4x + 1$

13. $4x^2 + 3x - 1$

15. $4x^2 - 5x + 1$

17. $6x^2 - 5x + 1$

19. $8x^2 - 7x - 1$

21. $5x^2 + 11x + 2$

23. $6x^2 + 7x + 2$

25. $7x^2 + 9x + 2$

27. $4x^2 + 9x + 2$

29. $3x^2 + 5x + 2$

B **31.** $3x^2 - 14x + 8$

33. $6x^2 + 7x - 5$

35. $6x^2 - 4x - 5$

37. $2m^2 - 3m - 20$

2. $2x^2 - 7x + 6$

4. $x^2 - 3x - 8$

6. $3x^2 - 5x - 2$

8. $2x^2 - 11x + 6$

10. $4x^2 + 5x + 1$

12. $8x^2 + 6x + 1$

14. $6x^2 + x - 1$

16. $4x^2 - 4x + 1$

18. $4x^2 - 3x - 1$

20. $6x^2 + 7x + 1$

22. $5x^2 + 7x + 2$

24. $6x^2 + 13x + 2$

26. $2x^2 + 5x + 2$

28. $3x^2 + 7x + 2$

30. $7x^2 + 15x + 2$

32. $2y^2 - 13y + 15$

34. $5x^2 - 8x - 4$

36. $5x^2 - 7x - 4$

38. $12x^2 + 16x - 3$

39. $3u^2 - 11u - 4$

41. $6u^2 - uv - 2v^2$

43. $3x^2 + 2xy - 3y^2$

45. $8x^2 + 6x - 9$

47. $4m^2 + 10mn - 6n^2$

49. $3u^2 - 8uv - 6v^2$

51. $4u^2 - 19uv + 12v^2$

53. $4x^2 - 7x - 2$

55. $6x^2 - x - 2$

57. $6x^2 - 4x - 2$

59. $5x^2 + 17x + 6$

61. $3x^2 + 5x + 2$

63. $8x^2 + 19x + 6$

65. $5x^2 + 7x - 6$

67. $9x^2 + 15x - 6$

69. $7x^2 + 11x - 6$

71. $3x^2 + 2x + 1$

73. $6x^2 - 7x + 8$

C **75.** $12x^2 - 40xy - 7y^2$

77. $18x^2 - 9xy - 20y^2$

79. $10x^2 - 19x + 6$

81. $8x^2 + 19x - 15$

40. $8u^2 + 2u - 1$

42. $6x^2 - 7xy - 5y^2$

44. $2x^2 - 3xy - 4y^2$

46. $6x^2 - 5x - 6$

48. $3u^2 + 7uv - 6v^2$

50. $4m^2 - 9mn - 6n^2$

52. $12x^2 - xy - 6y^2$

54. $4x^2 - 27x - 2$

56. $6x^2 - 11x - 2$

58. $4x^2 - 2x - 2$

60. $5x^2 + 13x + 6$

62. $3x^2 + 9x + 6$

64. $8x^2 + 26x + 6$

66. $5x^2 - 7x - 6$

68. $9x^2 - 25x - 6$

70. $7x^2 - 19x - 6$

72. $5x^2 - 4x + 3$

74. $4x^2 + 5x + 6$

76. $15x^2 + 17xy - 4y^2$

78. $15m^2 + 2mn - 24n^2$

80. $10x^2 + 11x - 6$

82. $8x^2 - 26x + 15$

For the value of c given, find all integers b such that $x^2 + bx + c$ can be factored:

83. $c = 12$ **84.** $c = 10$ **85.** $c = 8$

86. $c = 6$ **87.** $c = 16$ **88.** $c = 20$

For the value of b given, find all positive integers c between 1 and 15, inclusive, such that $x^2 + bx + c$ can be factored:

89. $b = 7$ **90.** $b = 5$ **91.** $b = 3$

92. $b = 9$ **93.** $b = 11$ **94.** $b = 12$

6-7 Higher-Degree Polynomials and More Factoring

- ♦ Sum and Difference of Two Cubes
- ♦ Combined Factoring Processes
- ♦ Factoring by Grouping
- ♦ A Factoring Strategy

In this last section on factoring we will consider some additional basic factoring forms, as well as problems that require a combination of the processes we have considered. Higher-degree polynomials are included.

♦ SUM AND DIFFERENCE OF TWO CUBES

We can verify, by direct multiplication of the right sides, the following factoring formulas for the sum and difference of two cubes:

Sum and Difference of Two Cubes

1. The sum of two cubes

$$A^3 + B^3 = (A + B)(A^2 - AB + B^2)$$

2. The difference of two cubes

$$A^3 - B^3 = (A - B)(A^2 + AB + B^2)$$

These formulas are used in the same way as the factoring formula for the difference of two squares. Notice that neither $A^2 - AB + B^2$ nor $A^2 + AB + B^2$ factors further using integer coefficients. You can check this using the ac test. Both formulas should be learned.

Notice also that in each formula the binomial factor $A + B$ or $A - B$ has the same addition or subtraction sign as the original sum or difference. The term AB in the other factor has the opposite sign.

$$\overbrace{A^3 + B^3}^{\substack{\text{Same} \\ \text{sign}}} = (A \underset{\text{Opposite sign}}{+} B)(A^2 \; - \; AB + B^2)$$

$$\overbrace{A^3 - B^3}^{\substack{\text{Same} \\ \text{sign}}} = (A \underset{\text{Opposite sign}}{-} B)(A^2 \; + \; AB + B^2)$$

Example 1 **Factoring Sums and Differences of Cubes**	Factor as far as possible, using integer coefficients: **(A)** $x^3 - 8$ **(B)** $x^3 + 27$ **(C)** $8x^3 + 1$

Solution　**(A)**　To factor
$$x^3 - 8$$

we recognize that $8 = 2^3$, so the polynomial can be rewritten as the difference of two cubes

$$x^3 - 2^3$$

Now we can apply the formula from the box with $A = x$, $B = 2$ to obtain

$$x^3 - 8 = x^3 - 2^3 = (x - 2)(x^2 + 2x + 4)$$

(B) $x^3 + 27 = x^3 + 3^3 = (x + 3)(x^2 - 3x + 9)$
(C) $8x^3 + 1 = (2x)^3 + 1^3 = (2x + 1)[(2x)^2 - 2x + 1]$
$\qquad\qquad = (2x + 1)(4x^2 - 2x + 1)$

Matched Problem 1　Factor as far as possible, using integer coefficients: $\quad (3x-1)(9x^2+3x+1)$

(A) $x^3 + 8$ **(B)** $x^3 - 27$ **(C)** $27x^3 - 1$

$\quad (x+2)(x^2-2x+4) \qquad (x-3)(x^2+3x+9)$

♦　**COMBINED FACTORING PROCESSES**

We now consider several examples that involve the removal of common factors as well as other factoring processes considered earlier. The factoring process generally will be simpler if we remove common factors first before proceeding to other methods.

$2x(9x^2-4)$
$(3x-2)(3x+2)$

Example 2 **Factoring**	Factor as far as possible, using integer coefficients: $2x(2x^2-7x^2+3x)$ **(A)** $4x^3 - 14x^2 + 6x$ **(B)** $18x^3 - 8x$ $2x(2x-1)(x-3)$ **(C)** $8x^3y + 20x^2y^2 - 12xy^3$ **(D)** $3y^3 + 6y^2 + 6y$

$4xy(2x^2+5xy-3y^2) \; (2x-y)(x+3y) \qquad 3(y^2+2y+2)$

Solution　**(A)** $4x^3 - 14x^2 + 6x$ Remove common factors.
$\qquad = 2x(2x^2 - 7x + 3)$ Factor $2x^2 - 7x + 3$ by trial and error or the ac
$\qquad = 2x(2x - 1)(x - 3)$ test.

(B) $18x^3 - 8x$	Remove common factors.
$= 2x(9x^2 - 4)$	Factor the difference of two squares.
$= 2x(3x - 2)(3x + 2)$	Factoring is complete.
(C) $8x^3y + 20x^2y^2 - 12xy^3$	Remove common factors.
	Factor $2x^2 + 5xy - 3y^2$ by trial and error or the *ac* test.
$= 4xy(2x^2 + 5xy - 3y^2)$	
$= 4xy(2x - y)(x + 3y)$	
(D) $3y^3 + 6y^2 + 6y$	Remove common factors.
$= 3y(y^2 + 2y + 2)$	Cannot be factored further using integer coefficients.

Matched Problem 2   Factor as far as possible, using integer coefficients:

(A) $3x^3 - 15x^2y + 18xy^2$ **(B)** $3x^3 - 48x$

(C) $3x^3y + 3x^2y - 36xy$ **(D)** $4x^3 + 12x^2 + 12x$

♦ FACTORING BY GROUPING

Occasionally, polynomial forms of a different nature than we considered in Section 6-3 can be factored by appropriate grouping of terms. The following example illustrates the process.

Example 3
Factoring by Grouping

Factor by grouping terms:

(A) $x^2 + xy + 2x + 2y$ **(B)** $x^2 - 2x - xy + 2y$

Solution

(A) $x^2 + xy + 2x + 2y$	Group the first two and last two terms.
$= (x^2 + xy) + (2x + 2y)$	Remove common factors.
$= x(x + y) + 2(x + y)$	Since each term has the common factor $(x + y)$, we can complete the factoring.
$= (x + y)(x + 2)$	Notice that the factors are not first-degree polynomials of the same type.
(B) $x^2 - 2x - xy + 2y$	
$= (x^2 - 2x) - (xy - 2y)$	Be careful of signs here.
$= x(x - 2) - y(x - 2)$	
$= (x - 2)(x - y)$	

Matched Problem 3   Factor by grouping terms:

(A) $x^2 - xy + 5x - 5y$ **(B)** $x^2 + 4x - xy - 4y$

♦ A FACTORING STRATEGY

There is no general procedure, or **algorithm,** for factoring all polynomials having factors with integer coefficients. However, the following strategy may be helpful:

General Strategy for Factoring Polynomials

1. Remove common factors (Section 6-3).
2. If the polynomial has two terms, look for a difference of two squares or a sum or difference of two cubes (Sections 6-4 and 6-7).
3. If the polynomial has three terms:
   (A) See if it is a perfect square (Section 6-4).
   (B) Try trial and error (Section 6-5).
   (C) Or use the ac test (Section 6-6).
4. If the polynomial has more than three terms, try grouping (Sections 6-3 and 6-7).

Factoring requires skill, creativity, and perseverance. Often the appropriate technique is not apparent immediately, and practice is necessary to develop your recognition of what might work on a given problem. Exercise 6-7 contains numerous problems for this purpose.

Answers to Matched Problems

1. **(A)** $(x + 2)(x^2 - 2x + 4)$ **(B)** $(x - 3)(x^2 + 3x + 9)$
   (C) $(3x - 1)(9x^2 + 3x + 1)$
2. **(A)** $3x(x - 2y)(x - 3y)$ **(B)** $3x(x - 4)(x + 4)$
   (C) $3xy(x - 3)(x + 4)$ **(D)** $4x(x^2 + 3x + 3)$
3. **(A)** $(x - y)(x + 5)$ **(B)** $(x + 4)(x - y)$

EXERCISE 6-7

Factor as far as possible using integer coefficients:

A
1. $6x^3 + 9x^2$
2. $8x^2 + 2x$
3. $u^4 + 6u^3 + 8u^2$
4. $m^5 + 8m^4 + 15m^3$
5. $x^3 - 5x^2 + 6x$
6. $x^3 - 7x^2 + 12x$
7. $x^2 - 4$
8. $x^2 - 1$
9. $4x^2 - 1$
10. $9x^2 - 4$
11. $u^2 + v^2$
12. $m^2 + 64$
13. $2x^2 - 8$
14. $3x^2 - 3$
15. $9x^2 - 16y^2$
16. $25x^2 - 1$
17. $6u^2v^2 - 3uv^3$
18. $2x^3y - 6x^2y^3$
19. $4x^3y - xy^3$
20. $x^3y - 9xy^3$
21. $3x^4 + 27x^2$
22. $2x^3 + 8x$
23. $9u^2 + 4v^2$
24. $x^2 + 16y^2$
25. $x^3 - 8y^3$
26. $27a^3 + b^3$
27. $8x^2y - 72y^3$
28. $9x^2 - 36x^2y^2$
29. $x^2y^2 - 9y^2$
30. $4xy^2 - 16x^3$

31. $7x^2 + 14x + 7$
32. $4x^2 - 24x + 36$
33. $6x^2 - 60x + 150$
34. $8x^2 + 32x + 32$

B
35. $6x^2 + 36x + 48$
36. $4x^2 - 28x + 48$
37. $3x^3 - 6x^2 + 15x$
38. $2x^3 - 2x^2 + 8x$
39. $12x^3 + 16x^2y - 16xy^2$
40. $9x^2y + 3xy^2 - 30y^3$
41. $x^2 + 3x + xy + 3y$
42. $xy + 2x + y^2 + 2y$
43. $x^2 - 3x - xy + 3y$
44. $x^2 - 5x + xy - 5y$
45. $2ac + bc - 6ad - 3bd$
46. $2ac + 4bc - ad - 2bd$
47. $2mu + 2nu - mv - nv$
48. $3wx + 6wy - xz - 2yz$
49. $x^3 - 8$
50. $x^3 + 1$
51. $x^3 + 27$
52. $8y^3 - 1$
53. $3m^4 + 12m^2$
54. $3x(2x - y) - 2y(2x - y)$
55. $x^2 - 7x + 6$
56. $u^2 - 4u + 6$
57. $4y^2 - 1$
58. $4x^2 - 20x + 25$
59. $6x^3 - 8x^2 - 8x$
60. $3x^2 + 10xy - 8y^2$

61. $u^2 - 5u - uv + 5v$ **62.** $3x - x^2 - 3y + xy$

63. $ac + 2bc - 3ad - 6db$

64. $wy + 2xy - 5wz - 10xz$

65. $2xz + 3yz + 2xw + 3yw$

66. $3ac - 3ad - 2bc + 2bd$

C **67.** $4x^3y + 14x^2y^2 + 6xy^3$

68. $3x^3y - 15x^2y^2 + 18xy^3$

69. $60x^2y^2 - 200xy^3 - 35y^4$

70. $60x^4 + 68x^3y - 16x^2y^2$

71. $6x^2y^2 - 3xy^3 + 3xy^2$

72. $18x^4y - 9x^3y - 3x^2y$

73. $x^3 + x + 5x^2 + 5$ **74.** $x^4 + x^2 + 4x + 4$

75. $16xy^3 + 2x^4$ **76.** $54 - 2x^3y^3$

77. $4x^2 + 10x - 6$ **78.** $9x^2 - 15x - 6$

79. $8x^2 - 36x - 20$ **80.** $6x^2 + 28x - 10$

Some higher-degree polynomials may be "disguised" variations of lower-degree polynomials that can be factored. For example,

$$x^4 + 3x^2 + 2 = (x^2 + 1)(x^2 + 2)$$

as can be verified by multiplication. Recognizing the factorization may be made easier with a substitution: Let $u = x^2$, so $x^4 + 3x^2 + 2$ becomes $u^2 + 3u + 2$. This last is factored by trial and error as

$$u^2 + 3u + 2 = (u + 1)(u + 2)$$
$$= (x^2 + 1)(x^2 + 2)$$

Factor as far as possible using integer coefficients:

81. $x^4 + 4x^2 + 4$ **82.** $x^6 + 3x^3 + 2$

83. $x^6 - 3x^3 - 4$ **84.** $x^4 + 8x^2 + 16$

85. $x^4 - 16$ **86.** $x^6 - 8$

87. $x^6 + 27$ **88.** $x^6 - 25$

89. $x^4 - 1$ **90.** $x^6 - 64$

91. $x^4 - y^4$ **92.** $x^6 - y^3$

93. $x^6 + 4x^3 + 4$ **94.** $x^8 + 10x^4 + 25$

95. $x^4 - 5x^2 + 6$ **96.** $x^4 + 9x^2 + 8$

97. $2x^6 + 28x^3 + 98$ **98.** $3x^4 - 54x^2 + 243$

  # 6-8 Solving Equations by Factoring

- ◆ Zero Property
- ◆ Solving Equations by Factoring
- ◆ Solutions and Graphs

If an equation has the form

Polynomial $= 0$

and the polynomial can be factored into the product of factors of degree 1, then the equation can be quickly solved by setting each factor equal to 0. The basis for this technique is a real-number property called the zero property.

◆ ZERO PROPERTY

The real numbers have the property that if a product of two or more numbers is zero, then at least one of the factors must be zero. That is, the product of nonzero numbers cannot be zero. More formally we have the following:

Zero Property

For real numbers a and b,

$$ab = 0 \qquad \text{if and only if} \qquad a = 0 \text{ or } b = 0$$

The "or" in the statement "$a = 0$ or $b = 0$" should be interpreted to allow the possibility that both a and b are 0. This is the common mathematical use of the word. The zero property can be used to solve certain types of equations. If an expression that can be factored is set equal to zero, we can solve the given equation by setting each factor equal to zero. For example, if

$$(x - 5)(x + 6) = 0$$

then $$x - 5 = 0 \qquad \text{or} \qquad x + 6 = 0$$

These last two equations have solutions $x = 5$ and $x = -6$, so these are also the solutions to $(x - 5)(x + 6) = 0$.

♦ **SOLVING EQUATIONS BY FACTORING**

Examples will illustrate the use of the zero property in solving equations by factoring.

Example 1
Solving an Equation by Factoring

Solve $x^2 + x - 20 = 0$ by factoring.

Solution

$$\begin{aligned} x^2 + x - 20 &= 0 \qquad \text{Factor the left side.} \\ (x - 4)(x + 5) &= 0 \qquad (x-4)(x+5) = 0 \text{ if and only} \\ &\qquad\qquad\quad \text{if } (x-4) = 0 \text{ or } (x+5) = 0. \end{aligned}$$

$$x - 4 = 0 \qquad \text{or} \qquad x + 5 = 0$$
$$\boxed{x = 4} \qquad\qquad\qquad \boxed{x = -5}$$

Check Substitute $x = 4$ and $x = -5$ into $x^2 + x - 20$:

$x = 4$: $4^2 + 4 - 20 = 16 + 4 - 20 = 0$

$x = -5$: $(-5)^2 + (-5) - 20 = 25 - 5 - 20 = 0$

Matched Problem 1 Solve $x^2 - 2x - 48 = 0$ by factoring.

It is important to recognize that this method depends on a particular property of 0 and that to use it we must have a *product equal to 0*. The method cannot be applied directly to a product like $ab = 12$. Many equations can be rewritten in the form of a product equal to 0, however, and then this method applies.

Example 2
Solving by Factoring

Solve $3x^2 = 4x$.

Solution

$$3x^2 = 4x$$ Since x might be 0, we cannot divide by x. Rewrite the equation as a polynomial set equal to 0.

$$3x^2 - 4x = 0$$ Factor.

$$x(3x - 4) = 0$$ $x(3x - 4) = 0$ if and only if $x = 0$ or $3x - 4 = 0$.

$$x = 0 \quad \text{or} \quad 3x - 4 = 0$$

$$\boxed{x = 0} \quad \text{or} \quad \boxed{x = \frac{4}{3}}$$

Check $x = 0$: $3(0)^2 \overset{?}{=} 4(0)$ $x = \frac{4}{3}$: $3\left(\frac{4}{3}\right)^2 \overset{?}{=} 4\left(\frac{4}{3}\right)$

$$0 \overset{\checkmark}{=} 0 \qquad\qquad \frac{16}{3} \overset{\checkmark}{=} \frac{16}{3}$$

Matched Problem 2   Solve $5z^2 = 3z$.

Example 3
Solving by Factoring

Solve $x^2 - x - 1 = 0$ by factoring, if possible, using integer coefficients.

Solution $x^2 - x - 1$ cannot be factored using integer coefficients; hence, another method, which we will consider later, must be used.

Matched Problem 3   Solve $x^2 - x - 3 = 0$ by factoring, if possible, using integer coefficients.

Example 4
Solving by Factoring

Solve $x(x + 3) = -2$.

Solution

$$x(x + 3) = -2$$ Rewrite by multiplying out and shifting all terms to the left side.

$$x^2 + 3x + 2 = 0$$

$$(x + 1)(x + 2) = 0$$

$$x + 1 = 0 \quad \text{or} \quad x + 2 = 0$$

$$\boxed{x = -1} \quad \text{or} \quad \boxed{x = -2}$$

Checking this solution is left to you.

Matched Problem 4   Solve $(x - 1)(x + 2) = 18$.

Example 5
A Geometry Problem

Find the dimensions of a rectangle with area 36 square inches and length 5 inches more than its width.

Solution

Let x be the width so that $x + 5$ is the length. See Figure 1.

$$\text{Area} = (\text{Width})(\text{Length})$$
$$36 = x(x + 5)$$
$$x(x + 5) = 36$$
$$x^2 + 5x - 36 = 0$$
$$(x + 9)(x - 4) = 0$$

$$x + 9 = 0 \qquad \text{or} \qquad x - 4 = 0$$
$$x = -9 \qquad\qquad\qquad x = 4 \text{ inches} \quad \text{Width}$$

$$x + 5 = 9 \text{ inches} \quad \text{Length}$$

Discard. It is not possible to have a negative width. In general, common sense must be used to discard answers that are not physically possible.

Solution: The rectangle is 4 inches wide by 9 inches long.

| A = 36 in.² | x |

$x + 5$

Figure 1

Check

Length is 5 inches longer than width: $9 - 4 = 5$

$$\text{Area} = 4 \cdot 9 = 36 \text{ square inches}$$

Matched Problem 5

Find the dimensions of a rectangle with area 24 square inches if its width is 2 inches less than its length.

♦ **SOLUTIONS AND GRAPHS**

The graph of an equation of the form

$$y = \text{Polynomial in } x$$

can be graphed by plotting points, just as we did for linear equations in Chapters 4 and 5. For example, to graph $y = x^2 - 4$, we would make a table of values and plot the corresponding points:

x	y
-4	12
-3	5
-2	0
-1	-3
0	-4
1	-3
2	0
3	5
4	12

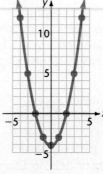

The resulting graph is called a *parabola,* and the lowest point on the graph its *vertex.*

The equation $x^2 - 4 = 0$ can be solved by factoring:

$$x^2 - 4 = 0$$
$$(x - 2)(x + 2) = 0$$

$$x - 2 = 0 \quad \text{or} \quad x + 2 = 0$$
$$x = 2 \quad \text{or} \quad x = -2$$

These solutions correspond exactly to the two points where the graph of $y = x^2 - 4$ crosses the x axis, since $y = 0$ represents the x axis. Thus, the solutions of $x^2 - 4 = 0$ are the x intercepts of the graph of $y = x^2 - 4$.

Example 6
Relating Solutions and Graphs

Graph the equation $y = x^2 - x - 6$ and solve the equation $x^2 - x - 6 = 0$. Relate the solutions of the equation to points on the graph.

Solution   We make a table of values and plot the corresponding points:

x	y
-4	14
-3	6
-2	0
-1	-4
0	-6
1	-6
2	-4
3	0
4	6

Solve $x^2 - x - 6 = 0$ by factoring:

$$x^2 - x - 6 = 0$$
$$(x - 3)(x + 2) = 0$$

$$x - 3 = 0 \quad \text{or} \quad x + 2 = 0$$
$$x = 3 \quad \text{or} \quad x = -2$$

These two solutions correspond to the two points $(-2, 0)$ and $(3, 0)$ where the parabola crosses the x axis. That is, the solutions of the equation $x^2 - x - 6 = 0$ are the x intercepts of the graph of the equation $y = x^2 - x - 6$.

Matched Problem 6   Graph the equation $y = x^2 - 3x - 4$ and solve the equation $x^2 - 3x - 4 = 0$. Relate the solutions to the equation to points on the graph.

Graphing parabolas and solving equations of the form $y = ax^2 + bx + c$ are considered more fully in Chapter 9.

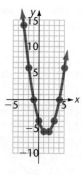

EXERCISE 6-8

A *Solve:*

1. $(x - 3)(x - 4) = 0$
2. $(x - 9)(x - 4) = 0$
3. $(x + 6)(x - 5) = 0$
4. $(x - 9)(x + 3) = 0$
5. $(x + 4)(3x - 2) = 0$
6. $(2x - 1)(x + 2) = 0$
7. $(4t + 3)(5t - 2) = 0$
8. $(2m + 3)(3m - 2) = 0$
9. $u(4u - 1) = 0$
10. $z(3z + 5) = 0$

Solve by factoring:

11. $x^2 - 6x + 5 = 0$
12. $x^2 - 5x + 6 = 0$
13. $x^2 - 4x + 3 = 0$
14. $x^2 - 8x + 15 = 0$
15. $x^2 - 4x - 12 = 0$
16. $x^2 + 4x - 5 = 0$
17. $x^2 - 3x = 0$
18. $x^2 + 5x = 0$
19. $4t^2 - 8t = 0$
20. $3m^2 + 12m = 0$
21. $x^2 - 25 = 0$
22. $x^2 - 36 = 0$
23. $x^2 + 6x + 9 = 0$
24. $x^2 + 16x + 64 = 0$
25. $x^2 - 8x + 16 = 0$
26. $x^2 - 10x + 25 = 0$
27. $x^2 + 6x + 8 = 0$
28. $x^2 + 7x + 12 = 0$
29. $x^2 - 3x - 28 = 0$
30. $x^2 - 2x - 48 = 0$

B *Solve each equation by factoring. If an equation cannot be solved by factoring, state this as your answer.*

31. $2x^2 = 3 - 5x$
32. $3x^2 = x + 2$
33. $3x(x - 2) = 2(x - 2)$
34. $2x(x - 1) = 3(x + 1)$
35. $4n^2 = 16n + 128$
36. $3m^2 + 12m = 36$
37. $3z^2 - 10z = 8$
38. $2y^2 + 15y = 8$
39. $3 = t^2 + 7t$
40. $y^2 = 5y - 2$
41. $\frac{u}{4}(u + 1) = 3$
42. $\frac{x^2}{2} = x + 4$
43. $y(y - 2) = 15$
44. $x(2x - 3) = 2$

45. $2x^2 + 2 = 5x$
46. $x^2 - 3x = 10$
47. $\frac{x^2}{4} = (x + 3)$
48. $\frac{x^2}{3} = 6(x - 4)$
49. $\frac{y}{2}(y - 5) = 12$
50. $\frac{x}{3}(x - 4) = 15$
51. $\frac{x}{4} \cdot \frac{x + 1}{3} = 6$
52. $\frac{x}{2} \cdot \frac{x + 1}{3} = 35$

C *Solve:*

53. $(x - 4)(x - 3) = 2$
54. $(x + 3)(x - 3) = 7$
55. $(x - 1)(x + 3) = 5$
56. $(x - 2)(x - 1) = 6$
57. $(x - 3)(x + 5) = 48$
58. $(x - 2)(x + 5) = 30$
59. $(x + 2)(x + 7) = 36$
60. $(x + 3)(x + 7) = 96$
61. $(x - 5)(x + 3) = 9$
62. $(x - 7)(x - 4) = 40$
63. $(x + 2)(x + 7) = -6$
64. $(x - 3)(x + 6) = -14$
65. $(2x - 3)(x - 4) = -3$
66. $(3x - 2)(x + 4) = 5$
67. $(4x - 3)(x + 2) = -6$
68. $(5x - 1)(x - 4) = 4$

For the given expression $ax^2 + bx + c$, graph the equation $y = ax^2 + bx + c$. Solve the equation $ax^2 + bx + c = 0$ and relate the solutions to the x intercepts of the graph.

69. $x^2 - 9$
70. $x^2 - 1$
71. $x^2 - 2x - 3$
72. $x^2 + 2x - 3$
73. $x^2 + 6x + 5$
74. $x^2 - 6x + 5$

APPLICATIONS

In Problems 75–78, find the dimensions of the rectangle having the given properties. The area of a rectangle is length times height. See the formulas inside the back cover of the text.

75. Area 33 square inches; width 8 inches less than the length

76. Area 104 square inches; length 5 inches more than the width

77. Area 96 square centimeters; length 6 times the width

78. Area 98 square centimeters; width half the length

In Problems 79–82, find the base and height of the triangle having the given properties. The area of a triangle is half the base times the height. See the formulas inside the back cover of the text.

79. Area 10 square feet; base 1 foot less than the height

80. Area 2 square feet; base 3 feet longer than the height

81. Area 9 square meters; base twice the height

82. Area 24 square centimeters; height one-third the base

In Problems 83–86, find the dimensions of the trapezoid having the given properties. The area of a trapezoid is half the height times the sum of the base and top. See the formulas inside the back cover of the text.

83. Area 12 square inches; top half the base and height half the top

84. Area 18 square inches; top half the base and height 1 inch more than the top

85. Area 48 square meters; height half the base and top two-thirds the height

86. Area 100 square meters; height twice the top and top one-third the base

87. A 4- by 3-inch rectangular piece of ceramic tile has the design shown in the figure. Both the vertical and horizontal stripes have the same width. What should this width be if the stripes take up 50% of the tile's area?

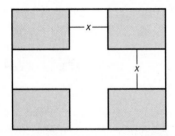

88. Repeat Problem 87 for a 5- by 4-inch piece of tile with a stripe taking up 60% of the tile's area.

89. An 8- by 10-inch photograph is to be surrounded by a matting of uniform width. See the figure. The photograph is to take up two-thirds of the total area of the mat and photo combined. How wide should the mat be?

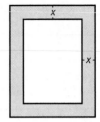

90. Repeat Problem 89 for a photograph which is 4- by 6-inches and which is to take up exactly half the area of the photo and mat combined.

91. A fruit grower is preparing a new section of his orchard for planting. For the particular variety of dwarf tree to be planted, he can expect a yield of 40 pounds per tree if he plants 20 trees. For each additional tree planted, the crowding will reduce the yield per tree by 1 pound. Thus, the total yield when x trees are added is $(20 + x)(40 - x) = 800 + 20x - x^2$. How many trees should the grower plant to get a total yield of 500 pounds of fruit?

92. The grower in Problem 91 also plans a blackberry patch. In the space available for his patch, he can plant 20 bushes and expect a yield of 14 pounds per bush. Each additional bush planted will reduce the yield by $\frac{1}{5}$ pound. Can the grower obtain a total yield of 405 pounds of blackberries?

93. An amateur California wine maker has room to plant 40 grapevines spaced so as to ultimately yield 8 liters of wine for each vine planted. He can plant the vines closer together but will lose 0.1 liter of wine per vine for each additional vine planted. Can the total amount of wine obtained be increased to 360 liters by planting more vines?

94. A farmer plans a small strawberry patch. He can plant 60 plants that will yield $\frac{1}{2}$ pound of strawberries per plant. Each additional plant included will decrease the yield per plant by $\frac{1}{200}$ of a pound. How many plants will provide a total yield of 32 pounds of strawberries?

6-9 Algebraic Long Division

♦ Long Division of Polynomials
♦ Solving a Cubic Equation

There are times when it is useful to find the quotient of two polynomials, and in particular to find the quotient of a polynomial divided by a first-degree polynomial. We will use a long-division process similar to that used in arithmetic. We will then show how the process can be used to solve some third-degree, or *cubic,* equations by factoring.

♦ **LONG DIVISION OF POLYNOMIALS**

Several examples will illustrate the long-division process.

Example 1
Polynomial
Long Division

Divide $2x^2 + 5x - 12$ by $x + 4$ and check.

Solution

$$x + 4 \overline{)2x^2 + 5x - 12}$$

Both polynomials are to be arranged in descending powers of the variable if this is not already done.

$$\begin{array}{r} 2x \\ x + 4 \overline{)2x^2 + 5x - 12} \end{array}$$

Divide the first term of the divisor into the first term of the dividend. That is, what must x be multiplied by so that the product is exactly $2x^2$? Answer: $2x$.

$$\begin{array}{r} 2x \\ x + 4 \overline{)2x^2 + 5x - 12} \\ \underline{2x^2 + 8x} \\ -3x - 12 \end{array}$$

Multiply the divisor by $2x$, line up like terms, subtract, and bring down -12 from above.

$$\begin{array}{r} 2x - 3 \\ x + 4 \overline{)2x^2 + 5x - 12} \\ \underline{2x^2 + 8x} \\ -3x - 12 \\ \underline{-3x - 12} \\ 0 \end{array}$$

Repeat the process until the degree of the remainder is less than that of the divisor or the remainder is 0.

Check $(x + 4)(2x - 3) = 2x^2 + 5x - 12$

Matched Problem 1 Divide $2x^2 + 7x + 3$ by $x + 3$ and check.

Example 2
Polynomial
Long Division

Divide $x^3 + 8$ by $x + 2$ and check.

Solution

$$
\begin{array}{r}
x^2 - 2x + 4 \\
x + 2\overline{)x^3 + 0x^2 + 0x + 8} \\
\underline{x^3 + 2x^2} \\
-2x^2 + 0x \\
\underline{-2x^2 - 4x} \\
4x + 8 \\
\underline{4x + 8} \\
0
\end{array}
$$

Insert, with 0 coefficients, any missing terms of lower degree than 3, and proceed as in Example 1.

To check, note that as the sum of two cubes,

$$x^3 + 8 = (x + 2)(x^2 - 2x + 4)$$

Matched Problem 2   Divide $(x^3 - 8)$ by $(x - 2)$ and check.

Example 3
Polynomial
Long Division

Divide $(3 - 7x + 6x^2)$ by $(3x + 1)$ and check.

Solution

$$
\begin{array}{r}
2x - 3 \\
3x + 1\overline{)6x^2 - 7x + 3} \\
\underline{6x^2 + 2x} \\
-9x + 3 \\
\underline{-9x - 3} \\
6 = R
\end{array}
$$
    (Remainder)

Arrange $3 - 7x + 6x^2$ in descending powers of x; then proceed as above until the degree of the remainder is less than the degree of the divisor.

Check   Just as in arithmetic, when there is a remainder, we check by adding the remainder to the product of the divisor and quotient. That is, we check that

$$\text{Polynomial} = \text{Divisor} \times \text{Quotient} + \text{Remainder}$$

Thus, in this example, we check

$$
\begin{aligned}
(3x + 1)(2x - 3) + 6 &\overset{?}{=} 6x^2 - 7x + 3 \\
6x^2 - 7x - 3 + 6 &\overset{?}{=} 6x^2 - 7x + 3 \\
6x^2 - 7x + 3 &\overset{\checkmark}{=} 6x^2 - 7x + 3
\end{aligned}
$$

Matched Problem 3   Divide $(2 + 6x^2 - x)$ by $(3x - 2)$ and check.

♦   **SOLVING A CUBIC EQUATION**

If we know, or could guess, one solution to a cubic equation, long division can help us find the others, as shown in the next example.

Example 4
Solving a Cubic
Equation by Factoring

Solve $x^3 - 4x^2 + 5x - 2 = 0$.

Solution   With the constant term in the expression being 2, we might guess 1 or 2 as a possible solution. Substituting $x = 2$ into $x^3 - 4x^2 + 5x - 2$, we obtain

$$2^3 - 4 \cdot 2^2 + 5 \cdot 2 - 2 = 8 - 16 + 10 - 2 = 0$$

so 2 is, in fact, a solution. This means that $x - 2$ is a factor of $x^3 - 4x^2 + 5x - 2$, although we will not prove this here. To find the other factors, we divide the cubic by $x - 2$:

$$
\begin{array}{r}
x^2 - 2x\ + 1 \\
x - 2 \overline{)x^3 - 4x^2 + 5x - 2} \\
\underline{x^3 - 2x^2} \\
-2x^2 + 5x \\
\underline{-2x^2 + 4x} \\
x - 2 \\
\underline{x - 2} \\
0
\end{array}
$$

Thus

$$x^3 - 4x^2 + 5x - 2 = (x - 2)(x^2 - 2x + 1)$$
$$= (x - 2)(x - 1)^2 = (x - 2)(x - 1)(x - 1)$$

The solutions to the equation are found by using the zero property. We set $x - 2 = 0$ and $x - 1 = 0$, and obtain $x = 2$ and $x = 1$.

In the solution to Example 4, 2 was a solution to $x^3 - 4x^2 + 5x - 2 = 0$, and $x - 2$ turned out to be a factor of the polynomial. This is true in general: **If r is a real number that yields 0 when substituted into the polynomial, then $x - r$ is a factor of the polynomial.**

Matched Problem 4   Solve $x^3 - 6x^2 + 11x - 6 = 0$, given that $x = 2$ is one solution.

Answers to   **1.** $2x + 1$ **2.** $x^2 + 2x + 4$ **3.** $2x + 1$, R = 4
Matched Problems   **4.** $x = 1, 2, 3$

EXERCISE 6-9

Divide using the long-division process. Check the answers.

A **1.** $(x^2 + 5x + 6) \div (x + 3)$

2. $(x^2 + 6x + 8) \div (x + 4)$

3. $(2x^2 + x - 6) \div (x + 2)$

4. $(3x^2 - 5x - 2) \div (x - 2)$

5. $(2x^2 - 3x - 4) \div (x - 3)$

6. $(3x^2 - 11x - 1) \div (x - 4)$

7. $(2m^2 + m - 10) \div (2m + 5)$

8. $(3y^2 + 5y - 12) \div (3y - 4)$

9. $(6x^2 + 5x - 6) \div (3x - 2)$

10. $(8x^2 - 14x + 3) \div (2x - 3)$

11. $(6x^2 + 11x - 12) \div (3x - 2)$

12. $(6x^2 + x - 13) \div (2x + 3)$

13. $(3x^2 + 13x - 12) \div (3x - 2)$

14. $(2x^2 - 7x - 1) \div (2x + 1)$

15. $(2x^2 - 9x - 35) \div (2x + 5)$

16. $(2x^2 + 9x + 10) \div (2x + 5)$

17. $(3x^2 + 11x - 20) \div (3x - 4)$

18. $(3x^2 - 13x + 12) \div (3x - 4)$

19. $(4x^2 + 21x - 18) \div (4x - 3)$

20. $(4x^2 - 23x + 15) \div (4x - 3)$

21. $(5x^2 - 13x - 6) \div (5x + 2)$

22. $(5x^2 + 22x + 8) \div (5x + 2)$

23. $(6x^2 + 35x - 6) \div (6x - 1)$

24. $(6x^2 - 49x + 8) \div (6x - 1)$

B 25. $(x^2 - 4) \div (x - 2)$ 26. $(y^2 - 9) \div (y + 3)$

27. $(m^2 - 7) \div (m - 3)$ 28. $(u^2 - 18) \div (u + 4)$

29. $(8c + 4 + 5c^2) \div (c + 2)$

30. $(4a^2 - 22 - 7a) \div (a - 3)$

31. $(9x^2 - 8) \div (3x - 2)$ 32. $(8x^2 + 7) \div (2x - 3)$

33. $(5y^2 - y + 2y^3 - 6) \div (y + 2)$

34. $(x - 5x^2 + 10 + x^3) \div (x + 2)$

35. $(x^3 - 1) \div (x - 1)$ 36. $(x^3 + 27) \div (x + 3)$

37. $(x^4 - 16) \div (x + 2)$ 38. $(x^5 + 32) \div (x - 2)$

39. $(3y - y^2 + 2y^3 - 1) \div (y + 2)$

40. $(3 + x^3 - x) \div (x - 3)$

41. $(x^2 + 12) \div (x + 1)$ 42. $(x^2 - 18) \div (x + 3)$

43. $(x^2 + 10x + 25) \div (x + 5)$

44. $(x^2 + 8x + 16) \div (x + 4)$

45. $(x^3 - 64) \div (x - 4)$ 46. $(x^3 + 27) \div (x - 3)$

47. $(4x^2 + 12x + 18) \div (2x + 9)$

48. $(6x^2 + 7x + 72) \div (3x + 8)$

49. $(8x^3 + 4x^2 + 2x + 1) \div (2x + 1)$

50. $(27x^3 + 9x^2 + 3x + 1) \div (3x + 1)$

C 51. $(4x^4 - 10x - 9x^2 - 10) \div (2x + 3)$

52. $(9x^4 - 2 - 6x - x^2) \div (3x - 1)$

53. $(16x - 5x^3 - 4 + 6x^4 - 8x^2) \div (2x - 4 + 3x^2)$

54. $(8x^2 - 7 - 13x + 24x^4) \div (3x + 5 + 6x^2)$

55. $(x^4 + 4x^3 - 8x^2 + 9x + 30) \div (x + 2)$

56. $(x^4 + 6x^3 + 4x^2 + 5x - 12) \div (x + 3)$

57. $(x^4 - 8x^2 + 16) \div (x^2 + 2)$

58. $(x^4 - 6x^2 + 9) \div (x^2 - 3)$

59. $(x^4 - 2x^3 + x - 2) \div (x + 1)$

60. $(x^4 - x^3 + 8x - 8) \div (x - 1)$

Solve the equation using the fact that the number given is one solution:

61. $x^3 - 8x^2 + 19x - 12 = 0; x = 4$

62. $x^3 + 5x^2 - x - 5 = 0; x = -5$

63. $x^3 + 6x^2 + 11x + 6 = 0; x = -3$

64. $x^3 - 7x - 6 = 0; x = 3$

65. $x^3 + 4x^2 - 31x - 70 = 0; x = -2$

66. $x^3 - 4x^2 - 17x + 60 = 0; x = 3$

67. $x^3 - 3x^2 - 6x + 8 = 0; x = 4$

68. $x^3 + 9x^2 + 26x + 24 = 0; x = -3$

CHAPTER SUMMARY

6-1 POLYNOMIAL ADDITION AND SUBTRACTION

A **polynomial** in one or two variables x and y is an algebraic expression constructed by adding or subtracting constants and terms of the form ax^n or $bx^m y^n$, where a and b are real number coefficients and m and n are positive integers. The **degree of a term** in a polynomial is the sum of the powers of the variable factors in the term; nonzero constants are assigned degree 0. The **degree of a polynomial** is the highest degree of its terms. The number 0 is a polynomial with no assigned degree. Polynomials with one, two, and three terms are called **monomials, binomials,** and **trinomials,** respectively. Polynomials are added by combining like terms. A second polynomial is subtracted from a first by adding the negative of the second to the first.

6-2 POLYNOMIAL MULTIPLICATION

Polynomials are multiplied by multiplying each term of the first by each term of the second. Binomials can be multiplied mentally by the FOIL method:

$$(A + B)(C + D) = AC + AD + BC + BD$$

First
Last   F   O   I   L
Inner
Outer

and squared by $(A + B)^2 = A^2 + 2AB + B^2$.

6-3 FACTORING OUT COMMON FACTORS

Common factors are factored out by using the distributive property. Grouping may lead to factoring out common factors.

6-4 FACTORING SOME BASIC FORMS OF SECOND-DEGREE POLYNOMIALS

The sum $A^2 + B^2$ of two squares cannot be factored unless there are common factors. The difference of two squares is factored

$$A^2 - B^2 = (A - B)(A + B)$$

A second-degree polynomial $x^2 + bx + c$ is a perfect square if c is the square of half of b. In this case the polynomial can be factored as $\left(x + \dfrac{b}{2}\right)^2$.

6-5 FACTORING SECOND-DEGREE POLYNOMIALS BY TRIAL AND ERROR

Some second-degree polynomials can be factored by trial and error. This method is effective when the constant term and the coefficient of the square term are not too large. The method is most useful when trying to factor $x^2 + bx + c$ into $(x + r)(x + s)$: we must find r and s with $r + s = b$ and $rs = c$.

6-6 FACTORING SECOND-DEGREE POLYNOMIALS SYSTEMATICALLY

Another method for factoring second-degree polynomials that are factorable in the integers uses the *ac* test. The **ac test** to factor $ax^2 + bx + c$ or $ax^2 + bxy + cy^2$ involves finding two integers p and q such that $pq = ac$ and $p + q = b$. If such a p and q exist the polynomial can be factored. If such a p and q do not exist, the polynomial cannot be factored in the integers. To factor the polynomial after finding p and q, rewrite the middle term as $px + qx$ or $pxy + qxy$, and factor by grouping.

6-7 HIGHER-DEGREE POLYNOMIALS AND MORE FACTORING

The sum and difference of two cubes are factored

$$A^3 + B^3 = (A + B)(A^2 - AB + B^2)$$

$$A^3 - B^3 = (A - B)(A^2 + AB + B^2)$$

A general strategy for factoring is

1. Remove common factors.
2. If the polynomial has two terms, look for a difference of two squares or a sum or difference of two cubes.
3. If the polynomial has three terms:
   (A) See if it is a perfect square.
   (B) Try trial and error.
   (C) Or use the *ac* test.
4. If the polynomial has more than three terms, try grouping.

6-8 SOLVING EQUATIONS BY FACTORING

An equation of the form Polynomial = 0, where the polynomial can be factored, can be solved by factoring using the **zero property:**

$$ab = 0 \qquad \text{if and only if} \qquad a = 0 \text{ or } b = 0$$

6-9 ALGEBRAIC LONG DIVISION

Algebraic long division can be used to find the quotient of two polynomials. Just as in arithmetic, if division of polynomial P by divisor D yields quotient Q and remainder R, then $P = DQ + R$.

CHAPTER REVIEW EXERCISE

Work through all the problems in this chapter review and check the answers in the back of the book. Answers to all review problems are there, and following each answer is a number in italics indicating the section in which that type of problem is discussed. Where weaknesses show up, review appropriate sections in the text.

A
1. Give the degree of the following monomials:
   (A) $7x^3$ (B) $3x^2y^5$ (C) $2xy^4z^6$

2. Give the degree of the following polynomials:
   (A) $3x^4 + 5x^2 - 8$
   (B) $6xy^5 - 10x^3y^5$
   (C) $2x^3 + 3y^3 - 4z^3$

3. Add $2x^2 + x - 3$, $2x - 3$, and $3x^2 + 2$.

4. Subtract $3x^2 - x - 2$ from $5x^2 - 2x + 5$.

5. Multiply $2x^2 - x + 3$ by $x^2 + 3x - 1$.

6. Divide using algebraic long division:
   $(6x^2 + 5x - 2) \div (2x - 1)$.

Perform the indicated operations and simplify:

7. $(3x - 2)(2x + 5)$

8. $(2u - 3v)(3u + 4v)$

9. $(2x^2 - 3x + 1) + (3x^3 - x^2 + 5)$

10. $(2x^2 - 3x + 1) - (3x^3 - x^2 + 5)$

11. $(4x^2 + 2x - 5) - (5x^2 - 4x + 2)$

12. $(4x^2 + 2x - 5) + (5x^2 - 4x + 2)$

Factor, if possible, using integer coefficients:

13. $x^2 - 9x + 14$
14. $3x^2 - 10x + 8$
15. $x^2 - 3x - 3$
16. $4x^2y - 6xy^2$
17. $x^3 - 5x^2 + 6x$
18. $4u^2 - 9$
19. $x(x - 1) + 3(x - 1)$
20. $m^2 + 4n^2$
21. $x^2 + 4x - 21$
22. $3x^2 - 11x - 4$
23. $2x^2 - 5x - 3$
24. $x^2 - x - 30$
25. $x^3 + 5x^2 + 6x$
26. $x^2 - 8x + 16$
27. $4x^2 - 3x - 1$
28. $x^2 + 8x + 7$
29. $3x^2 + 18x + 24$
30. $x^2 + 22x + 121$
31. $x^2 + 64$
32. $4x^2 - 100$
33. $x^2 + 10x + 24$
34. $5x^2 + 11x + 2$
35. $9x^4 - 64x^2$
36. $400 + 40x + x^2$

B *Solve the equation:*

37. $(x + 5)(x - 2) = 0$
38. $x(3x - 1) = 0$
39. $(x - 3)^2 = 0$
40. $x(x + 1)(x - 9) = 0$

Let P, Q, R, and S be the following polynomials:

$$P = 2x - 1 \qquad Q = 2x + 1$$
$$R = 2x^2 - 5x - 6 \qquad S = x^2 + x + 1$$

Find the indicated polynomial:

41. $R - PQ$

42. PQS

43. $PQ + R + S$

44. $(S - Q)(R - P)$

Divide using algebraic long division:

45. $(2x^2 - 7x - 1) \div (2x + 1)$

46. $(2 - 10x + 9x^3) \div (3x - 2)$

47. $(x^3 - x - 24) \div (x - 3)$

48. $(x^4 + 2x^3 + 4x^2) \div (x + 1)$

Perform the indicated operations and simplify:

49. $(2x - 3)(2x^2 - 3x + 2)$

50. $(9x^2 - 4)(3x^2 + 7x - 6)$

51. $(a + b)(a^2 - ab + b^2)$

52. $[(3x^2 - x + 1) - (x^2 - 4)] - [(2x - 5)(x + 3)]$

53. $(x^2 - 5x + 7)(x^2 + 7x - 5)$

54. $(x^3 - x) - [x - (x^2 - x)]$

55. $(x - 1)(x - 2)(x - 3)$

56. $(x - 1)(x - 2) + (x - 3)(x - 4)$

Factor, if possible, using integer coefficients:

57. $3u^2 - 12$

58. $2x^2 - xy - 3y^2$

59. $x^2 - xy + y^2$

60. $6y^3 + 3y^2 - 45y$

61. $2x^3 - 4x^2y - 10xy^2$

62. $12x^3y + 27xy^3$

63. $x^2(x - 1) - 9(x - 1)$

64. $4x^2 + 10x - 3$

65. $x^3 - 125y^3$

66. $x^4 + 6x^2 + 9$

67. $8x^2 - 47x - 6$

68. $x^8 - y^4$

69. $6x^2 - 47x - 8$

70. $x^3 - z^3$

71. $xy + xz + y^2 + yz$

72. $6x^2 - 5x - 6$

Factor, using integer coefficients, by grouping:

73. $x^2 - xy + 4x - 4y$

74. $x^2 + xy - 3x - 3y$

75. $2u^2 - 3u + 6u - 9$

76. $6x^2 + 4x - 3x - 2$

77. $xy + x + z + yz$

78. $x^3 + x + x^4 + x^2$

In Problems 79–82, solve the equation by factoring:

79. $x^2 + 3x - 4 = 0$

80. $2x^2 + 3x = 2$

81. $x^2 - 11x = 0$

82. $12 + 8x + x^2 = 0$

C **83.** Simplify:
$$[-2xy(x^2 - 4y^2)] - [-2xy(x - 2y)(x + 2y)].$$

84. Divide $2x^4 - 14x^2 + 20x + 4$ by $2x^2 + 6$ using long division.

Factor, if possible, using integer coefficients:

85. $12x^2 + 18x - 12$

86. $12x^2 + 7x - 12$

87. $12x^2 + 40x + 12$

88. $12x^2 + 32x - 12$

89. $x^3 + 216$

90. $x^4 + x^2$

91. $x^2y^2 - x^2z^2$

92. $xy^3 - x^4$

93. $36x^3y + 24x^2y^2 - 45xy^3$

94. $12u^4 - 12u^3v - 20u^3v^2$

95. $6ac + 4bc - 12ad - 8bd$

96. $12ux - 15vx - 4u + 5v$

97. $8x^3 + 1$

98. Find all integers b such that $2x^2 + bx - 4$ can be factored into two first-degree factors using integer coefficients.

99. Solve $x^3 - 4x^2 - 4x + 16 = 0$, given that $x = 4$ is one solution.

100. Find the dimensions of a rectangle with area 48 square inches and length 10 inches more than half the width.

CHAPTER PRACTICE TEST

The following practice test is provided for you to test your knowledge of the material in this chapter. You should try to complete it in 50 minutes or less. The answers in the back of the book indicate the section in the text that covers the material in the question. Actual tests in your class may vary from this practice test in difficulty, length, or emphasis, depending on the goals of your course or instructor.

In Problems 1–5, let P, Q, and R be the following polynomials:

$$P = 2x - 5 \qquad Q = x - 4 \qquad R = 3x^2 + x - 8$$

Find the indicated polynomials:

1. $P + Q + R$ **2.** $R - P$ **3.** RQ

4. $PQ - R$ **5.** PQR

In Problems 6–15, factor the polynomials using integer coefficients. If a polynomial cannot be so factored, indicate this.

6. $6xy + 10x^2$ **7.** $4x^2 - 9$

8. $x^2 - 12x + 36$ **9.** $x^2 - 2x - 15$

10. $64x^3 - 1$ **11.** $x^2 + 16$

12. $6x^2 - 5x - 4$ **13.** $6x^2 - 9x - 6$

14. $ax + by + bx + ay$ **15.** $x^2 + 11x + 12$

16. Divide $x^3 + 2x^2 + 3x + 90$ by $x + 5$ using long division.

Solve:

17. $(x - 1)(x + 2) = 0$ **18.** $x^2 + 4x - 45 = 0$

7

Fractional Forms

Polynomials involve only the operations of addition, subtraction, and multiplication of variables and constants. In this chapter, we extend the kinds of algebraic expressions we consider to include those in which division may involve variables in a denominator.

Photo reference: see Exercise 7-4, Problem 70.

7-1   Reducing to Lowest Terms

- ♦ Rational Expressions
- ♦ Reducing to Lowest Terms

In this section we will introduce the simplest algebraic expressions that include division involving a variable. These are fractional forms in which both the numerator and the denominator are polynomials.

♦ RATIONAL EXPRESSIONS

Fractional forms in which the numerator and denominator are polynomials are called **rational expressions.** For example:

$$\frac{1}{x} \qquad \frac{2}{x-3} \qquad \frac{y-3}{4y^2-5x+6} \qquad \frac{x^2-8xy+y^2}{x^2+y^2}$$

are all rational expressions.

In Chapter 3 we worked with simple fractional forms. In this chapter we will use the same basic ideas established there on more complex fractional forms.

The **fundamental principle of fractions**

$$\frac{ak}{bk} = \frac{a}{b} \qquad b,\, k \neq 0$$

will play an important role in our work.

♦ REDUCING TO LOWEST TERMS

If the numerator and denominator in a quotient of two polynomials contain a common factor, it may be divided out using the fundamental principle of fractions. The process that converts ak/bk to a/b is usually denoted by

$$\frac{ak}{bk} = \frac{a}{b} \qquad \text{or} \qquad \frac{\overset{1}{\cancel{ak}}}{\underset{1}{\cancel{bk}}} = \frac{a}{b} \qquad \text{or} \qquad \frac{\overset{a}{\cancel{ak}}}{\underset{b}{\cancel{bk}}} = \frac{a}{b}$$

Recall that removing a common factor from a numerator and denominator is equivalent to dividing the numerator and denominator by this same common factor. If all common factors are removed from the numerator and the denominator, the resulting rational expression is said to be reduced to **lowest terms.** The language exactly parallels that introduced for fractions in Chapter 3. Several examples should make the process clear. In the examples it is important to keep in mind:

Common factors divide out.   Remember: factors are multiplied.

Common terms do not divide out.   Remember: terms are added or subtracted.

<table>
<tr><td>

Example 1
Reducing to
Lowest Terms

</td><td>

Reduce to lowest terms by eliminating all common factors from numerator and denominator:

(A) $\dfrac{5(x + 3)}{2x(x + 3)}$ **(B)** $\dfrac{6x^2 - 3x}{3x}$ **(C)** $\dfrac{x^2y - xy^2}{x^2 - xy}$

(D) $\dfrac{4x^3 + 10x^2 - 6x}{2x^3 - 18x}$ **(E)** $\dfrac{2 - x}{x - 2}$

</td></tr>
</table>

Solution **(A)** $\dfrac{5(x + 3)}{2x(x + 3)} = \dfrac{5(\overset{1}{\cancel{x + 3}})}{2x(\underset{1}{\cancel{x + 3}})}$ Divide out common factors.

$$= \dfrac{5}{2x}$$

(B) $\dfrac{6x^2 - 3x}{3x} = \dfrac{\overset{1}{\cancel{3x}}(2x - 1)}{\underset{1}{\cancel{3x}}}$ Factor the top, then divide out common factors.

$$= 2x - 1$$ Note the *term* $3x$ cannot be divided out in $\dfrac{6x^2 - 3x}{3x}$.

(C) $\dfrac{x^2y - xy^2}{x^2 - xy} = \dfrac{\overset{1}{\cancel{xy}}(\overset{1}{\cancel{x - y}})}{\underset{1}{\cancel{x}}(\underset{1}{\cancel{x - y}})}$ Factor the top and bottom, then divide out common factors.

$$= y$$

(D) $\dfrac{4x^3 + 10x^2 - 6x}{2x^3 - 18x} = \dfrac{\overset{1}{\cancel{2x}}(2x^2 + 5x - 3)}{\underset{1}{\cancel{2x}}(x^2 - 9)}$

$$= \dfrac{(2x - 1)(\overset{1}{\cancel{x + 3}})}{(x - 3)(\underset{1}{\cancel{x + 3}})}$$

$$= \dfrac{2x - 1}{x - 3}$$

(E) $\dfrac{2 - x}{x - 2} = \dfrac{-(\overset{1}{\cancel{x - 2}})}{\underset{1}{\cancel{x - 2}}} = -1$

In Example 1(E), we used the result first introduced in Section 2-6:

$$-(a - b) = -a + b = b - a$$

Matched Problem 1 Reduce to lowest terms by eliminating all common factors from numerator and denominator:

(A) $\dfrac{3x(x^2 + 2)}{2(x^2 + 2)}$ **(B)** $\dfrac{4m}{8m^2 - 4m}$ **(C)** $\dfrac{x^2 - 3x}{x^2y - 3xy}$

(D) $\dfrac{2x^3 - 8x}{4x^3 - 14x^2 + 12x}$ **(E)** $\dfrac{y - x}{x^2 - xy}$

**Example 2
Reducing to
Lowest Terms**

Reduce to lowest terms by eliminating all common factors from numerator and denominator:

(A) $\dfrac{x^2 + 7x + 12}{x^2 - x - 12}$ (B) $\dfrac{x^3 - 9x}{x^2 + 3x}$

Solution   (A) $\dfrac{x^2 + 7x + 12}{x^2 - x - 12} = \dfrac{(x + 3)(x + 4)}{(x + 3)(x - 4)} = \dfrac{x + 4}{x - 4}$

(B) $\dfrac{x^3 - 9x}{x^2 + 3x} = \dfrac{x(x + 3)(x - 3)}{x(x + 3)} = x - 3$

Matched Problem 2

Reduce to lowest terms by eliminating all common factors from numerator and denominator:

(A) $\dfrac{x^2 - 2x - 15}{x^2 + 5x + 6}$ (B) $\dfrac{x^2 + 2x}{x^3 + 4x^2 + 4x}$

**Answers to
Matched Problems**

1. (A) $\dfrac{3x}{2}$ (B) $\dfrac{1}{2m - 1}$ (C) $\dfrac{1}{y}$ (D) $\dfrac{x + 2}{2x - 3}$ (E) $-\dfrac{1}{x}$

2. (A) $\dfrac{x - 5}{x + 2}$ (B) $\dfrac{1}{x + 2}$

EXERCISE 7-1

Eliminate all common factors from numerator and denominator:

A 1. $\dfrac{2x^2}{6x}$ 2. $\dfrac{9y}{3y^3}$ 3. $\dfrac{6y^2}{10y^5}$

4. $\dfrac{8x^4}{18x}$ 5. $\dfrac{A}{A^2}$ 6. $\dfrac{B^2}{B}$

7. $\dfrac{4xy^4}{14x^2y}$ 8. $\dfrac{16x^3y^2}{50xy^4}$ 9. $\dfrac{x + 3}{(x + 3)^2}$

10. $\dfrac{(y - 1)^2}{y - 1}$ 11. $\dfrac{8(y - 5)^2}{2(y - 5)}$ 12. $\dfrac{4(x - 1)}{12(x - 1)^3}$

13. $\dfrac{2x^2(x + 7)}{6x(x + 7)^3}$ 14. $\dfrac{15y^3(x - 9)^3}{5y^4(x - 9)^2}$ 15. $\dfrac{22a^2(b + 3)^4}{26a(b + 3)^5}$

16. $\dfrac{15(x - 5)^3y^4}{12(x - 5)^4y^3}$ 17. $\dfrac{x^2 - 2x}{2x - 4}$ 18. $\dfrac{2x^2 - 10x}{4x - 20}$

19. $\dfrac{x^3 + 4x^2}{x(x + 4)}$ 20. $\dfrac{3x^2 + 9}{x^3 + 3x}$ 21. $\dfrac{x^4 - x^3}{2x^2 - 2x}$

22. $\dfrac{x^5 - 3x^3}{2x^4 - 6x^2}$ 23. $\dfrac{9y - 3y^2}{3y}$ 24. $\dfrac{2x^2 - 4x}{2x}$

25. $\dfrac{6x^4}{3x^4 - 9x^2}$ 26. $\dfrac{2x^3}{6x^4 - 4x^2}$ 27. $\dfrac{m^2 - mn}{m^2n - mn^2}$

28. $\dfrac{a^2b + ab^2}{ab + b^2}$ 29. $\dfrac{ab - a^2b^2}{a^3b^3 - a^2b^2}$ 30. $\dfrac{x^4y^4 - x^2y^2}{xy - x^3y^3}$

31. $\dfrac{(2x - 1)(2x + 1)}{3x(2x + 1)}$ 32. $\dfrac{(x + 3)(2x + 5)}{2x^2(2x + 5)}$

33. $\dfrac{(2x + 1)(x - 5)}{(3x - 7)(x - 5)}$ 34. $\dfrac{(3x + 2)(x + 9)}{(2x - 5)(x + 9)}$

35. $\dfrac{(x + 2)(x - 3)}{(x - 3)(x - 2)}$ 36. $\dfrac{(x + 5)(x - 2)}{(x - 2)(x + 4)}$

B 37. $\dfrac{x^2 + 5x + 6}{2x^2 + 6x}$ 38. $\dfrac{x^2 + 6x + 8}{3x^2 + 12x}$

39. $\dfrac{x^2 - 9}{x^2 + 6x + 9}$ 40. $\dfrac{x^2 - 4}{x^2 + 4x + 4}$

41. $\dfrac{x^2 - 4x + 4}{x^2 - 5x + 6}$ 42. $\dfrac{x^2 - 6x + 9}{x^2 - 5x + 6}$

43. $\dfrac{2x^2 + 5x - 3}{4x^2 - 1}$ 44. $\dfrac{9x^2 - 4}{3x^2 + 7x - 6}$

45. $\dfrac{x^2 - 1}{x^2 - x}$ 46. $\dfrac{x^2 - 4}{x^2 + 2x}$

47. $\dfrac{y^2 + 3y}{y^2 - 9}$ 48. $\dfrac{y^2 - 1}{(y - 1)^2}$

49. $\dfrac{9x^2 - 3x + 6}{3}$ 50. $\dfrac{2 - 6x - 4x^2}{2}$

51. $\dfrac{10 + 5m - 15m^2}{5m}$ 52. $\dfrac{12t^2 + 4t - 8}{4t}$

53. $\dfrac{6x}{3x^2 - 27}$ 54. $\dfrac{12}{4x^2 + 8}$

55. $\dfrac{4m^3n - 2m^2n^2 + 6mn^3}{2mn}$

56. $\dfrac{6x^3y - 12x^2y^2 - 9xy^3}{3xy}$

57. $\dfrac{3x^2y^3}{3x^3y + 6x^2y^2 + 3xy^3}$

58. $\dfrac{2ab}{4a^3 - 8a^2b + 4ab^2}$

59. $\dfrac{x^2 - x - 6}{x - 3}$

60. $\dfrac{x^2 + 2x - 8}{x - 2}$

61. $\dfrac{4x^2 - 9y^2}{4x^2y + 6xy^2}$

62. $\dfrac{a^2 - 16b^2}{4ab - 16b^2}$

63. $\dfrac{x^3 - 4xy^2}{x^2 + 2xy}$

64. $\dfrac{a^3 + 3a^2b}{a^4 - 9a^2b^2}$

65. $\dfrac{8 - y}{y - 8}$

66. $\dfrac{5 - m}{m - 5}$

67. $\dfrac{(x - 3)(x - 5)}{(5 - x)(x + 3)}$

68. $\dfrac{(4 - x)(x - 2)}{(2 - x)(x + 4)}$

69. $\dfrac{y^2 - xy}{x - y}$

70. $\dfrac{n - m}{m^2 - mn}$

71. $\dfrac{3 - x}{x^2 - x - 6}$

72. $\dfrac{2 - y}{y^2 - 4}$

73. $\dfrac{x^2 - 5x + 6}{2x - x^2}$

74. $\dfrac{x^2 + 3x - 10}{2x - x^2}$

75. $\dfrac{x^2 - 4x - 12}{12 - 8x + x^2}$

76. $\dfrac{x^2 + x - 20}{x^2 - 8x + 16}$

C 77. $\dfrac{x^2 - xy + 2x - 2y}{x^2 - y^2}$

78. $\dfrac{u^2 + uv - 2u - 2v}{u^2 + 2uv + v^2}$

79. $\dfrac{x^2y - 8xy + 15y}{xy - 3y}$

80. $\dfrac{m^3 + 7m^2 + 10m}{m^2 + 5m}$

81. $\dfrac{6x^3 + 28x^2 - 10x}{12x^3 - 4x^2}$

82. $\dfrac{12x^3 - 78x^2 - 42x}{16x^4 + 8x^3}$

83. $\dfrac{x^3 - 8}{x^2 - 4}$

84. $\dfrac{y^3 + 27}{2y^3 - 6y^2 + 18y}$

85. $\dfrac{2x^2 + 3x + 1}{x^2 + 3x + 2}$

86. $\dfrac{3x^2 - x - 2}{x^2 - 2x + 1}$

87. $\dfrac{2x^2 + 3x - 2}{x^2 + 2x + 1}$

88. $\dfrac{3x^2 - x - 2}{3x^2 + 5x + 2}$

89. $\dfrac{2x^2 + 3x + 1}{2x^2 - 5x - 3}$

90. $\dfrac{x^2 + x - 2}{2x^2 + 3x - 2}$

91. $\dfrac{2a^2b - 2ab}{a^2b^2 - ab^2 + a^2b - ab}$

92. $\dfrac{3xy^2 - 3xy}{x^2y^2 - x^2y - xy^2 + xy}$

7-2 **Multiplication and Division**

♦ Multiplication
♦ Division

The earlier treatment of multiplication and division of rational numbers in Chapter 3 extends naturally to real-number fractions and rational forms in general.

♦ **MULTIPLICATION**

The definition of multiplication for fractions extends to multiplying quotients of real numbers.

> **Multiplication of Real Fractions**
>
> For any real numbers a, b, c, and d (b, $d \neq 0$),
>
> $$\frac{a}{b} \cdot \frac{c}{d} = \frac{a \cdot c}{b \cdot d} \qquad \frac{3}{4} \cdot \frac{x}{2} = \frac{3x}{8}$$

This definition, coupled with the fundamental principle of real fractions,

$$\frac{ak}{bk} = \frac{a}{b} \qquad b, k \neq 0$$

provides the basic tool for multiplying and reducing rational expressions to lowest terms.

Example 1
Multiplication

Multiply and reduce to lowest terms:

(A) $\dfrac{2x^2y}{3z} \cdot \dfrac{3z^2}{4xy}$ **(B)** $(x^2 - 1) \cdot \dfrac{x + 2}{x + 1}$ **(C)** $\dfrac{x^2 - 1}{x^2 + 5x + 6} \cdot \dfrac{x + 2}{x + 1}$

Solution

(A)
$$\frac{2x^2y}{3z} \cdot \frac{3z^2}{4xy} = \frac{(2x^2y)(3z^2)}{(3z)(4xy)}$$

$$= \frac{6x^2yz^2}{12xyz} \qquad \text{The numerator } 6x^2yz^2 \text{ and denominator } 12xyz \text{ have common factor } 6xyz.$$

$$= \frac{(6xyz)xz}{(6xyz)2}$$

$$= \frac{xz}{2}$$

Instead of multiplying out and then removing the common factor $6xyz$, we could have recognized from the original product that both the numerator and denominator would contain 2, 3, x, y, and z as factors. Consequently, these common factors could have been eliminated before multiplying:

$$\frac{\overset{1 \cdot x \cdot 1}{\cancel{2} \cdot \cancel{x^2} \cdot \cancel{y}}}{\underset{1 \cdot 1}{\cancel{3} \cdot \cancel{z}}} \cdot \frac{\overset{1 \cdot z}{\cancel{3} \cdot \cancel{z^2}}}{\underset{2 \cdot 1 \cdot 1}{\cancel{4} \cdot \cancel{x} \cdot \cancel{y}}} = \frac{xz}{2}$$

(B) $(x^2 - 1) \cdot \dfrac{x + 2}{x + 1}$ Factor all numerators and denominators and divide out common factors. Then multiply.

$$= \frac{(x - 1)\overset{1}{\cancel{(x + 1)}}}{1} \cdot \frac{x + 2}{\underset{1}{\cancel{x + 1}}}$$

$$= \frac{(x - 1)(x + 2)}{1}$$

$$= (x - 1)(x + 2)$$

(C) $\dfrac{x^2 - 1}{x^2 + 5x + 6} \cdot \dfrac{x + 2}{x + 1} = \dfrac{(x - 1)\overset{1}{\cancel{(x + 1)}}}{\underset{1}{\cancel{(x + 2)}}(x + 3)} \cdot \dfrac{\overset{1}{\cancel{x + 2}}}{\underset{1}{\cancel{x + 1}}}$

$$= \frac{x - 1}{x + 3}$$

Matched Problem 1   Multiply and reduce to lowest terms:

     (A) $\dfrac{3a^2b^2}{5c} \cdot \dfrac{10c^2}{9ab}$ **(B)** $\dfrac{x+2}{x+3} \cdot (x^2-9)$ **(C)** $\dfrac{x+2}{x+3} \cdot \dfrac{x^2+7x+12}{x^2-4}$

♦ **DIVISION**

The definition of division of real fractions is the same as given in Chapter 3 for quotients of integers: $A \div B = Q$ means $B \cdot Q = A$. The same rule of inverting and multiplying then applies.

Division of Real Fractions

For any real numbers a, b, c, and d (b, d, $c \neq 0$),

$$\underset{\text{Divisor}}{\dfrac{a}{b} \div \dfrac{c}{d}} = \dfrac{a}{b} \cdot \underset{\text{Reciprocal of divisor}}{\dfrac{d}{c}} \qquad \dfrac{x}{2} \div \dfrac{4}{3} = \dfrac{x}{2} \cdot \dfrac{3}{4} = \dfrac{3x}{8}$$

That is, to divide one fraction by another, multiply by the reciprocal of the divisor.

Example 2
Division

Divide and reduce to lowest terms:

     (A) $\dfrac{2x^2y}{3z} \div \dfrac{4xy}{3z^2}$ **(B)** $(x^2-4) \div \dfrac{x-2}{x+3}$ **(C)** $\dfrac{x^2-x}{x+1} \div \dfrac{x^2-2x+1}{x^2+2x+1}$

Solution   **(A)** $\dfrac{2x^2y}{3z} \div \dfrac{4xy}{3z^2} = \dfrac{2x^2y}{3z} \cdot \dfrac{3z^2}{4xy}$ Invert and multiply. The product is the same as Example 1(A).

$$= \dfrac{xz}{2}$$

     (B) $(x^2-4) \div \dfrac{x-2}{x+3} = (x-2)(x+2) \cdot \dfrac{x+3}{x-2}$

$$= (x+2)(x+3)$$

     (C) $\dfrac{x^2-x}{x+1} \div \dfrac{x^2-2x+1}{x^2+2x+1} = \dfrac{x(x-1)}{x+1} \cdot \dfrac{(x+1)(x+1)}{(x-1)(x-1)}$

$$= \dfrac{x(x+1)}{x-1}$$

Matched Problem 2   Divide and reduce to lowest terms:

     (A) $\dfrac{2a^2b}{3c^2} \div \dfrac{6ab}{5c}$ **(B)** $\dfrac{x-1}{x+2} \div (x^2-1)$

     (C) $\dfrac{x^2+3x+2}{x^2-9} \div \dfrac{x+2}{x^2+4x+3}$

Answers to Matched Problems

1. (A) $\dfrac{2abc}{3}$ (B) $(x+2)(x-3)$ (C) $\dfrac{x+4}{x-2}$

2. (A) $\dfrac{5a}{9c}$ (B) $\dfrac{1}{(x+1)(x+2)}$ (C) $\dfrac{(x+1)^2}{x-3}$

EXERCISE 7-2

Perform the indicated operations and simplify:

A

1. $\dfrac{15}{16} \cdot \dfrac{24}{27}$

2. $\dfrac{6}{7} \cdot \dfrac{28}{9}$

3. $\dfrac{36}{8} \div \dfrac{9}{4}$

4. $\dfrac{4}{6} \div \dfrac{24}{8}$

5. $\dfrac{-12}{25} \div \dfrac{16}{15}$

6. $\dfrac{14}{9} \div \dfrac{-35}{12}$

7. $\dfrac{-8}{27} \cdot \dfrac{20}{33}$

8. $\dfrac{-10}{21} \cdot \dfrac{14}{25}$

9. $\dfrac{y^4}{3u^5} \cdot \dfrac{2u^3}{3y}$

10. $\dfrac{6x^3y}{7u} \cdot \dfrac{14u^3}{12xy}$

11. $\dfrac{uvw}{5xyz} \div \dfrac{5vy}{uwxz}$

12. $\dfrac{3c^2d}{a^3b^3} \div \dfrac{3a^3b^3}{cd}$

13. $\dfrac{-3xy}{z^2} \cdot \dfrac{-xz}{15y^2}$

14. $\dfrac{3x^2y}{4z^3} \cdot \dfrac{-2xz^2}{9xy^2}$

15. $\dfrac{-8ac}{5b^2} \div \dfrac{4ab}{25c}$

16. $\dfrac{-2bc^2}{3a^2} \div \dfrac{-ab}{6c^2}$

17. $\dfrac{x+3}{2x^2} \cdot \dfrac{4x}{x+3}$

18. $\dfrac{3x^2y}{x-y} \cdot \dfrac{x-y}{6xy}$

19. $\dfrac{a^2-a}{a-1} \cdot \dfrac{a+1}{a}$

20. $\dfrac{x+3}{x^3+3x^2} \cdot \dfrac{x^3}{x-3}$

21. $\dfrac{x^2-9}{4x^2} \cdot \dfrac{6x^3}{x^2-6x+9}$

22. $\dfrac{6y^4}{y^2-4} \cdot \dfrac{y^2+2y}{4y^2}$

23. $\dfrac{y^2+3y+2}{5y^2} \div \dfrac{(y+2)^2}{15y}$

24. $\dfrac{x^2+3x+2}{2x^3} \div \dfrac{(x+1)^2}{10x}$

25. $\dfrac{4x}{x-4} \div \dfrac{8x^2}{x^2-6x+8}$

26. $\dfrac{x-2}{4y} \div \dfrac{x^2+x-6}{12y^2}$

B

27. $\dfrac{d^5}{3a} \div \left(\dfrac{d^2}{6a^2} \cdot \dfrac{a}{4d^3} \right)$

28. $\left(\dfrac{d^5}{3a} \div \dfrac{d^2}{6a^2} \right) \cdot \dfrac{a}{4d^3}$

29. $\dfrac{3x^3}{y^4} \cdot \left(\dfrac{-4z}{5} \cdot \dfrac{10y^2}{12xz} \right)$

30. $\dfrac{6z^2}{x^3} \cdot \left(\dfrac{2y}{3xz} \cdot \dfrac{-x^4}{8yz} \right)$

31. $\dfrac{-5x^2}{z^5} \div \left(\dfrac{10z}{3x} \cdot \dfrac{6y^2}{xz^3} \right)$

32. $\dfrac{2x^3}{3y^3} \div \left(\dfrac{-6y}{5z^2} \cdot \dfrac{15z^4}{9x} \right)$

33. $\dfrac{3x^3}{y^4} \div \left(\dfrac{-4z}{5} \div \dfrac{10y^2}{12xz} \right)$

34. $\dfrac{6z^2}{x^3} \cdot \left(\dfrac{2y}{3xz} \div \dfrac{-x^4}{8yz} \right)$

35. $\dfrac{-5x^2}{z^5} \cdot \left(\dfrac{10z}{3x} \div \dfrac{6y^2}{xz^3} \right)$

36. $\dfrac{2x^3}{3y^3} \cdot \left(\dfrac{-6y}{5z^2} \div \dfrac{15z^4}{9x} \right)$

37. $\dfrac{2x^2+4x}{12x^2y} \cdot \dfrac{6x}{x^2+6x+8}$

38. $\dfrac{6x^2}{4x^2y-12xy} \cdot \dfrac{x^2+x-12}{3x^2+12x}$

39. $\dfrac{2y^2+7y+3}{4y^2-1} \div (y+3)$

40. $(t^2-t-12) \div \dfrac{t^2-9}{t^2-3t}$

41. $\dfrac{x^2+5x+4}{x^2-1} \div (x+4)$

42. $\dfrac{x^2+5x+6}{x^2-4} \div (x+3)$

43. $(x-5) \div \dfrac{x^2-1}{x^2-6x+5}$

44. $(x+4) \div \dfrac{x^2-9}{x^2+7x+12}$

45. $\dfrac{x^2-6x+9}{x^2-x-6} \div \dfrac{x^2+2x-15}{x^2+2x}$

46. $\dfrac{m+n}{m^2-n^2} \div \dfrac{m^2-mn}{m^2-2mn+n^2}$

47. $-(x^2-4) \cdot \dfrac{3}{x+2}$

48. $-(x^2-3x) \cdot \dfrac{x-2}{x-3}$

49. $\dfrac{a+b}{a-b} \cdot (a^2-2ab+b^2)$

50. $\dfrac{p-q}{p+q} \cdot (p^2+2pq+q^2)$

51. $\dfrac{a^2 - 3a - 10}{a^2 + a - 2} \cdot \dfrac{a^2 - 1}{a - 5}$

52. $\dfrac{a^2 + 2a - 15}{a^2 + a - 12} \cdot \dfrac{a^2 - 25}{a + 4}$

53. $\dfrac{2 - m}{2m - m^2} \cdot \dfrac{m^2 + 4m + 4}{m^2 - 4}$

54. $\dfrac{x^2 - 9}{x^2 + 5x + 6} \cdot \dfrac{x + 2}{x - 3}$

C 55. $\left(\dfrac{x^2 - xy}{xy + y^2} \div \dfrac{x^2 - y^2}{x^2 + 2xy + y^2} \right) \div \dfrac{x^2 - 2xy + y^2}{x^2 y + xy^2}$

56. $\dfrac{x^2 - xy}{xy + y^2} \div \left(\dfrac{x^2 - y^2}{x^2 + 2xy + y^2} \div \dfrac{x^2 - 2xy + y^2}{x^2 y + xy^2} \right)$

57. $\left(\dfrac{xy^2 - xz^2}{3y - 3z} \div \dfrac{y^2 + 2yz + z^2}{x^2} \right) \cdot \dfrac{x^2}{y + z}$

58. $\left(\dfrac{xy^2 - xz^2}{3y - 3z} \cdot \dfrac{y^2 + 2yz + z^2}{x^2} \right) \div \dfrac{x^2}{y + z}$

59. $\dfrac{a^2 - 4a + 3}{a^2 - 4} \cdot \left(\dfrac{a^2 + 4a + 4}{a^2 - 1} \div \dfrac{a^2 - 2a - 3}{(a + 1)^2} \right)$

60. $\dfrac{x^2 + 3x + 2}{x^2 + 7x + 12} \cdot \left(\dfrac{x^2 + 5x + 4}{x^2 + 5x + 6} \div \dfrac{x^2 + 2x + 1}{x^2 + 6x + 9} \right)$

61. $\left(\dfrac{x^2 + x - 2}{x^2 + x - 20} \div \dfrac{x^2 - 4x + 3}{x^2 - 3x - 4} \right) \cdot \dfrac{x^2 + 6x + 5}{x^2 - 9}$

62. $\left(\dfrac{a^2 + 4a + 3}{a^2 + 6a + 8} \cdot \dfrac{a^2 + 5a + 6}{a^2 + 6a + 5} \right) \div \dfrac{a^2 + 9a + 20}{a^2 + 6a + 9}$

In Problems 63–66, determine all values of x for which the two expressions represent the same real numbers:

63. $\dfrac{x^2 - 1}{x - 1}$ and $x + 1$

64. $\dfrac{x^2 - x - 6}{x - 3}$ and $x + 2$

65. $\dfrac{x^2 + x - 12}{x + 4}$ and $x - 3$

66. $\dfrac{x^2 - 4}{x + 2}$ and $x - 2$

67. Try to evaluate the following quotient without using a calculator. [*Hint:* Let $x = 108,641$, so $108,642 = x + 1$, and so on.]

$$\dfrac{(108,641)^2 - (108,643)^2}{(108,642)(108,646) - (108,644)^2}$$

68. Repeat Problem 67 for

$$\dfrac{(302,754)^2 - (302,757)^2}{(302,751)(302,755) - (302,754)^2}$$

7-3 Addition and Subtraction

♦ Addition and Subtraction
♦ Sign Changes and Multiplication by -1

We will add and subtract rational expressions exactly as we did fractions in Chapter 3. You may want to review Section 3-4.

♦ ADDITION AND SUBTRACTION

To add or subtract real fractions, we proceed in the same way we did for rational numbers. If the denominators are the same, we add or subtract the numerators:

Addition and Subtraction of Real Fractions

For any real numbers a, b, and c ($b \neq 0$),

1. $\dfrac{a}{b} + \dfrac{c}{b} = \dfrac{a + c}{b}$ $\qquad \dfrac{x}{2} + \dfrac{3}{2} = \dfrac{x + 3}{2}$

2. $\dfrac{a}{b} - \dfrac{c}{b} = \dfrac{a - c}{b}$ $\qquad \dfrac{2x}{3} - \dfrac{x}{3} = \dfrac{2x - x}{3} = \dfrac{x}{3}$

If the denominators are not the same, we can use the fundamental principle of fractions to change the fractions to equivalent fractions with the same denominator. We can use any common denominator, but the least common denominator (LCD) will save effort in reducing the resulting sum or difference. The **LCD** in this context means the same as it did in Chapter 3, the least common multiple of all the denominators. We may be able to find the LCD by inspection. If not, we can proceed exactly as we did in Section 3-4:

Finding the Least Common Denominator (LCD)

Step 1   Factor each denominator completely, using integer coefficients.

Step 2   The LCD must contain each *different* factor that occurs in all the denominators to the highest power it occurs in any one denominator.

Example 1
Addition and Subtraction with a Common Denominator

Combine into a single fraction and reduce to lowest terms:

(A) $\dfrac{x-3}{2(x+1)} + \dfrac{x-5}{2(x+1)}$ **(B)** $\dfrac{4x-5}{x(x-1)} - \dfrac{x-2}{x(x-1)}$

Solution   **(A)** $\dfrac{x-3}{2(x+1)} + \dfrac{x-5}{2(x+1)}$ Since the denominators are equal, we can add using Property 1 from the rules for adding real fractions.

$\qquad = \dfrac{(x-3)+(x-5)}{2(x+1)}$

$\qquad = \dfrac{x-3+x-5}{2(x+1)}$ Simplify the numerator.

$\qquad = \dfrac{2x-8}{2(x+1)}$ Factor the numerator to reduce to lowest terms.

$\qquad = \dfrac{2(x-4)}{2(x+1)}$ Remove the common factor 2 from the numerator and denominator.

$\qquad = \dfrac{x-4}{x+1}$

(B) $\dfrac{4x-5}{x(x-1)} - \dfrac{x-2}{x(x-1)}$

$\qquad = \dfrac{(4x-5)-(x-2)}{x(x-1)}$ Using parentheses at this stage will minimize sign errors. The arrows indicate where an error commonly occurs.

$\qquad = \dfrac{4x-5-x+2}{x(x-1)}$ Simplify the numerator.

$$= \frac{3x - 3}{x(x - 1)}$$ Factor the numerator.

$$= \frac{3(x - 1)}{x(x - 1)}$$

$$= \frac{3}{x}$$

Matched Problem 1 Combine into a single fraction and reduce to lowest terms:

(A) $\dfrac{x^2 - 4x + 1}{x(x + 2)} + \dfrac{x^2 - 1}{x(x + 2)}$ **(B)** $\dfrac{3x + 5}{3(x + 4)} - \dfrac{x - 3}{3(x + 4)}$

Example 2
Adding with Different Denominators

Combine into a single fraction and reduce to lowest terms:

$$\frac{1}{4(x + 1)} + \frac{1}{2(x^2 - 1)}$$

Solution Find the LCD of the denominators:

$$4(x + 1) = 2^2(x + 1)$$

$$2(x^2 - 1) = 2(x + 1)(x - 1)$$

$$\begin{aligned} \text{LCD} &= 2^2(x + 1)(x - 1) \\ &= 4(x^2 - 1) \end{aligned}$$

The LCD contains each different factor to the highest power it occurs in any denominator.

Use the fundamental principle of fractions to change each fraction to one with the LCD as denominator:

$$\frac{1}{4(x + 1)} = \frac{(x - 1)}{4(x + 1)(x - 1)} = \frac{x - 1}{4(x^2 - 1)}$$

$$\frac{1}{2(x^2 - 1)} = \frac{2}{2 \cdot 2(x^2 - 1)} = \frac{2}{4(x^2 - 1)}$$

Now we can add the fractions:

$$\begin{aligned} \frac{1}{4(x + 1)} + \frac{1}{2(x^2 - 1)} &= \frac{x - 1}{4(x^2 - 1)} + \frac{2}{4(x^2 - 1)} \\ &= \frac{x - 1 + 2}{4(x^2 - 1)} = \frac{x + 1}{4(x^2 - 1)} \\ &= \frac{x + 1}{4(x + 1)(x - 1)} \\ &= \frac{1}{4(x - 1)} \end{aligned}$$

As we indicated before, we could have used any common denominator to combine the fractions. For example, here we could have used the product $4(x + 1) \cdot 2(x^2 - 1)$. If we had done so, however, more common factors would have had to be removed to reduce the result to lowest terms. Try using $4(x + 1) \cdot (x^2 - 1)$ as the common denominator to see what happens.

Matched Problem 2   Combine into a single fraction and reduce to lowest terms:

$$\frac{3x + 1}{5(x^2 - 2x - 3)} - \frac{1}{10(x + 1)}$$

Example 3
Subtracting with Different Denominators

Combine into a single fraction and reduce to lowest terms:

$$\frac{3}{(x - 1)(x + 2)} - \frac{5}{(x + 2)(x - 3)}$$

Solution $\dfrac{3}{(x - 1)(x + 2)} - \dfrac{5}{(x + 2)(x - 3)}$

The LCD is $(x - 1)(x + 2)(x - 3)$. Use the fundamental principle of fractions to convert each fraction so the denominator is the LCD.

$$= \frac{3(x - 3)}{(x - 1)(x + 2)(x - 3)} - \frac{5(x - 1)}{(x - 1)(x + 2)(x - 3)}$$

$$= \frac{3(x - 3) - 5(x - 1)}{(x - 1)(x + 2)(x - 3)}$$

$$= \frac{3x - 9 - 5x + 5}{(x - 1)(x + 2)(x - 3)}$$

$$= \frac{-2x - 4}{(x - 1)(x + 2)(x - 3)}$$

$$= \frac{-2(x + 2)}{(x - 1)(x + 2)(x - 3)} = \frac{-2}{(x - 1)(x - 3)}$$

Matched Problem 3   Combine into a single fraction and reduce to lowest terms:

$$\frac{2}{(x + 1)(x + 3)} + \frac{3}{(x + 1)(x - 2)}$$

◆ **SIGN CHANGES AND MULTIPLICATION BY −1**

The relationship

$$-(a - b) = b - a$$

was introduced in Section 2-6 and discussed again in Section 7-1. It is often useful in combining fractions. Keep in mind that the relation results from multiplying by −1:

$$-(a - b) = (-1)(a - b)$$
$$= (-1)a - (-1)b$$
$$= -a - (-b)$$
$$= -a + b$$
$$= b - a$$

Example 4
Sign Changes

Replace the question mark with an appropriate algebraic expression:

(A) $x - 3 = -(?)$ **(B)** $2x - 5 = -(?)$ **(C)** $-(x - 1) = (?) - x$

Solution **(A)** $x - 3 = -(3 - x)$
(B) $2x - 5 = -(5 - 2x)$
(C) $-(x - 1) = 1 - x$

Matched Problem 4

Replace the question mark with an appropriate algebraic expression:

(A) $a - 2 = -(?)$ **(B)** $3 - a = -(?)$ **(C)** $-(2a - 3) = (?) - 2a$

Example 5
Using $-(a - b) = b - a$

Combine into a single fraction and reduce to lowest terms:

$$\frac{x}{x - 1} + \frac{2}{1 - x}$$

Solution $\dfrac{x}{x - 1} + \dfrac{2}{1 - x} = \dfrac{x}{x - 1} + \dfrac{2}{-(x - 1)}$ Recognize $1 - x = -(x - 1)$.

$= \dfrac{x}{x - 1} - \dfrac{2}{x - 1}$ Recall $\dfrac{a}{-b} = -\dfrac{a}{b}$

$= \dfrac{x - 2}{x - 1}$

Matched Problem 5

Combine into a single fraction and reduce to lowest terms:

$$\frac{3}{y - 3} + \left(\frac{-y}{3 - y}\right) \quad = \frac{3}{y-3} + \frac{y}{y-3} = \frac{3+y}{y-3}$$

Answers to Matched Problems

1. (A) $\dfrac{2(x - 2)}{x + 2}$ **(B)** $\dfrac{2}{3}$ **2.** $\dfrac{1}{2(x - 3)}$ **3.** $\dfrac{5}{(x - 2)(x + 3)}$

4. (A) $2 - a$ **(B)** $a - 3$ **(C)** 3 **5.** $\dfrac{y + 3}{y - 3}$

EXERCISE 7-3

Combine into single fractions and reduce to lowest terms:

A 1. $\dfrac{3}{2x} - \dfrac{1}{2x}$ **2.** $\dfrac{2}{3x} + \dfrac{1}{3x}$

3. $\dfrac{5}{3x^2} + \dfrac{1}{3x^2}$ **4.** $\dfrac{5}{2x^2} - \dfrac{1}{2x^2}$

5. $\dfrac{2x}{3y} + \dfrac{x}{3y}$ **6.** $\dfrac{5x}{2y} - \dfrac{x}{2y}$

7. $\dfrac{x}{4y} - \dfrac{3x}{4y}$ **8.** $\dfrac{x}{6y} + \dfrac{5x}{6y}$

9. $\dfrac{2x}{x + 3} + \dfrac{6}{x + 3}$ **10.** $\dfrac{4}{x + 3} - \dfrac{1}{x + 3}$

11. $\dfrac{4}{x-1} - \dfrac{3}{x-1}$

12. $\dfrac{2x}{x-1} + \dfrac{-2}{x-1}$

47. $\dfrac{5}{x^2+9x+14} - \dfrac{4}{x^2+10x+21}$

13. $\dfrac{x}{x-1} - \dfrac{1}{x-1}$

14. $\dfrac{x}{x+3} + \dfrac{3}{x+3}$

48. $\dfrac{3}{x^2+x-2} - \dfrac{4}{x^2+2x-3}$

15. $\dfrac{x}{x+1} + \dfrac{1}{x+1}$

16. $\dfrac{x}{x-3} - \dfrac{3}{x-3}$

49. $\dfrac{3}{x(x+1)} - \dfrac{6}{x(x+2)}$

17. $\dfrac{1}{x+1} - \dfrac{1}{x}$

18. $\dfrac{1}{x-2} - \dfrac{1}{x}$

50. $\dfrac{1}{x(x+1)} + \dfrac{2}{x(x+2)}$

19. $\dfrac{3}{x-2} - \dfrac{2}{x}$

20. $\dfrac{4}{x+1} - \dfrac{3}{x}$

51. $\dfrac{1}{x^2-4} - \dfrac{2}{x^2-4x+4}$

21. $\dfrac{1}{z} + \dfrac{2}{z+3}$

22. $\dfrac{1}{z} - \dfrac{2}{z+3}$

52. $\dfrac{2}{x^2-4} + \dfrac{1}{x^2-4x+4}$

23. $\dfrac{3}{y+1} - \dfrac{2}{y+2}$

24. $\dfrac{3}{y+1} + \dfrac{2}{y+2}$

53. $\dfrac{-1}{x^2-5x+6} + \dfrac{6}{x^2-9}$

B 25. $\dfrac{3}{x+1} - \dfrac{6}{x^2-1}$

26. $\dfrac{3}{x+1} + \dfrac{6}{x^2-1}$

54. $\dfrac{1}{x^2-5x+6} - \dfrac{3}{x^2-9}$

27. $\dfrac{1}{x} + \dfrac{x}{x(x+1)}$

28. $\dfrac{1}{x} - \dfrac{x}{x(x+1)}$

55. $1 + \dfrac{2}{x} + \dfrac{3}{x^2}$ **56.** $3 + \dfrac{2}{x} + \dfrac{1}{x^2}$

29. $\dfrac{x}{x-1} - \dfrac{x+1}{x}$

30. $\dfrac{x}{x+4} - \dfrac{x-4}{x}$

57. $3 - \dfrac{2}{x} + \dfrac{1}{x^2}$ **58.** $5 + \dfrac{3}{x} - \dfrac{1}{x^2}$

31. $\dfrac{x}{x+1} - \dfrac{x}{x-1}$

32. $\dfrac{x}{x-2} - \dfrac{x}{x+2}$

59. $\dfrac{2}{x+1} + \dfrac{1}{x^2-1} - \dfrac{3}{x-1}$

33. $x - \dfrac{x^2+1}{x}$

34. $x + \dfrac{4-x^2}{x}$

60. $\dfrac{1}{x-2} + \dfrac{3}{x^2-5x+6} - \dfrac{5}{x-3}$

35. $\dfrac{x^2-4}{x+3} - x$

36. $\dfrac{1-x^2}{x+2} + x$

61. $\dfrac{2}{x+4} - \dfrac{3}{x^2+5x+4} + \dfrac{4}{x+1}$

37. $1 - \dfrac{y}{y-1}$

38. $1 + \dfrac{y}{y-1}$

62. $\dfrac{4}{x+2} - \dfrac{3}{x^2-4} + \dfrac{2}{x-2}$

39. $\dfrac{1}{(x-1)(x-2)} + \dfrac{2}{(x-1)(x-3)}$

63. $\dfrac{-1}{x^2+3x+2} + \dfrac{7}{x^2+5x+4}$

40. $\dfrac{2}{(x-1)(x-2)} - \dfrac{1}{(x-1)(x-3)}$

64. $\dfrac{1}{x^2+3x+2} - \dfrac{3}{x^2+5x+4}$

41. $\dfrac{3}{(x+4)(x-1)} - \dfrac{2}{(x+4)(x-3)}$

C 65. $\dfrac{3}{x-1} + \dfrac{2}{x+1} - \dfrac{6}{x^2-1}$

42. $\dfrac{4}{(x+3)(x+2)} + \dfrac{1}{(x+3)(x-1)}$

66. $\dfrac{2}{x-1} - \dfrac{3}{x+1} + \dfrac{x-5}{x^2-1}$

43. $\dfrac{x}{(x-1)(x+2)} - \dfrac{1}{x}$ **44.** $\dfrac{2x}{(x+3)(x-4)} - \dfrac{2}{x}$

67. $\dfrac{1}{x(x-1)} + \dfrac{1}{x^2-1} + \dfrac{2}{x(x+1)}$

45. $\dfrac{-2}{x^2+5x+6} + \dfrac{4}{x^2+6x+8}$

68. $\dfrac{-3}{x(x+1)} - \dfrac{2}{x+1} + \dfrac{3}{x}$

46. $\dfrac{5}{x^2+3x-4} + \dfrac{-2}{x^2+6x+8}$

69. $\dfrac{2}{x+3} + \dfrac{3}{x-3} + \dfrac{x+15}{x^2-9}$

70. $\dfrac{-3}{x+2} + \dfrac{1}{x-2} + \dfrac{x-10}{x^2-4}$

71. $\dfrac{2}{x+3} + \dfrac{4}{x-2} - \dfrac{x-7}{x^2+x-6}$

72. $\dfrac{3}{x+2} - \dfrac{1}{x+1} + \dfrac{x+7}{x^2+3x+2}$

73. $\dfrac{1}{y-1} + \dfrac{2}{1-y^2}$

74. $\dfrac{-y}{y^2-1} - \dfrac{1}{1-y}$

75. $\dfrac{-2}{y^2-1} + \dfrac{2}{1-y} - \dfrac{1}{1+y}$

76. $\dfrac{4}{x^2-4} - \dfrac{2}{2-x} + \dfrac{1}{2+x}$

77. $\left(\dfrac{1}{x} + \dfrac{1}{y}\right) \div (x+y)$

78. $\left(\dfrac{y}{x} - \dfrac{x}{y}\right) \div (x-y)$

79. $\left(\dfrac{y}{x} + \dfrac{x}{y}\right) \div \dfrac{x+y}{xy}$

80. $\left(\dfrac{1}{x} - \dfrac{1}{y}\right) \div (y-x)$

81. $\left(\dfrac{1}{x-1} - \dfrac{1}{x+2}\right) \div \dfrac{3}{x^2-4}$

82. $\left(\dfrac{2}{x-3} - \dfrac{3}{x+1}\right) \div \dfrac{x^2-22x+121}{x^2-9}$

83. $\left(\dfrac{4}{x-2} - \dfrac{2}{x+3}\right) \div \dfrac{x^2+2x-8}{x^2-4}$

84. $\left(\dfrac{1}{x+1} - \dfrac{1}{x-3}\right) \div \dfrac{4}{x^2-1}$

7-4 Solving Equations Involving Fractional Forms

- Equations with Constants in Denominators
- Equations with Variables in Denominators
- Applications
- Solutions and Graphs

Equations involving fractional forms are generally solved by eliminating the fractions. Special care must be taken, however, in the case where a variable occurs in a denominator. We deal with such equations in this section and also consider again the relationship between an equation of the form

$$(\text{Expression in } x) = 0$$

and the graph of the related equation

$$y = \text{Expression in } x$$

♦ EQUATIONS WITH CONSTANTS IN DENOMINATORS

We have already considered equations involving constants in denominators, such as

$$\frac{2}{3} - \frac{x-4}{2} = \frac{5x}{6}$$

in Section 3-5 and found we could easily convert such an equation into an equivalent equation with integer coefficients by multiplying both sides by the LCM of the denominators—in this case 6:

$$6 \cdot \frac{2}{3} - 6 \cdot \frac{(x-4)}{2} = 6 \cdot \frac{5x}{6} \qquad \text{Multiply each side by 6 to clear fractions.}$$

Since 6 is the LCM of the denominators, each denominator will divide into 6 exactly, leaving

$$4 - 3(x - 4) = 5x$$

and we finish the solution in a few steps:

$$\downarrow$$

$$4 - 3x + 12 = 5x \qquad \text{A sign error frequently occurs where the arrow points.}$$

$$-3x + 16 = 5x$$

$$-8x = -16$$

$$x = 2$$

♦ EQUATIONS WITH VARIABLES IN DENOMINATORS

If an equation involves a variable in one or more denominators, such as

$$\frac{3}{4} + \frac{1}{x} = \frac{5}{2x}$$

we proceed in essentially the same way as above so long as we are careful:

We must avoid any value that makes a denominator in the equation 0.

**Example 1
Solving Fractional
Form Equations**

Solve $\dfrac{3}{4} + \dfrac{1}{x} = \dfrac{5}{2x}$.

$$3x + 4 = 10$$
$$3x = 6$$
$$x = 2$$

Solution

$$\frac{3}{4} + \frac{1}{x} = \frac{5}{2x} \qquad x \neq 0$$

If 0 turns up later as an apparent solution, then it must be discarded, because we cannot divide by 0. Since $x \neq 0$, we can multiply both sides by $4x$, the LCM of the denominators.

$$4x \cdot \frac{3}{4} + 4x \cdot \frac{1}{x} = 4x \cdot \frac{5}{2x}$$

All denominators divide out.

$$3x + 4 = 10$$

$$3x = 6$$

$$x = 2$$

Matched Problem 1   Solve $\dfrac{2}{3} + \dfrac{1}{x} = \dfrac{7}{3x}$.

$$2x + 3 = 7$$
$$2x = 4$$
$$x = 2$$

Example 2
Solving Fractional
Form Equations

Solve $\dfrac{2x}{x-1} - 3 = \dfrac{8-6x}{x-1}$.

Solution

$$\dfrac{2x}{x-1} - 3 = \dfrac{8-6x}{x-1} \qquad x \neq 1$$

If 1 turns up later as an apparent solution, it must be rejected, since we cannot divide by $1 - 1 = 0$. Multiply by $(x - 1)$, the LCM of the denominators. Also, it is a good idea to place all binomial numerators and denominators in parentheses to avoid multiplication errors.

$$(x-1)\dfrac{2x}{(x-1)} - 3(x-1) = (x-1)\dfrac{(8-6x)}{(x-1)}$$

$$2x - 3(x-1) = 8 - 6x$$

$$2x - 3x + 3 = 8 - 6x$$

$$-x + 3 = 8 - 6x$$

$$5x = 5$$

$$x = 1 \qquad\qquad\qquad x \text{ cannot equal 1.}$$

The original equation has no solution. Results such as $x = 1$ in this case are called **extraneous solutions.**

Matched Problem 2 Solve $\dfrac{3x}{x-4} - 5 = \dfrac{4+2x}{x-4}$.

Example 3
Solving Fractional
Form Equations

Solve $3x + \dfrac{21}{2} = \dfrac{6}{x}$.

Solution

$$3x + \dfrac{21}{2} = \dfrac{6}{x}$$

If 0 turns up later as an apparent solution, it must be rejected, since we cannot substitute 0 for x in $6/x$. Multiply both sides by $2x$, the LCM of the denominators.

$$2x \cdot 3x + 2x \cdot \dfrac{21}{2} = 2x \cdot \dfrac{6}{x}$$

Eliminate denominators by reducing fractions.

$$6x^2 + 21x = 12$$

Write as a polynomial set equal to 0.

$$6x^2 + 21x - 12 = 0$$

Since the coefficient of each term is divisible by 3, multiply each side by $\frac{1}{3}$ to simplify further.

$$\frac{1}{3} \cdot 6x^2 + \frac{1}{3} \cdot 21x - \frac{1}{3} \cdot 12 = \frac{1}{3} \cdot 0 \qquad \text{Eliminate denominators by reducing fractions.}$$

$$2x^2 + 7x - 4 = 0 \qquad \text{Factor, if possible.}$$

$$(2x - 1)(x + 4) = 0 \qquad \text{Solve using the zero property.}$$

$$2x - 1 = 0 \qquad \text{or} \qquad x + 4 = 0$$

$$2x = 1 \qquad\qquad\qquad x = -4$$

$$x = \frac{1}{2}$$

Checking the solution is left to you.

Matched Problem 3   Solve $5x = \dfrac{5}{2} + \dfrac{15}{x}$.

♦ **APPLICATIONS**

Applications will sometimes lead naturally to an equation with a variable in the denominator.

Example 4
A Proportion Problem

Suppose you are told that the ratio of women to men in a college is $3:5$ and there are 2,640 students enrolled. How many women are in the college?

Solution   Let $x =$ Number of women. Then the number of men is $2{,}640 - x$, and the ratio of women to men is

$$\frac{x}{2{,}640 - x}$$

Thus,

$$\frac{x}{2{,}640 - x} = \frac{3}{5} \qquad \text{Cross multiply.}$$

$$5x = 3(2{,}640 - x)$$

$$5x = 7{,}920 - 3x$$

$$8x = 7{,}920$$

$$x = \frac{7{,}920}{8} = 990 \text{ women}$$

Check   The ratio of women to men is $\dfrac{990}{2{,}640 - 990} = \dfrac{990}{1{,}650} = \dfrac{3}{5}$.

Matched Problem 4   If in a college the ratio of men to women is $2:3$ and there are 2,450 students enrolled, how many men are in the school?

Example 5 Application	A baseball player begins the week with 31 hits in 100 at bats, for a batting average (hits/at bats) of .310. During the week he gets 11 hits and raises his average to exactly .350. How many times did the player bat during the week?

Solution   Let x be the number of at bats during the week. The player's overall average at the end of the week is

$$\frac{\text{Total hits}}{\text{Total at bats}} = \frac{31 + 11}{100 + x} = \frac{42}{100 + x} = .350$$

Thus

$$.35(100 + x) = 42$$
$$35 + .35x = 42$$
$$.35x = 7$$
$$x = \frac{7}{.35} = 20 \text{ at bats}$$

Check $\dfrac{42}{100 + 20} = \dfrac{42}{120} = .350$

Matched Problem 5   A basketball player has made 18 out of 24 free throws for a free throw percentage of 0.75, that is, 75%. Several games later she has made 10 more free throws and raised her free throw percentage to 80%. How many more free throws did she attempt?

♦ SOLUTIONS AND GRAPHS

The graph of an equation in which y is set equal to a fractional form involving x can be found by plotting points, just as we plotted straight lines in Chapters 4 and 5, and parabolas in Chapter 6. However, the graphs will look substantially different when there are values of x that make the denominator 0. Consider, for example, the graph of

$$y = \frac{1 - 2x}{x}$$

Since we cannot allow the variable x to equal 0, the graph will have two separate parts, one to the left of the y axis and one to the right of it. We make a table of values for both halves, including values of x near 0, and plot the corresponding points:

x	y
−5	−2.20
−4	−2.25
−3	−2.33
−2	−2.50
−1	−3.00
$-\frac{1}{2}$	−4.00
$-\frac{1}{10}$	−12.00
0	Not defined
$\frac{1}{10}$	8.00
$\frac{1}{2}$	0.00
1	−1.00
2	−1.50
3	−1.67
4	−1.75
5	−1.80

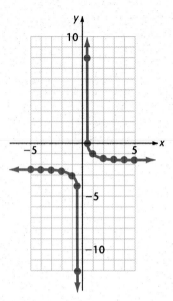

Figure 1

Notice that the graph crosses the x axis at $x = \frac{1}{2}$ and that solving the equation

$$\frac{1 - 2x}{x} = 0$$

also yields $x = \frac{1}{2}$:

$$\frac{1 - 2x}{x} = 0$$

$$1 - 2x = 0$$

$$-2x = -1$$

$$x = \frac{-1}{-2} = \frac{1}{2}$$

Thus, the solution of the equation $\dfrac{1 - 2x}{x} = 0$ is the x intercept of the graph of the

equation $y = \dfrac{1 - 2x}{x}$.

Answers to Matched Problems
1. $x = 2$ **2.** No solution; $x = 4$ is extraneous
3. $x = -\frac{3}{2}, 2$ **4.** 980   **5.** 11

EXERCISE 7-4

Solve:

A **1.** $\dfrac{2}{x} - \dfrac{1}{3} = \dfrac{5}{x}$

2. $\dfrac{1}{2} - \dfrac{2}{x} = \dfrac{3}{x}$

3. $\dfrac{5}{6} - \dfrac{1}{y} = \dfrac{2}{3y}$

4. $\dfrac{1}{x} + \dfrac{2}{3} = \dfrac{1}{2}$

5. $\dfrac{1}{x} + \dfrac{1}{3} = \dfrac{7}{3x}$

6. $\dfrac{1}{x} + \dfrac{1}{5} = \dfrac{11}{6x}$

7. $\dfrac{3}{2x} + \dfrac{1}{x} = \dfrac{5}{4}$

8. $\dfrac{2}{3x} + \dfrac{1}{x} = \dfrac{1}{3}$

9. $\dfrac{4}{3x} - \dfrac{5}{6x} = \dfrac{1}{2}$

10. $\dfrac{5}{2x} - \dfrac{3}{x} = \dfrac{-1}{4}$

11. $\dfrac{1}{x} + \dfrac{1}{4} = \dfrac{3}{x} - \dfrac{3}{20}$

12. $\dfrac{1}{x} - \dfrac{1}{6} = \dfrac{3}{2x} - \dfrac{5}{12}$

13. $\dfrac{2}{3} - \dfrac{4}{x} = \dfrac{1}{x} - \dfrac{1}{21}$

14. $\dfrac{2}{5} + \dfrac{3}{x} = \dfrac{1}{x} + \dfrac{4}{x}$

15. $3 - \dfrac{4}{x} = \dfrac{x+2}{x}$

16. $4 + \dfrac{3}{x} = \dfrac{4x+3}{x}$

17. $\dfrac{2}{3x} + \dfrac{1}{2} = \dfrac{4}{x} + \dfrac{4}{3}$

18. $\dfrac{1}{m} - \dfrac{1}{9} = \dfrac{4}{9} - \dfrac{2}{3m}$

19. $\dfrac{1}{2t} + \dfrac{1}{8} = \dfrac{2}{t} - \dfrac{1}{4}$

20. $\dfrac{4}{3k} - 2 = \dfrac{k+4}{6k}$

B 21. $\dfrac{9}{L+1} - 1 = \dfrac{12}{L+1}$

22. $\dfrac{7}{y-2} - \dfrac{1}{2} = 3$

23. $\dfrac{3}{2x-1} + 4 = \dfrac{6x}{2x-1}$

24. $\dfrac{5x}{x+5} = 2 - \dfrac{25}{x+5}$

25. $\dfrac{3N}{N-2} - \dfrac{9}{4N} = 3$

26. $\dfrac{2E}{E-1} = 2 + \dfrac{5}{2E}$

27. $5 + \dfrac{2x}{x-3} = \dfrac{6}{x-3}$

28. $\dfrac{6}{x-2} = 3 + \dfrac{3x}{x-2}$

29. $y = \dfrac{9}{y}$

30. $\dfrac{t}{2} = \dfrac{2}{t}$

31. $3 + \dfrac{4x}{x-5} = \dfrac{5x+3}{x-5}$

32. $2 + \dfrac{6x}{x-1} = \dfrac{7x+2}{x-1}$

33. $5 - \dfrac{2x}{x-3} = \dfrac{x-1}{x-3}$

34. $6 - \dfrac{3x}{x+4} = \dfrac{6x+18}{x+4}$

35. $\dfrac{8}{x+4} = \dfrac{7x+8}{x^2+5x+4}$

36. $\dfrac{3}{x-4} = \dfrac{4x-5}{x^2-4x}$

37. $\dfrac{3x-2}{x^2+2x} = \dfrac{2}{x+2}$

38. $\dfrac{x+1}{x^2-3x+2} = \dfrac{4}{x-1}$

39. $\dfrac{5}{x-3} = \dfrac{33-x}{x^2-6x+9}$

40. $\dfrac{D^2+2}{D^2-4} = \dfrac{D}{D-2}$

41. $\dfrac{n-5}{6n-6} = \dfrac{1}{9} - \dfrac{n-3}{4n-4}$

42. $\dfrac{1}{3} - \dfrac{s-2}{2s+4} = \dfrac{s+2}{3s+6}$

43. $\dfrac{3}{4} + \dfrac{x+3}{x-1} = \dfrac{6x+3}{2x-2}$

44. $\dfrac{7}{3} - \dfrac{2x-1}{x+2} = \dfrac{5x+1}{3x+6}$

45. $\dfrac{5x-3}{4x+28} = \dfrac{1}{3} + \dfrac{x+3}{2x+14}$

46. $\dfrac{25x+12}{20x-10} = \dfrac{2}{5} + \dfrac{x+3}{2x-1}$

47. $y = \dfrac{15}{y-2}$

48. $2x - 3 = \dfrac{2}{x}$

49. $2 + \dfrac{2}{x^2} = \dfrac{5}{x}$

50. $1 - \dfrac{3}{x} = \dfrac{10}{x^2}$

C 51. $\dfrac{2}{x-2} = 3 - \dfrac{5}{2-x}$

52. $\dfrac{3x}{x-4} - 2 = \dfrac{3}{4-x}$

53. $5 - \dfrac{x}{x-3} = \dfrac{4x-1}{3-x}$

54. $1 - \dfrac{2x}{x-1} = \dfrac{x+1}{1-x}$

55. $\dfrac{x+2}{3x-1} = \dfrac{3x}{1-3x} + 2$

56. $\dfrac{x-3}{4x-5} = \dfrac{10}{5-4x} + 2$

57. $\dfrac{10}{x^2-x} + \dfrac{5}{x^2+x} + \dfrac{8}{x^2-1} = \dfrac{120}{x^3-x}$

58. $\dfrac{12}{x^2-2x} + \dfrac{20}{x^2+2x} + \dfrac{30}{x^2-4} = \dfrac{170}{x^3-4x}$

59. $\dfrac{5x-22}{x^2-6x+9} - \dfrac{11}{x^2-3x} - \dfrac{5}{x} = 0$

60. $\dfrac{1}{x^2-x-2} - \dfrac{3}{x^2-2x-3} = \dfrac{1}{x^2-5x+6}$

61. $\dfrac{1}{x^2-1} - \dfrac{1}{x^2+4x+3} + \dfrac{1}{x^2+2x-3} = 0$

62. $\dfrac{1}{x^2-9x+20} - \dfrac{1}{x^2-8x+15}$
$+ \dfrac{1}{x^2-7x+12} = 0$

APPLICATIONS

63. *Sports* For a basketball team the ratio of three-point shots attempted to two-point shots attempted is 2:9. During the season the team attempted 1,848 total shots. How many three-point shots did it attempt?

64. *Sales* At a fast-food restaurant, the ratio of diet drinks sold to regular drinks sold is 1.5 to 4. If the restaurant sells 22,000 drinks in a month, how many diet drinks did it sell?

65. *Personnel* In an insurance company, the ratio of male to female agents is 8:7. If the company has 4,830 agents in total, how many are women?

66. **Education**   At a university, the ratio of full-time students to part-time students is 5:8. If the university's total enrollment is 16,042, how many part-time students are enrolled?

67. **Sports**   A baseball player has 118 hits in 412 at bats, for a batting average of approximately .286. How many consecutive hits will raise his average to exactly .300?

68. **Sports**   A basketball player has made 35 of 46 free throws, or approximately 76.1%. How many consec-

utive free throws made will raise her percentage to exactly 80%?

69. **Sports**   If a baseball team has won 42 out of its first 78 games, how many consecutive games must it win to reach a winning percentage of 60%?

70. **Safety**   A construction company has been accident-free for 145 of the first 163 workdays in a year. How many consecutive accident-free days must it experience to raise the percentage of workdays that are accident-free to 90%?

7-5 Formulas and Equations with Several Variables

Literal equations were introduced in Section 2-7. There, we noted that solving for one variable in terms of the others often results in fractional forms. We can now deal more fully with such forms. In the following examples, formulas involving several variables are changed to isolate a particular variable. The remaining variables are treated as constants in the solution process. The process is the same as we have used for single-variable equations: first, remove any fractions by multiplying by the LCM of the denominators as in Sections 3-5 and 7-4; then solve by isolating the variable as in Section 2-7.

Example 1
Solving a Literal Equation

Solve the formula $c = wrt/1,000$ for t. The formula gives the cost of using an electrical appliance, where $w =$ Power in watts, $r =$ Rate per kilowatt-hour, $t =$ Time in hours.

Solution

$$c = \frac{wrt}{1,000}$$ Start with the given formula. Multiply both sides of the equation by 1,000 to remove fractions.

$$1,000c = wrt$$ Divide both sides by wr, the coefficient of t.

$$\frac{1,000c}{wr} = t$$

It is not necessary that the variable t be isolated on the left side of the equation, since the final form is equivalent to $t = 1,000c/wr$. If desired, equations can be reversed at any step by using the symmetric property of equality. Usually, the variable is isolated on the left side of the equation, but this is only tradition, not a mathematical necessity.

Matched Problem 1   Solve the formula $c = wrt/1,000$ for w.

Example 2
Solving a Literal
Equation

Solve the formula $A = P + Prt$ for r. This is the simple interest formula.

Solution

$$A = P + Prt$$

$$P + Prt = A$$

$$Prt = A - P$$

$$\frac{Prt}{Pt} = \frac{A - P}{Pt}$$

$$r = \frac{A - P}{Pt}$$

Reverse the equation using the symmetric property; then perform operations to isolate r on the left side.

Matched Problem 2

Solve the formula $A = P + Prt$ for t.

Example 3
Solving a Literal
Equation

Solve the formula $A = P + Prt$ for P.

Solution

$$A = P + Prt$$

$$P + Prt = A$$

If we write $P = A - Prt$, we have not solved for P. To solve for P is to isolate P entirely on the left side with a coefficient of 1.

In general, if the variable we are solving for appears on both sides of an equation, we have not solved for it.

Since P is a common factor to both terms on the left, we factor P out and complete the problem:

$$P(1 + rt) = A$$

$$\frac{P(1 + rt)}{(1 + rt)} = \frac{A}{(1 + rt)}$$

$$P = \frac{A}{1 + rt}$$

Divide both sides by $(1 + rt)$ to isolate P.

Note that P appears only on the left side.

Matched Problem 3

Solve $A = ac + ab$ for a.

Example 4
Solving a Literal
Equation

Solve $K = \dfrac{3t^2}{s}$ for s.

Solution

$$K = \dfrac{3t^2}{s}$$ Multiply both sides by s to remove fractions.

$$sK = 3t^2$$ Divide both sides by K to isolate s.

$$s = \dfrac{3t^2}{K}$$

Matched Problem 4   Solve $M = \dfrac{c + d}{e}$ for e.

Example 5
Solving a Literal
Equation

Solve $\dfrac{1}{a} = \dfrac{1}{b} + \dfrac{1}{c}$ for a.

Solution

$$\dfrac{1}{a} = \dfrac{1}{b} + \dfrac{1}{c}$$ Multiply both sides by abc to clear fractions.

$$abc \cdot \dfrac{1}{a} = abc \cdot \dfrac{1}{b} + abc \cdot \dfrac{1}{c}$$

$$bc = ac + ab$$ Reverse the equation.

$$ac + ab = bc$$ Factor out a.

$$a(c + b) = bc$$ Divide both sides by $(c + b)$.

$$a = \dfrac{bc}{c + b}$$

Matched Problem 5   Solve $\dfrac{1}{a} = \dfrac{1}{b} + \dfrac{1}{c}$ for b.

Example 6
Solving a Literal
Equation

Solve $y = \dfrac{x + 2}{3x - 1}$ for x in terms of y.

Solution

$$y = \dfrac{x + 2}{3x - 1}$$ Remove fractions by multiplying each side by $3x - 1$.

$$y(3x - 1) = x + 2$$ Simplify the left side.

$$3yx - y = x + 2$$ Shift terms containing x to the left side.

$$3yx - y - x = 2$$

$$3yx - x = 2 + y \qquad \text{Factor out } x \text{ on the left side.}$$
$$(3y - 1)x = 2 + y$$
$$x = \frac{2 + y}{3y - 1}$$

Matched Problem 6   Solve $y = \dfrac{2x - 1}{x + 3}$ for x in terms of y.

Answers to Matched Problems

1. $w = \dfrac{1,000c}{rt}$ **2.** $t = \dfrac{A - P}{Pr}$ **3.** $a = \dfrac{A}{c + b}$

4. $e = \dfrac{c + d}{M}$ **5.** $b = \dfrac{ac}{c - a}$ **6.** $x = \dfrac{3y + 1}{2 - y}$

EXERCISE 7-5

A *Solve for x:*

1. $y = 3x - 8$ **2.** $y = -2x - 7$

3. $y = -5x + 2$ **4.** $y = 6x + 5$

5. $y - 3 = 2x + 4$ **6.** $y + 5 = -3x + 2$

7. $y + 4 = -3x - 7$ **8.** $y - 6 = 2x - 9$

9. $y = \dfrac{xz}{18}$ **10.** $z = \dfrac{-xy}{222}$

11. $y = \dfrac{-1}{xz}$ **12.** $z = \dfrac{1}{xy}$

13. $y = 3x + 4z$ **14.** $z = 4y - 3x$

15. $y = \dfrac{x + 1}{z}$ **16.** $z = \dfrac{x - 3}{2z}$

B *The formulas and equations in Problems 17–50 are used in the sciences and mathematics. Some are included inside the back cover of this text. Solve for the indicated variable.*

17. Solve $A = P + I$ for I.   *Simple interest*

18. Solve $R = R_1 + R_2$ for R_2.   *Electric circuits— resistance in series*

19. Solve $d = rt$ for r.   *Distance–rate–time*

20. Solve $d = 1,100t$ for t.   *Sound distance in air*

21. Solve $I = Prt$ for t.   *Simple interest*

22. Solve $C = 2\pi r$ for r.   *Circumference of a circle*

23. Solve $C = \pi D$ for π.   *Circumference of a circle*

24. Solve $e = mc^2$ for m.   *Mass–energy equation*

25. Solve $ax + b = 0$ for x.   *Linear equation in one variable*

26. Solve $p = 2a + 2b$ for a.   *Perimeter of a rectangle*

27. Solve $s = 2t - 5$ for t.   *Slope–intercept form for a line*

28. Solve $y = mx + b$ for m.   *Slope–intercept form for a line*

29. Solve $3x - 4y - 12 = 0$ for y.   *Linear equation in two variables*

30. Solve $Ax + By + C = 0$ for y.   *Linear equation in two variables*

31. Solve $I = E/R$ for E.   *Electric circuits— Ohm's law*

32. Solve $m = b/a$ for b.   *Optics—magnification*

33. Solve the formula in Problem 31 for R.

34. Solve the formula in Problem 32 for a.

35. Solve $A = 2l + 2b$ for l.   *Perimeter of a parallelogram*

36. Solve the equation in Problem 35 for b.

37. Solve $A = \frac{1}{2}h(B + b)$ for b.   *Area of a trapezoid*

38. Solve the equation in Problem 37 for B.

39. Solve $A = \frac{1}{2}Dd$ for D.   *Area of a kite*

40. Solve the equation in Problem 39 for d.

41. Solve $C = 100B/L$ for B.   *Anthropology— cephalic index*

42. Solve IQ = $(100)(MA)/(CA)$ for (MA).   *Psychology— intelligence quotient*

43. Solve the formula in Problem 41 for L.

44. Solve the formula in Problem 42 for (CA).

45. Solve $F = G(mM/d^2)$ *Gravitational force*
    for m.   *between two masses*

46. Solve the formula in Problem 45 for G.

47. Solve $P = M - Mdt$ for d.   *Simple discount*

48. Solve the formula in Problem 47 for t.

49. Solve $C = \frac{5}{9}(F - 32)$ for F.   *Celsius–Fahrenheit*

50. Solve $F = \frac{9}{5}C + 32$ for C.   *Fahrenheit–Celsius*

C **51.** Solve $b = c - 3cd$ for c.

52. Solve $p = q + 2qr$ for q.

53. Solve $y = xz - 2x$ for x.

54. Solve $z = xy - 3y$ for y.

55. Solve $a + b = ab + bc$ for b.

56. Solve $p - q = qr - pq$ for p.

57. Solve $\dfrac{1}{x} = \dfrac{1}{y} - \dfrac{1}{z}$ for x.

58. Solve $\dfrac{1}{x} = \dfrac{1}{y} - \dfrac{1}{z}$ for z.

59. Solve $y = 1 + \dfrac{1}{x + 1}$ for x.

60. Solve $y = 1 - \dfrac{1}{1 - x}$ for x.

61. Solve $y = \dfrac{3x + 1}{2x - 1}$ for x in terms of y.

62. Solve $y = \dfrac{x - 3}{3x - 2}$ for x in terms of y.

63. Solve $a = \dfrac{b + 2}{2b - 1}$ for b in terms of a.

64. Solve $a = \dfrac{3b - 2}{b + 1}$ for b in terms of a.

65. Solve $x = \dfrac{y - 3z}{2y + z}$ for y.

66. Solve $x = \dfrac{y - 3z}{2y + z}$ for z.

The formulas and equations in Problems 67–70 are used in the sciences and business. Solve for the indicated variable.

67. Solve $\dfrac{1}{f} = \dfrac{1}{a} + \dfrac{1}{b}$ for f.   *Optics—focal length*

68. Solve $\dfrac{1}{R} = \dfrac{1}{R_1} + \dfrac{1}{R_2}$ for R_1.   *Electrical resistance*

69. Solve $P = M - Mdt$ for M *Simple discount*

70. Solve $V = P - Pdt$ for P *Depreciation*

  ## 7-6 Complex Fractions

- ♦ Simplifying by Using Division
- ♦ Simplifying by Using the Fundamental Principle of Fractions
- ♦ Which Method?
- ♦ Solving Equations Involving Complex Fractions

A fraction form with fractions in its numerator or denominator is called a **complex fraction.** It is often necessary to represent a complex fraction as a *simple fraction.* In all cases we will consider, a **simple fraction** means the quotient of two polynomials or two integers. The process does not involve any new concepts. It is a matter of applying old concepts appropriately.

♦ SIMPLIFYING BY USING DIVISION

All fractions, simple or complex, represent division problems. To simplify, it is often easiest just to divide.

Example 1
Simplifying by Dividing

Express as a simple fraction:

(A) $\dfrac{\dfrac{3}{5}}{\dfrac{2}{3}}$ (B) $\dfrac{\dfrac{a}{2b}}{\dfrac{a^2}{b}}$

Solution

(A) $\dfrac{\dfrac{3}{5}}{\dfrac{2}{3}} = \dfrac{3}{5} \div \dfrac{2}{3} = \dfrac{3}{5} \cdot \dfrac{3}{2} = \dfrac{9}{10}$

(B) $\dfrac{\dfrac{a}{2b}}{\dfrac{a^2}{b}} = \dfrac{a}{2b} \div \dfrac{a^2}{b} = \dfrac{a}{2b} \cdot \dfrac{b}{a^2} = \dfrac{1}{2a}$

Matched Problem 1

Express as a simple fraction:

(A) $\dfrac{\dfrac{3}{4}}{\dfrac{5}{8}}$ (B) $\dfrac{\dfrac{xy}{z^2}}{\dfrac{x}{yz}}$

Example 2
Simplifying by Dividing

Express as a simple fraction:

$$\dfrac{\dfrac{x}{x^2 - 1}}{\dfrac{x^2}{x + 1}}$$

Solution

$$\dfrac{\dfrac{x}{x^2 - 1}}{\dfrac{x^2}{x + 1}} = \dfrac{x}{x^2 - 1} \div \dfrac{x^2}{x + 1} = \dfrac{x}{x^2 - 1} \cdot \dfrac{x + 1}{x^2}$$

$$= \dfrac{x}{(x + 1)(x - 1)} \cdot \dfrac{x + 1}{x^2} = \dfrac{1}{x(x - 1)}$$

Matched Problem 2

Express as a simple fraction:

$$\dfrac{\dfrac{a^2}{a^2 - 2a - 3}}{\dfrac{a}{a^2 - 9}}$$

♦ SIMPLIFYING BY USING THE FUNDAMENTAL PRINCIPLE OF FRACTIONS

The fundamental principle of fractions

$$\frac{a}{b} = \frac{ka}{kb} \qquad b, k \neq 0$$

can also be used to simplify complex fractions. Consider, for example, the complex fraction simplified in Example 2:

$$\frac{\dfrac{x}{x^2 - 1}}{\dfrac{x^2}{x + 1}}$$

If we multiply the numerator and denominator by $k = x^2 - 1 = (x - 1)(x + 1)$, the fractions in the numerator and denominator will disappear:

$$\frac{(x^2 - 1)\dfrac{x}{x^2 - 1}}{(x^2 - 1)\dfrac{x^2}{x + 1}} = \frac{x}{(x - 1)x^2} = \frac{1}{(x - 1)x}$$

We chose the multiplier $x^2 - 1$ because it is the LCD of the fractions in the numerator and denominator.

Example 3
Simplifying a Complex Fraction

Express as a simple fraction:

(A) $\dfrac{1 - \dfrac{1}{x}}{x - \dfrac{1}{x}}$ **(B)** $\dfrac{\dfrac{a}{b} - \dfrac{b}{a}}{\dfrac{1}{b} + \dfrac{1}{a}}$

Solution   **(A)** $\dfrac{1 - \dfrac{1}{x}}{x - \dfrac{1}{x}} = \dfrac{x\left(1 - \dfrac{1}{x}\right)}{x\left(x - \dfrac{1}{x}\right)}$ Multiply top and bottom by x, the LCD of the fractions in the numerator and denominator.

$$= \frac{x - 1}{x^2 - 1}$$ Factor the denominator and reduce to lowest terms.

$$= \frac{x - 1}{(x - 1)(x + 1)}$$

$$= \frac{1}{x + 1}$$

(B) $\dfrac{\dfrac{a}{b} - \dfrac{b}{a}}{\dfrac{1}{b} + \dfrac{1}{a}} = \dfrac{ab\left(\dfrac{a}{b} - \dfrac{b}{a}\right)}{ab\left(\dfrac{1}{b} + \dfrac{1}{a}\right)}$ 　　The LCD of the fractions in the numerator and denominator is ab.

$$= \dfrac{a^2 - b^2}{a + b}$$ 　　Reduce to lowest terms.

$$= a - b$$

Matched Problem 3 　Express as a simple fraction:

(A) $\dfrac{4 - \dfrac{1}{x^2}}{1 - \dfrac{1}{2x}}$ 　　　　**(B)** $\dfrac{\dfrac{1}{a} - \dfrac{1}{b}}{b - \dfrac{a^2}{b}}$

♦ **WHICH METHOD?**

When internal fractions within a fraction have denominators that contain more than one term as in Example 2, the division method is usually easier to use. If internal fractions have single-term denominators as in Example 3, using the fundamental principle of fractions to clear denominators is generally easier.

♦ **SOLVING EQUATIONS INVOLVING COMPLEX FRACTIONS**

Example 4
Solving Equations Involving Complex Fractions

Solve:

(A) $\dfrac{\dfrac{x}{x^2 - 1}}{\dfrac{x^2}{x + 1}} = \dfrac{1}{x}$ 　　　　**(B)** $\dfrac{\dfrac{x}{x^2 - 1}}{\dfrac{x^2}{x + 1}} = \dfrac{1}{x - 1}$

Solution **(A)** $\dfrac{\dfrac{x}{x^2 - 1}}{\dfrac{x^2}{x + 1}} = \dfrac{1}{x}$ 　　Simplify the complex fraction on the left side of the equation. This is the same fraction seen in Example 2, so we use the answer obtained there. Note that $x \neq 0$, $x \neq \pm 1$.

$$\dfrac{1}{x(x - 1)} = \dfrac{1}{x}$$ 　　Multiply by $x(x - 1)$, the LCD.

$$1 = x - 1$$

$$2 = x$$

that is,

$$x = 2$$

(B) $\dfrac{\dfrac{x}{x^2 - 1}}{\dfrac{x^2}{x + 1}} = \dfrac{1}{x - 1}$ Proceed as in Example 4(A). Once again, $x \neq 0$ and $x \neq \pm 1$.

$$\dfrac{1}{x(x - 1)} = \dfrac{1}{x - 1}$$

$$1 = x \qquad \text{Since } x \text{ cannot equal 1, 1 is an extraneous solution.}$$

Therefore, the equation has no solution.

Matched Problem 4 Solve:

(A) $\dfrac{4 - \dfrac{1}{x^2}}{1 - \dfrac{1}{2x}} = \dfrac{1}{x}$ **(B)** $\dfrac{4 - \dfrac{1}{x^2}}{1 - \dfrac{1}{2x}} = \dfrac{2}{x}$

Answers to Matched Problems **1. (A)** $\dfrac{6}{5}$ **(B)** $\dfrac{y^2}{z}$ **2.** $\dfrac{a(a + 3)}{a + 1}$ **3. (A)** $\dfrac{2(2x + 1)}{x}$ **(B)** $\dfrac{1}{a(a + b)}$

4. (A) $-\dfrac{1}{4}$ **(B)** No solution

EXERCISE 7-6

Write the expression as a simple fraction reduced to lowest terms:

A **1.** $\dfrac{\dfrac{2}{3}}{\dfrac{4}{5}}$ **2.** $\dfrac{\dfrac{3}{4}}{\dfrac{3}{8}}$ **3.** $\dfrac{\dfrac{7}{12}}{\dfrac{2}{3}}$ **4.** $\dfrac{\dfrac{4}{5}}{\dfrac{8}{15}}$

5. $\dfrac{\dfrac{-5}{9}}{\dfrac{15}{33}}$ **6.** $\dfrac{\dfrac{-12}{35}}{\dfrac{18}{5}}$ **7.** $\dfrac{\dfrac{x^2 y}{z}}{\dfrac{xy^2}{z^2}}$ **8.** $\dfrac{\dfrac{x^2 y^2}{z^2}}{\dfrac{xy^3}{z}}$

9. $\dfrac{\dfrac{a^4}{b^3 c^2}}{\dfrac{a^2}{b^3 c^4}}$ **10.** $\dfrac{\dfrac{a^3}{b^3 c}}{\dfrac{a}{bc}}$ **11.** $\dfrac{\dfrac{a}{bc}}{\dfrac{b}{ac}}$ **12.** $\dfrac{\dfrac{xy^2}{z^3}}{\dfrac{x}{yz}}$

13. $\dfrac{\dfrac{xy}{zw}}{\dfrac{-wxy}{z}}$ **14.** $\dfrac{\dfrac{-x^2 y}{zw}}{\dfrac{w^2 xy}{z^2}}$

15. $\dfrac{\dfrac{x^2 - y^2}{x}}{\dfrac{x - y}{y}}$ **16.** $\dfrac{\dfrac{a^2 - b^2}{ab}}{\dfrac{a + b}{ab}}$

17. $\dfrac{\dfrac{x + 1}{x^2}}{\dfrac{x^2 + 2x + 1}{x}}$ **18.** $\dfrac{\dfrac{x^2}{x - 1}}{\dfrac{x^3}{x^2 - 2x + 1}}$

B **19.** $\dfrac{x - \dfrac{1}{x}}{x - 1}$ **20.** $\dfrac{\dfrac{x - 1}{x}}{1 - \dfrac{1}{x}}$

21. $\dfrac{\dfrac{x^2 - y^2}{xy}}{1 - \dfrac{y}{x}}$ **22.** $\dfrac{\dfrac{1}{a} - \dfrac{1}{b}}{\dfrac{b}{a} - \dfrac{a}{b}}$

23. $\dfrac{1 - \dfrac{1}{a^2}}{1 + \dfrac{1}{a}}$ **24.** $\dfrac{\dfrac{a}{b} - \dfrac{b}{a}}{\dfrac{1}{a} + \dfrac{1}{b}}$

25. $\dfrac{\dfrac{1}{x} - \dfrac{1}{y}}{\dfrac{1}{x^2} - \dfrac{1}{y^2}}$ **26.** $\dfrac{a - \dfrac{b^2}{a}}{\dfrac{1}{a} - \dfrac{1}{b}}$

27. $\dfrac{1 - \dfrac{b}{a}}{1 - \dfrac{a}{b}}$ **28.** $\dfrac{1 + \dfrac{a}{b}}{1 + \dfrac{b}{a}}$

29. $\dfrac{\dfrac{x-1}{x-3}}{x^2-4x+3}$

30. $\dfrac{\dfrac{x+1}{x+5}}{x^2+5x+4}$

31. $\dfrac{\dfrac{x-2}{x+2}}{\dfrac{x^2-x-2}{x^2+x-2}}$

32. $\dfrac{\dfrac{x+3}{x-3}}{\dfrac{x^2+x-6}{x^2-x-6}}$

33. $\dfrac{\dfrac{x^2-1}{x^2+x-12}}{\dfrac{x^2+5x+4}{x^2-4x+3}}$

34. $\dfrac{\dfrac{x^2-x-2}{x^2+5x+6}}{\dfrac{x^2-4}{x^2+4x+3}}$

35. $\dfrac{\dfrac{x^2+2x-8}{x^2+4x+3}}{\dfrac{x^2+6x+8}{x^2-4x-5}}$

36. $\dfrac{\dfrac{x^2+3x-4}{x^2-x-6}}{\dfrac{x^2+5x+4}{x^2-5x+6}}$

37. $\dfrac{\dfrac{m^2}{m^2+m-6}}{\dfrac{m^3}{m^2-9}}$

38. $\dfrac{\dfrac{2x^2y}{x^2+xy-2y^2}}{\dfrac{4xy^2}{x^2-4y^2}}$

39. $\dfrac{\dfrac{n^2-m^2}{m^2-mn}}{\dfrac{-m^3}{m^2+mn}}$

40. $\dfrac{\dfrac{bc-ac}{a^2-7a+12}}{\dfrac{-(a-b)}{a^2-a-6}}$

Solve:

41. $\dfrac{\dfrac{x-2}{x+2}}{\dfrac{x^2-x-2}{x^2+x-2}} = \dfrac{2}{x+1}$

42. $\dfrac{\dfrac{x-2}{x+2}}{\dfrac{x^2-x-2}{x^2+x-2}} = \dfrac{-2}{x+1}$

43. $\dfrac{\dfrac{x+3}{x-3}}{\dfrac{x^2+x-6}{x^2-x-6}} = \dfrac{4}{x-2}$

44. $\dfrac{\dfrac{x+3}{x-3}}{\dfrac{x^2+x-6}{x^2-x-6}} = \dfrac{-4}{x-2}$

45. $\dfrac{\dfrac{x^2}{x^2+x-6}}{\dfrac{x^3}{x^2-9}} = \dfrac{-1}{x^2-2x}$

46. $\dfrac{\dfrac{x^2}{x^2+x-6}}{\dfrac{x^3}{x^2-9}} = \dfrac{-3}{x^2-2x}$

C *Write the expression as a simple fraction reduced to lowest terms:*

47. $\dfrac{\dfrac{x^2}{x+y}-x}{\dfrac{y^2}{x+y}-y}$

48. $\dfrac{\dfrac{x}{x+y}-1}{\dfrac{x^3}{x^2-y^2}-x}$

49. $\dfrac{\dfrac{x^3-y^3}{x^3y^3}}{\dfrac{x-y}{x^2y^2}}$

50. $\dfrac{\dfrac{a^2-b^2}{a^3+b^3}}{\dfrac{a^3-b^3}{a^2+ab+b^2}}$

51. $\dfrac{\dfrac{x^2+2xy+y^2}{x^2-2xy+y^2}}{\dfrac{x^3+y^3}{x^2-y^2}}$

52. $\dfrac{\dfrac{x^2-2xy+y^2}{x^2+2xy+y^2}}{\dfrac{x^2-y^2}{x^3-y^3}}$

53. $\dfrac{\dfrac{x}{x+1}-\dfrac{x}{x-1}}{\dfrac{x+1}{x-1}-\dfrac{x-1}{x+1}}$

54. $\dfrac{\dfrac{a-b}{a}-\dfrac{a+b}{b}}{\dfrac{a+b}{a}+\dfrac{a-b}{b}}$

55. $1+\dfrac{1}{1+\dfrac{1}{x}}$

56. $1-\dfrac{a}{a-\dfrac{1}{a}}$

57. $\dfrac{1}{1-\dfrac{1}{1+x}}$

58. $\dfrac{1}{1+\dfrac{1}{1-x}}$

Solve:

59. $1+\dfrac{1}{1+\dfrac{1}{x}} = \dfrac{1}{x+1}$

60. $1+\dfrac{1}{1+\dfrac{1}{x}} = \dfrac{-1}{x+1}$

61. $1+\dfrac{x}{x-\dfrac{1}{x}} = \dfrac{x}{x^2-1}$

62. $1+\dfrac{x}{x-\dfrac{1}{x}} = \dfrac{2x^2}{x^2-1}$

APPLICATIONS

63. *Domestic* If one pump can empty a pool in r minutes and a second can empty it in s minutes, then the two working together can empty it in

$$t = \dfrac{1}{\dfrac{1}{r}+\dfrac{1}{s}} \text{ minutes}$$

Express t as a simple fraction.

64. *Domestic* If one person can water-seal a wood deck in x hours and a second person can do the same job in y hours, then the two working together can seal the deck in

$$h = \cfrac{1}{\cfrac{1}{x} + \cfrac{1}{y}} \text{ hours}$$

Express h as a simple fraction.

65. *Sports* A baseball pitcher who has allowed R earned runs in I innings pitched will have an earned run average, ERA, given by

$$\text{ERA} = \frac{R}{I/9}$$

Express ERA as a simple fraction.

66. *Sports* A soccer goalkeeper who has allowed g goals scored while playing m minutes will have a goals against average, GAA, of

$$\text{GAA} = \frac{g}{m/90}$$

Express GAA as a simple fraction.

CHAPTER SUMMARY

7-1 REDUCING TO LOWEST TERMS

A **rational expression** is a fraction form in which the numerator and denominator are polynomials. The **fundamental principle of real fractions**

$$\frac{ak}{bk} = \frac{a}{b}$$

is used with rational expressions to **reduce to lower terms.**

7-2, 7-3 MULTIPLICATION AND DIVISION, ADDITION AND SUBTRACTION

Multiplication, division, addition, and subtraction are performed as for rational numbers:

$$\frac{a}{b} \cdot \frac{c}{d} = \frac{ac}{bd} \qquad \frac{a}{b} \div \frac{c}{d} = \frac{a}{b} \cdot \frac{d}{c} \qquad \frac{a}{b} + \frac{c}{b} = \frac{a+c}{b} \qquad \frac{a}{b} - \frac{c}{b} = \frac{a-c}{b}$$

To add or subtract rational expressions with different denominators, use the fundamental principle to change the form of the fractions so that the denominators are the same. The **least common denominator (LCD)** is generally used. To find the LCD, factor each denominator and include each factor to the highest power at which it occurs in any denominator. In simplifying fractions, the identity $a - b = -(b - a)$ may be useful.

7-4 SOLVING EQUATIONS INVOLVING FRACTIONAL FORMS

Equations with only constants in the denominators are solved by first removing fractions, through multiplication of both sides by the LCM of the denominators. Equations involving variables in the denominators are solved the same way, but values that make a denominator 0 must be avoided.

7-5 FORMULAS AND EQUATIONS WITH SEVERAL VARIABLES

Formulas involving several variables may be solved for one variable in terms of the other; the process is the same as for one-variable equations.

7-6 COMPLEX FRACTIONS

A **complex fraction** is a fractional form with fractions in its numerator or denominator. It may be reduced to a **simple fraction,** the quotient of two polynomials, either by using the fundamental principle of fractions or by using division.

CHAPTER REVIEW EXERCISE

Work through all the problems in this chapter review and check the answers in the back of the book. Answers to all review problems are there, and following each answer is a number in italics indicating the section in which that type of problem is discussed. Where weaknesses show up, review appropriate sections in the text.

A *Reduce to lowest terms:*

1. $\dfrac{6x^3}{21x}$

2. $\dfrac{3(x-3)^2}{42(x-3)}$

3. $\dfrac{8x+16}{20(x+2)}$

4. $\dfrac{pq^2 + p^2q}{p^2 + pq}$

5. $\dfrac{(x+3)(x-3)}{(x-3)(x+4)}$

6. $\dfrac{x^2 - 1}{x+1}$

Perform the indicated operations and reduce to lowest terms:

7. $1 + \dfrac{2}{3x}$

8. $\dfrac{2}{x} - \dfrac{1}{6x} + \dfrac{1}{3}$

9. $\dfrac{3x^2(x-3)}{6y} \cdot \dfrac{8y^3}{9(x-3)^2}$

10. $(d-2)^2 \div \dfrac{d^2 - 4}{d-2}$

11. $\dfrac{2}{3x-1} - \dfrac{1}{2x}$

12. $\dfrac{x}{x-4} - 1$

13. $\dfrac{\frac{1}{4}}{\frac{5}{6}}$

14. $\dfrac{4\frac{2}{3}}{1\frac{1}{2}}$

15. $\dfrac{1}{x} - \dfrac{1}{x^2} + \dfrac{1}{x^3}$

16. $\dfrac{x}{x-1} - \dfrac{x-1}{x}$

17. $\dfrac{x}{x-1} \div \dfrac{x-1}{x}$

18. $\dfrac{x-1}{x} \cdot \dfrac{3x^2}{x^2 - 1}$

19. $\dfrac{\frac{1}{x}}{\frac{2}{x^2}}$

20. $\dfrac{\frac{x^2}{x-1}}{\frac{x^3}{x-1}}$

21. $\dfrac{x^2}{x+2} \cdot \dfrac{x-1}{x+2} \cdot \dfrac{(x+2)^2}{x^3}$

22. $\dfrac{3}{x} - \left(\dfrac{4}{x^2} + \dfrac{3}{x}\right)$

In Problems 23–27, solve the equation:

23. $\dfrac{2}{3} - \dfrac{2}{x} = \dfrac{4}{x}$

24. $\dfrac{x}{x-1} + 2 = \dfrac{4}{x-1}$

25. $\dfrac{x+1}{x-1} = \dfrac{1}{2}$

26. $\dfrac{2}{3m} - \dfrac{1}{4m} = \dfrac{1}{12}$

27. $\dfrac{3x}{x-5} - 8 = \dfrac{15}{x-5}$

28. Solve the formula for the area of a triangle, $A = \dfrac{bh}{2}$, for b.

29. Solve the equation $y = \dfrac{1}{3}x + \dfrac{3}{4}$ for x.

30. A baseball player's batting average A is the quotient of hits h divided by times at bat b, that is, $A = h/b$. Solve the formula for b.

B *Reduce to lowest terms:*

31. $\dfrac{x^2 + 6x - 7}{4x - 4}$

32. $\dfrac{x^2 - 4}{x^2 - 3x + 2}$

33. $\dfrac{5x^2 + 20x + 15}{5x^2 - 5}$

34. $\dfrac{a^2 - b^2}{b^2 + 2ba + a^2}$

35. $\dfrac{3 - z}{z^2 - 9}$

36. $\dfrac{-2pqr + 3qrs}{2p - 3s}$

Perform the indicated operations and reduce to lowest terms:

37. $\dfrac{y-2}{y^2 - 4y + 4} \div \dfrac{y^2 + 2y}{y^2 + 4y + 4}$

38. $\dfrac{6x^4 - 6x^3}{3xy} \cdot \dfrac{xy + y}{x^4 - x^2}$

39. $\dfrac{x+1}{x+2} - \dfrac{x+2}{x+3}$

40. $\dfrac{1}{2x^2} + \dfrac{1}{3x(x-1)}$

41. $\dfrac{m+2}{m-2} - \dfrac{m^2 + 4}{m^2 - 4}$

42. $\dfrac{1}{x^2 - y^2} - \dfrac{1}{x^2 - 2xy + y^2}$

43. $\dfrac{x - \dfrac{1}{x}}{1 - \dfrac{1}{x^2}}$

44. $\dfrac{\dfrac{x}{y} - \dfrac{y}{x}}{\dfrac{x}{y} + 1}$

45. $\dfrac{x}{x+5} - \dfrac{x+3}{x}$

46. $\dfrac{1}{2} + \dfrac{1}{x+2} + \dfrac{1}{x-2}$

47. $\dfrac{a^2 + 5a + 4}{a^2 + 5a + 6} \div \dfrac{a^2 + a}{a^2 + 3a}$

48. $\dfrac{b^2 + b}{b^2 + 5b + 6} \cdot \dfrac{b^2 + 7b + 12}{b^2 + 4b}$

49. $x - \dfrac{x^2}{x^2 + 1}$

50. $\dfrac{3}{x^3} \cdot \dfrac{x^2}{6} \cdot \dfrac{9}{x} \cdot \dfrac{1}{12}$

51. $\dfrac{\dfrac{3x^2 + 6x}{x^2 - 9}}{3x^2}$

52. $\dfrac{\dfrac{x^2 - x}{x^3 + x^2}}{x^2 + 2x + 1}$

In Problems 53–58, solve the equation:

53. $\dfrac{3x}{x - 2} - 4 = \dfrac{14 - 4x}{x - 2}$

54. $\dfrac{2x}{x - 1} - 3 = \dfrac{7 - 3x}{x - 1}$

55. $\dfrac{1}{x} + \dfrac{1}{x + 1} + \dfrac{1}{3x} = \dfrac{3}{x + 1}$ **56.** $\dfrac{x + 1}{x} = \dfrac{x}{x + 1}$

57. $\dfrac{5}{2x + 3} - 5 = \dfrac{-5x}{2x + 3}$ **58.** $\dfrac{3}{x} - \dfrac{2}{x + 1} = \dfrac{1}{2x}$

59. Solve the formula $S = \dfrac{n(a + L)}{2}$ for L.

60. Solve the formula $M = xA + A$ for A.

61. Solve the formula for the surface area of a rectangular solid, $S = 2ab + 2bc + 2ac$, for a.

62. A softball player has 11 hits in 38 times at bat. How many consecutive hits will raise her batting average, hits/at bats, to .400?

63. If $\frac{1}{2}$ is added to the reciprocal of a number, the sum is 2. Find the number by setting up an equation and solving.

C *Reduce to lowest terms:*

64. $\dfrac{x^3 + 8x^2 + 15x}{5x^2 + 15x}$ **65.** $\dfrac{a^3 - 8}{a^2 - 7a + 10}$

66. $\dfrac{p^3 + 3p^2 + 2p}{p^3 - p^2 - 2p}$

Perform the indicated operations and reduce to lowest terms:

67. $\dfrac{y^2 - y - 6}{y^2 + 4y + 4} \div \dfrac{3 - y}{2 + y}$ **68.** $\dfrac{2x + 4}{2x - y} + \dfrac{2x - y}{y - 2x}$

69. $\dfrac{\dfrac{3}{x - 1} - 3}{\dfrac{2}{x - 1} + 2}$ **70.** $2 - \dfrac{2}{2 - \dfrac{2}{x}}$

71. $\dfrac{1 + \dfrac{1}{x}}{\dfrac{1}{x}} - \dfrac{1}{\dfrac{1}{x}}$

In Problems 72–74, solve the equation:

72. $5 - \dfrac{2x}{3 - x} = \dfrac{6}{x - 3}$

73. $y = \dfrac{3x + 1}{2x - 3}$ for x in terms of y

74. $1 = \dfrac{1}{a} + \dfrac{1}{b}$ for a.

75. A basketball team on a winning streak has a record of 34 wins and 16 losses. When the winning streak began, the team had won 60% of the games played to that point. How many consecutive games has the team now won?

CHAPTER PRACTICE TEST

The following practice test is provided for you to test your knowledge of the material in this chapter. You should try to complete it in 50 minutes or less. The answers in the back of the book indicate the section in the text that covers the material in the question. Actual tests in your class may vary from this practice test in difficulty, length, or emphasis, depending on the goals of your course or instructor.

Reduce to lowest terms:

1. $\dfrac{18x^3}{30x}$ **2.** $\dfrac{4x - 10}{6x - 15}$

3. $\dfrac{x^2 - 4}{x - 2}$ **4.** $\dfrac{x + 6}{x^2 + 5x - 6}$

5. $\dfrac{x^2 + 4x + 3}{x^2 + 2x - 3}$ **6.** $\dfrac{2x^2 - 4xy + 2y^2}{x^2 - y^2}$

Perform the indicated operations and reduce to lowest terms:

7. $\dfrac{1}{x} + \dfrac{1}{x^2}$ **8.** $\dfrac{3}{x + 1} - \dfrac{3}{(x + 1)^2}$

9. $\dfrac{x}{x + 1} + \dfrac{1 - x^2}{x}$ **10.** $\dfrac{x^3}{x^2 - 1} \div \dfrac{x}{x + 1}$

11. $\dfrac{x^2 + 2x + 1}{x - 1} \cdot \dfrac{x^2 - 2x + 1}{x + 1}$

12. $\dfrac{\dfrac{3x^2}{5y^2}}{\dfrac{9x^3}{20y^4}}$

13. $\dfrac{x}{1 - \dfrac{x}{x+1}}$

In Problems 14–17, solve the equation:

14. $\dfrac{3}{8} + \dfrac{5}{x} = \dfrac{3x+1}{2x}$

15. $\dfrac{12}{17} = \dfrac{x-5}{x+5}$

16. $\dfrac{1}{2x} + \dfrac{2}{3x} + \dfrac{5}{6x} = \dfrac{2}{x}$

17. $\dfrac{3}{x+4} - 5 = \dfrac{x+7}{x+4}$

18. Solve $y = 2 + \dfrac{2}{x}$ for x.

19. An archer has hit 12 bull's-eyes in her first 15 practice shots. How many consecutive bull's-eyes will raise her percentage of bull's-eyes to 90%?

20. In a large company, the ratio of female to male executives is $3:7$. If the company has 840 executives employed, how many additional female executives are needed to raise the ratio to $5:6$?

CUMULATIVE REVIEW EXERCISE—CHAPTERS 1–7

This set of exercises reviews the major concepts and techniques of Chapters 1–7. Work through all the problems, and check the answers in the back of the book. Answers to all review problems are there, and following each answer is a number in italics indicating the section in which that type of problem is discussed. Where weaknesses show up, review appropriate sections in the text.

A *Find, evaluate, or calculate the quantity requested:*

1. The least common multiple of 20, 45, and 84

2. The least common multiple of x, $3x - 6$, and $x^2 - 4$

3. $(-3)(4) - (-6) \div 2 + |-8|$

4. The larger of $-7/11$ and $-5/8$

5. The sum of $3x^2 - 2x + 1$ and $2x^2 + 3x + 4$

6. The difference $(5x^3 - 2x + 4) - (-x^3 + 6x^2 - 4)$

7. The product of $2x^2 + 7x - 4$ and $x + 3$

8. The quotient of $x^4 - 2x^3 + 2x - 1$ divided by $x + 1$

9. The product $(2x - 3y)(4x + 5y)$

10. The sum $\dfrac{x}{4} + \dfrac{x+2}{3}$

11. The difference $\dfrac{x^2}{2} - \dfrac{x-4}{5}$

12. The product $\dfrac{6x^2y}{5} \cdot \dfrac{10x}{3y}$

13. The quotient $\dfrac{ab}{2c^2} \div \dfrac{2a^2b}{c}$

14. The quotient $\dfrac{\dfrac{6}{x^2}}{\dfrac{3}{2x}}$

15. The slope and y intercept of the line $y = 3x - 5$

16. The equation of the line having slope -2 and passing through the point $(4, 1)$

17. A number such that 70% of the number is 84

Rewrite the given expression or quantity in the form requested:

18. 280 as a product of primes

19. $6x^2y - 5xy^2 + 4xy - 3xy^2 + 2x^2y - xy$ with like terms combined

20. $4a^2b - 8ab + 12ab^2$ with all common factors factored out

21. $x^2 + 6x + 9$ in completely factored form

22. $\dfrac{3}{5x}$ in higher terms, with denominator $15xy^2$

23. $\dfrac{12x(x+2)}{3x^2 - 12}$ in lowest terms

24. $2x^2 - 18$ in completely factored form

25. $5x^2 + 50$ in completely factored form

26. $x^2 - 9x + 20$ in completely factored form

27. $2x^2 + 10xy - xy - 5y^2$ in completely factored form

28. $\dfrac{x^2 + 2x + 1}{x^2 - 1}$ in lowest terms

Solve the equation, inequality, or system:

29. $7 - 3x = 6 - 2(4 - 5x)$

30. $\dfrac{3}{5}x - 5 = \dfrac{6-x}{5}$

31. $\dfrac{3}{5x} + 5 = \dfrac{4(x-10)}{5x}$

32. $x^2 - 3x + 2 = 0$

33. $\dfrac{1}{x} + \dfrac{1}{3x} + \dfrac{1}{5x} = \dfrac{12x - 1}{15x}$

34. $x + 2y = 13$
$x - 3y = -2$

35. $3x + 5y = 1$
$2x + \ \ y = 10$

36. $-5 < 4 - x \le 3$

37. $y = 3x + 10$; solve for x

38. $a = \dfrac{2}{b} - \dfrac{c}{2}$; solve for b

Solve graphically:

39. $0 \le y \le 5$
$y \ge x - 2$

40. $x - \ \ y = 4$
$2x + 3y = 33$

Graph:

41. The line having equation $y = -2x + 3$

42. The line through $(-3, -4)$ having slope 2

43. $y > x - 3$

Set up the appropriate equation, or equations, and solve:

44. At a college bookstore the ratio of new textbooks purchased to used texts purchased is $5:7$. If there were 15,204 books purchased during the year, how many of these were new books?

45. Insulin comes in a solution that contains 100 units of insulin per milliliter. How many milliliters must be administered to a patient if he is to receive 12 units of insulin?

46. A schoolteacher receives raises of 6.2%, 5.4%, and 10% in 3 successive years. Her salary after the last raise is $27,827. What was her salary, to the nearest dollar, before the first raise?

47. A solution of 90% alcohol is to be diluted with pure water to obtain 1,000 milliliters of a 36% alcohol solution. How much water should be used?

48. On a long stretch of straight tracks, two trains pass each other going in opposite directions. One is traveling at 48 miles per hour and the other at 62 miles per hour. How long will it be before the trains are 55 miles apart?

49. A shipment of 180 boxes of oranges and grapefruit from a Florida grower weighs 8,500 pounds. Each box of oranges weighs 50 pounds, and each box of grapefruit weighs 45 pounds. How many boxes of each are in the shipment?

B *Find, evaluate, or calculate the quantity requested:*

50. The quotient of $x^5 + x^3 - 3x^2 - 2x + 3$ divided by $x^2 - 1$

51. The sum $\dfrac{3}{x - 1} + \dfrac{3}{x + 1}$

52. The value of $-3x^2 + |y - 5| - x \div y^2$ for $x = 8$ and $y = 2$

53. The least common denominator for $\dfrac{1}{x}, \dfrac{1}{x + 3}$, and $\dfrac{x - 3}{x^2}$

54. The equation of the line through the points $(2, -5)$ and $(-4, 3)$

55. The quotient $\dfrac{x^2 + 2x + 1}{x^2 - x - 2} \div \dfrac{x + 1}{x^2 - 4}$ reduced to lowest terms

56. The product $\dfrac{x^3 - x}{x + 1} \cdot \dfrac{1}{x^2 - x}$ reduced to lowest terms

57. The complex fraction $\dfrac{\dfrac{x}{x - 5}}{\dfrac{x^2}{x^2 - 4x - 5}}$ written as a simple fraction in lowest terms

Translate the expression or statement into an algebraic expression, equality, or inequality with x as the variable. The reciprocal of x is 1/x.

58. 6 more than the reciprocal of a number

59. The reciprocal of 6 more than a number

60. Eighty more than a number is more than 8 times the number.

61. Three less than a number is at least 3 times the number.

62. The sum of the reciprocals of three consecutive integers beginning with x

63. The sum of 5 times a number and 4 times its reciprocal is the quotient of 11 divided by the product of 16 and the number.

Rewrite the given expression or quantity in the form requested:

64. $2x^2 + 7x - 15$ in completely factored form

65. $4x^2 - 15x - 4$ in completely factored form

66. $x^3 - 64$ in completely factored form

67. $x^4 - 1$ in completely factored form

68. $3x^3 + 24$ in completely factored form

69. $\{1 - [1 - (1 - a)]\} \cdot \{a - [a - (a - 1)]\}$ without grouping symbols, multiplied out, and with like terms combined

70. $11x + 7y = 30$ in the form $y = mx + b$

71. $\dfrac{\dfrac{x}{1-x}}{1 - \dfrac{1}{1-x}}$ as a simple fraction in reduced form

72. $3xy - 4yz + 4wz - 3xw$ in completely factored form

Solve the equation, inequality, or system:

73. $3x - 5y = 6$
$\quad\; 4x - 2y = 6$

74. $x - 5 \le \dfrac{x}{5} - 1$

75. $\dfrac{2}{x-3} + 4 = \dfrac{2x}{x-3}$

76. $\dfrac{2}{x+3} + 4 = \dfrac{-10x}{x+3}$

77. $xy = x + y$; solve for x

78. $\dfrac{y}{x^2} - \dfrac{1}{x} = \dfrac{z}{x}$; solve for x

Solve graphically:

79. $6x + 4y \ge 12$
$\qquad x \ge 1$
$\qquad y \ge 1$

80. $6x + 4y \le 12$
$\qquad 4x + 6y \le 12$
$\qquad x \ge 0$
$\qquad y \ge 0$

Set up the appropriate equation, or equations, and solve:

81. The apothecary system of measurement is still used in some medical situations. The basic unit for measuring weight is the *grain*. A grain is equivalent to 60 milligrams in the metric system. How many grains are in 500 milligrams?

82. A company makes two models of wooden outdoor benches. Each bench of model A requires 18 board feet of lumber, and each of model B requires 20 board feet. Each bench of model A requires 2 hours of labor, while each of model B requires only 1.5. If the company used 2,100 board feet of lumber and 190 hours of labor producing benches, how many of each kind were produced?

83. Two workers for a political candidate are to address envelopes for a mailing. One worker can address 70 envelopes per hour. The other can address 80 per hour but starts the task 15 minutes after his colleague. How long will it be before the two have addressed the same number of envelopes?

84. The restaurant reviewer for the local paper is required to leave a tip of 15% for meals eaten in the course of her job. Tips are to be figured on the cost

of the meal, excluding the 6% tax charged. If the total cost of meal, tip, and tax is $22.99, what was the tip?

85. The average number y of rooms that a hotel can rent per night at a rate of x dollars is given by the equation $y = -2x + 240$. If the hotel is averaging 150 rooms rented per night, what rate is it charging?

C *Rewrite the given expression or quantity in the form requested:*

86. $3x^2 + 10x - 8$ in completely factored form

87. $2x^2 + 5x - 12$ in completely factored form

88. $x^2 + 2x + 3$ in completely factored form

89. $x^6 - 1$ in completely factored form

90. $x^3 - xy^2 + 2x^2y - 2y^3$ in completely factored form

91. $\dfrac{1 - \dfrac{1}{1 - \dfrac{1}{1-x}}}{x - \dfrac{x}{1 - \dfrac{1}{1+x}}}$ as a simple fraction in reduced form

Solve:

92. $x^2 + 2x + 1 = 4$

93. $3x^2 + 5x - 12 = 0$

94. $\dfrac{2}{x-2} + \dfrac{3}{x-1} = \dfrac{5x-8}{x^2-3x+2}$

95. $\dfrac{2}{x-2} + \dfrac{3}{x-1} = \dfrac{3x-6}{x^2-3x+2}$

96. $\dfrac{1}{x-1} + \dfrac{1}{x+1} = \dfrac{y}{x^2-1}$; solve for x

97. $\dfrac{2}{x-2} + \dfrac{3}{x-1} = \dfrac{x-4}{x^2-3x+2}$

Set up the appropriate equation, or equations, and solve:

98. An investor has $12,000 invested in two bonds. One bond pays 9% interest annually, and the other pays 11%. If the investor receives $1,152 in interest each year, how much is invested in each bond?

99. The average number y of rooms a hotel can rent per night at a rate of x dollars is given by an equation of the form $y = mx + b$. If the hotel can rent an average of 100 rooms when the rate is $70 and 120 rooms when the rate is $60, find m and b.

100. A 2-liter quantity of 60% alcohol solution is to be strengthened to an 85% solution by replacing some of it with pure alcohol. How much must be replaced?

8

Exponents and Radicals

Natural number exponents were introduced in Chapter 1. In this chapter, the exponent concept is extended to integers, and the rules for manipulating expressions involving exponents are expanded.

Square roots and the radical notation $\sqrt{}$ were reviewed in Section 4-1. This chapter deals with square roots in more detail. Sums, differences, products, and quotients that involve radicals are considered, as are equations that involve square roots.

Photo reference: see Exercise 8-3, Problem 56.

8-1   Natural Number Exponents

♦   Five Exponent Laws
♦   Summary and Use of the Five Laws

Earlier we defined a number raised to a natural number power. Recall:

Natural Number Exponent

For a natural number n,

$$a^n = a \cdot a \cdot \ \cdots \ a \qquad n \text{ factors of } a \qquad (1)$$
$$2^5 = 2 \cdot 2 \cdot 2 \cdot 2 \cdot 2 = 32 \qquad 5 \text{ factors of } 2$$

We then introduced the first law of exponents: If m and n are positive integers and a is a real number, then $a^m a^n = a^{m+n}$. By now you have used this law over and over again, in particular in multiplying polynomials. More complicated algebraic forms involving exponents are frequently encountered. Four other exponent laws combined with the first law provide efficient tools for simplifying and changing these forms. In this section we will review the first law and discuss the four additional laws.

♦   **FIVE EXPONENT LAWS**

In the following discussion, m and n are natural numbers and a and b are real numbers, excluding division by 0, of course.

From the definition of a natural number exponent in Equation (1)

$$a^3 a^4 = \underbrace{(a \cdot a \cdot a)}_{\substack{3 \\ \text{factors}}}\underbrace{(a \cdot a \cdot a \cdot a)}_{\substack{4 \\ \text{factors}}}$$
$$\underbrace{}_{\substack{3+4 \\ \text{factors}}}$$

$$= (a \cdot a \cdot a \cdot a \cdot a \cdot a \cdot a) = a^{3+4} = a^7$$

In general:

Law 1

$$a^m a^n = a^{m+n} \qquad 2^3 \cdot 2^2 = \underbrace{2 \cdot 2 \cdot 2 \cdot 2 \cdot 2}_{\substack{2^3 \quad\ \ 2^2}} = 2^5$$
$$a^5 a^2 = a^{5+2} = a^7$$

Example 1 **Using Law 1**	Rewrite, using law 1: $x^7 x^4$.

Solution $x^7 x^4 \boxed{= x^{7+4}} = x^{11}$

Matched Problem 1   Rewrite, using law 1:

(A) $a^6 a^3$ **(B)** $x^{10} x^8$

Again, from the definition of exponent in Equation (1)

$$(a^3)^4 = a^3 \cdot a^3 \cdot a^3 \cdot a^3$$

$$\overset{\text{4 groups of 3 factors each}}{= (a \cdot a \cdot a)(a \cdot a \cdot a)(a \cdot a \cdot a)(a \cdot a \cdot a)}$$

$$\overset{4 \cdot 3 \text{ factors}}{= (a \cdot a \cdot a \cdot a \cdot a \cdot a \cdot a \cdot a \cdot a \cdot a \cdot a \cdot a)}$$

$$= a^{4 \cdot 3} = a^{12}$$

This generalizes to:

Law 2

$$(a^n)^m = a^{mn} \qquad (2^3)^2 \boxed{= (2 \cdot 2 \cdot 2)(2 \cdot 2 \cdot 2)} = 2^6$$

$$(a^5)^3 = a^{3 \cdot 5} = a^{15}$$

Example 2 **Using Law 2**	Rewrite, using law 2: $(x^4)^7$.

Solution $(x^4)^7 \boxed{= x^{7 \cdot 4}} = x^{28}$

Matched Problem 2   Rewrite, using law 2:

(A) $(y^2)^5$ **(B)** $(u^4)^6$

Using the definition of exponent in Equation (1) and the commutative property for multiplication:

$$(ab)^4 = \overset{\overset{\text{4}}{\overbrace{\text{factors of } (ab)}}}{(ab)(ab)(ab)(ab)} = \overset{\overset{\text{4}}{\overbrace{\text{factors of } a}}}{(a \cdot a \cdot a \cdot a)} \overset{\overset{\text{4}}{\overbrace{\text{factors of } b}}}{(b \cdot b \cdot b \cdot b)}$$

$$= a^4 b^4$$

In general, it can be shown that:

Law 3

$$(ab)^m = a^m b^m$$ $(2 \cdot 3)^3 = (2 \cdot 3)(2 \cdot 3)(2 \cdot 3) = 2^3 \cdot 3^3$

$(ab)^3 = a^3 b^3$

Example 3
Using Law 3

Rewrite, using law 3:

(A) $(xy)^4$ **(B)** $(2x^2y^3)^3$

Solution   **(A)** $(xy)^4 = x^4 y^4$

(B) $(2x^2y^3)^3 = (2^1 x^2 y^3)^3 = 2^{3 \cdot 1} x^{3 \cdot 2} y^{3 \cdot 3} = 2^3 x^6 y^9 = 8x^6 y^9$

Matched Problem 3

Rewrite, using law 3:

(A) $(uv)^5$ **(B)** $(3a^3 b^4)^4$

Using the definition of exponent in Equation (1) still one more time,

$$\left(\frac{a}{b}\right)^5 = \overbrace{\frac{a}{b} \cdot \frac{a}{b} \cdot \frac{a}{b} \cdot \frac{a}{b} \cdot \frac{a}{b}}^{\substack{5 \\ \text{factors of } a/b}} \cdot = \frac{\overbrace{a \cdot a \cdot a \cdot a \cdot a}^{5 \text{ factors of } a}}{\underbrace{b \cdot b \cdot b \cdot b \cdot b}_{5 \text{ factors of } b}} = \frac{a^5}{b^5}$$

In general:

Law 4

$$\left(\frac{a}{b}\right)^m = \frac{a^m}{b^m}$$ $\left(\frac{2}{3}\right)^2 = \frac{2}{3} \cdot \frac{2}{3} = \frac{2 \cdot 2}{3 \cdot 3} = \frac{2^2}{3^2}$

$\left(\frac{a}{b}\right)^3 = \frac{a^3}{b^3}$

Example 4
Using Law 4

Rewrite, using law 4:

(A) $\left(\dfrac{x}{y}\right)^5$ **(B)** $\left(\dfrac{2u^2}{v^3}\right)^3$

Solution   **(A)** $\left(\dfrac{x}{y}\right)^5 = \dfrac{x^5}{y^5}$

(B) $\left(\dfrac{2u^2}{v^3}\right)^3 = \left(\dfrac{2^1 u^2}{v^3}\right)^3 = \dfrac{2^{3\cdot 1} u^{3\cdot 2}}{v^{3\cdot 3}} = \dfrac{2^3 u^6}{v^9} = \dfrac{8u^6}{v^9}$

Matched Problem 4   Rewrite, using law 4:

(A) $\left(\dfrac{w}{z}\right)^7$ **(B)** $\left(\dfrac{3x^3}{y^2}\right)^4$

We now look at the quotient form a^m/a^n. Consider the following three special cases:

1. $\dfrac{a^7}{a^3} = \dfrac{a \cdot a \cdot a \cdot a \cdot a \cdot a \cdot a}{a \cdot a \cdot a}$

$= \dfrac{(a \cdot a \cdot a)(a \cdot a \cdot a \cdot a)}{(a \cdot a \cdot a)} = a^{7-3} = a^4$

2. $\dfrac{a^3}{a^3} = \dfrac{a \cdot a \cdot a}{a \cdot a \cdot a} = 1$

3. $\dfrac{a^4}{a^7} = \dfrac{a \cdot a \cdot a \cdot a}{a \cdot a \cdot a \cdot a \cdot a \cdot a \cdot a}$

$= \dfrac{(a \cdot a \cdot a \cdot a)}{(a \cdot a \cdot a \cdot a)(a \cdot a \cdot a)} = \dfrac{1}{a^{7-4}} = \dfrac{1}{a^3}$

In general, it can be shown that:

Law 5

$$\dfrac{a^m}{a^n} = \begin{cases} a^{m-n} & \text{if } m > n \\[2mm] 1 & \text{if } m = n \\[2mm] \dfrac{1}{a^{n-m}} & \text{if } n > m \end{cases}$$

$\dfrac{2^3}{2^2} = \dfrac{2 \cdot 2 \cdot 2}{2 \cdot 2} = 2^{3-2} = 2$

$\dfrac{2^2}{2^2} = 1$

$\dfrac{2^2}{2^3} = \dfrac{2 \cdot 2}{2 \cdot 2 \cdot 2} = \dfrac{1}{2^{3-2}} = \dfrac{1}{2}$

$\dfrac{x^8}{x^3} = x^{8-3} = x^5 \qquad \dfrac{x^8}{x^8} = 1 \qquad \dfrac{x^3}{x^8} = \dfrac{1}{x^5}$

Example 5
Using Law 5

Rewrite, using law 5:

(A) $\dfrac{u^5}{u}$ **(B)** $\dfrac{d^3}{d^3}$ **(C)** $\dfrac{y^4}{y^6}$

Solution   **(A)** $\dfrac{u^5}{u} = u^{5-1} = u^4$ **(B)** $\dfrac{d^3}{d^3} = 1$ **(C)** $\dfrac{y^4}{y^6} = \dfrac{1}{y^{6-4}} = \dfrac{1}{y^2}$

Matched Problem 5   Rewrite, using law 5:

(A) $\dfrac{x^7}{x^2}$ **(B)** $\dfrac{c^4}{c^4}$ **(C)** $\dfrac{w}{w^4}$

♦ **SUMMARY AND USE OF THE FIVE LAWS**

We have given plausible arguments for each of the five exponent laws. Formal proofs of these laws are beyond the scope of this course.

It is very important to observe and remember:

CAUTION **The laws of exponents involve products and quotients, not sums and differences.**

Many mistakes are made in algebra by people applying a law of exponents to sums or differences. For example, $(ab)^2 = a^2b^2$, but $(a + b)^2 \neq a^2 + b^2$. Rather, $(a + b)^2 = a^2 + 2ab + b^2$. The exponent laws are summarized here for convenient reference.

Laws of Exponents

1. $a^m a^n = a^{m+n}$
2. $(a^n)^m = a^{mn}$
3. $(ab)^m = a^m b^m$
4. $\left(\dfrac{a}{b}\right)^m = \dfrac{a^m}{b^m}, \ b \neq 0$

5. $\dfrac{a^m}{a^n} = \begin{cases} a^{m-n} & \text{if } m > n \\[2mm] 1 & \text{if } m = n \\[2mm] \dfrac{1}{a^{n-m}} & \text{if } n > m, \ a \neq 0 \end{cases}$

Example 6
Using the Laws of Exponents

Rewrite, using the laws of exponents:

(A) $x^{11}x^7$ (B) $(y^{10})^3$ (C) $(xy)^8$

(D) $\left(\dfrac{x}{y}\right)^8$ (E) $\dfrac{y^3}{y^9}$ (F) $\dfrac{y^{12}}{y^4}$

Solution (A) $x^{11}x^7 \boxed{= x^{11+7}} = x^{18}$ (B) $(y^{10})^3 \boxed{= y^{10 \cdot 3}} = y^{30}$

(C) $(xy)^8 = x^8 y^8$ (D) $\left(\dfrac{x}{y}\right)^8 = \dfrac{x^8}{y^8}$

(E) $\dfrac{y^3}{y^9} \boxed{= \dfrac{1}{y^{9-3}}} = \dfrac{1}{y^6}$ (F) $\dfrac{y^{12}}{y^4} \boxed{= y^{12-4}} = y^8$

Matched Problem 6

Rewrite, using the laws of exponents:

(A) $y^6 y^{13}$ (B) $(y^4)^{10}$ (C) $(xy)^{10}$

(D) $\left(\dfrac{x}{y}\right)^{10}$ (E) $\dfrac{x^{10}}{x^5}$ (F) $\dfrac{y^5}{y^{15}}$

Example 7
Using the Laws of Exponents

Rewrite, using the laws of exponents:

(A) $(x^3y^2)^5$ **(B)** $\left(\dfrac{x^3}{y^2}\right)^4$ **(C)** $\dfrac{3x^3y^2}{6x^4y}$

Solution   **(A)** $(x^3y^2)^5 \boxed{= (x^3)^5(y^2)^5 = x^{3\cdot5}y^{2\cdot5}} = x^{15}y^{10}$

(B) $\left(\dfrac{x^3}{y^2}\right)^4 \boxed{= \dfrac{(x^3)^4}{(y^2)^4} = \dfrac{x^{3\cdot4}}{y^{2\cdot4}} = \dfrac{x^{12}}{y^8}}$

(C) $\dfrac{3x^3y^2}{6x^4y} \boxed{= \dfrac{3}{6}\cdot\dfrac{x^3}{x^4}\cdot\dfrac{y^2}{y} = \dfrac{1}{2}\cdot\dfrac{1}{x}\cdot y} = \dfrac{y}{2x}$

Matched Problem 7   Rewrite, using the laws of exponents:

(A) $(w^7z^4)^3$ **(B)** $\left(\dfrac{w^7}{z^4}\right)^5$ **(C)** $\dfrac{8w^3z^5}{2w^6z^3}$

Knowing the rules of the game of chess doesn't make you a good chess player; similarly, memorizing the laws of exponents doesn't necessarily make you good at using them. To acquire skill in their use, you must use these laws in a fairly large variety of problems. Exercise 8-1 should help you acquire this skill.

Answers to Matched Problems

1. (A) a^9 **(B)** x^{18} **2. (A)** y^{10} **(B)** u^{24}

3. (A) u^5v^5 **(B)** $81a^{12}b^{16}$ **4. (A)** $\dfrac{w^7}{z^7}$ **(B)** $\dfrac{81x^{12}}{y^8}$

5. (A) x^5 **(B)** 1 **(C)** $\dfrac{1}{w^3}$

6. (A) y^{19} **(B)** y^{40} **(C)** $x^{10}y^{10}$ **(D)** $\dfrac{x^{10}}{y^{10}}$ **(E)** x^5 **(F)** $\dfrac{1}{y^{10}}$

7. (A) $w^{21}z^{12}$ **(B)** $\dfrac{w^{35}}{z^{20}}$ **(C)** $\dfrac{4z^2}{w^3}$

EXERCISE 8-1

Rewrite, using the laws of exponents:

A 1. x^5x^8 **2.** $y^{11}y^6$ **3.** y^9y^6 **4.** $x^{12}x^3$

5. x^5x **6.** x^8x **7.** $(x^4)^4$ **8.** $(x^6)^6$

9. $(x^3)^5$ **10.** $(y^4)^6$ **11.** $(y^5)^3$ **12.** $(x^4)^5$

13. $(xy)^8$ **14.** $(xy)^9$ **15.** x^7y^7 **16.** x^5y^5

17. x^4y^4 **18.** x^6y^6 **19.** $(xy)^5$ **20.** $(xy)^7$

21. $\dfrac{x^4}{y^4}$ **22.** $\dfrac{x^6}{y^6}$ **23.** $\left(\dfrac{x}{y}\right)^2$ **24.** $\left(\dfrac{x}{y}\right)^4$

25. $\dfrac{x^{10}}{x^4}$ **26.** $\dfrac{y^4}{y^{11}}$ **27.** $\dfrac{x^5}{x^{13}}$ **28.** $\dfrac{y^9}{y^2}$

29. $\dfrac{y^{10}}{y^{14}}$ **30.** $\dfrac{x^{14}}{x^7}$ **31.** $\dfrac{y^9}{y^3}$ **32.** $\dfrac{x^{16}}{x^{24}}$

33. $10^{13} \times 10^2$ **34.** $10^{11} \times 10^4$

35. $(10^3)^4$ **36.** $(10^5)^3$

37. $(2 \times 10^3)(3 \times 10^8)$ **38.** $(4 \times 10^5)(2 \times 10^5)$

39. $(4 \times 10^6) \div (2 \times 10^3)$ **40.** $(8 \times 10^5) \div (4 \times 10^2)$

41. $(6 \times 10^8) \div (4 \times 10^2)$ **42.** $(8 \times 10^6) \div (6 \times 10^3)$

43. $(-1)^5$ **44.** $(-1)^4$ **45.** $(-1)^6$

46. $(-1)^7$ **47.** $(-2)^3$ **48.** $(-3)^3$

49. $(-3)^2$ **50.** $(-2)^2$ **51.** $(x^5y^2)^4$

B 52. $(2xy^2)^5$ **53.** $(3a^2b)^3$ **54.** $(a^2b^3)^2$

55. $(2x^4)(3x^3)$

56. $\dfrac{6x^4}{4x^3}$

57. $\dfrac{3a^4}{6a^8}$

58. $(3a^2)(4a^3)$

59. $\dfrac{12a^4b^2}{4ab^3}$

60. $(x^2y)^3(xy^3)^2$

61. $(ab^4)^2(a^2b)^3$

62. $\dfrac{2x^3y^2}{6xy^3}$

63. $5(xy^3)^4$

64. $(3x^2y)^2$

65. $(5xy^3)^4$

66. $3(x^2y)^2$

67. $(-xy^2)^2$

68. $(-x^2y)^3$

69. $-(xy^2)^2$

70. $-(x^2y)^3$

71. -2^4

72. -3^3

73. $(-3)^3$

74. $(-2)^4$

75. $-x^4$

76. $(-x)^4$

77. $(-x)^5$

78. $-x^5$

C **79.** $(2xy^3z^4)^4$

80. $(3a^4b^2c)^3$

81. $(3a^4)(2a^5)(a^6)$

82. $(2x^3)(x^7)(3x^{11})$

83. $(3a)^4(2a^5)^2a^6$

84. $(2x)^3x^7(3x^8)^2$

85. $\dfrac{(2a^2b^3)^3}{(4ab)^2}$

86. $\dfrac{(3x^2y)^4}{(3xy^2)^3}$

87. $(-ab^2)^4(-a^2b)$

88. $(-x^2y)^3(-xy^3)^2$

89. $\dfrac{(a^5b^2)^2}{(-a^4b^3)^3}$

90. $\dfrac{(-a^4b)^2}{(a^2b^3)^3}$

91. $\dfrac{(-a^4b)^3(ab^5)}{-a^8b^8}$

92. $\dfrac{(x^2y)(x^4y^2)}{-x^4y^5}$

8-2 Integer Exponents

- ◆ Zero Exponents
- ◆ Negative Exponents
- ◆ Summary

The fifth law of exponents suggests the following pattern:

$$\frac{x^3}{x^1} = x^{3-1} = x^2$$

$$\frac{x^3}{x^2} = x^{3-2} = x^1$$

$$1 = \frac{x^3}{x^3} = x^{3-3} = x^0$$

$$\frac{1}{x} = \frac{x^3}{x^4} = x^{3-4} = x^{-1}$$

$$\frac{1}{x^2} = \frac{x^3}{x^5} = x^{3-5} = x^{-2}$$

The pattern suggests that x^0 should be defined to be 1 and that x^{-n} should be defined to be $1/x^n$. We will extend the exponent concept to 0 and negative integers in this manner. All five exponent laws will still hold when exponents are extended in this way to include zero and negative integers.

◆ ZERO EXPONENTS

We will define a^0 to be 1 for any $a \neq 0$. If the laws of exponents are to hold, then this must be true. For example,

$$a^0a^3 = a^{0+3} = a^3 = 1 \cdot a^3$$

so a^0 must be 1. The expression 0^0 is not defined. Any possible definition of 0^0 will lead to difficulties with the exponent laws.

Definition of Zero Exponent

For all real numbers $a \neq 0$,

$$a^0 = 1 \qquad 12^0 = 1$$

0^0 is not defined.

Thus, all the following are equal to 1:

$$7^0 = 1 \qquad 1{,}238^0 = 1 \qquad \pi^0 = 1 \qquad z^0 = 1 \text{ (for } z \neq 0)$$

$$(a^2 b^7 c^{11})^0 = 1 \text{ (for } a, b, c \neq 0)$$

In Section 6-1, we defined the degree of a nonzero constant term a in a polynomial to be 0. If you think of a as ax^0, you can now see why this is done.

♦ NEGATIVE EXPONENTS

We will define a^{-n} to be $1/a^n$ for $a \neq 0$ and n a natural number. Again, this definition is forced upon us if the exponent laws are to hold. For example, if $a \neq 0$, then

$$a^{-3} \cdot a^3 = a^{-3+3} = a^0 = 1$$

so by the definition of division, a^{-3} must be $\dfrac{1}{a^3}$.

Definition of Negative-Integer Exponents

If n is a positive integer and a is a nonzero real number, then

$$a^{-n} = \frac{1}{a^n} \qquad a^{-2} = \frac{1}{a^2}$$

Notice that, since

$$\frac{1}{a^{-3}} = \frac{1}{\dfrac{1}{a^3}} = 1 \cdot \frac{a^3}{1} = a^3$$

both of the following statements are true:

$$a^{-3} = \frac{1}{a^3} \qquad \text{and} \qquad a^3 = \frac{1}{a^{-3}}$$

In other words, we can move a cube back and forth between numerator and denominator simply by changing the sign of the exponent. More generally we have

For $a \neq 0$ and an integer n,

$$a^{-n} = \frac{1}{a^n} \quad \text{and} \quad a^n = \frac{1}{a^{-n}}$$

Thus we can shift any integer power back and forth between numerator and denominator by changing the sign of the exponent. This is true, however, only for *factors* in the numerator and denominator, not for *terms*. Thus, for example,

$$\frac{1}{2^{-3} + 5} \neq \frac{2^3}{5}$$

Example 1
Rewriting Exponent Forms

Rewrite, using positive exponents only:

(A) x^{-4} **(B)** $\dfrac{1}{x^{-4}}$ **(C)** 10^{-4} **(D)** $\dfrac{x^{-4}}{y^{-8}}$

Solution   **(A)** $x^{-4} = \dfrac{1}{x^4}$ **(B)** $\dfrac{1}{x^{-4}} = x^4$

(C) $10^{-4} = \dfrac{1}{10^4} = 0.0001$ **(D)** $\dfrac{x^{-4}}{y^{-8}} = x^{-4} \cdot \dfrac{1}{y^{-8}} = \dfrac{1}{x^4} \cdot y^8 = \dfrac{y^8}{x^4}$

Matched Problem 1

Rewrite, using positive exponents only:

(A) x^{-7} **(B)** $\dfrac{1}{x^{-7}}$ **(C)** 10^{-7} **(D)** $\dfrac{x^{-7}}{y^{-4}}$

All five laws of exponents remain true, with the definition of exponent extended to the integers. The fifth law of exponents can now be restated in a more compact form:

Law 5

For a real number $a \neq 0$; m and n integers

$$\frac{a^m}{a^n} = a^{m-n} = \frac{1}{a^{n-m}} \qquad \frac{2^3}{2^2} = 2^{3-2} = 2 \qquad \frac{2^2}{2^2} = 2^0 = 1 \qquad \frac{2^2}{2^3} = 2^{-1} = \frac{1}{2}$$

Example 2
Rewriting Exponent Forms

Rewrite, using negative integers only:

(A) $\dfrac{3^3}{3^5}$ **(B)** $\dfrac{8^{-3}}{8^4}$

Rewrite, using positive exponents only:

(C) $\dfrac{3^3}{3^5}$ **(D)** $\dfrac{8^{-3}}{8^4}$

Solution **(A)** $\dfrac{3^3}{3^5} = 3^{3-5} = 3^{-2}$

(B) $\dfrac{8^{-3}}{8^4} = 8^{-3-4} = 8^{-7}$

(C) Using part (A), $\dfrac{3^3}{3^5} = 3^{-2} = \dfrac{1}{3^2}$

(D) Using part (B), $\dfrac{8^{-3}}{8^4} = 8^{-7} = \dfrac{1}{8^7}$

Matched Problem 2 Rewrite, using positive exponents only:

(A) $\dfrac{5^2}{5^6}$ **(B)** $\dfrac{x^{-2}}{x^6}$

Rewrite, using negative exponents only:

(C) $\dfrac{5^2}{5^6}$ **(D)** $\dfrac{x^{-2}}{x^6}$

◆ SUMMARY

Tables 1 and 2 provide a summary of all our work on exponents to this point.

Table 1 **Integer Exponents—Definition of a^p (p an integer, a a real number)**

1. If p is a positive integer, then

$$a^p = a \cdot a \cdot \,\cdots\, a \qquad p \text{ factors of } a \qquad 3^4 = 3 \cdot 3 \cdot 3 \cdot 3 = 81$$

2. If $p = 0$, then

$$a^p = 1 \qquad a \neq 0 \qquad\qquad 3^0 = 1$$

3. If p is a negative integer, then

$$a^p = \dfrac{1}{a^{-p}} \qquad a \neq 0 \qquad\qquad 3^{-4} = \dfrac{1}{3^{-(-4)}} = \dfrac{1}{3^4}$$

Table 2 **Integer Exponents—Laws (n and m integers, a and b real numbers)**

1. $a^m a^n = a^{m+n}$ $2^3 2^4 = 2^7$

2. $(a^n)^m = a^{mn}$ $(2^4)^3 = 2^{12}$

3. $(ab)^m = a^m b^m$ $(2 \cdot 5)^3 = 2^3 5^3$

4. $\left(\dfrac{a}{b}\right)^m = \dfrac{a^m}{b^m}$ $b \neq 0$ $\left(\dfrac{2}{5}\right)^3 = \dfrac{2^3}{5^3}$

5. $\dfrac{a^m}{a^n} = a^{m-n} = \dfrac{1}{a^{n-m}}$ $a \neq 0$ $\dfrac{2^3}{2^4} = 2^{3-4} = \dfrac{1}{2^{4-3}}$

Rewrite, using positive exponents only:

Example 3
Rewriting
Exponent Forms

(A) $x^{-3}x^8$

(B) $(x^{-3}y^2)^{-4}$

(C) $\left(\dfrac{x^{-1}}{x^{-2}}\right)^{-3}$

(D) $\dfrac{2x^2y^{-1}}{4x^{-2}y^{-3}}$

(E) $\dfrac{10^{-5} \times 10^2}{10^3 \times 10^{-8}}$

(F) $\left(\dfrac{x^{-2}y^2}{z^{-3}}\right)^{-2}$

Solution **(A)** $x^{-3}x^8 \boxed{= x^{-3+8}} = x^5$

$\dfrac{x^{12}}{y^8}$ $\dfrac{x^3}{x^6} = \dfrac{1}{x^3}$

(B) $(x^{-3}y^2)^{-4} \boxed{= (x^{-3})^{-4}(y^2)^{-4} = x^{(-3)(-4)}y^{2(-4)}}$

$$= x^{12}y^{-8}$$

$$= x^{12} \cdot \dfrac{1}{y^8}$$

$$= \dfrac{x^{12}}{y^8}$$

(C) $\left(\dfrac{x^{-1}}{x^{-2}}\right)^{-3} \boxed{= \dfrac{(x^{-1})^{-3}}{(x^{-2})^{-3}} = \dfrac{x^{(-1)(-3)}}{x^{(-2)(-3)}}} = \dfrac{x^3}{x^6} = \dfrac{1}{x^3}$

or $\left(\dfrac{x^{-1}}{x^{-2}}\right)^{-3} \boxed{= (x^{-1-(-2)})^{-3} = (x^{-1+2})^{-3}}$

$$= (x^1)^{-3} = \dfrac{1}{x^3}$$

(D) $\dfrac{2x^2y^{-1}}{4x^{-2}y^{-3}} \boxed{= \dfrac{1}{2} \cdot \dfrac{x^2}{x^{-2}} \cdot \dfrac{y^{-1}}{y^{-3}} = \dfrac{1}{2}x^{2+(+2)}y^{-1+(+3)}}$

$$= \dfrac{1}{2}x^4y^2$$

or, changing to positive exponents first,

$$\dfrac{2x^2y^{-1}}{4x^{-2}y^{-3}} = \dfrac{1}{2} \cdot \boxed{\dfrac{x^2x^2y^3}{y} = \dfrac{1}{2} \cdot x^{2+2}y^{3-1}} = \dfrac{1}{2}x^4y^2$$

(E) $\dfrac{10^{-5} \times 10^2}{10^3 \times 10^{-8}} = \dfrac{10^{-5+2}}{10^{3-8}} = \dfrac{10^{-3}}{10^{-5}} = 10^{-3-(-5)} = 10^2 = 100$

(F) $\left(\dfrac{x^{-2}y^2}{z^{-3}}\right)^{-2} \boxed{= \dfrac{x^{(-2)(-2)}y^{2(-2)}}{z^{(-3)(-2)}}} = \dfrac{x^4y^{-4}}{z^6} = \dfrac{x^4}{y^4z^6}$

Matched Problem 3 Rewrite, using positive exponents only:

(A) $x^{-4}x^{11}$ **(B)** $(x^{-1}y^3)^{-2}$ **(C)** $\left(\dfrac{x^{-4}}{x^{-3}}\right)^{-2}$

(D) $\dfrac{3x^{-4}y^{-2}}{9xy^{-6}}$ **(E)** $\dfrac{10^8 \times 10^{-4}}{10^{-6} \times 10^2}$ **(F)** $\left(\dfrac{x^{-1}y^2}{xz^{-2}}\right)^{-1}$

Writing expressions using only positive exponents makes it easier to compare two expressions, for example, your answer and the answer in the text, and for this reason answers to the exercises in this section are requested in such form. There are many instances, however, when it will be preferable to leave negative exponents in an expression. The next section will show such a situation.

It is important to remember that the laws of exponents involve products and quotients, not sums and differences. Consider the following:

I. $(ab)^2 = a^2b^2$ but $(a + b)^2 \neq a^2 + b^2$

For example,

$$(2 \cdot 3)^2 = 6^2 = 36 = 2^2 3^2 \quad \text{but} \quad (2 + 3)^2 = 5^2 = 25 \neq 2^2 + 3^2$$

II. $(ab)^{-1} = a^{-1}b^{-1}$ but $(a + b)^{-1} \neq a^{-1} + b^{-1}$

For example,

$$(2 \cdot 3)^{-1} = \frac{1}{2 \cdot 3} = \frac{1}{6} = \frac{1}{2} \cdot \frac{1}{3} = 2^{-1}3^{-1}$$

On the other hand,

$$(2 + 3)^{-1} = 5^{-1} = \frac{1}{5} \quad \text{and} \quad 2^{-1} + 3^{-1} = \frac{1}{2} + \frac{1}{3} = \frac{5}{6} \quad \text{so} \quad (2 + 3)^{-1} \neq 2^{-1} + 3^{-1}$$

 CAUTION **In general, $(a + b)^n \neq a^n + b^n$.**

The expressions $(a + b)^n$ and $a^n + b^n$ will be equal only for very special values of a, b, or n, such as a or $b = 0$, or $n = 1$.

Example 4
Rewriting
Exponent Forms

Rewrite, using positive exponents only:

(A) $(2^{-1} + 3^{-1})^2$ $\qquad 2^1 + 3^1 = 5$

(B) $(a^{-1} + b^{-1})^2$ $\qquad a + b$

Solution **(A)** $(2^{-1} + 3^{-1})^2 = \left(\dfrac{1}{2} + \dfrac{1}{3}\right)^2 = \left(\dfrac{5}{6}\right)^2 = \dfrac{25}{36}$

(B) $(a^{-1} + b^{-1})^2 = \left(\dfrac{1}{a} + \dfrac{1}{b}\right)^2 = \left(\dfrac{b + a}{ab}\right)^2 = \dfrac{(b + a)^2}{(ab)^2}$

$$= \dfrac{b^2 + 2ab + a^2}{a^2b^2}$$

Matched Problem 4 Rewrite, using positive exponents only:

      (A) $2^{-2} + 3^{-2}$ **(B)** $a^{-2} + b^{-2}$

Answers to Matched Problems

1. (A) $\dfrac{1}{x^7}$ **(B)** x^7 **(C)** 0.000 000 1   **(D)** $\dfrac{y^4}{x^7}$

2. (A) $\dfrac{1}{5^4}$ **(B)** $\dfrac{1}{x^8}$ **(C)** 5^{-4} **(D)** x^{-8}

3. (A) x^7 **(B)** $\dfrac{x^2}{y^6}$ **(C)** x^2 **(D)** $\dfrac{y^4}{3x^5}$

          (E) 10^8 **(F)** $\dfrac{x^2}{y^2z^2}$

4. (A) $\dfrac{13}{36}$ **(B)** $\dfrac{a^2 + b^2}{a^2b^2}$

EXERCISE 8-2

A *Rewrite, using positive exponents only:*

1. 8^0 **2.** 10^0 **3.** 313^0

4. $(-201)^0$ **5.** 2^{-3} **6.** 3^{-2}

7. $\dfrac{1}{3^{-3}}$ **8.** $\dfrac{1}{2^{-2}}$ **9.** $(-2)^{-2}$

10. $(-2)^{-3}$ **11.** $(-3)^{-3}$ **12.** $(-3)^{-2}$

13. a^{-5} **14.** b^{-6} **15.** $\dfrac{1}{b^{-7}}$

16. $\dfrac{1}{a^{-8}}$ **17.** $\dfrac{3^{-2}}{3^5}$ **18.** $\dfrac{4^{-1}}{4^3}$

19. $\dfrac{4^{-8}}{4^{-5}}$ **20.** $\dfrac{3^{-9}}{3^{-6}}$ **21.** $\dfrac{x^7}{x^{-4}}$

22. $\dfrac{y^9}{y^{-5}}$ **23.** $\dfrac{y^{-3}}{y^6}$ **24.** $\dfrac{y^{-5}}{y^8}$

25. $3^5 3^{-7}$ **26.** $2^6 2^{-4}$

27. $3^8 3^{-4}$ **28.** $2^4 2^{-7}$

29. $10^{12} \times 10^{-4}$ **30.** $10^{-5} \div 10^{-1}$

31. $10^{-1} \div 10^5$ **32.** $10^3 \div 10^{-7}$

33. $(10^{-2})^{-2}$ **34.** $(10^{-3})^{-1}$

35. $(x^{-2})^3$ **36.** $(y^{-3})^3$

37. $(y^{-2})^{-4}$ **38.** $(x^{-3})^{-2}$

39. $(x^{-1}y^2)^2$ **40.** $(x^2y^{-1})^3$

B 41. $(x^{-2}y)^{-3}$ **42.** $(x^2y^{-3})^{-1}$

43. $\left(\dfrac{a^{-2}}{b}\right)^{-1}$ **44.** $\left(\dfrac{a^2}{b^{-1}}\right)^{-3}$

45. $\dfrac{x^{-1}y^4}{x^{-5}y^{-3}}$ **46.** $\dfrac{x^4y^{-1}}{x^{-1}y^{-4}}$

47. $\dfrac{10^{-1} \times 10^4}{10^6 \times 10^{-5}}$ **48.** $\dfrac{10^8 \times 10^{-2}}{10^7 \times 10^{-11}}$

49. $(a^2a^{-5})^{-2}$ **50.** $(x^{-1}x^4)^{-3}$

51. $\left(\dfrac{x^{-2}}{x^{-4}}\right)^{-1}$ **52.** $\left(\dfrac{a^{-3}}{a^{-2}}\right)^{-2}$

53. $(3a^2b^{-1})^{-2}$ **54.** $(2a^{-1}b^3)^{-1}$

55. $\dfrac{1}{(3x^{-3}y^2)^{-1}}$ **56.** $\dfrac{1}{(2xy^{-3})^{-2}}$

57. $\left(\dfrac{a^{-1}}{3b}\right)^{-1}$ **58.** $\left(\dfrac{a}{2b^{-1}}\right)^{-2}$

C 59. $(3^{-1} + 4^{-1})^{-1}$ **60.** $(2^{-1} + 3^{-1})^{-1}$

61. $(2^{-1} + 3^{-1})^{-2}$ **62.** $(3^{-1} + 4^{-1})^{-2}$

63. $(10^{-2} + 10^{-3})^{-1}$ **64.** $(10^{-1} + 10^{-2})^{-1}$

65. $\left(\dfrac{2xy^{-1}}{3x^{-2}y}\right)^{-1}$ **66.** $\left(\dfrac{4x^{-1}y^2}{3xy^{-1}}\right)^{-1}$

67. $\left(\dfrac{3x^{-2}y^{-1}}{2xy}\right)^{-2}$ **68.** $\left(\dfrac{6a^{-3}b^{-4}}{3a^2b^{-1}}\right)^{-2}$

69. $\left(\dfrac{10a^3b^{-2}}{5a^{-1}b^2}\right)^{-1}$ **70.** $\left(\dfrac{9a^{-1}b^{-1}}{6a^{-2}b^{-2}}\right)^{-3}$

71. $\dfrac{10a^{-1}10^{-3}a^2}{10^{-1}a^4}$ **72.** $\dfrac{10^{-2}a10^{-3}a^{-1}}{10^4a^{-6}}$

73. $(a^{-1} + b^{-1})^{-1}$ **74.** $(a^{-1} + b^{-1})^{-2}$

75. $(x^{-2} - y^{-2})^{-2}$ **76.** $(x^{-1} - y^{-1})^{-1}$

77. $p^{-2} + q^{-2}$ **78.** $r^{-2} - s^{-2}$

79. $a^{-1} - b^{-1}$ **80.** $a^{-1} + b^{-1}$

81. $(x^{-1} + y^{-1})^2$ **82.** $(x^{-1} - y^{-1})^2$

8-3 Scientific Notation

♦ Scientific Notation
♦ Arithmetic in Scientific Notation
♦ Application

Writing and working with very large or very small numbers such as 100,000,000,000 or 0.000 000 000 000 1 can be awkward. In particular, your calculator can accept only a limited number of digits in a number before converting it to another form. Yet work in science, economics, and other areas often requires dealing with numbers that are very, very large or very, very small. For example:

1. The assets of the largest banks in the world are in excess of **400,000,000,000** dollars.
2. The energy of subatomic particles is measured in units called electron volts, which are equal to **0.000 000 000 000 000 000 160 2** joule, a joule being the standard unit of energy.

It is convenient to represent numbers of such large or small magnitudes in a notation called *scientific notation,* which we introduce in this section. This is also the notation to which your calculator likely converts numbers too large or too small to display.

♦ SCIENTIFIC NOTATION

To write a number in **scientific notation** means to write it as the product of a number between 1 and 10, excluding 10, and an integer power of 10. Any decimal fraction, however large or small, can be so represented. For example:

$$4 = 4 \times 10^0 \qquad 0.2 = 2 \times 10^{-1}$$
$$43 = 4.3 \times 10^1 \qquad 0.02 = 2 \times 10^{-2}$$
$$435 = 4.35 \times 10^2 \qquad 0.002 = 2 \times 10^{-3}$$
$$4,351 = 4.351 \times 10^3 \qquad 0.0002 = 2 \times 10^{-4}$$

Recall how powers of 10 are related to decimals:

$$100,000 = 10^5$$
$$10,000 = 10^4$$
$$1,000 = 10^3$$
$$100 = 10^2$$
$$10 = 10^1$$
$$1 = 10^0$$

$$0.1 = 10^{-1}$$
$$0.01 = 10^{-2}$$
$$0.001 = 10^{-3}$$
$$0.0001 = 10^{-4}$$
$$0.000\ 01 = 10^{-5}$$

and so on.

The following rule will help you shift the decimal point when converting to scientific notation. Remember that we are shifting the decimal point to obtain a number that has one digit, not equal to zero, to the left of the decimal.

Converting Positive Numbers to Scientific Notation

1. If the number is greater than or equal to 10, the number of places the decimal point is shifted left appears as a positive exponent.
2. If the number is less than 10 but greater than or equal to 1, the decimal place is not shifted at all, and the exponent used is 0.
3. If the number is less than 1, the number of places the decimal point is shifted right appears as a negative exponent.

Thus,

$$3{,}450{,}000 = 3{,}450\ 000 \times 10^6 = 3.45 \times 10^6$$

6 Places left   Positive exponent

$$3.45 = 3{,}45 \times 10^0$$

0 Places shifted   Zero exponent

and

$$0.000\ 034\ 5 = 000\ 003{,}45 \times 10^{-5} = 3.45 \times 10^{-5}$$

5 Places right   Negative exponent

Work is easily checked by converting back to standard notation. Remember that multiplying by 10 shifts the decimal point to the right and dividing by 10, that is, multiplying by 10^{-1}, shifts it left:

$$3.45 \times 10^6 = 3{,}450{,}000 = 3{,}450{,}000$$

Shift right 6 places

$$3.45 \times 10^{-5} = 0.00003{,}45 = 0.000\ 034\ 5$$

Shift left 5 places

Example 1 **Converting to** **Scientific Notation**	Write in scientific notation: **(A)** 380 **(B)** 10,400 **(C)** 0.065 **(D)** 0.000 41

Solution   **(A)** $380 = 3\,8\,0 = 3.8 \times 10^2$

(B) $10,400 = 1\,0\,4\,0\,0 = 1.04 \times 10^4$

(C) $0.065 = 0\,0\,6\,5 = 6.5 \times 10^{-2}$

(D) $0.000\,41 = 0\,0\,0\,0\,4\,1 = 4.1 \times 10^{-4}$

Matched Problem 1   Write in scientific notation:

(A) 65,000   **(B)** 880   **(C)** 0.0045   **(D)** 0.000 006

Your calculator probably displays numbers using an "E" or a space to indicate scientific notation with the power of 10 following the E or space. Thus, it most likely represents 4,351,000,000 as "4.351E+9" or "4.351 9," meaning 4.351×10^9, and 0.000 000 4351 as "4.351E−7" or "4.351 −7," meaning 4.351×10^{-7}.

Example 2 **Converting to** **Standard Notation**	Write in standard decimal notation: **(A)** $6.031\,08 \times 10^8$ **(B)** $6.031\,08 \times 10^{-8}$

Solution   **(A)** $6.031\,08 \times 10^8 = 6\,0\,3\,1\,0\,8\,0\,0\,0 = 603,108,000$

(B) $6.031\,08 \times 10^{-8} = 0\,0\,0\,0\,0\,0\,0\,0\,6\,0\,3\,1\,0\,8$
$= 0.000\,000\,060\,310\,8$

Matched Problem 2   Write in standard decimal notation:

(A) $7.209\,46 \times 10^{-9}$ **(B)** $7.209\,46 \times 10^9$

It is important to keep in mind that $10^6 = 10 \cdot 10^5$, so 10^6 is *10 times* as large as 10^5. Similarly, 10^8 is 10^3, or *1,000*, *times* as large as 10^5, and so on. For example, the mass of the Earth is roughly 6×10^{24} kilograms, and the mass of Saturn is roughly 6×10^{26} kilograms. Thus, Saturn is roughly 100 times as massive as the Earth.

♦ **ARITHMETIC IN SCIENTIFIC NOTATION**

Scientific notation can be helpful in handling certain types of arithmetic calculations that involve very large or very small numbers. It is particularly helpful with products and quotients.

Example 3 **Complicated Arithmetic** **in Scientific Notation**	Evaluate, using scientific notation:

$$\frac{(600,000)(0.000\ 15)}{(0.02)(2,250,000,000)}$$

Solution

$$\frac{(600,000)(0.000\ 15)}{(0.02)(2,250,000,000)} = \frac{6 \times 10^5 \times 1.5 \times 10^{-4}}{2 \times 10^{-2} \times 2.25 \times 10^9}$$

$$= \frac{6 \times 1.5}{2 \times 2.25} \cdot \frac{10^5 \times 10^{-4}}{10^{-2} \times 10^9}$$

$$= \frac{9}{4.5} \cdot \frac{10^1}{10^7} = 2 \cdot \frac{1}{10^6}$$

$$= 2 \times 10^{-6}$$

Matched Problem 3   Evaluate, using scientific notation:

$$\frac{(0.000\ 08)(5,000,000,000)}{(2,500,000)(0.000\ 002)}$$

Calculating sums and differences of numbers in scientific notation can be simplified somewhat by factoring out powers of 10. The application that follows uses this approach to find the sum of several large numbers expressed in scientific notation.

♦   **APPLICATION**

The following table gives the masses of the nine planets in our solar system:

Planet	Mass in Kilograms
Mercury	3.303×10^{23}
Venus	4.870×10^{24}
Earth	5.970×10^{24}
Mars	6.42×10^{23}
Jupiter	1.899×10^{27}
Saturn	5.686×10^{26}
Uranus	8.66×10^{25}
Neptune	1.030×10^{26}
Pluto	1.3×10^{22}

Jupiter is not only the largest planet, it is significantly larger than the other eight combined. To see this, first add the masses of the other eight planets:

$3.303 \times 10^{23} + 4.870 \times 10^{24} + 5.970 \times 10^{24} + 6.42 \times 10^{23} + 5.686 \times 10^{26} + 8.66 \times 10^{25} + 1.030 \times 10^{26} + 1.3 \times 10^{22}$

The sum will be more manageable if we factor out 10^{22}, the largest power of 10 common to all terms, and then convert the remaining factors to standard notation in order to add. This gives

$$10^{22}(3.303 \times 10 + 4.870 \times 10^2 + 5.970 \times 10^2 + 6.42 \times 10 + 5.686 \times 10^4$$
$$+ 8.66 \times 10^3 + 1.030 \times 10^4 + 1.3)$$

$$= 10^{22}(33.03 + 487.0 + 597.0 + 64.2 + 56,860 + 8,660 + 10,300 + 1.3)$$

$$= 10^{22}(77,002.53) = 10^{22}(7.700\ 253 \times 10^4)$$

$$= 7.700\ 253 \times 10^{26}$$

Now take the ratio of the mass of Jupiter to the total of the other eight planets:

$$\frac{\text{Mass of Jupiter}}{\text{Mass of other eight}} = \frac{1.899 \times 10^{27}}{7.700\ 253 \times 10^{26}} = \frac{1.899}{7.700253} \times 10$$

$$= 0.2466 \times 10 \approx 2.5$$

That is, Jupiter is almost 2.5 times larger than the other eight planets combined.

Answers to Matched Problems
1. **(A)** 6.5×10^4 **(B)** 8.8×10^2 **(C)** 4.5×10^{-3} **(D)** 6×10^{-6}
2. **(A)** $0.000\ 000\ 007\ 209\ 46$ **(B)** $7,209,460,000$ 3. 8×10^4, or $80,000$

EXERCISE 8-3

A *Write in scientific notation:*

1. 2,000
2. 4,500
3. 108,000
4. 3,000,000
5. 412
6. 965
7. 48
8. 72
9. 0.63
10. 0.082
11. 0.046
12. 0.55
13. 0.000 44
14. 0.000 06
15. 0.000 095
16. 0.0008
17. 0.000 003
18. 0.000 002 5
19. 0.000 001 8
20. 0.000 007
21. 1
22. 10
23. 6,350,000,000
24. 1,125,000,000,000
25. 0.000 000 000 845
26. 0.000 000 000 65
27. 0.000 000 480 5
28. 0.000 000 200 6

Write in standard decimal notation:

29. 3×10^4
30. 2×10^5
31. 4×10^{-3}
32. 6×10^{-4}
33. 8×10^{-2}
34. 7×10^{-5}
35. 3.4×10^6
36. 5.1×10^5
37. 4.8×10^8
38. 1.5×10^{-6}
39. 2.4×10^{-4}
40. 1.6×10^{-5}
41. 3.2×10^{-6}
42. 7.5×10^{-2}
43. 5.25×10^4
44. 4.68×10^5
45. 4.15×10^{-3}
46. 8.24×10^{-4}

Simplify and express answers in scientific notation:

47. $(3 \times 10^4)(5 \times 10^{-3})$
48. $(4 \times 10^5)(3 \times 10^{-7})$
49. $(6 \times 10^4)(4 \times 10^{-6})$
50. $(2 \times 10^8)(8 \times 10^5)$
51. $\dfrac{32 \times 10^4}{2 \times 10^5}$
52. $\dfrac{28 \times 10^6}{2 \times 10^4}$
53. $\dfrac{66 \times 10^8}{6 \times 10^5}$
54. $\dfrac{45 \times 10^3}{3 \times 10^6}$

B *Simplify, if necessary, and write in scientific notation:*

55. 3,664,000,000 miles, the distance from Pluto to the sun

56. 491,920,000,000 kilometers, the distance from Earth to the star Alpha Cephei

57. The ratio 35,960,000 miles : 3,664,000,000 miles, the ratio of the distance of the sun from Mercury to its distance from Pluto

58. The ratio 150,000,000 kilometers : 5,900,000,000 kilometers, the ratio of the sun's distance from Earth to its distance from Pluto

The assets of the largest banks in four U.S. cities, as of the end of 1989, were as follows:

Citibank, New York	$161,988,000,000
Bank of America, San Francisco	$86,712,000,000
Security Pacific, Los Angeles	$54,313,683,000
First National, Chicago	$36,410,165,000

In Problems 59–64, find the quantity requested and write the answer in scientific notation:

59. The assets of Citibank

60. The assets of First National

61. The amount by which the assets of Citibank exceed those of Bank of America

62. The amount by which the assets of First National trail those of Security Pacific

63. The ratio of the assets of First National to the assets of Citibank

64. The ratio of the assets totaling $24,000,000 for a small community bank to the assets of Security Pacific

Very small time intervals, such as those used to measure times in computer operations, are recorded in units such as milliseconds, microseconds, and nanoseconds:

$$1,000 \; milliseconds = 10^3 \; milliseconds$$
$$= 1 \; second$$
$$1,000,000 \; microseconds = 10^6 \; microseconds$$
$$= 1 \; second$$
$$1,000,000,000 \; nanoseconds = 10^9 \; nanoseconds$$
$$= 1 \; second$$

In Problems 65–70, convert the given time to seconds and express the answer in scientific notation:

65. 1 microsecond

66. 1 nanosecond

67. 5.6 nanoseconds

68. 3.2 microseconds

69. 0.01 millisecond

70. 0.03 microsecond

C *Convert each numeral to scientific notation:*

71. 1,500,000,000,000,000 kilograms, the estimated coal reserves of the United States

72. 10,500,000,000,000,000 kilowatt-hours, the energy equivalent to the U.S. coal reserves

73. 0.000 000 000 000 000 000 000 000 000 910 8 kilograms, the mass of an electron

74. 0.000 000 000 000 000 000 000 000 001 675 kilograms, the mass of a neutron

Convert each numeral to scientific notation and simplify. Express the answer in scientific notation and as a decimal fraction

75. $\dfrac{(21,000,000)(6,000)}{0.0063}$

76. $\dfrac{(0.000\ 48)(2,000,000,000)}{(0.016)(60,000)}$

77. $\dfrac{(3,600)(0.000\ 002)}{(0.0018)(8,000,000)}$ **78.** $\dfrac{(0.027)(2,000,000)}{0.0009}$

79. $\dfrac{(4,200)(0.000\ 004)}{(0.0007)(1,200,000)}$

80. $\dfrac{(0.000\ 035)(6,000,000,000)}{(140,000)(0.005)}$

81. $\dfrac{(32,000,000,000)(9,000,000)}{(0.000\ 000\ 002\ 4)(0.000\ 12)}$

82. $\dfrac{(1,500,000,000)(0.000\ 000\ 2)}{(0.01)(3,000,000)}$

83. $(1,000,000)^2(0.000\ 004)^{-1}(0.002)^{-2}$

84. $(0.0004)^2(2,000,000)^{-2}(400)^{-1}$

Air quality standards establish maximum amounts of pollutants considered acceptable in the air. The amounts are frequently given in parts per million, ppm. One ppm represents 1/1,000,000 of the volume, just as 1% represents 1/100 of the volume. Use ratios to rewrite each of the standards given in Problems 85–88 in (A) scientific notation, (B) decimal form, (C) percent form. For example,

$$30 \; ppm = \frac{30}{1,000,000} = \frac{30}{10^6} = \textbf{(A)} \; 30 \times 10^{-6}$$
$$= \textbf{(B)} \; 0.000\ 03 = \textbf{(C)} \; 0.003\%$$

85. 9 ppm, the standard for carbon monoxide, when averaged over an 8-hour period

86. 0.03 ppm, the standard for sulfur oxides, when averaged over a year

87. 0.05 ppm, the standard for nitrogen dioxide, when averaged over a year

88. 35 ppm, the standard for carbon monoxide, when averaged over a 1-hour period

8-4 Square Roots and Radicals

♦ Square Roots
♦ Square Root Properties
♦ Simplest Radical Form

Square roots were considered in Section 4-1, and the radical notation $\sqrt{a}$ was used there to denote the *positive* square root of the number a. In this section we consider square roots in more detail. We also look at some properties of radicals that allow us to write expressions involving radicals in ways that make comparisons easier. The properties suggest a connection between radicals and exponents that is developed in detail in subsequent courses.

♦ SQUARE ROOTS

Recall the following about square roots:

x is a **square root** of y if $x^2 = y$.

-3 and 3 are square roots of 9, since $(-3)^2 = 9$ and $3^2 = 9$.

$\sqrt{y}$ denotes the positive square root of y.

$\sqrt{9} = 3$

$-\sqrt{y}$ denotes the negative square root of y.

$-\sqrt{9} = -3$

$\sqrt{y}$ is an irrational number unless y is one of the perfect squares 1, 4, 9, . . .

$\sqrt{2}, \sqrt{3}, \sqrt{5}, \sqrt{6}$ are irrational numbers.

The symbol $\sqrt{}$ is called a **radical,** or radical sign, and the expression contained within the radical sign is called the **radicand.**

How many square roots of a real number are there? The following result answers this question:

Number of Square Roots

(A) Every positive real number has exactly two real square roots, each the opposite or negative of the other.
(B) Negative real numbers have no real-number square roots, since no real number squared can be negative.
(C) The square root of 0 is 0.

Example 1
Square Roots

Evaluate, if possible:

(A) $\sqrt{4}$ **(B)** $-\sqrt{4}$ **(C)** $\sqrt{-4}$ **(D)** $\sqrt{0}$

Solution   **(A)** $\sqrt{4} = 2$ **(B)** $-\sqrt{4} = -2$
             (C) $\sqrt{-4}$ is not a real number.   **(D)** $\sqrt{0} = 0$

Matched Problem 1   Evaluate, if possible:

(A) $\sqrt{49}$ (B) $-\sqrt{49}$ (C) $\sqrt{-49}$ (D) $\sqrt{0}$

♦   **SQUARE ROOT PROPERTIES**

Consider the following examples involving square roots:

(A) $\sqrt{6^2} = \sqrt{36} = 6$.

(B) $\sqrt{4}\sqrt{36} = 2 \cdot 6 = 12$ and $\sqrt{4 \cdot 36} = \sqrt{144} = 12$, so that $\sqrt{4}\sqrt{36} = \sqrt{4 \cdot 36}$.

(C) $\dfrac{\sqrt{36}}{\sqrt{4}} = \dfrac{6}{2} = 3$ and $\sqrt{\dfrac{36}{4}} = \sqrt{9} = 3$, so that $\dfrac{\sqrt{36}}{\sqrt{4}} = \sqrt{\dfrac{36}{4}}$

These examples suggest the following general properties:

Properties of Radicals

For nonnegative real numbers a and b:

1. $\sqrt{a^2} = a$ $\qquad\qquad\qquad$ $\sqrt{3^2} = 3$

2. $\sqrt{a}\sqrt{b} = \sqrt{ab}$ $\qquad\qquad$ $\sqrt{2}\sqrt{3} = \sqrt{2 \cdot 3}$

3. $\dfrac{\sqrt{a}}{\sqrt{b}} = \sqrt{\dfrac{a}{b}}$ $\quad b \neq 0$ $\qquad$ $\dfrac{\sqrt{2}}{\sqrt{3}} = \sqrt{\dfrac{2}{3}}$

Example 2
Using the Properties of Radicals

Simplify, using the properties of radicals:

(A) $\sqrt{5}\sqrt{10}$ (B) $\dfrac{\sqrt{32}}{\sqrt{8}}$ (C) $\sqrt{\dfrac{7}{4}}$

Solution   (A) Apply Property 2 to $\sqrt{5}\sqrt{10}$:

$$\sqrt{5}\sqrt{10} = \sqrt{5 \cdot 10}$$
$$= \sqrt{50} \qquad\qquad \text{Try to factor 50 in a way that involves perfect squares.}$$
$$= \sqrt{25 \cdot 2} \qquad\quad \text{Apply Property 2.}$$
$$= \sqrt{25}\sqrt{2}$$
$$= 5\sqrt{2}$$

(B) Apply Property 3:

$$\frac{\sqrt{32}}{\sqrt{8}} = \sqrt{\frac{32}{8}} = \sqrt{4} = 2$$

(C) $\sqrt{\dfrac{7}{4}} = \dfrac{\sqrt{7}}{\sqrt{4}} = \dfrac{\sqrt{7}}{2}$ or $\dfrac{1}{2}\sqrt{7}$

Matched Problem 2   Simplify, using the properties of radicals:

(A) $\sqrt{3}\sqrt{6}$ **(B)** $\dfrac{\sqrt{18}}{\sqrt{2}}$ **(C)** $\sqrt{\dfrac{11}{9}}$

The properties of radicals appear quite similar to three properties for exponents:

$$(a^m)^n = a^{mn}$$

$$a^m b^m = (ab)^m$$

$$\frac{a^m}{b^m} = \left(\frac{a}{b}\right)^m$$

If $\sqrt{a}$ were to be written as a power a^p of a, what might p be? If $\sqrt{a} = a^p$, then we would have

$$a^1 = a = \sqrt{a}\sqrt{a}$$
$$= a^p a^p$$
$$= a^{2p}$$

Thus, we would need $2p = 1$, that is, $p = \frac{1}{2}$. Now, if we write $\sqrt{a}$ as $a^{1/2}$, the properties of radicals become

$$(a^2)^{1/2} = a$$

$$a^{1/2} b^{1/2} = (ab)^{1/2}$$

$$\frac{a^{1/2}}{b^{1/2}} = \left(\frac{a}{b}\right)^{1/2}$$

which is exactly what the laws of exponents would give us. It is in fact possible to write radicals as exponents in this way. This topic is explored further in subsequent courses.

♦   **SIMPLEST RADICAL FORM**

Arithmetic expressions involving radicals can be evaluated using calculators. Algebraic expressions containing radicals can be changed to a variety of equivalent forms to aid in their evaluation or to compare two expressions that appear quite different. One form that is often useful is called the *simplest radical form.*

Simplest Radical Form for Square Roots

An algebraic expression that contains square root radicals is in **simplest radical form** if all three of the following conditions are satisfied:

1. No radicand, when expressed in completely factored form, contains a factor raised to a power greater than 1. $\sqrt{x^3}$ violates this condition.

2. No radical appears in a denominator. $\dfrac{3}{\sqrt{5}}$ violates this condition.

3. No fraction appears within a radical. $\sqrt{\dfrac{2}{3}}$ violates this condition.

It should be understood that in some circumstances forms other than the simplest radical form may be more useful. The situation dictates the choice.

Example 3
Simplest Radical Form

Change to simplest radical form. All variables represent positive real numbers.

(A) $\sqrt{72}$ **(B)** $\sqrt{8x^3}$ $6\sqrt{2}$ $2x\sqrt{2x}$

Solution   **(A)** $\sqrt{72} = \sqrt{6^2 \cdot 2}$ Violates Condition 1, since 6 is raised to a power greater than 1. Use Property 2: $\sqrt{ab} = \sqrt{a}\sqrt{b}$

$$= \sqrt{6^2}\sqrt{2} \qquad \sqrt{a^2} = a \quad (a \ge 0)$$

$$= 6\sqrt{2}$$

(B) $\sqrt{8x^3} = \sqrt{(2^2x^2)(2x)}$ Violates Condition 1. Separate $8x^3$ into a perfect square part (2^2x^2) and what is left over ($2x$); then use radical Property 2: $\sqrt{ab} = \sqrt{a}\sqrt{b}$.

$$= \sqrt{2^2x^2}\sqrt{2x} \qquad \sqrt{a^2} = a \quad (a \ge 0)$$

$$= 2x\sqrt{2x}$$

Matched Problem 3

Change to simplest radical form. All variables represent positive real numbers.

(A) $\sqrt{32}$ **(B)** $\sqrt{18y^3}$

Example 4
Simplest Radical Form

Change to simplest radical form. All variables represent positive real numbers.

(A) $\dfrac{3x}{\sqrt{3}}$ **(B)** $\sqrt{\dfrac{x}{2}}$ $\dfrac{\sqrt{9x}}{3}$

Solution   **(A)** $\dfrac{3x}{\sqrt{3}}$ has a radical in the denominator; hence, it violates Condition 2. To remove the radical from the denominator we multiply top and bottom by $\sqrt{3}$ to obtain $\sqrt{3^2}$ in the denominator:

$$\frac{3x}{\sqrt{3}} = \frac{3x}{\sqrt{3}} \cdot \frac{\sqrt{3}}{\sqrt{3}}$$

$$= \frac{3x\sqrt{3}}{\sqrt{3^2}}$$

$$= \frac{3x\sqrt{3}}{3} = x\sqrt{3}$$

(B) $\sqrt{\dfrac{x}{2}}$ has a fraction within the radical; hence, it violates Condition 3. To remove the fraction from the radical, we multiply the top and bottom of $x/2$ inside the radical by 2 to make the denominator a perfect square:

$$\sqrt{\frac{x}{2}} = \sqrt{\frac{2 \cdot x}{2 \cdot 2}}$$

$$= \sqrt{\frac{2x}{2^2}}$$

$$= \frac{\sqrt{2x}}{\sqrt{2^2}} = \frac{\sqrt{2x}}{2}$$

Matched Problem 4   Change to simplest radical form. All variables represent positive real numbers.

(A) $\dfrac{2x}{\sqrt{2}}$ (B) $\sqrt{\dfrac{y}{3}}$

The first property of radicals, $\sqrt{a^2} = a$, is true only for *nonnegative* values of a. It is not true when a is negative. Problems 85 and 86 in Exercise 8-4 suggest how $\sqrt{a^2}$ should be interpreted if we allow a to take on any real value. All the other problems in Exercise 8-4 restrict variables to nonnegative real numbers so that $\sqrt{a^2} = a$ will hold true.

Answers to Matched Problems

1. (A) 7   (B) -7 (C) Not a real number   (D) 0
2. (A) $3\sqrt{2}$ (B) 3   (C) $\dfrac{\sqrt{11}}{3}$ or $\dfrac{1}{3}\sqrt{11}$
3. (A) $4\sqrt{2}$ (B) $3y\sqrt{2y}$ 4. (A) $x\sqrt{2}$ (B) $\dfrac{\sqrt{3y}}{3}$ or $\dfrac{1}{3}\sqrt{3y}$

EXERCISE 8-4

All variables represent positive real numbers unless stated to be otherwise.

A *Simplify, using the properties of radicals, and express each answer in simplest radical form:*

1. $\sqrt{16}$ 2. $\sqrt{25}$ 3. $-\sqrt{81}$ 4. $-\sqrt{49}$
5. $\sqrt{x^2}$ 6. $\sqrt{y^2}$ 7. $\sqrt{9m^2}$ 8. $\sqrt{4u^2}$
9. $\sqrt{3}\sqrt{27}$ 10. $\sqrt{2}\sqrt{8}$
11. $\sqrt{216}\sqrt{6}$ 12. $\sqrt{1{,}000}\sqrt{10}$
13. $\sqrt{2} \div \sqrt{8}$ 14. $\sqrt{216} \div \sqrt{6}$
15. $\sqrt{1{,}000} \div \sqrt{10}$ 16. $\sqrt{3} \div \sqrt{27}$
17. $\sqrt{6}\sqrt{12}$ 18. $\sqrt{10}\sqrt{14}$
19. $\sqrt{21}\sqrt{3}$ 20. $\sqrt{15}\sqrt{5}$
21. $\sqrt{10} \div \sqrt{14}$ 22. $\sqrt{21} \div \sqrt{3}$
23. $\sqrt{15} \div \sqrt{5}$ 24. $\sqrt{6} \div \sqrt{12}$
25. $\sqrt{a}\sqrt{a^3}$ 26. $\sqrt{x^5}\sqrt{x}$
27. $\sqrt{x^5} \div \sqrt{x}$ 28. $\sqrt{a} \div \sqrt{a^3}$
29. $\sqrt{2x}\sqrt{6x}$ 30. $\sqrt{6a}\sqrt{3a}$
31. $\sqrt{6a} \div \sqrt{3a}$ 32. $\sqrt{2x} \div \sqrt{6x}$
33. $\sqrt{8}$ 34. $\sqrt{18}$ 35. $\sqrt{x^3}$ 36. $\sqrt{m^3}$
37. $\sqrt{18y^3}$ 38. $\sqrt{8x^3}$ 39. $\sqrt{\tfrac{1}{4}}$ 40. $\sqrt{\tfrac{1}{9}}$
41. $-\sqrt{\tfrac{4}{9}}$ 42. $-\sqrt{\tfrac{9}{16}}$ 43. $\dfrac{1}{\sqrt{x^2}}$ 44. $\dfrac{1}{\sqrt{y^2}}$
45. $\dfrac{1}{\sqrt{3}}$ 46. $\dfrac{1}{\sqrt{5}}$ 47. $\sqrt{\tfrac{1}{3}}$ 48. $\sqrt{\tfrac{1}{5}}$

49. $\dfrac{1}{\sqrt{x}}$ 50. $\dfrac{1}{\sqrt{y}}$ 51. $\sqrt{\dfrac{1}{x}}$ 52. $\sqrt{\dfrac{1}{y}}$

B 53. $\sqrt{25x^2y^4}$ 54. $\sqrt{49x^4y^2}$ 55. $\sqrt{4x^5y^3}$
56. $\sqrt{9x^3y^5}$ 57. $\sqrt{8x^7y^6}$ 58. $\sqrt{18x^8y^5}$

59. $\dfrac{1}{\sqrt{3y}}$ 60. $\dfrac{1}{\sqrt{2x}}$ 61. $\dfrac{4xy}{\sqrt{2y}}$
62. $\dfrac{6x^2}{\sqrt{3x}}$ 63. $\dfrac{2x^2y}{\sqrt{3xy}}$ 64. $\dfrac{3a}{\sqrt{2ab}}$
65. $\sqrt{\tfrac{2}{3}}$ 66. $\sqrt{\tfrac{3}{5}}$ 67. $\sqrt{\dfrac{3m}{2n}}$
68. $\sqrt{\dfrac{6x}{7y}}$ 69. $\sqrt{\dfrac{4a^3}{3b}}$ 70. $\sqrt{\dfrac{9m^5}{2n}}$

Simplify and express in simplest radical form. Evaluate both the original expression and the answer using a calculator.

71. $\sqrt{6}\sqrt{3}$ 72. $\sqrt{2}\sqrt{6}$ 73. $\sqrt{\tfrac{1}{5}}$
74. $\sqrt{\tfrac{1}{3}}$ 75. $\dfrac{\sqrt{33}}{\sqrt{2}}$ 76. $\dfrac{\sqrt{23}}{\sqrt{5}}$

C *In Problems 77–84, rewrite the expression in simplest radical form:*

77. $\dfrac{\sqrt{2x}\sqrt{5}}{\sqrt{20x}}$ 78. $\dfrac{\sqrt{6}\sqrt{8x}}{\sqrt{3x}}$ 79. $\sqrt{a^2 + b^2}$
80. $\sqrt{m^2 + n^2}$ 81. $\sqrt{x^4 - 2x^2}$ 82. $\sqrt{m^3 + 4m^2}$
83. $\sqrt{x^2 + 2x + 1}$ for $x \geq -1$
84. $\sqrt{a^2 + 4a + 4}$ for $a \geq -2$

85. Is $\sqrt{x^2} = \begin{cases} x & \text{when } x \geq 0 \\ -x & \text{when } x < 0 \end{cases}$ true for $x = 4$?
For $x = -4$?

86. Is $\sqrt{x^2} = |x|$ true for $x = 4$? For $x = -4$?

87. Find the fallacy in the following "proof" that all real numbers are equal: If m and n are any real numbers, then

$$(m - n)^2 = (n - m)^2$$

$$m - n = n - m$$

$$2m = 2n$$

$$m = n$$

88. If $x^2 = y^2$, does it necessarily follow that $x = y$? [*Hint:* Can you find a pair of numbers that make the first equation true but the second equation false?]

APPLICATIONS

*Problems 89 and 90 use the **Pythagorean theorem**. Recall that this theorem states that in a right triangle, the square of the length of the hypotenuse is equal to the sum of the squares of the lengths of the sides.*

$$c^2 = a^2 + b^2$$

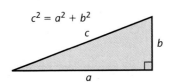

89. *Sight Distance* The accompanying figure shows how to find the distance d to the horizon from an altitude h above the earth's surface. From the Pythagorean theorem, $(h + R)^2 = R^2 + d^2$, or, equivalently,

$d = \sqrt{(h + R)^2 - R^2}$. Use $R = 3,950$ miles as the radius of the earth to find d when h is $\frac{1}{2}$ mile.

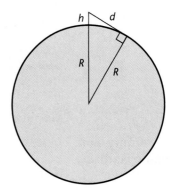

90. *Pythagorean Theorem* If a right triangle has a hypotenuse of 25 centimeters and one side is 7 centimeters, how long is the other side?

91. *Inventory* Under certain conditions, the optimum order size S for an inventory item that has annual demand D, order cost C, and annual holding cost H is given by

$$S = \sqrt{\frac{2DC}{H}}$$

Find the optimum order size for an item with annual demand of 72,000 cases, order cost of \$20 per order, and annual holding cost of \$1.25 per case.

92. *Pendulum* The period T of a pendulum of length L is given by the formula

$$T = 2\pi\sqrt{\frac{L}{g}}$$

where g is the acceleration of gravity. Use $g = 32$ to find the period of a pendulum 3 feet long.

8-5 Sums and Differences Involving Radicals

Algebraic expressions can often be simplified by combining terms that contain exactly the same radical forms. We proceed in essentially the same way that we do when we combine like terms. You will recall that the distributive law played a central role in this process. Remember that we wrote

$$3x + 5x = (3 + 5)x = 8x$$

and concluded that we could combine like terms by adding their numerical coefficients. We have a similar rule for radicals:

Combining Radical Expressions

Two terms involving the same radical expression can be combined into a single term by adding the numerical coefficients of the expression.

Example 1
Adding and Subtracting Radicals

Simplify:

(A) $3\sqrt{2} + 5\sqrt{2}$ **(B)** $2\sqrt{m} - 7\sqrt{m}$ **(C)** $3\sqrt{x} - 2\sqrt{5} + 4\sqrt{x} - 7\sqrt{5}$

Solution

(A) $3\sqrt{2} + 5\sqrt{2} = (3 + 5)\sqrt{2} = 8\sqrt{2}$

(B) $2\sqrt{m} - 7\sqrt{m} = (2 - 7)\sqrt{m} = -5\sqrt{m}$

(C) $3\sqrt{x} - 2\sqrt{5} + 4\sqrt{x} - 7\sqrt{5} = 3\sqrt{x} + 4\sqrt{x} - 2\sqrt{5} - 7\sqrt{5}$

$$= 7\sqrt{x} - 9\sqrt{5}$$

Matched Problem 1

Simplify:

(A) $2\sqrt{3} + 4\sqrt{3}$ **(B)** $3\sqrt{x} - 5\sqrt{x}$ **(C)** $2\sqrt{y} - 3\sqrt{7} + 4\sqrt{y} - 2\sqrt{7}$

Occasionally terms containing radicals can be combined after they have been expressed in simplest radical form.

6.2

Example 2
Adding and Subtracting Radicals

Express in simplest radical form and simplify:

(A) $4\sqrt{8} - 2\sqrt{18}$ **(B)** $2\sqrt{12} - \sqrt{\dfrac{1}{3}}$ $12\sqrt{3} - \dfrac{\sqrt{3}}{3}$ $\dfrac{11\sqrt{3}}{3}$

Solution

(A) $4\sqrt{8} - 2\sqrt{18} = 4 \cdot \sqrt{4} \cdot \sqrt{2} - 2 \cdot \sqrt{9} \cdot \sqrt{2}$
$$= 4 \cdot 2 \cdot \sqrt{2} - 2 \cdot 3 \cdot \sqrt{2}$$
$$= 8\sqrt{2} - 6\sqrt{2}$$
$$= 2\sqrt{2}$$

(B) $2\sqrt{12} - \sqrt{\dfrac{1}{3}} = 2 \cdot \sqrt{4} \cdot \sqrt{3} - \sqrt{\dfrac{1}{3} \cdot \dfrac{3}{3}}$

$$= 4\sqrt{3} - \frac{\sqrt{3}}{3} \qquad \text{Note that } \frac{\sqrt{3}}{3} = \frac{1}{3}\sqrt{3}$$

$$= \left(4 - \frac{1}{3}\right)\sqrt{3}$$

$$= \frac{11}{3}\sqrt{3} \qquad \text{or} \qquad \frac{11\sqrt{3}}{3}$$

Matched Problem 2   Express in simplest radical form and simplify:

$$\text{(A) } 5\sqrt{3} - 2\sqrt{12} \qquad \text{(B) } 3\sqrt{8} - \sqrt{\frac{1}{2}}$$

Answers to Matched Problems

1. (A) $6\sqrt{3}$ (B) $-2\sqrt{x}$ (C) $6\sqrt{y} - 5\sqrt{7}$

2. (A) $\sqrt{3}$ (B) $\dfrac{11\sqrt{2}}{2}$ or $\dfrac{11}{2}\sqrt{2}$

EXERCISE 8-5

Simplify by combining as many terms as possible. All variables represent positive real numbers. Use exact radical forms only.

A
1. $6\sqrt{3} + 4\sqrt{3}$
2. $4\sqrt{2} + 5\sqrt{2}$
3. $6\sqrt{x} - 3\sqrt{x}$
4. $12\sqrt{m} - 3\sqrt{m}$
5. $4\sqrt{7} - 3\sqrt{5}$
6. $2\sqrt{3} + 5\sqrt{2}$
7. $5\sqrt{x} - \sqrt{x}$
8. $8\sqrt{a} - 3\sqrt{a}$
9. $2\sqrt{a} + 7\sqrt{a}$
10. $3\sqrt{x} + 2\sqrt{x}$
11. $\sqrt{x} - 6\sqrt{x}$
12. $\sqrt{a} - 4\sqrt{a}$
13. $3\sqrt{5} - \sqrt{5} + 2\sqrt{5}$
14. $4\sqrt{7} - 6\sqrt{7} + \sqrt{7}$
15. $\sqrt{a} + 3\sqrt{a} - 5\sqrt{a}$
16. $\sqrt{b} - 2\sqrt{b} + 3\sqrt{b}$
17. $4\sqrt{2} - \sqrt{3} + 2\sqrt{2}$
18. $2\sqrt{5} + 6\sqrt{4} - 4\sqrt{5}$
19. $3\sqrt{x} + 2\sqrt{y} - \sqrt{x}$
20. $5\sqrt{x} - 3\sqrt{y} - \sqrt{y}$
21. $\sqrt{7} - 2\sqrt{5} + 3\sqrt{7} + 7\sqrt{5}$
22. $\sqrt{3} - 2\sqrt{2} + 3\sqrt{3} + \sqrt{2}$
23. $\sqrt{a} + 2\sqrt{b} - 3\sqrt{b} + 4\sqrt{a}$
24. $-\sqrt{x} - 3\sqrt{y} + 5\sqrt{x} - 7\sqrt{y}$

B
25. $\sqrt{12} - \sqrt{3}$
26. $\sqrt{50} + 3\sqrt{2}$
27. $\sqrt{18} + \sqrt{2}$
28. $3\sqrt{8} - 2\sqrt{2}$
29. $10\sqrt{20} + 5\sqrt{80}$
30. $\sqrt{63} - 3\sqrt{28}$
31. $\sqrt{8} + 2\sqrt{27}$
32. $2\sqrt{12} + 3\sqrt{18}$
33. $\sqrt{4x} - \sqrt{9x}$
34. $\sqrt{8mn} + 2\sqrt{18mn}$
35. $\sqrt{8a} + \sqrt{18a}$
36. $\sqrt{12x} - \sqrt{27x}$
37. $\sqrt{48x} - \sqrt{75x}$
38. $\sqrt{20a} + \sqrt{45a}$
39. $\sqrt{24} - \sqrt{12} + 3\sqrt{3}$
40. $\sqrt{8} - \sqrt{20} + 4\sqrt{2}$
41. $\sqrt{2x} + \sqrt{8x} + \sqrt{18x}$
42. $\sqrt{3a} + \sqrt{12a} + \sqrt{27a}$
43. $\sqrt{x} + \sqrt{x^3}$
44. $\sqrt{x} - \sqrt{x^5}$
45. $\sqrt{a} - \sqrt{4a^3} + \sqrt{9a}$
46. $2\sqrt{x^3} + \sqrt{4x} - 5\sqrt{x}$

C
47. $\sqrt{125x^7}$
48. $\sqrt{64a^5}$
49. $\sqrt{27x^3y}$
50. $\sqrt{8xy^5}$
51. $\sqrt{x^3y^4z^5}$
52. $\sqrt{x^4y^5z^6}$
53. $\sqrt{32ab^2c^3}$
54. $\sqrt{12a^5b^4c^3}$
55. $\sqrt{\frac{2}{3}} - \sqrt{\frac{3}{2}}$
56. $\sqrt{\frac{1}{8}} + \sqrt{8}$
57. $\sqrt{\frac{xy}{2}} + \sqrt{8xy}$
58. $\sqrt{\frac{3uv}{2}} - \sqrt{24uv}$
59. $\sqrt{12} - \sqrt{\frac{1}{2}}$
60. $\sqrt{\frac{3}{5}} + 2\sqrt{20}$
61. $\sqrt{\frac{1}{2}} + \frac{\sqrt{2}}{2} + \sqrt{8}$
62. $\frac{\sqrt{3}}{3} + 2\sqrt{\frac{1}{3}} + \sqrt{12}$
63. $\frac{\sqrt{x}}{4} + \sqrt{\frac{x}{4}}$
64. $\sqrt{\frac{x}{9}} - \frac{\sqrt{x}}{9}$
65. $\sqrt{\frac{8x}{25}} - \frac{7\sqrt{2x}}{25}$
66. $\frac{\sqrt{18x}}{36} + \sqrt{\frac{2x}{36}}$

8-6 Products and Quotients Involving Radicals

♦ Products
♦ Quotients—Rationalizing Denominators

The properties of radicals allowed us to simplify some products and quotients involving radicals, for example,

$$\sqrt{4}\sqrt{9} = \sqrt{36} = 6 \quad \text{and} \quad \frac{\sqrt{18}}{\sqrt{2}} = \sqrt{\frac{18}{2}} = \sqrt{9} = 3$$

We now consider several types of products and quotients that involve radicals in more complicated ways. The distributive law again plays a central role in our approach to simplifying such expressions.

In the examples in this section, all variables represent positive real numbers.

♦ PRODUCTS

The following examples illustrate the treatment of radicals in products.

Example 1
Multiplying Out

Multiply and simplify:

(A) $\sqrt{2}(\sqrt{2} - 3)$ (B) $\sqrt{x}(\sqrt{x} + 5)$ (C) $(\sqrt{5} - 2)(\sqrt{5} + 4)$
(D) $(\sqrt{x} - 3)(\sqrt{x} + 5)$ (E) $(\sqrt{a} + \sqrt{b})^2$

Solution (A) $\sqrt{2}(\sqrt{2} - 3) \;\boxed{= \sqrt{2}\sqrt{2} - 3\sqrt{2}} = 2 - 3\sqrt{2}$

(B) $\sqrt{x}(\sqrt{x} + 5) \;\boxed{= \sqrt{x}\sqrt{x} + 5\sqrt{x}} = x + 5\sqrt{x}$

(C) $(\sqrt{5} - 2)(\sqrt{5} + 4)$ You can use the FOIL method to multiply out this product.

$$\boxed{= \sqrt{5}\sqrt{5} + 4\sqrt{5} - 2\sqrt{5} - 8}$$
$$= 5 + 2\sqrt{5} - 8$$
$$= 2\sqrt{5} - 3$$

(D) $(\sqrt{x} - 3)(\sqrt{x} + 5) \;\boxed{= \sqrt{x}\sqrt{x} - 3\sqrt{x} + 5\sqrt{x} - 15}$
$$= x + 2\sqrt{x} - 15$$

(E) $(\sqrt{a} + \sqrt{b})^2$ *Note:* $(\sqrt{a} + \sqrt{b})^2 \neq a + b.$
$$= (\sqrt{a})^2 + 2\sqrt{a}\sqrt{b} + (\sqrt{b})^2$$
$$= a + 2\sqrt{ab} + b$$

Matched Problem 1

Multiply and simplify:

(A) $\sqrt{3}(2 - \sqrt{3})$ (B) $\sqrt{y}(2 + \sqrt{y})$ (C) $(\sqrt{3} - 1)(\sqrt{3} + 4)$
(D) $(\sqrt{y} + 2)(\sqrt{y} - 5)$ (E) $(\sqrt{x} - \sqrt{y})^2$

Example 2
Evaluating a Polynomial

Show that $(2 - \sqrt{3})$ is a solution of the equation $x^2 - 4x + 1 = 0$.

Solution
$$x^2 - 4x + 1 \overset{?}{=} 0 \qquad \text{Substitute } 2 - \sqrt{3} \text{ for } x.$$
$$(2 - \sqrt{3})^2 - 4(2 - \sqrt{3}) + 1 = 4 - 4\sqrt{3} + 3 - 8 + 4\sqrt{3} + 1$$
$$= 0$$

Matched Problem 2   Show that $(2 + \sqrt{3})$ is a solution of $x^2 - 4x + 1 = 0$.

♦ **QUOTIENTS—RATIONALIZING DENOMINATORS**

The next examples illustrate the treatment of radicals in quotients.

Example 3
Reducing to
Lowest Terms

Reduce to lowest terms $\dfrac{6 - \sqrt{12}}{10}$.

Solution

$$\frac{6 - \sqrt{12}}{10} = \frac{6 - \sqrt{4 \cdot 3}}{10} \qquad \text{Be careful: } \frac{6 - \sqrt{12}}{10} \neq \frac{\overset{3}{\cancel{6}} - \sqrt{12}}{\underset{5}{\cancel{10}}}$$

We cannot cancel *terms*.

$$= \frac{6 - 2\sqrt{3}}{10} = \frac{\overset{1}{\cancel{2}}(3 - \sqrt{3})}{\underset{5}{\cancel{10}}} = \frac{3 - \sqrt{3}}{5}$$

Matched Problem 3   Reduce to lowest terms $\dfrac{12 - \sqrt{32}}{8}$.

Recall that to express $\dfrac{\sqrt{2}}{\sqrt{3}}$ in simplest radical form, we multiplied the numerator and denominator by $\sqrt{3}$ to clear the denominator of the radical:

$$\frac{\sqrt{2}}{\sqrt{3}} = \frac{\sqrt{2} \cdot \sqrt{3}}{\sqrt{3} \cdot \sqrt{3}} = \frac{\sqrt{6}}{3}$$

The denominator is thus converted to a rational number. The process of converting irrational denominators to rational forms is called **rationalizing the denominator.** How can we rationalize the binomial denominator in

$$\frac{1}{\sqrt{3} - \sqrt{2}}$$

Multiplying the numerator and denominator by $\sqrt{3}$ or $\sqrt{2}$ will not help. You should try it. However, recall the product

$$(a - b)(a + b) = a^2 - b^2$$

This suggests that if we multiply the numerator and denominator by the denominator, but with the middle sign changed, we can obtain squares of each term in the denominator. Thus,

$$\frac{1}{\sqrt{3} - \sqrt{2}} = \frac{1(\sqrt{3} + \sqrt{2})}{(\sqrt{3} - \sqrt{2})(\sqrt{3} + \sqrt{2})}$$

$$= \frac{\sqrt{3} + \sqrt{2}}{(\sqrt{3})^2 - (\sqrt{2})^2}$$

$$= \frac{\sqrt{3} + \sqrt{2}}{3 - 2} = \sqrt{3} + \sqrt{2}$$

The expressions $\sqrt{3} + \sqrt{2}$ and $\sqrt{3} - \sqrt{2}$ are called **conjugates** of each other.

Example 4
Rationalizing the Denominator

Rationalize denominators and simplify:

(A) $\dfrac{\sqrt{2}}{\sqrt{6} - 2}$ (B) $\dfrac{\sqrt{x} - \sqrt{y}}{\sqrt{x} + \sqrt{y}}$

Solution

(A) $\dfrac{\sqrt{2}}{\sqrt{6} - 2} = \dfrac{\sqrt{2}(\sqrt{6} + 2)}{(\sqrt{6} - 2)(\sqrt{6} + 2)} = \dfrac{\sqrt{12} + 2\sqrt{2}}{6 - 4}$

$$= \frac{2\sqrt{3} + 2\sqrt{2}}{2} = \frac{2(\sqrt{3} + \sqrt{2})}{2} = \sqrt{3} + \sqrt{2}$$

(B) $\dfrac{\sqrt{x} - \sqrt{y}}{\sqrt{x} + \sqrt{y}} = \dfrac{(\sqrt{x} - \sqrt{y})(\sqrt{x} - \sqrt{y})}{(\sqrt{x} + \sqrt{y})(\sqrt{x} - \sqrt{y})}$

$$= \frac{x - 2\sqrt{xy} + y}{x - y}$$

Matched Problem 4

Rationalize denominators and simplify:

(A) $\dfrac{\sqrt{2}}{\sqrt{2} + 3}$ (B) $\dfrac{\sqrt{x} + \sqrt{y}}{\sqrt{x} - \sqrt{y}}$

Answers to Matched Problems

1. (A) $2\sqrt{3} - 3$ (B) $2\sqrt{y} + y$ (C) $3\sqrt{3} - 1$
   (D) $y - 3\sqrt{y} - 10$ (E) $x - 2\sqrt{xy} + y$
2. $(2 + \sqrt{3})^2 - 4(2 + \sqrt{3}) + 1 = 4 + 4\sqrt{3} + 3 - 8 - 4\sqrt{3} + 1 = 0$
3. $\dfrac{3 - \sqrt{2}}{2}$ 4. (A) $\dfrac{2 - 3\sqrt{2}}{-7}$ or $\dfrac{-2 + 3\sqrt{2}}{7}$ (B) $\dfrac{x + 2\sqrt{xy} + y}{x - y}$

EXERCISE 8-6

A *Multiply and simplify where possible:*

1. $4(\sqrt{5} + 2)$ 2. $3(\sqrt{3} - 4)$ 3. $2(5 - \sqrt{2})$
4. $5(3 - \sqrt{5})$ 5. $\sqrt{2}(\sqrt{2} + 3)$ 6. $\sqrt{3}(\sqrt{3} + 2)$
7. $\sqrt{5}(\sqrt{5} - 4)$ 8. $\sqrt{6}(\sqrt{6} + 4)$
9. $\sqrt{5}(3 - \sqrt{5})$ 10. $\sqrt{3}(2 - \sqrt{3})$
11. $\sqrt{a}(\sqrt{a} + 1)$ 12. $\sqrt{x}(5 - \sqrt{x})$

B
13. $\sqrt{y}(2 - \sqrt{y})$ 14. $\sqrt{z}(\sqrt{z} + 6)$
15. $\sqrt{6}(\sqrt{2} - 1)$ 16. $\sqrt{3}(5 + \sqrt{6})$
17. $\sqrt{2}(\sqrt{6} - \sqrt{2})$ 18. $\sqrt{20}(\sqrt{5} - 1)$
19. $(\sqrt{3} - 4)(\sqrt{3} + 1)$ 20. $(\sqrt{5} + 2)(\sqrt{5} - 3)$
21. $(\sqrt{6} + 1)(3 - \sqrt{6})$ 22. $(\sqrt{7} - 4)(2 + \sqrt{3})$
23. $(\sqrt{x} - 3)(\sqrt{x} + 4)$ 24. $(\sqrt{x} + 3)(\sqrt{x} - 1)$
25. $(\sqrt{x} + 2)(3 + \sqrt{x})$ 26. $(\sqrt{x} - 5)(2 + \sqrt{x})$
27. $(\sqrt{x} + 4)(6 - \sqrt{x})$ 28. $(\sqrt{x} - 1)(5 - \sqrt{x})$

29. $(\sqrt{3} - 2)^2$ **30.** $(\sqrt{2} - 3)^2$

31. $(\sqrt{5} + 4)^2$ **32.** $(\sqrt{6} + 1)^2$

33. $(\sqrt{x} - 1)^2$ **34.** $(\sqrt{x} - 3)^2$

35. $(\sqrt{x} + 2)^2$ **36.** $(\sqrt{x} + 4)^2$

37. $(4 + \sqrt{x})^2$ **38.** $(5 - \sqrt{x})^2$

39. $(1 - \sqrt{x})^2$ **40.** $(3 + \sqrt{x})^2$

41. $(\sqrt{3} + 2)(\sqrt{3} - 2)$ **42.** $(\sqrt{5} - 3)(\sqrt{5} + 3)$

43. $(\sqrt{x} - 4)(\sqrt{x} + 4)$ **44.** $(\sqrt{x} + 1)(\sqrt{x} - 1)$

45. $(6 - \sqrt{x})(6 + \sqrt{x})$ **46.** $(2 + \sqrt{x})(2 - \sqrt{x})$

47. $(3\sqrt{5} + 5)(5\sqrt{3} + 3)$ **48.** $(2\sqrt{6} + 1)(6\sqrt{2} + 1)$

49. $(4\sqrt{x} - 3)(5\sqrt{x} + 6)$ **50.** $(6\sqrt{x} + 5)(4\sqrt{x} - 3)$

51. $(2\sqrt{2} - 5)(3\sqrt{2} + 2)$ **52.** $(4\sqrt{3} - 1)(3\sqrt{3} - 2)$

53. $(3\sqrt{x} - 2)(2\sqrt{x} - 3)$ **54.** $(4\sqrt{y} - 2)(3\sqrt{y} + 1)$

Show that the given number is a solution to the given equation:

55. $3 - \sqrt{2}$, $x^2 - 6x + 7 = 0$

56. $3 + \sqrt{2}$, $x^2 - 6x + 7 = 0$

57. $1 + \sqrt{3}$, $x^2 - 2x - 2 = 0$

58. $1 - \sqrt{3}$, $x^2 - 2x - 2 = 0$

59. $4 - \sqrt{5}$, $x^2 - 8x + 11 = 0$

60. $4 + \sqrt{5}$, $x^2 - 8x + 11 = 0$

61. $2 + \sqrt{6}$, $x^2 - 4x - 2 = 0$

62. $2 - \sqrt{6}$, $x^2 - 4x - 2 = 0$

Reduce by removing common factors from numerator and denominator:

63. $\dfrac{8 + 4\sqrt{2}}{12}$ **64.** $\dfrac{6 - 2\sqrt{3}}{6}$ **65.** $\dfrac{-3 - 6\sqrt{5}}{9}$

66. $\dfrac{-4 + 2\sqrt{7}}{4}$ **67.** $\dfrac{6 - \sqrt{18}}{3}$ **68.** $\dfrac{10 + \sqrt{8}}{2}$

69. $\dfrac{2 - \sqrt{12}}{2}$ **70.** $\dfrac{\sqrt{20} + 8}{2}$

C *Rationalize denominators and simplify:*

71. $\dfrac{1}{\sqrt{11} + 3}$ **72.** $\dfrac{1}{\sqrt{5} + 2}$ **73.** $\dfrac{2}{\sqrt{5} + 1}$

74. $\dfrac{4}{\sqrt{6} - 2}$ **75.** $\dfrac{\sqrt{y}}{\sqrt{y} + 3}$ **76.** $\dfrac{\sqrt{x}}{\sqrt{x} - 2}$

77. $\dfrac{\sqrt{3} + 2}{\sqrt{3} - 2}$ **78.** $\dfrac{\sqrt{2} - 1}{\sqrt{2} + 2}$ **79.** $\dfrac{\sqrt{x} + 2}{\sqrt{x} - 3}$

80. $\dfrac{\sqrt{a} - 3}{\sqrt{a} + 2}$ **81.** $\dfrac{\sqrt{5} + \sqrt{3}}{\sqrt{5} - \sqrt{3}}$ **82.** $\dfrac{\sqrt{x}}{\sqrt{x} - \sqrt{y}}$

83. $\dfrac{1 + \sqrt{3}}{\sqrt{3} - \sqrt{2}}$ **84.** $\dfrac{\sqrt{7}}{\sqrt{7} + \sqrt{2}}$

$8\text{-}7$ **Radical Equations**

- ♦ Solving Radical Equations
- ♦ Solutions and Graphs

An equation that involves a variable within a radical sign is called a **radical equation.** Some examples are

$$\sqrt{x - 3} = 5 \qquad 2\sqrt{x} = x - 3 \qquad \sqrt{x + 7} = \sqrt{x^2 + 5}$$

In this section we solve such equations and also consider again the relationship between solutions and points on a related graph.

♦ SOLVING RADICAL EQUATIONS

The key to solving radical equations is the following property of real numbers:

If $a = b$, then $a^2 = b^2$.

Restated, this says, if both sides of an equation are squared, then every solution of the original equation is also a solution to the resulting squared equation. To solve a radical equation, we will square both sides of the equation to remove the radicals.

Example 1
Solving a Radical Equation

Solve $\sqrt{x - 3} = 5$.

Solution

$$\sqrt{x - 3} = 5 \qquad \text{Square both sides.}$$

$$(\sqrt{x - 3})^2 = 5^2 \qquad \text{Recall } (\sqrt{a})^2 = a. \text{ This is the definition of } \sqrt{a}.$$

$$x - 3 = 25$$

$$x = 28$$

Check $\quad \sqrt{28 - 3} = \sqrt{25} = 5$ as desired.

Matched Problem 1   Solve $\sqrt{3x + 1} = 4$.

Example 2
Solving a Radical Equation

Solve $5\sqrt{x} = x + 6$.

Solution

$$5\sqrt{x} = x + 6 \qquad \text{Square both sides.}$$

$$(5\sqrt{x})^2 = (x + 6)^2 \qquad \text{Simplify.}$$

$$25x = x^2 + 12x + 36 \qquad \text{Simplify.}$$

$$0 = x^2 - 13x + 36 \qquad \text{Factor.}$$

$$0 = (x - 9)(x - 4) \qquad \text{Solve by setting each factor equal to 0.}$$

$$x = 9 \qquad \text{or} \qquad x = 4$$

Check $\quad x = 9: 5\sqrt{x} = 5\sqrt{9} = 5 \cdot 3 = 15 \qquad$ We substitute 9 into the left side.

$$x + 6 = 9 + 6 = 15 \qquad \text{We substitute 9 into the right side.}$$

Therefore, 9 is a solution, since the two sides are equal when 9 is substituted for x.

$$x = 4: 5\sqrt{x} = 5\sqrt{4} = 5 \cdot 2 = 10$$

$$x + 6 = 4 + 6 = 10$$

Therefore, 4 is also a solution.

Matched Problem 2   Solve $5\sqrt{x} = x + 4$.

The following example shows why it is important to check all solutions.

Example 3 **Solving a Radical** **Equation**	Solve $2\sqrt{x} = x - 3$.

Solution

$$2\sqrt{x} = x - 3 \qquad \text{Square both sides.}$$
$$4x = x^2 - 6x + 9 \qquad \text{Simplify.}$$
$$0 = x^2 - 10x + 9 \qquad \text{Solve by factoring.}$$
$$0 = (x - 9)(x - 1)$$
$$x = 9 \qquad \text{or} \qquad x = 1$$

Check

$$x = 9: \ 2\sqrt{x} = 2\sqrt{9} = 2 \cdot 3 = 6$$
$$x - 3 = 6$$

Therefore, 9 is a solution.

$$x = 1: \ 2\sqrt{x} = 2\sqrt{1} = 2 \cdot 1 = 2$$
$$x - 3 = -2$$

Therefore, 1 is not a solution.

   In this example, one of the two possible solutions turns out not to be a solution. This is because the converse of the property "$a = b$ implies $a^2 = b^2$" is not true. Squaring both sides of an equation may result in extra solutions that do not satisfy the original equation. These are called **extraneous solutions.** Each solution of the original is included in the solutions of the square, but the converse is not necessarily true; the square equation may have solutions that are not solutions to the original equation. In summary:

> Squaring both sides of an equation will not lose any solutions of the original equation, but may introduce extraneous solutions that do not satisfy the original equation.

Matched Problem 3   Solve $\sqrt{x} = x - 12$.

$x = x^2 - 24x + 144$

$x^2 - 25x + 144$

$\dfrac{16}{9}$

$\overline{54}$

◆ SOLUTIONS AND GRAPHS

Consider the equation $\sqrt{x} = 2$, or, equivalently, $\sqrt{x} - 2 = 0$. We can graph the related equation $y = \sqrt{x} - 2$ by plotting points:

x	y
0	-2
1	-1
4	0
9	1
16	2
25	3

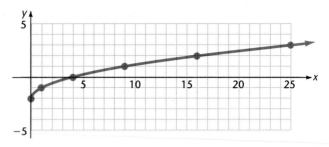

Figure 1

Notice that we chose only nonnegative values for x, since the square root of a negative number is not real. Also, we chose perfect square values of x to make the calculation of the square root easy. The solution to the equation $\sqrt{x} = 2$ is $x = 4$, corresponding to the point where the graph crosses the x axis.

Example 4
Relating Solutions and Graphs

Graph the equation $y = \sqrt{x - 3} - 5$ and solve the equation $\sqrt{x - 3} - 5 = 0$. Relate the solutions of the equation to points on the graph.

Solution

We make a table of values and plot the corresponding points. Notice that the smallest value of x that is allowed is $x = 3$.

x	y
3	-5
4	-4
7	-3
12	-2
19	-1
28	0
39	1
52	2

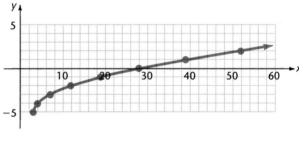

Figure 2

Rewriting the equation yields:

$$\sqrt{x - 3} - 5 = 0$$
$$\sqrt{x - 3} = 5$$

This is the equation solved in Example 1, so the solution is $x = 28$, corresponding to the point where the graph crosses the x axis, that is, the x intercept.

Matched Problem 4

Graph the equation $y = \sqrt{3x + 1} - 4$ and solve the equation $\sqrt{3x + 1} - 4 = 0$. Relate the solutions of the equation to points on the graph.

Graphs can also illustrate how squaring both sides of an equation can introduce extraneous roots. Consider the equation from Example 3:

$$2\sqrt{x} = x - 3$$

Rewriting the equation as $2\sqrt{x} - x + 3 = 0$ and graphing the related equation $y = 2\sqrt{x} - x + 3$ would lead to the graph in Figure 3. The solution found in Example 3, $x = 9$, corresponds to the point where the graph crosses the x axis.

However, if we square both sides of the original equation

$$2\sqrt{x} = x - 3$$

we obtain

$$4x = (x - 3)^2 = x^2 - 6x + 9$$

or $x^2 - 10x + 9 = 0$. The graph of $y = x^2 - 10x + 9$ is shown in Figure 4. Notice that this graph crosses the x axis at *two* points, $x = 1$ and $x = 9$. The first point does not correspond to a solution of the original equation.

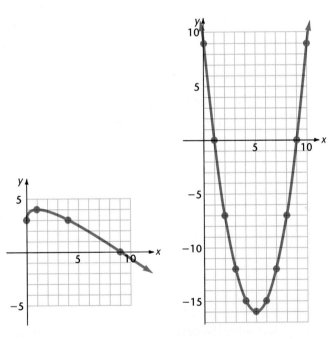

Figure 3   **Figure 4**

Answers to   **1.** $x = 5$ **2.** $x = 1$ and $x = 16$ **3.** $x = 16$ ($x = 9$ is extraneous)
Matched Problems   **4.** $x = 5$ is the x intercept

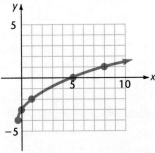

EXERCISE 8-7

Solve:

A **1.** $\sqrt{2x + 6} = 4$ **2.** $\sqrt{2x - 6} = 4$

  3. $\sqrt{2x - 1} = 3$ **4.** $\sqrt{2x + 1} = 3$

  5. $\sqrt{2x + 1} = -1$ **6.** $\sqrt{2x - 1} = -3$

  7. $\sqrt{15 - 2x} = 3$ **8.** $\sqrt{21 - 5x} = 4$

  9. $\sqrt{33 - 8x} = 5$ **10.** $\sqrt{49 - 2x} = 6$

 11. $\sqrt{3 - 11x} = 5$ **12.** $\sqrt{-1 - 3x} = 3$

 13. $-\sqrt{1 + x} = 2$ **14.** $-\sqrt{5 - x} = 2$

 15. $\sqrt{x + 5} = \sqrt{3x - 9}$ **16.** $\sqrt{5 - x} = \sqrt{x + 1}$

 17. $\sqrt{2x - 3} = \sqrt{x + 2}$ **18.** $\sqrt{x - 2} = \sqrt{10 - x}$

B **19.** $\sqrt{x} = x$ **20.** $\sqrt{x} = -x$

 21. $3\sqrt{x} = x$ **22.** $\sqrt{2x} = \dfrac{1}{2}x$

 23. $\sqrt{3x} = \dfrac{1}{3}x$ **24.** $2\sqrt{x} = x$

 25. $6\sqrt{x} = x + 5$ **26.** $7\sqrt{x} = x + 10$

 27. $3\sqrt{x} = x + 2$ **28.** $4\sqrt{x} = x + 3$

 29. $4\sqrt{x} = x + 4$ **30.** $6\sqrt{x} = x + 9$

 31. $5\sqrt{x} = x + 4$ **32.** $5\sqrt{x} = x + 6$

 33. $8\sqrt{x} = x + 15$ **34.** $7\sqrt{x} = x + 12$

 35. $6\sqrt{x} = x + 8$ **36.** $9\sqrt{x} = x + 20$

 37. $\sqrt{x} = x - 2$ **38.** $\sqrt{x} = x - 6$

 39. $2\sqrt{x} = x - 3$ **40.** $2\sqrt{x} = x - 8$

 41. $5\sqrt{x} = 6 - x$ **42.** $4\sqrt{x} = x - 12$

 43. $2\sqrt{x} = 15 - x$ **44.** $2\sqrt{x} = 8 - x$

 45. $3\sqrt{x} = 4 - x$ **46.** $3\sqrt{x} = 10 - x$

C **47.** $8\sqrt{x - 5} = x + 10$ **48.** $\sqrt{x + 4} = x - 2$

 49. $\sqrt{x + 3} = x + 1$ **50.** $6\sqrt{x - 5} = x + 3$

 51. $5\sqrt{x + 4} = x - 2$ **52.** $2\sqrt{x + 9} = x + 1$

 53. $2\sqrt{x - 5} = 5 - x$ **54.** $3\sqrt{x - 5} = 5 - x$

 55. $\sqrt{x - 1} = x - 3$ **56.** $\sqrt{x + 5} = x - 1$

 57. $2\sqrt{x + 2} = x - 1$ **58.** $3\sqrt{x + 5} = x + 5$

 59. $1 + \sqrt{x} = x - 1$ **60.** $3 + \sqrt{2x} = x - 1$

 61. $4 + \sqrt{3x} = x - 2$ **62.** $5 + \sqrt{x} = x + 3$

 63. $\sqrt{x - 3} = \sqrt{4x}$ **64.** $\sqrt{x + 1} = \sqrt{3 - 2x}$

 65. $\sqrt{6 - 3x} = \sqrt{1 - 2x}$ **66.** $\sqrt{5x} = \sqrt{2x - 6}$

Graph the equation that results from setting y equal to the given expression. Solve the equation that results from setting the given expression equal to 0. Relate the solutions to the x intercepts of the graph.

 67. $\sqrt{2x - 1} - 3$ **68.** $\sqrt{2x + 1} - 3$

 69. $2 + \sqrt{1 + x}$ **70.** $2 + \sqrt{5 - x}$

 71. $\sqrt{x + 5} - \sqrt{3x - 9}$ **72.** $\sqrt{5 - x} - \sqrt{x + 1}$

CHAPTER SUMMARY

8-1, 8-2 NATURAL NUMBER EXPONENTS, INTEGER EXPONENTS

The definition of a^p for an integer p is given by:

Definition of a^p (p an integer, a a real number)

---

1. If p is a positive integer, then $a^p = a \cdot a \cdot \cdots \cdot a$ (p factors of a).

2. If $p = 0$, then $a^p = 1$, $a \neq 0$.

3. If p is a negative integer, then $a^p = \dfrac{1}{a^{-p}}$, $a \neq 0$.

The **five basic exponent laws** can be summarized as follows:

Laws of Exponents (n and m integers, a and b real numbers)

1. $a^m a^n = a^{m+n}$
2. $(a^n)^m = a^{mn}$
3. $(ab)^m = a^m b^m$
4. $\left(\dfrac{a}{b}\right)^m = \dfrac{a^m}{b^m}$ $\qquad b \neq 0$
5. $\dfrac{a^m}{a^n} = a^{m-n} = \dfrac{1}{a^{n-m}}$ $\qquad a \neq 0$

8-3 SCIENTIFIC NOTATION

A number is written in **scientific notation** when it is expressed as a product of a number between 1 and 10, excluding 10, and an integer power of 10.

8-4 SQUARE ROOTS AND RADICALS

A number b is a **square root** of the number a if $b^2 = a$. A positive real number a has exactly two square roots: one positive denoted by $\sqrt{a}$ and one negative denoted by $-\sqrt{a}$. Negative real numbers have no real-number square roots; the square root of 0 is 0. The square root **radical** ($\sqrt{}$) satisfies these properties for a and b nonnegative real numbers:

1. $\sqrt{a^2} = a$
2. $\sqrt{a}\sqrt{b} = \sqrt{ab}$
3. $\dfrac{\sqrt{a}}{\sqrt{b}} = \sqrt{\dfrac{a}{b}}$ $\qquad b \neq 0$

An algebraic expression that contains square root radicals is in **simplest radical form** if all three of the following conditions are satisfied:

1. No **radicand** (the expression within the radical sign) when expressed in completely factored form contains a factor raised to a power greater than 1. ($\sqrt{x^3}$ violates this condition.)
2. No radical appears in a denominator. ($3/\sqrt{5}$ violates this condition.)
3. No fraction appears within a radical. ($\sqrt{2/3}$ violates this condition.)

8-5 SUMS AND DIFFERENCES INVOLVING RADICALS

Two terms involving the same radical expression can be combined into a single term by adding the numerical coefficients; for example, $3\sqrt{2} + 4\sqrt{2} = 7\sqrt{2}$.

8-6 PRODUCTS AND QUOTIENTS INVOLVING RADICALS

The process of converting denominators involving radicals to rational forms is called **rationalizing the denominator.** The product

$$(a - b)(a + b) = a^2 - b^2$$

is useful in rationalizing binomial denominators involving square root radicals.

8-7 RADICAL EQUATIONS

A **radical equation** involves a variable within a radical. Such equations are solved by removing radicals by squaring both sides of the equation. This operation may introduce **extraneous solutions.**

CHAPTER REVIEW EXERCISE

Work through all the problems in this chapter review and check the answers in the back of the book. Answers to all review problems are there, and following each answer is a number in italics indicating the section in which that type of problem is discussed. Where weaknesses show up, review appropriate sections in the text.

All variables in these problems represent positive real numbers.

A *Evaluate each expression. Write the answer as an integer or fraction without exponents, or as a decimal fraction.*

1. 3^4

2. 3^{-2}

3. $\left(\dfrac{1}{3}\right)^0$

4. $\dfrac{1}{3^{-2}}$

5. $-\sqrt{25}$

6. 3.0×10^{-5}

7. 2.34×10^7

8. $\dfrac{3^{-2}}{3}$

9. $(3^{-2})^{-1}$

10. $10^{-21}10^{19}$

11. $(25^{-5})(25^5)$

12. $(3 \times 10^4)(2 \times 10^{-6})$

13. $(2 \times 10^4)^{-2}$

14. $(3 \times 10^4) \div (2 \times 10^{-6})$

Rewrite each expression in a simpler form using only positive exponents and putting the answer in simplest radical form where necessary:

15. $\left(\dfrac{2x^2}{3y^3}\right)^2$

16. $(x^2y^{-3})^{-1}$

17. $-\sqrt{25}$

18. $\sqrt{4x^2y^4}$

19. $\sqrt{\dfrac{25}{y^2}}$

20. $4\sqrt{x} - 7\sqrt{x}$

21. $\sqrt{5}(\sqrt{5} + 2)$

22. x^4x^7

23. $(x^4)^7$

24. $\dfrac{x^4}{x^7}$

25. $(x^4x^7)^2$

26. $\left(\dfrac{x^4}{y^7}\right)^2$

27. x^4x^{-7}

28. $(x^4)^{-7}$

29. $\dfrac{x^4}{x^{-7}}$

30. $\left(\dfrac{x^{-4}}{y^7}\right)^2$

31. $(x^4y^{-7})^2$

32. $\dfrac{x^0}{y^{-7}}$

33. $\dfrac{x^4}{y^0}$

34. $\dfrac{1}{\sqrt{3}}$

35. $\dfrac{2}{\sqrt{x}}$

36. $\sqrt{3x}\sqrt{12x}$

Solve:

37. $\sqrt{x} = 4$

38. $\sqrt{1-x} = -1$

39. $\sqrt{1+x} = 1$

40. $\sqrt{1-x} = 1$

41. $\sqrt{x+2} = 3$

42. $\sqrt{x-3} = 3$

Write in scientific notation:

43. 65,430,000,000,000

44. 0.000 000 000 078 9

B *Rewrite each expression in a simpler form using only positive exponents and putting the answer in simplest radical form where necessary:*

45. $\dfrac{1}{(2x^2y^{-3})^{-2}}$

46. $\dfrac{3m^4n^{-7}}{6m^2n^{-2}}$

47. $\sqrt{36x^4y^7}$

48. $\dfrac{1}{\sqrt{2y}}$

49. $\sqrt{\dfrac{3x}{2y}}$

50. $\sqrt{\tfrac{2}{3}} + \sqrt{\tfrac{3}{2}}$

51. $(\sqrt{3} - 1)(\sqrt{3} + 2)$

52. $(-3ab^{-2})^2$

53. $x^{-2}x^2$

54. $\left(\dfrac{-x^2}{y}\right)^{-3}$

55. $\left(\dfrac{a^{-2}}{3b^4}\right)^{-1}$

56. $\sqrt{12} + \sqrt{75}$

57. $\sqrt{18x} + \sqrt{8x^3}$

Evaluate each expression and write the answer in scientific notation:

58. $\dfrac{(480,000)(0.005)}{1,200,000}$

59. $\dfrac{18 \times 10^{-5}}{12 \times 10^4}$

60. $\dfrac{15 \times 10^4}{6 \times 10^{-5}}$

In Problems 61–64, solve the equation:

61. $2\sqrt{x} = x + 1$

62. $2\sqrt{x} = x$

63. $2\sqrt{x} = 2x$

64. $\sqrt{x+4} = x - 2$

65. Show that $-1 + \sqrt{3}$ is a solution to $x^2 + 2x - 2 = 0$.

C *Rewrite each expression in a simpler form using only positive exponents and putting the answer in simplest radical form where necessary:*

66. $\left(\dfrac{9m^3n^{-3}}{3m^{-2}n^2}\right)^{-2}$

67. $(x^{-1} + y^{-1})^{-1}$

68. $\dfrac{\sqrt{8m^3n^4}}{\sqrt{12m^2}}$

69. $\dfrac{\sqrt{3}}{\sqrt{3} - \sqrt{2}}$

70. $\dfrac{\sqrt{x}-2}{\sqrt{x}+2}$ **71.** $\sqrt{4x^4+16x^2}$

72. $\dfrac{x^{-1}10^{-2}x^3 10^4}{10^{-5}x^6 10^7 x^{-8}}$ **73.** $\left(\dfrac{2a^{-2}b^3 c^{-4}}{3a^{-3}b^4 c^{-5}}\right)^{-1}$

74. $\sqrt{\dfrac{2a^{-2}b^3 c^{-4}}{3a^{-6}b^7 c^{-8}}}$

In Problems 75–78, solve the equation:

75. $2\sqrt{x}=x-3$ **76.** $\sqrt{3x}=\dfrac{1}{2}$

77. $1+\sqrt{2x}=-3+x$ **78.** $\sqrt{x+5}=x-1$

79. Because the volume of mercury increases almost linearly with temperature over a fairly wide temperature range, mercury is often used in thermometers. If 1 cubic centimeter of mercury at 0°C is heated to a temperature of T°C, its volume is given by the formula

$$V = 1 + (1.8 \times 10^{-4})T$$

Find the volume of the sample at (2×10^2)°C as a decimal fraction.

80. If a is a square root of b, then does $a^2 = b$ or does $b^2 = a$?

CHAPTER PRACTICE TEST

The following practice test is provided for you to test your knowledge of the material in this chapter. You should try to complete it in 50 minutes or less. The answers in the back of the book indicate the section in the text that covers the material in the question. Actual tests in your class may vary from the practice test in difficulty, length, or emphasis, depending on the goals of your course or instructor.

All variables represent positive real numbers.

Evaluate each expression. Write the answer as an integer or fraction without exponents, or as a decimal fraction:

1. $3^0 \times 4^1 \times 5^{-2}$ **2.** 3.45×10^{-6}

3. 3.45×10^6 **4.** $-\sqrt{100}$

Rewrite each expression in a simpler form using only positive exponents and putting the answer in simplest radical form where necessary:

5. $(x^2)^3 x^4$ **6.** $\left(\dfrac{x}{y^2}\right)^3 \left(\dfrac{y}{x^2}\right)^4$

7. $\left(\dfrac{a^2}{b^{-3}c^4}\right)^{-1}$ **8.** $\left(\dfrac{1}{x}+y^{-1}\right)^{-1}$

9. $\sqrt{25x^2 y^3}$ **10.** $\sqrt{20x}+3\sqrt{5x}$

11. $(xy^2)^{-2}x^2$ **12.** $\left(\dfrac{x^2}{y}\right)^{-1}(xy)^{-1}$

13. $\dfrac{1}{\sqrt{5x}}$ **14.** $(\sqrt{x}-5)(\sqrt{x}-1)$

15. $\dfrac{\sqrt{x}-5}{\sqrt{x}+1}$

Rewrite in scientific notation:

16. $7,654,000,000,000,000$

17. $0.000\,000\,000\,045\,67$

Solve:

18. $\sqrt{x-1}=5$ **19.** $4\sqrt{x-5}=x-1$

20. $\sqrt{x+1}=x-5$

Quadratic Equations

First-degree, or linear, equations in one variable were introduced in Section 2-7. Each such equation can be transformed into the form

$$ax + b = 0 \qquad a \neq 0 \qquad \text{First-degree equation}$$

We have solved many equations of this type and found that they always have a single solution.

In this chapter we will consider second-degree equations in one variable, also called *quadratic equations.* A **quadratic equation** in one variable is any equation that can be written in the form

$$ax^2 + bx + c = 0 \qquad a \neq 0 \qquad \text{Quadratic equation}$$

where x is a variable and a, b, and c are constants. We will refer to this form as the **standard form** for the quadratic equation. The equations

$$2x^2 - 3x + 5 = 0 \qquad \text{and} \qquad 15 = 180t - 16t^2$$

are both quadratic equations, since they are either in the standard form or can be transformed into this form.

Problems that give rise to quadratic equations are many and varied. For example, to find the dimensions of a rectangle with an area of 36 square inches and length 5 inches more than its width (Figure 1), we are led to the equation

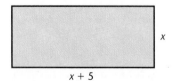

$$x(x + 5) = 36 \qquad w \cdot l = \text{Area}$$
$$x^2 + 5x = 36$$

Figure 1   or $$x^2 + 5x - 36 = 0$$

Photo reference: see Exercise 9-1, Problem 92.

We actually have at hand all the tools we need to solve equations of this type—it is a matter of putting this material together in the right way. This is exactly what we will do in this chapter.

9-1 Solution by Factoring and by Square Root

♦ Solution by Factoring
♦ Solution by Square Root

In this section, we will consider two elementary methods of solving quadratic equations. The methods, however, apply only in certain situations. The first, solving a second-degree equation by factoring, was introduced in Section 6-8 and is therefore covered very briefly here. The second method involves solving equations of the form $x^2 = a$. More general methods are discussed in Sections 9-2 and 9-3.

♦ SOLUTION BY FACTORING

If the coefficients a, b, and c in the quadratic equation

$$ax^2 + bx + c = 0$$

are such that $ax^2 + bx + c$ can be written as the product of two first-degree factors with integer coefficients, then the quadratic equation can be quickly and easily solved by the method of factoring, as was done in Section 6-8.

Example 1
Solving by Factoring

Solve $x^2 - 9x + 20 = 0$.

Solution

$$x^2 - 9x + 20 = 0$$
$$(x - 4)(x - 5) = 0$$
$$x - 4 = 0 \quad \text{or} \quad x - 5 = 0$$
$$x = 4 \quad \text{or} \quad x = 5$$

Matched Problem 1   Solve $x^2 - 7x - 30 = 0$.

♦ SOLUTION BY SQUARE ROOT

Some quadratic equations can be solved simply by taking square roots. This method applies when a squared term is equated to a positive constant. A key step in the process uses this result:

For $a \geq 0$, if $x^2 = a$, then $x = \pm\sqrt{a}$.

The method is very fast, and it leads to a general method for solving all quadratic equations, as will be seen in the next section. A few examples should make the process clear.

Example 2 **Solving by Taking** **Square Roots**	Solve $x^2 - 9 = 0$.

Solution

$$x^2 - 9 = 0 \qquad \text{Notice that the first-degree term is missing.}$$

$$x^2 = 9 \qquad \text{What number squared is 9?}$$

$$\boxed{x = \pm\sqrt{9}} \qquad \text{Short for } \sqrt{9} \text{ or } -\sqrt{9}$$

$$x = \pm 3$$

Since the constant 9 is a perfect square, we could also have solved this equation by factoring:

$$x^2 - 9 = 0$$
$$(x - 3)(x + 3) = 0$$
$$x = 3 \quad \text{or} \quad x = -3$$

You can check the solutions: $3^2 - 9 = 0$ and $(-3)^2 - 9 = 0$. In subsequent examples the checking is left to you.

Matched Problem 2 Solve $x^2 - 16 = 0$.

Example 3 **Solving by Taking** **Square Roots**	Solve $x^2 - 7 = 0$.

Solution

$$x^2 - 7 = 0$$
$$x^2 = 7 \qquad \text{What number squared is 7?}$$
$$x = \pm\sqrt{7}$$

Matched Problem 3 Solve $x^2 - 8 = 0$.

Example 4
Solving by Taking Square Roots

Solve $2x^2 - 3 = 0$.

Solution

$$2x^2 - 3 = 0$$
$$2x^2 = 3$$
$$x^2 = \frac{3}{2}$$
$$x = \pm\sqrt{\frac{3}{2}} = \pm\frac{\sqrt{6}}{2}$$

Matched Problem 4   Solve $3x^2 - 2 = 0$.

$\pm\frac{2}{3}$ $\pm\frac{\sqrt{6}}{3}$

Example 5
Solving by Taking Square Roots

Solve $(x - 2)^2 - 16 = 0$.

Solution

$$(x - 2)^2 - 16 = 0$$ $x-2 = \sqrt{16}$ $x-2 = \pm 4$ $x = 2 \pm 4$
$$(x - 2)^2 = 16$$ Solve for $(x - 2)$ first; then solve for x.
$$x - 2 = \pm 4$$
$$\boxed{x = 2 \pm 4}$$ The notation 2 ± 4 means $2 + 4$ or $2 - 4$.
$$x = 6, -2$$

Matched Problem 5   Solve $(x + 3)^2 - 25 = 0$.

$x+3 = \pm 5$

3 ± 5 $8, -2$

Example 6
Solving by Taking Square Roots

Solve $(x + \frac{1}{2})^2 - \frac{5}{4} = 0$.

Solution

$$\left(x + \frac{1}{2}\right)^2 - \frac{5}{4} = 0$$ Solve for $x + \frac{1}{2}$ first, then solve for x.

$$\left(x + \frac{1}{2}\right)^2 = \frac{5}{4}$$

$$x + \frac{1}{2} = \pm\sqrt{\frac{5}{4}}$$

$$x + \frac{1}{2} = \pm\frac{\sqrt{5}}{2}$$

$$x = -\frac{1}{2} \pm \frac{\sqrt{5}}{2}$$

$$x = \frac{-1 \pm \sqrt{5}}{2}$$

Matched Problem 6   Solve $(x - \frac{1}{3})^2 - \frac{7}{9} = 0$.

Example 7
An Equation with
No Solution

Solve $x^2 = -9$.

Solution   There is no solution in the real numbers, since no real number squared is negative.

Matched Problem 7   Solve $x^2 + 4 = 0$. $x = \sqrt{4}$ $x = \pm 2$

The introduction to Chapter 1 posed the following problem:

A rectangular field is twice as long as it is wide and has an area of 6,962 square yards. What are its dimensions?

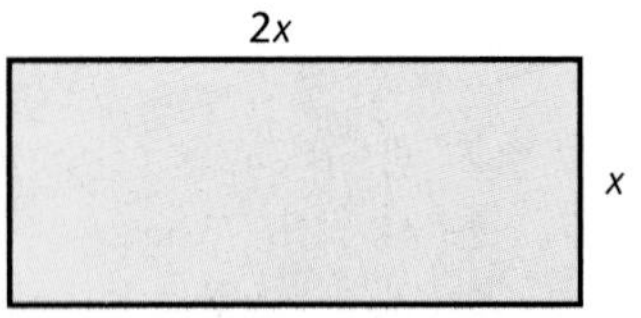

We can now restate the problem in algebraic terms and quickly obtain the solution. Let $x =$ Width of the field, so $2x =$ Length. The area is $x \cdot 2x$, so

$$2x^2 = 6,962$$
$$x^2 = 3,481$$
$$x = \pm\sqrt{3,481} = \pm 59$$

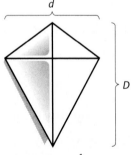

2x

x

Figure 2

Since the problem requires a positive width, the only solution that makes sense physically is $x = 59$. Thus, the field is 59 yards by 118 yards.

Example 8
An Area Problem

The area of a kite as shown in Figure 3 is $\frac{1}{2}Dd$. Find the height D of the kite if the ratio $D:d$ is $4:3$ and the total area of the kite is 294 square inches.

Solution   Since $D:d = 4:3$,

d

D

Area of kite = $\frac{1}{2}$ product
of the diagonals

Figure 3

$$\frac{D}{d} = \frac{4}{3}$$
$$3D = 4d$$
$$\frac{3}{4}D = d$$

Now substitute this expression for d into the equation for the area:

$$A = \frac{1}{2}Dd$$

$$294 = \frac{1}{2}D\left(\frac{3}{4}D\right) = \frac{3}{8}D^2$$

$$\frac{8 \cdot 294}{3} = D^2$$

$$D^2 = 784$$

$$D = \pm\sqrt{784} = \pm 28$$

Since D must be positive, we discard -28, so the solution is $D = 28$ inches.

Check   If $D = 28$ and $D/d = \frac{4}{3}$, then $d = 21$ inches and the area is $\frac{1}{2} \cdot 28 \cdot 21 = 294$ square inches as desired.

Matched Problem 8   Find the width d for a kite with $D : d = 5 : 3$ and total area 2,430 square centimeters.

Answers to Matched Problems

1. $x = -3, 10$ **2.** $x = \pm 4$ **3.** $x = \pm\sqrt{8}$ or $\pm 2\sqrt{2}$

4. $x = \pm\sqrt{\dfrac{2}{3}}$ or $\pm\dfrac{\sqrt{6}}{3}$ **5.** $x = -8, 2$ **6.** $x = \dfrac{1 \pm \sqrt{7}}{3}$

7. No real solutions   **8.** 54 centimeters

EXERCISE 9-1

A *Solve by factoring:*

1. $x^2 - 8x + 15 = 0$

2. $x^2 + 2x - 15 = 0$

3. $x^2 + 3x - 28 = 0$

4. $x^2 - 3x + 18 = 0$

5. $x^2 - x - 42 = 0$

6. $x^2 + 9x + 8 = 0$

7. $x^2 + 4x - 21 = 0$

8. $x^2 - 3x - 70 = 0$

9. $x^2 + 11x + 28 = 0$

10. $x^2 + 10x + 24 = 0$

11. $x^2 + 16x - 36 = 0$

12. $x^2 - 9x - 36 = 0$

13. $x^2 - 15x + 36 = 0$

14. $x^2 - 10x - 24 = 0$

15. $x^2 + 10x - 24 = 0$

16. $x^2 - 10x + 24 = 0$

17. $x^2 + 12x + 36 = 0$

18. $x^2 + 5x - 24 = 0$

19. $x^2 - 14x + 24 = 0$

20. $x^2 - 2x - 24 = 0$

21. $x^2 + 5x - 36 = 0$

22. $x^2 - 12x + 36 = 0$

Solve for all real solutions by the square root method:

23. $x^2 = 16$

24. $x^2 = 49$

25. $m^2 - 64 = 0$

26. $n^2 - 25 = 0$

27. $x^2 = 3$

28. $y^2 = 2$

29. $u^2 - 5 = 0$

30. $x^2 - 11 = 0$

31. $a^2 = 18$

32. $y^2 = 8$

33. $x^2 - 12 = 0$

34. $n^2 - 27 = 0$

35. $x^2 = \frac{4}{9}$

36. $y^2 = \frac{9}{16}$

37. $9x^2 = 4$

38. $16y^2 = 9$

B **39.** $9x^2 - 4 = 0$

40. $16y^2 - 9 = 0$

41. $25x^2 - 4 = 0$

42. $9x^2 - 1 = 0$

43. $4t^2 - 3 = 0$

44. $9x^2 - 7 = 0$

45. $2x^2 - 5 = 0$

46. $3m^2 - 7 = 0$

47. $3m^2 - 1 = 0$

48. $5n^2 - 1 = 0$

49. $(y - 2)^2 = 9$

50. $(x - 3)^2 = 4$

51. $(x + 2)^2 = 25$

52. $(y + 3)^2 = 16$

53. $(y - 2)^2 = 3$

54. $(y - 3)^2 = 5$

55. $(x - \frac{1}{2})^2 = \frac{9}{4}$

56. $(x - \frac{1}{3})^2 = \frac{4}{9}$

57. $(x - 3)^2 = -4$

58. $(t + 1)^2 = -9$

59. $(x - \frac{3}{2})^2 = \frac{4}{9}$

60. $(y + \frac{5}{2})^2 = \frac{5}{2}$

C *Solve by factoring:*

61. $2x^2 - 5x - 3 = 0$

62. $3x^2 + x - 2 = 0$

63. $3x^2 + 5x - 2 = 0$

64. $2x^2 + 5x + 3 = 0$

65. $4x^2 + 7x - 2 = 0$

66. $3x^2 - x - 4 = 0$

67. $2x^2 - 3x + 1 = 0$

68. $2x^2 + 7x - 4 = 0$

69. $2x^2 - 5x + 3 = 0$

70. $2x^2 - 3x - 2 = 0$

71. $2x^2 + 3x + 1 = 0$

72. $2x^2 + 3x - 2 = 0$

73. $2x^2 + 5x - 3 = 0$

74. $2x^2 - 9x + 4 = 0$

75. $2x^2 + x - 1 = 0$

76. $2x^2 - x - 1 = 0$

77. $3x^2 - 5x - 2 = 0$

78. $3x^2 + 7x - 6 = 0$

79. $3x^2 - x - 4 = 0$

80. $3x^2 + 2x - 8 = 0$

Solve the equation for the indicated variable:

81. $a^2 + b^2 = c^2$; b *Pythagorean theorem*

82. $k = \frac{1}{2}mv^2$; v *Kinetic energy*

83. $(h + R)^2 = R^2 + d^2$; d *Sight distance*

84. $T^2g = 4\pi^2L$; T *Pendulum*

85. $S^2H = 2DC$; S *Inventory*

86. $A = \pi r^2$; r *Area of circle*

APPLICATIONS

87. *Geometry* If the side of a square is increased by 10%, the area of the resulting square becomes 484 square meters. What was the area of the original square?

88. *Geometry* If the side of a square is increased by 40%, the area of the resulting square becomes 441 square meters. What was the area of the original square?

89. *Geometry* A rectangular field that is 3 times as long as it is wide has area 36,300 square yards. What are the dimensions of the field?

90. *Geometry* A rectangular field that is 4 times as long as it is wide has area 16,900 square yards. What are the dimensions of the field?

91. *Physics* The force F in pounds on a flat surface with area a square feet that results from a wind blowing at v miles per hour is given approximately by $F = 0.003v^2a$. What wind velocity will result in a force of 729 pounds on a flat surface of size 270 square feet?

92. *Physics* One method of measuring the velocity of water in a river is to use an L-shaped tube as indicated in the accompanying figure. Torricelli's law in physics tells us that the height (in feet) that the water is pushed up into the tube above the surface is related to the water's velocity (in feet per second) by the formula $v^2 = 2gh$, where g is approximately 32 feet per second per second. [*Note:* The device can also be used as a simple speedometer for a boat.] How fast is a stream flowing if $h = 0.5$ foot? Find the answer to two decimal places.

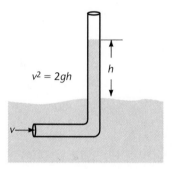

   # 9-2 **Solution by Completing the Square**

♦ Completing the Square
♦ Solution of Quadratic Equations by Completing the Square

The factoring and square root methods discussed in the last section are fast and easy to use when they apply. Unfortunately, many quadratic equations will not yield to either method as stated. For example, the simple-looking polynomial in

$$x^2 + 4x - 6 = 0$$

cannot be factored using integer coefficients, and the square root method is not applicable either. The equation requires a new method, if it can be solved at all.

In this section we will discuss a method, called **solution by completing the square,** that will work for all quadratic equations. In the next section we will use this method to develop a general formula that will be used in the future whenever the methods of the preceding section fail.

The method of completing the square is based on the process of transforming the standard quadratic equation

$$ax^2 + bx + c = 0 \tag{1}$$

into the form

$$(x + A)^2 = B \tag{2}$$

where A and B are constants. This last equation can be solved easily (assuming $B \geq 0$) by the square root method discussed in the last section. That is,

$$(x + A)^2 = B$$
$$x + A = \pm\sqrt{B}$$
$$x = -A \pm \sqrt{B}$$

♦ **COMPLETING THE SQUARE**

In Section 6-4, we learned to recognize a perfect square binomial:

$$(A + B)^2 = A^2 + 2AB + B^2$$
$$(x + B)^2 = x^2 + 2Bx + B^2 \qquad {\scriptstyle (x + 5)^2 = x^2 + 10x + 25}$$

In the second form, we note that the last term, B^2, is the square of half the coefficient of x in the middle term. Thus, to complete the square for $x^2 + bx$, we add a third term, $(b/2)^2$, which is the square of half the coefficient of x. For example, to complete the square for $x^2 + 4x$, we add the square $(\frac{1}{2} \cdot 4)^2$ to obtain $x^2 + 4x + 4$. This result is summarized in the following box:

> **Completing the Square**
>
> To **complete the square** of a quadratic of the form
>
> $$x^2 + bx$$
>
> add the square of half the coefficient of x. That is, add
>
> $$\left(\frac{b}{2}\right)^2$$
>
> Then $\qquad x^2 + bx + \left(\frac{b}{2}\right)^2 = \left(x + \frac{b}{2}\right)^2$
>
> The coefficient of x^2 must be 1 for this rule to apply.

Example 1
Completing the Square

Complete the square and factor the resulting expression:

(A) $x^2 + 8x$ **(B)** $x^2 - 7x$

Solution

(A) $x^2 + 8x + \left(\dfrac{8}{2}\right)^2 = x^2 + 8x + 16 = (x + 4)^2$

$$\textbf{(B)} \quad x^2 - 7x + \left(\frac{-7}{2}\right)^2 = x^2 - 7x + \frac{49}{4} = \left(x - \frac{7}{2}\right)^2$$

Matched Problem 1   Complete the square and factor the resulting expression:

   (A) $x^2 - 3x$ **(B)** $x^2 + 6x$

♦   **SOLUTION OF QUADRATIC EQUATIONS BY COMPLETING THE SQUARE**

Solving quadratic equations by the method of completing the square is best illustrated by examples. In this course we are going to be interested only in real-number solutions.

Example 2
Solving by Completing the Square

Solve $x^2 + 10x - 3 = 0$.

Solution

$$x^2 + 10x - 3 = 0$$
   Add 3 to both sides of the equation to leave the left side in the form $x^2 + Bx$.

$$x^2 + 10x = 3$$
   Complete the square on the left side. The same amount must be added to the right side to preserve the equality.

$$x^2 + 10x + 25 = 3 + 25$$
   Factor the left side as a perfect square.

$$(x + 5)^2 = 28$$
   Solve by square roots.

$$x + 5 = \pm\sqrt{28}$$
$$x = -5 \pm \sqrt{28}$$
$$= -5 \pm 2\sqrt{7}$$

Check
$$(-5 + 2\sqrt{7})^2 + 10(-5 + 2\sqrt{7}) - 3$$
$$= 25 - 20\sqrt{7} + 28 - 50 + 20\sqrt{7} - 3 = 0$$

Checking the solution $-5 - 2\sqrt{7}$ is left to you.

Matched Problem 2   Solve $x^2 + 6x + 4 = 0$.

Example 3
Solving by Completing the Square

Solve $2x^2 + 6x - 3 = 0$.

Solution

$$2x^2 + 6x - 3 = 0$$
   Divide both sides by 2 to change the left side to a form where the coefficient of x^2 is 1.

$$x^2 + 3x - \frac{3}{2} = 0$$
   Proceed as in Example 2.

$$x^2 + 3x = \frac{3}{2} \qquad\qquad \left(\frac{3}{2}\right)^2 = \frac{9}{4}$$

$$x^2 + 3x + \frac{9}{4} = \frac{3}{2} + \frac{9}{4}$$

$$\left(x + \frac{3}{2}\right)^2 = \frac{15}{4}$$

$$x + \frac{3}{2} = \pm\sqrt{\frac{15}{4}}$$

$$x = -\frac{3}{2} \pm \sqrt{\frac{15}{4}}$$

$$= \frac{-3}{2} \pm \frac{\sqrt{15}}{2}$$

$$= \frac{-3 \pm \sqrt{15}}{2}$$

Matched Problem 3 Solve $3x^2 + 4x - 1 = 0$.

The method of solving a quadratic equation by completing the square may be summarized as follows:

Solving a Quadratic Equation by Completing the Square

1. Write the quadratic equation in standard form:

$$ax^2 + bx + c = 0$$

2. Divide both sides by a to obtain a standard form with the coefficient of x^2 equal to 1, say,

$$x^2 + Bx + C = 0$$

3. Subtract the constant C from each side:

$$x^2 + Bx = -C$$

4. Complete the square of $x^2 + Bx$ by adding $(B/2)^2$ to each side to obtain

$$x^2 + Bx + \left(\frac{B}{2}\right)^2 = -C + \left(\frac{B}{2}\right)^2$$

$$\left(x + \frac{B}{2}\right)^2 = -C + \left(\frac{B}{2}\right)^2$$

5. Solve by the method of square roots.

Answers to Matched Problems

1. **(A)** $x^2 - 3x + \dfrac{9}{4} = \left(x - \dfrac{3}{2}\right)^2$ **(B)** $x^2 + 6x + 9 = (x + 3)^2$

2. $x = -3 \pm \sqrt{5}$ 3. $x = \dfrac{-2 \pm \sqrt{7}}{3}$

EXERCISE 9-2

A *Complete the square and factor:*

1. $x^2 - 8x$
2. $x^2 + 4x$
3. $x^2 + 10x$
4. $x^2 - 12x$
5. $x^2 - 10x$
6. $x^2 - 2x$
7. $x^2 - 6x$
8. $x^2 + 2x$
9. $x^2 - 4x$
10. $x^2 + 12x$

Solve by the method of completing the square:

11. $x^2 - 8x + 6 = 0$
12. $x^2 + 4x - 3 = 0$
13. $x^2 + 10x - 1 = 0$
14. $x^2 - 12x + 10 = 0$
15. $x^2 - 10x + 15 = 0$
16. $x^2 - 2x - 5 = 0$
17. $x^2 - 6x + 6 = 0$
18. $x^2 + 2x - 1 = 0$
19. $x^2 - 4x + 2 = 0$
20. $x^2 + 12x + 25 = 0$

B *Complete the square and factor:*

21. $x^2 + 7x$
22. $x^2 + 5x$
23. $x^2 + 3x$
24. $x^2 - x$
25. $x^2 - 5x$
26. $x^2 + x$
27. $x^2 + 9x$
28. $x^2 - 9x$
29. $x^2 - 11x$
30. $x^2 + 11x$

Solve by the method of completing the square:

31. $x^2 + 7x - 5 = 0$
32. $x^2 + 5x + 3 = 0$
33. $x^2 + 3x + 1 = 0$
34. $x^2 - x - 3 = 0$
35. $x^2 - 5x - 3 = 0$
36. $x^2 + x - 1 = 0$
37. $x^2 - 9x + 12 = 0$
38. $x^2 + 9x + 10 = 0$
39. $x^2 - 11x + 11 = 0$
40. $x^2 + 11x + 22 = 0$

C 41. $2x^2 + 4x - 5 = 0$
42. $2x^2 + 6x - 1 = 0$
43. $2x^2 + 5x - 4 = 0$
44. $2x^2 + x - 6 = 0$
45. $x^2 + 12x + 30 = 0$
46. $x^2 + 5x + 5 = 0$
47. $3x^2 - 6x + 1 = 0$
48. $3x^2 + 2x - 1 = 0$
49. $x^2 + 2x = 3$
50. $2x^2 - 5x = 3$
51. $2x^2 - 5x = -3$
52. $4x^2 + 3x = 1$
53. $4x^2 - 8x + 3 = 0$
54. $6x^2 - 9x + 3 = 0$
55. $6x^2 - x - 1 = 0$
56. $4x^2 - 9x + 5 = 0$
57. $4x^2 + 4x - 5 = 0$
58. $6x^2 + 7x - 5 = 0$
59. $6x^2 + 3x - 3 = 0$
60. $4x^2 + 2x - 15 = 0$
61. $4x^2 - 4x - 3 = 0$
62. $6x^2 - 11x + 4 = 0$
63. $6x^2 - 7x + 2 = 0$
64. $4x^2 - x - 1 = 0$

9-3 The Quadratic Formula

- ♦ Deriving the Quadratic Formula
- ♦ Use of the Quadratic Formula
- ♦ The Discriminant
- ♦ Which Method?

We now apply the method of completing the square developed in the last section to obtain a general formula for solving any quadratic equation.

♦ DERIVING THE QUADRATIC FORMULA

We apply the steps for the method of completing the square to the general quadratic equation in standard form

$$ax^2 + bx + c = 0 \qquad a \neq 0$$

1. Write the equation in standard form: we are given the equation in standard form

$$ax^2 + bx + c = 0$$

2. Divide both sides by a to obtain a standard form with coefficient of x^2 equal to 1:

$$x^2 + \frac{b}{a}x + \frac{c}{a} = 0$$

3. Subtract the constant term from both sides:

$$x^2 + \frac{b}{a}x = -\frac{c}{a}$$

4. Complete the square of $x^2 + \dfrac{b}{a}x$ by adding the square of half the coefficient of x, that is, the square of $\dfrac{1}{2}$ of $\dfrac{b}{a}$, to both sides:

$$x^2 + \frac{b}{a}x + \left(\frac{b}{2a}\right)^2 = -\frac{c}{a} + \left(\frac{b}{2a}\right)^2$$

5. Solve by the method of square roots:

$$\left(x + \frac{b}{2a}\right)^2 = -\frac{c}{a} + \frac{b^2}{4a^2}$$

$$= \frac{-4ac + b^2}{4a^2} = \frac{b^2 - 4ac}{4a^2}$$

$$x + \frac{b}{2a} = \pm\sqrt{\frac{b^2 - 4ac}{4a^2}}$$

$$= \pm\frac{\sqrt{b^2 - 4ac}}{\sqrt{4a^2}}$$

$$= \pm\frac{\sqrt{b^2 - 4ac}}{2a}$$

We need to note two things at this stage. If $b^2 - 4ac < 0$, the equation has no solution in the set of real numbers. Assume, therefore, that $b^2 - 4ac \geq 0$. Also, $\sqrt{4a^2}$ is either $2a$ or $-2a$, but we need use only $2a$, since there is already a $\pm$ sign in the expression.* Continuing:

---

* If a is positive, $\sqrt{4a^2} = 2a$ and

$$\pm\frac{\sqrt{b^2 - 4ac}}{\sqrt{4a^2}} = \pm\frac{\sqrt{b^2 - 4ac}}{2a}$$

If a is negative, $\sqrt{4a^2} = -2a$ and

$$\pm\frac{\sqrt{b^2 - 4ac}}{\sqrt{4a^2}} = \mp\frac{\sqrt{b^2 - 4ac}}{2a} = \pm\frac{\sqrt{b^2 - 4ac}}{2a}$$

$$x = -\frac{b}{2a} \pm \frac{\sqrt{b^2 - 4ac}}{2a}$$

$$x = \frac{-b \pm \sqrt{b^2 - 4ac}}{2a}$$

This last equation, the solution to $ax^2 + bx + c = 0$, is called the **quadratic formula.**

Quadratic Formula

The solutions to $ax^2 + bx + c = 0$ are given by

$$x = \frac{-b \pm \sqrt{b^2 - 4ac}}{2a} \qquad a \neq 0 \qquad \text{Quadratic Formula}$$

♦ USE OF THE QUADRATIC FORMULA

The quadratic formula yields the solution to any quadratic equation. However, when $b^2 - 4ac < 0$, the quadratic formula does not define a real number. We will limit our interest in this course to solutions that are real numbers. The following examples show us how to use the quadratic formula. The first example was worked in the last section by completing the squares.

Example 1
Applying the
Quadratic Formula

Solve $2x^2 + 6x - 3 = 0$ using the quadratic formula.

Solution $2x^2 + 6x - 3 = 0$ The equation is in standard form, with $a = 2$, $b = 6$, and $c = -3$.

$$x = \frac{-b \pm \sqrt{b^2 - 4ac}}{2a} \qquad \text{Substitute for } a, b, \text{ and } c \text{ in the formula.}$$

$$x = \frac{-6 \pm \sqrt{6^2 - 4 \cdot 2(-3)}}{2 \cdot 2}$$

$$= \frac{-6 \pm \sqrt{36 + 24}}{4} = \frac{-6 \pm \sqrt{60}}{4} = \frac{-6 \pm 2\sqrt{15}}{4}$$

$$= \frac{2(-3 \pm \sqrt{15})}{4} = \frac{-3 \pm \sqrt{15}}{2}$$

Matched Problem 1   Solve $3x^2 + 4x - 1 = 0$ using the quadratic formula.

Example 2
Applying the
Quadratic Formula

Solve $3x^2 + \frac{1}{2}x - 1 = 0$.

Solution $3x^2 + \dfrac{1}{2}x - 1 = 0$ It will be easier to clear the equations of fractions first. Multiply by 2, the LCD.

$6x^2 + x - 2 = 0$ Here $a = 6$, $b = 1$, $c = -2$.

$$x = \frac{-1 \pm \sqrt{1 - 4 \cdot 6(-2)}}{2 \cdot 6}$$

$$= \frac{-1 \pm \sqrt{49}}{12}$$

$$= \frac{-1 \pm 7}{12}$$

$$x = \frac{1}{2}, \frac{-2}{3}$$

Matched Problem 2   Solve $2x^2 + \frac{1}{3}x - 1 = 0$.

Example 3
Applying the
Quadratic Formula

Solve $16x^2 - 24x + 9 = 0$.

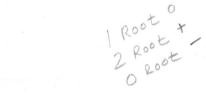

Solution   Here $a = 16$, $b = -24$, and $c = 9$.

$$x = \frac{24 \pm \sqrt{(-24)^2 - 4 \cdot 16 \cdot 9}}{2 \cdot 16}$$

$$= \frac{24 \pm \sqrt{576 - 576}}{32}$$

$$= \frac{24}{32} = \frac{3}{4}$$

Matched Problem 3   Solve $25x^2 - 40x + 16 = 0$.

Example 4
Applying the
Quadratic Formula

Solve $x^2 + x + 1 = 0$.

Solution   Here $a = 1$, $b = 1$, and $c = 1$.

$$x = \frac{-1 \pm \sqrt{1 - 4 \cdot 1 \cdot 1}}{2 \cdot 1} = \frac{-1 \pm \sqrt{-3}}{2}$$

Since the expression under the radical is negative, it does not represent a real number. This equation has no real solution.

Matched Problem 4   Solve $x^2 - 2x + 3 = 0$.

♦ **THE DISCRIMINANT**

Notice that in Examples 1 and 2, the equation had two real roots and the radical expression $\pm\sqrt{b^2 - 4ac}$ accounted for this. In Example 3, on the other hand, the equation had only one root, and the radical expression was equal to 0. Finally, in Example 4, the equation had no real root because the quantity in the radical was negative. The quantity $b^2 - 4ac$ thus provides an indicator for the number of real roots of a quadratic equation, discriminating among the three possibilities of 0, 1, or 2 roots. The expression $b^2 - 4ac$ is therefore called the **discriminant.**

Discriminant Test

$$ax^2 + bx + c = 0 \qquad a, b, c \text{ real numbers, } a \neq 0$$

$b^2 - 4ac$	roots
Positive	Two real roots
Zero	One real root
Negative	No real roots

If the discriminant $b^2 - 4ac$ is a perfect square, the two real roots are rational numbers. If the discriminant is positive, but not a perfect square, the two roots are irrational numbers.

Example 5
Applying the Discriminant

Apply the discriminant test to determine the number of real roots of the equation:

(A) $2x^2 - 3x + 1 = 0$ (B) $3x^2 + 5x + 4 = 0$
(C) $25x^2 + 10x + 1 = 0$

Solution   (A) $2x^2 - 3x + 1 = 0$
The discriminant $b^2 - 4ac$ here is equal to

$$(-3)^2 - 4 \cdot 2 \cdot 1 = 9 - 8 = 1$$

so the equation has two real roots. Check that they are 1 and $\frac{1}{2}$.
(B) $3x^2 + 5x + 4 = 0$
The discriminant $b^2 - 4ac$ here is equal to

$$5^2 - 4 \cdot 3 \cdot 4 = -23$$

so the equation has no real roots.

(C) $25x^2 + 10x + 1 = 0$

The discriminant $b^2 - 4ac$ here is equal to

$$10^2 - 4 \cdot 25 \cdot 1 = 0$$

so the equation has one real root. Check that it is $-\frac{1}{5}$.

Matched Problem 5 Apply the discriminant test to determine the number of real roots of the equation:

(A) $25x^2 - 30x + 9 = 0$ **(B)** $9x^2 + 3x + 1 = 0$
(C) $4x^2 + 5x + 1 = 0$

♦ WHICH METHOD?

The quadratic formula may be used to solve *any* quadratic equation. On the other hand, whenever the square root method or the factoring method readily produces results, these methods are generally faster. It may not be obvious when these methods are preferable, but there are two such circumstances that are easily recognized.

1. Any quadratic equation of the form

$$ax^2 + c = 0 \qquad \text{Note that the } bx \text{ term is missing.}$$

can always be solved (if solutions exist in the real numbers) by the square root method, since the equation is equivalent to $x^2 = -c/a$.
2. Any equation of the form

$$ax^2 + bx = 0 \qquad \text{Note that the } c \text{ term is missing.}$$

can always be solved by factoring, since $ax^2 + bx = x(ax + b)$.

In addition, with practice you will be able to recognize the factors of other quadratic expressions and use this to solve the corresponding equations. However, if in doubt, just use the quadratic formula. This strategy can be summarized as follows:

Solving $ax^2 + bx + c = 0$

1. If $b = 0$, use the square root method.
2. If $c = 0$ or $ax^2 + bx + c$ is easily factored, use the factoring method.
3. Otherwise, use the quadratic formula.

It is important to realize that the quadratic formula can *always* be used and will produce the same results as either of the other methods. For example, let us solve

$$2x^2 + 7x - 15 = 0$$

in two ways.

Suppose you observe that the polynomial can be factored as $(2x - 3)(x + 5)$. Then,

$$(2x - 3)(x + 5) = 0$$

$$2x - 3 = 0 \quad \text{or} \quad x + 5 = 0$$

$$x = \tfrac{3}{2} \quad \text{or} \quad x = -5$$

Suppose you had used the quadratic formula instead.

$$x = \frac{-b \pm \sqrt{b^2 - 4ac}}{2a} \qquad a = 2,\ b = 7,\ c = -15$$

$$x = \frac{-(7) \pm \sqrt{7^2 - 4(2)(-15)}}{2(2)}$$

$$x = \frac{-7 \pm \sqrt{169}}{4} = \frac{-7 \pm 13}{4}$$

$$x = \tfrac{3}{2}, -5$$

The quadratic formula produces the same result as the factoring method, as it should, but with a little more work.

Answers to Matched Problems

1. $x = \dfrac{-2 \pm \sqrt{7}}{3}$ 2. $x = \dfrac{-1 \pm \sqrt{73}}{12}$ 3. $x = \dfrac{4}{5}$ 4. No real roots

5. **(A)** One **(B)** No real roots **(C)** Two

EXERCISE 9-3

A *Specify the constants a, b, and c for each quadratic equation when written in the standard form $ax^2 + bx + c = 0$:*

1. $x^2 + 4x + 2 = 0$ 2. $x^2 + 8x + 3 = 0$

3. $x^2 - 3x - 2 = 0$ 4. $x^2 - 6x - 8 = 0$

5. $3x^2 - 2x + 1 = 0$ 6. $2x^2 - 5x + 3 = 0$

7. $2u^2 = 1 - 3u$ 8. $m = 1 - 3m^2$

9. $2x^2 - 5x = 0$ 10. $3y^2 - 5 = 0$

Solve by use of the quadratic formula:

11. $x^2 + 4x + 2 = 0$ 12. $x^2 + 8x + 3 = 0$

13. $x^2 - 3x - 2 = 0$ 14. $x^2 - 6x - 8 = 0$

15. $3x^2 - 2x + 1 = 0$ 16. $2x^2 - 5x + 3 = 0$

17. $2u^2 = 1 - 3u$ 18. $m = 1 - 3m^2$

19. $2x^2 - 5x = 0$ 20. $3y^2 - 5 = 0$

21. $y^2 - 6y - 3 = 0$ 22. $y^2 - 10y - 3 = 0$

B 23. $3t + t^2 = 1$ 24. $x^2 = 1 - x$

25. $2x^2 - 6x + 3 = 0$ 26. $2x^2 - 4x + 1 = 0$

27. $3m^2 = 1 - m$ 28. $3u + 2u^2 = 1$

29. $x^2 = 2x - 3$ 30. $x^2 + 8 = 4x$

31. $2x = 3 + \dfrac{3}{x}$ 32. $x + \dfrac{2}{x} = 6$

33. $m^2 = \dfrac{8m - 1}{5}$ 34. $x^2 = 3x + \dfrac{1}{2}$

Apply the discriminant test to determine the number of real roots, and if real roots exist, solve by use of the quadratic formula:

35. $x^2 + 8x + 17 = 0$ 36. $64x^2 - 48x + 9 = 0$

37. $36x^2 + 12x + 1 = 0$ 38. $5x^2 + 7x + 2 = 0$

39. $3x^2 - 7x + 3 = 0$ 40. $6x^2 - 11x + 6 = 0$

41. $9x^2 - 8x + 2 = 0$ 42. $100x^2 + 140x + 49 = 0$

43. $16x^2 - 24x + 9 = 0$ 44. $2x^2 + 7x + 5 = 0$

45. $x^2 + 5x - 20 = 0$ 46. $x^2 - 5x - 30 = 0$

47. $x^2 - 8x + 10 = 0$ 48. $x^2 + 6x + 7 = 0$

49. $3x^2 + 15x + 17 = 0$ 50. $4x^2 - 25x + 35 = 0$

51. $5x^2 - 12x + 7 = 0$ 52. $6x^2 + 10x + 3 = 0$

53. $x^2 + 5x + 7 = 0$ **54.** $x^2 - 5x + 8 = 0$

55. $4x^2 - 20x + 25 = 0$ **56.** $8x^2 + 4x + 2 = 0$

C *The following problems are mixed as to type. Use the most efficient method to find all real solutions to each equation.*

57. $x^2 - x - 6 = 0$ **58.** $x^2 + 2x - 8 = 0$

59. $x^2 + 7x = 0$ **60.** $x^2 = 3x$

61. $x^2 + 20x + 100 = 0$ **62.** $x^2 - 18x + 81 = 0$

63. $x^2 + 5x + 10 = 0$ **64.** $x^2 - 5x + 9 = 0$

65. $x^2 - 5x - 14 = 0$ **66.** $x^2 + 5x - 66 = 0$

67. $2x^2 = 32$ **68.** $3x^2 - 27 = 0$

69. $x^2 + 2x - 2 = 0$ **70.** $m^2 - 3m - 1 = 0$

71. $2x^2 = 4x$ **72.** $2y^2 + 3y = 0$

73. $2x^2 + x - 2 = 0$ **74.** $3x^2 - x - 1 = 0$

75. $x^2 + 8x + 16 = 0$ **76.** $x^2 - 4x + 4 = 0$

77. $x^2 - 2x = 1$ **78.** $x^2 - 2 = 2x$

79. $u^2 = 3u - \frac{3}{2}$ **80.** $t^2 = \frac{3}{2}(t + 1)$

81. $M = M^2$ **82.** $t(t - 3) = 0$

83. $6y = \dfrac{1 - y}{y}$ **84.** $2x + 1 = \dfrac{6}{x}$

85. $I^2 - 50 = 0$ **86.** $72 = u^2$

87. $(B - 2)^2 = 3$ **88.** $(u + 3)^2 = 5$

89. $x^2 + 4 = 0$ **90.** $(x - 2)^2 = -9$

91. $\dfrac{24}{n} = 12n - 28$ **92.** $3x = \dfrac{84 - 9x}{x}$

93. $x = 4 + \dfrac{5}{x}$ **94.** $x - 6 = \dfrac{7}{x}$

95. $x = \dfrac{1}{x + 3} - 1$ **96.** $x - 2 = \dfrac{1}{x + 2} - 4$

97. $3u^2 = \sqrt{3}u + 2$ **98.** $t^2 - \sqrt{5}t - 11 = 0$

99. $\dfrac{24}{10 + x} + 1 = \dfrac{24}{10 - x}$

100. $\dfrac{1.2}{x - 1} + \dfrac{1.2}{x} = 1$

9-4 **Graphing Quadratic Equations**

♦ Graphing $y = x^2$ and $y = -x^2$
♦ Graphing Equations of the Form $y = ax^2 + bx + c$
♦ Graphs and Solutions to Quadratic Equations

In Chapter 4, we defined the solution set of the linear equation

$$y = mx + b$$

to be the set of all pairs of real numbers (x, y) that make the equation true. The graph of the equation is the graph of all these ordered pairs and turns out always to be a straight line. In this section, we extend these ideas to quadratic equations

$$y = ax^2 + bx + c$$

A **solution** to such an equation is again a pair of real numbers (x, y) that make the equation true. The **solution set** for the equation is the set of all such pairs, and the **graph of the equation** is the graph consisting of all solution pairs. The graph is a curve called a **parabola.** A parabola has interesting geometric properties that are studied more thoroughly in later courses.

♦ GRAPHING $y = x^2$ AND $y = -x^2$

To graph $y = x^2$ we make a table of some of its values and plot them:

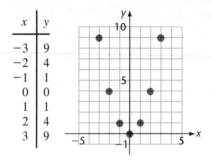

x	y
-3	9
-2	4
-1	1
0	0
1	1
2	4
3	9

Figure 1

The graph is not a straight line. If we plot more values, the graph fills in to give us a parabola that looks like this:

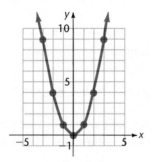

Figure 2

This is the graph of $y = x^2$.

To graph $y = -x^2$, we make a table of values as we did above, and plot the points:

x	y
-3	-9
-2	-4
-1	-1
0	0
1	-1
2	-4
3	-9

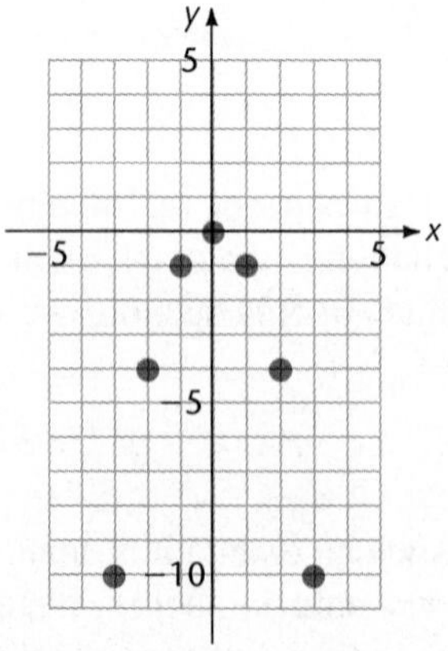

Figure 3

Notice that all the y values are simply the negatives of those obtained before for $y = x^2$ and the effect is just to turn our previous parabola upside down. The same is true of the complete graph:

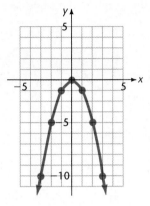

Figure 4

The graphs of $y = x^2$ and $y = -x^2$ exhibit the features of parabolas that will enable us to graph all quadratic equations $y = ax^2 + bx + c$:

1. The shape of the curve is $\cup$ for $y = +x^2$ and $\cap$ for $y = -x^2$. Either curve is called a **parabola.**
2. The parabola has a highest or lowest point called the **vertex** of the parabola. For both $y = x^2$ and $y = -x^2$, the vertex is $(0, 0)$.

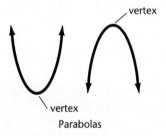

Parabolas

Figure 5

♦ **GRAPHING EQUATIONS OF THE FORM**
$y = ax^2 + bx + c$

All equations of the form $y = ax^2 + bx + c$ have graphs that are parabolas and appear similar to either $y = x^2$ or $y = -x^2$. However, the vertex of the parabola may be moved away from the origin, and the graph may appear elongated or flattened in comparison with $y = x^2$ and $y = -x^2$. The following result lets us graph $y = ax^2 + bx + c$ fairly efficiently. The examples that follow will illustrate the process.

Graphing $y = ax^2 + bx + c$

1. If $a > 0$, the basic shape of the graph is the same as $y = x^2$. If $a < 0$, the basic shape of the graph is the same as $y = -x^2$. The graph looks like

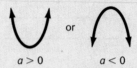

$a > 0$ $a < 0$

2. The vertex of the parabola, that is, the highest or lowest point on the curve, occurs at

$$x = \frac{-b}{2a}$$

The y value can be obtained by substituting this value for x into the equation $y = ax^2 + bx + c$. Begin to graph the parabola by plotting this point.

3. Plotting the points corresponding to a few values of x to each side of the vertex will give us a good picture of the parabola. Plot enough points that the shape becomes apparent.

Example 1

Graphing
$y = ax^2 + bx + c$

Graph:

(A) $y = 3x^2$ **(B)** $y = -x^2 + 3$ **(C)** $y = x^2 - 4x + 4$

Solution   **(A)** $y = 3x^2$

1. Here a, the coefficient of x^2, is 3, which is greater than 0. Therefore, the shape of the graph is like that of $y = x^2$. That is, the graph looks like

2. The lowest point on the graph has coordinates

$$x = \frac{-b}{2a} = \frac{0}{6} = 0 \quad \text{and} \quad y = 3(0)^2 = 0$$

Plot the vertex at the origin.

3. Make a table for a few values of x to each side of 0. Plot these points. The shape of the curve, shown in Figure 6, is now apparent.

x	y
−3	27
−2	12
−1	3
0	0
1	3
2	12
3	27

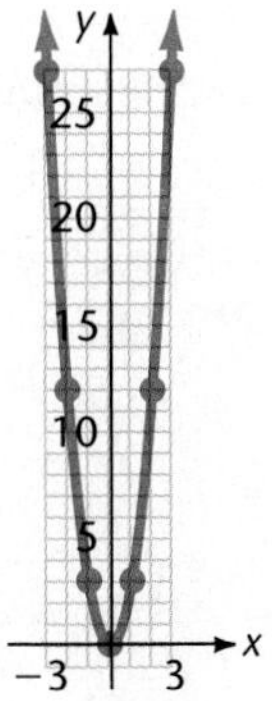

Figure 6

(B) $y = -x^2 + 3$

1. $a = -1$, so the parabola has a shape like $y = -x^2$. That is, the shape is

x	y
−3	−6
−2	−1
−1	2
1	2
2	−1
3	−6

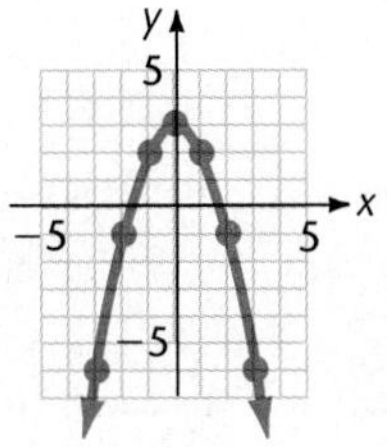

Figure 7

2. The vertex is at $x = \dfrac{-b}{2a} = \dfrac{0}{-2} = 0$ and $y = -(0)^2 + 3 = 3$. Plot the point $(0, 3)$.

3. Make a table for a few values of x to each side of 0. Plot these points and sketch the curve. See Figure 7.

(C) $y = x^2 - 4x + 4$

1. $a = 1$, so the parabola has a shape like $y = x^2$. That is, the shape is

x	y
−2	16
−1	9
0	4
1	1
3	1
4	4
5	9
6	16

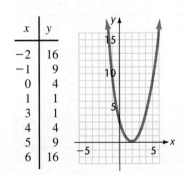

Figure 8

2. The vertex is at

$$x = \frac{-b}{2a} = \frac{4}{2} = 2 \qquad \text{and} \qquad y = (2)^2 - 4(2) + 4 = 0$$

Plot the point $(2, 0)$.

3. Make a table for a few values of x to each side of 2. Plot these points. Sketch the curve as shown in Figure 8.

Matched Problem 1 Graph:

(A) $y = -4x^2$ **(B)** $y = x^2 - 2$ **(C)** $y = x^2 + 4x + 4$

Example 2
Graphing
$y = ax^2 + bx + c$

Graph:

(A) $y = x^2 + x - 2$ **(B)** $y = -4x^2 - 9x + 9$

Solution   **(A)** $y = x^2 + x - 2$

1. The shape of the graph is the same as $y = x^2$; that is, the graph looks like

2. The vertex is

x	y
1	0
0	-2
-1	-2
-2	0

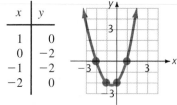

$$x = \frac{-b}{2a} = \frac{-1}{2} = -\frac{1}{2} \text{ and } y = \left(-\frac{1}{2}\right)^2 + \left(-\frac{1}{2}\right) - 2 = -\frac{9}{4}$$

Plot the point $(-\frac{1}{2}, -\frac{9}{4})$.

3. We will plot a couple of points to each side of $x = -\frac{1}{2}$ to locate the graph completely. See Figure 9.

Figure 9

(B) $y = -4x^2 - 9x + 9$

1. The shape of the graph is the same as $y = -x^2$; that is, the graph looks like

2. The vertex is

x	y
0	9
-1	14
-2	12
-3	0

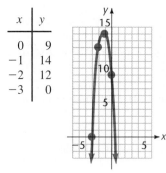

$$x = \frac{-b}{2a} = \frac{9}{-8} = -\frac{9}{8} \quad \text{and} \quad y = -4\left(-\frac{9}{8}\right)^2 - 9\left(-\frac{9}{8}\right) + 9 = 14\frac{1}{16}$$

Plot the point $(-\frac{9}{8}, 14\frac{1}{16})$.

3. We will plot a couple of points to each side of $x = -\frac{9}{8}$ to locate the graph completely. See Figure 10.

Figure 10

Matched Problem 2   Graph:

(A) $y = x^2 - x - 6$ **(B)** $y = -x^2 + x - 2$

◆ GRAPHS AND SOLUTIONS TO QUADRATIC EQUATIONS

If the graph of $y = ax^2 + bx + c$ crosses the x axis, then at that value of x, the value of y is 0. That is, such a value of x is a solution to

$$ax^2 + bx + c = 0$$

In Example 2(A) above, the graph of $y = x^2 + x - 2$ crossed the x axis at $x = -2$ and $x = 1$, the roots of the quadratic equation $x^2 + x - 2 = 0$. In Example 2(B), the graph of $y = -4x^2 - 9x + 9$ shows one root, $x = -3$, both in the table and on the graph. The graph also shows there is a second root, apparently between 0 and 1. The root can be found to be $x = \frac{3}{4}$ by factoring or using the quadratic formula.

In general, a quadratic equation has 0, 1, or 2 real roots. These cases correspond to the following situations geometrically:

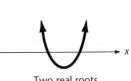

No real roots
(y never equals 0)

One real root
($y = 0$ at one point)

Two real roots
($y = 0$ at two points)

or to similar pictures with the graphs upside down.

Answers to Matched Problems

1. (A)

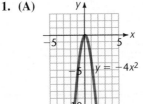

$y = -4x^2$

(B)

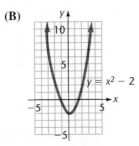

$y = x^2 - 2$

(C)

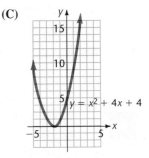

$y = x^2 + 4x + 4$

2. (A)

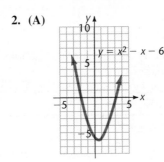

$y = x^2 - x - 6$

(B)

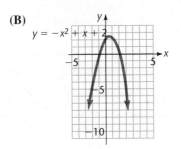

$y = -x^2 + x + 2$

EXERCISE 9-4

Graph:

A
1. $y = 5x^2$
2. $y = 6x^2$
3. $y = -5x^2$

4. $y = -6x^2$
5. $y = -4x^2$
6. $y = -2x^2$

7. $y = 2x^2$
8. $y = 4x^2$
9. $y = \frac{1}{5}x^2$

10. $y = \frac{1}{6}x^2$
11. $y = -\frac{1}{5}x^2$
12. $y = -\frac{1}{6}x^2$

13. $y = \frac{3}{2}x^2$
14. $y = -\frac{3}{2}x^2$
15. $y = -\frac{2}{3}x^2$

16. $y = \frac{2}{3}x^2$
17. $y = x^2 - 1$

18. $y = x^2 - 3$
19. $y = x^2 + 2$

20. $y = x^2 + 4$
21. $y = -x^2 + 1$

22. $y = -x^2 - 1$
23. $y = -x^2 - 3$

24. $y = -x^2 + 2$
25. $y = 2x^2 + 3$

26. $y = 4x^2 + 2$
27. $y = -2x^2 + 8$

28. $y = -3x^2 + 12$
29. $y = 3x^2 - 3$

30. $y = 2x^2 - 2$
31. $y = -3x^2 - 3$

32. $y = -2x^2 - 2$

B **33.** $y = x^2 - 3x$ **34.** $y = x^2 + 5x$

35. $y = 4x^2 - 8x$ **36.** $y = 3x^2 + 12x$

37. $y = x^2 - 25$ **38.** $y = x^2 - 36$

39. $y = x^2 + 7x$ **40.** $y = x^2 + 2x$

41. $y = x^2 - 4x$ **42.** $y = x^2 - 6x$

43. $y = -x^2 + 5x$ **44.** $y = -x^2 + 3x$

45. $y = 4 - x^2$ **46.** $y = 9 - x^2$

47. $y = x^2 - 6x + 5$ **48.** $y = x^2 - 5x + 6$

49. $y = x^2 - 4x + 3$ **50.** $y = x^2 - 8x + 15$

51. $y = x^2 - 6x + 10$ **52.** $y = x^2 + 4x - 5$

53. $y = x^2 + 5x - 6$ **54.** $y = x^2 + 3x - 4$

55. $y = x^2 - 2x - 8$ **56.** $y = x^2 - 4x - 12$

57. $y = x^2 - 5x - 6$ **58.** $y = x^2 - 6x + 8$

59. $y = x^2 + 6x + 8$ **60.** $y = x^2 + 5x + 6$

61. $y = x^2 - 4x + 4$ **62.** $y = x^2 + 6x + 9$

C **63.** $y = (x - 4)^2$ **64.** $y = (x - 5)^2$

65. $y = (x + 4)^2$ **66.** $y = (x + 5)^2$

67. $y = -(x - 1)^2$ **68.** $y = -(x - 2)^2$

69. $y = -(x + 1)^2$ **70.** $y = (x + 2)^2$

71. $y = x^2 - 3x + \dfrac{9}{4}$ **72.** $y = x^2 - x + \dfrac{1}{4}$

73. $y = x^2 - \dfrac{2}{3}x + \dfrac{1}{9}$ **74.** $y = x^2 - \dfrac{1}{3}x + \dfrac{1}{36}$

9-5 Applications

- ◆ Literal Equations
- ◆ Applications

We conclude this chapter by solving literal equations for a squared variable and by considering other applications.

◆ **LITERAL EQUATIONS**

We have previously solved literal equations for one variable in terms of the others. The formulas involved only the first power of the variable to be isolated. If the formula involves the square of that variable, the methods for solving quadratic equations may be applied.

Example 1
Solving a Quadratic Literal Equation

Solve for the indicated variable in terms of the other variables:

(A) $V = \pi r^2 h$ for r (volume of a cylinder, r positive)

(B) $S = \dfrac{n}{2}(n + 1)$ for n (sum of first n positive integers)

Solution   **(A)** $V = \pi r^2 h$ Divide by πh, the coefficient of r^2.

$\dfrac{V}{\pi h} = r^2$ Solve by the square root method.

$r = \sqrt{\dfrac{V}{\pi h}}$ Since r is positive, only the positive square root is used.

(B) $S = \dfrac{n}{2}(n + 1)$ Clear fractions.

$2S = n(n + 1)$ Multiply out.

$2S = n^2 + n$

$0 = n^2 + n - 2S$ Apply the quadratic formula with $a = 1$, $b = 1$, $c = -2S$.

$n = \dfrac{-1 \pm \sqrt{1 + 8S}}{2}$ Since n must be positive to make sense in our example, use only the positive solution.

$n = \dfrac{-1 + \sqrt{1 + 8S}}{2}$

Matched Problem 1   Solve for the indicated variable in terms of the other variables:

(A) $A = \pi(R + r)(R - r)$ for R (area of a disk of radius R with a hole of radius r, all variables positive)

(B) $c^2 = a^2 + b^2$ for a (Pythagorean theorem, all variables positive)

♦ **APPLICATIONS**

Many real-world problems lead directly to quadratic equations for their solutions. Quadratic equations may have two solutions, so it is important to check both of them in the original problem to see if one or both must be rejected. It is often the case that only one of the solutions will make sense in the context of the original application.

To get started, our first example is a straightforward word problem involving numbers. The second example is a geometric problem that is slightly more involved. Remember to draw figures, make diagrams, write down related formulas, and so on. Use scratch paper to try out ideas.

Example 2
A Number Problem

If 4 times the reciprocal of a number is subtracted from the original number, the difference is 3. Find the number.

Solution   Let x = Number. Then

$$x - 4 \cdot \frac{1}{x} = 3$$ Write an equation.

$$x \cdot x - x \cdot 4 \cdot \frac{1}{x} = x \cdot 3$$ Clear fractions ($x \neq 0$).

$$x^2 - 4 = 3x$$ Convert to standard form.

$$x^2 - 3x - 4 = 0$$ Solve by one of the methods discussed in earlier sections.

$$(x + 1)(x - 4) = 0$$ Factoring works.

$x + 1 = 0$ or $x - 4 = 0$

$x = -1$ $x = 4$ Both answers satisfy the original conditions, as you can check.

Matched Problem 2   The sum of a number and 4 times its reciprocal is 5. Find the number.

Example 3
A Geometry Problem

A painting measuring 6 by 8 inches has a frame of uniform width with a total area equal to the area of the painting. How wide is the frame? Give the answer in simplest radical form and as a decimal fraction to two decimal places.

Solution   Let x be the width of the frame.

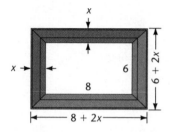

Figure 1

$$\left(\begin{array}{c}\text{Total area of picture}\\\text{and frame}\end{array}\right) = \left(\begin{array}{c}\text{Twice the area of}\\\text{the picture}\end{array}\right)$$

$$(6 + 2x)(8 + 2x) = 2(6 \cdot 8)$$

$$48 + 28x + 4x^2 = 96$$

$$x^2 + 7x - 12 = 0$$

$$x = \frac{-b \pm \sqrt{b^2 - 4ac}}{2a} \qquad a = 1, \, b = 7, \, c = -12$$

$$x = \frac{-7 \pm \sqrt{7^2 - 4(1)(-12)}}{2(1)}$$

$$x = \frac{-7 \pm \sqrt{97}}{2} \qquad \text{Check to see if one of the answers must be rejected.}$$

The negative answer must be rejected, since it has no meaning relative to the original problem; hence,

$$x = \frac{-7 + \sqrt{97}}{2} \approx 1.42 \text{ inches}$$

Matched Problem 3   If the length and width of a rectangle 4 by 2 inches are each increased equally by a certain amount, the area of the new rectangle will be twice that of the original rectangle. What are the dimensions, to two decimal places, of the new rectangle?

Example 4
A Rate-Time Problem

A fishing boat takes 3 hours longer to go 24 kilometers up a river than to return. If the boat cruises at 6 kilometers per hour in still water, what is the rate of the current?

Solution   Let c = Rate of current.

$$\text{Time up} = \text{Time down} + 3$$

$$\frac{\text{Distance up}}{\text{Rate up}} = \frac{\text{Distance down}}{\text{Rate down}} + 3 \qquad \text{Recall the rate-time formula } d = rt. \text{ Thus, } t = d/r.$$

$$\frac{24}{6 - c} = \frac{24}{6 + c} + 3 \qquad \text{Multiply both sides by } (6 - c)(6 + c).$$

$$24(6 + c) = 24(6 - c) + 3(36 - c^2)$$

$$144 + 24c = 144 - 24c + 108 - 3c^2$$

$$3c^2 + 48c - 108 = 0$$
$$c^2 + 16c - 36 = 0$$
$$(c - 2)(c + 18) = 0$$

$$c - 2 = 0 \quad \text{or} \quad c + 18 = 0$$
$$c = 2 \qquad \qquad \cancel{c = -18}$$
Reject

Rate of current = 2 kilometers per hour

Check

Rate up = 6 − 2 = 4 kilometers per hour

Rate down = 6 + 2 = 8 kilometers per hour

$$\text{Time up} = \frac{24}{4} = 6 \text{ hours}$$

$$\text{Time down} = \frac{24}{8} = 3 \text{ hours}$$

Thus, it took 3 hours longer to go up than to go down.

Matched Problem 4   If in Example 4 it takes the boat 2 hours longer to go the 24 kilometers up the river than to return, and the boat travels at 5 kilometers per hour in still water, what is the rate of the current?

Answers to Matched Problems

1. **(A)** $R = \sqrt{\dfrac{A + \pi r^2}{\pi}}$ **(B)** $a = \sqrt{c^2 - b^2}$ 2. 1 or 4

3. 5.12 by 3.12 inches   4. 1 kilometer per hour

EXERCISE 9-5

These problems are not grouped from easy (A) to difficult or theoretical (C). They are grouped somewhat according to type. The most difficult problems are marked with two stars (**) and the moderately difficult problems with one star (*). The easier problems are not marked.

LITERAL EQUATIONS

Solve for the indicated letter in terms of the other letters:

1. $d = \frac{1}{2}gt^2$ for t (positive)

2. $A = \pi r(r + l)$ for r (positive)

3. $A = P(1 + r)^2$ for r (positive)

4. $P = EI - RI^2$ for I

For each problem set up an appropriate equation and solve:

NUMBER PROBLEMS

5. Find a positive number that is 56 less than its square.

6. Find two consecutive positive even integers whose product is 168.

7. Find all numbers with the property that when the number is added to itself, the sum is the same as when the number is multiplied by itself.

8. Find two numbers such that their sum is 21 and their product is 104.

9. The sum of a number and its reciprocal is $\frac{17}{4}$. Find the number or numbers.

10. Find all numbers such that 6 times the reciprocal of the number is 1 less than the original number.

11. The sum of the first n positive integers is given by $[n(n + 1)]/2$. Find n if the sum is 120.

12. The sum of the first n odd integers is given by n^2. Find n if the sum is 121.

13. The sum of the first n even integers is given by $n(n + 1)$. Find n if the sum is 110.

14. The sum of the first n multiples of 3 is given by $[3n(n + 1)]/2$. Find n if the sum is 165.

BUSINESS AND ECONOMICS

15. If P dollars is invested at r percent compounded annually, at the end of 2 years it will grow to $A = P(1 + r)^2$. At what interest rate will $100 grow to $144 in 2 years? [*Note:* $A = 144$ and $P = 100$.]

16. Repeat Problem 15 for $1,000 growing to $1,210 in 2 years.

★ 17. Cost equations for manufacturing companies are often quadratic in nature. (At very high or very low outputs the costs are more per unit because of inefficiency of plant operation at these extremes.) If the cost equation for manufacturing a certain pharmaceutical drug is $C = x^2 - 10x + 31$, where C is the cost of manufacturing x units per week (both x and C are in thousands), find the output x for a $15,000 weekly cost.

★ 18. Repeat Problem 17 for a weekly cost of $6,000.

★ 19. The manufacturing company in Problem 17 sells its pharmaceuticals for $3 per unit. Thus, its revenue equation is $R = 3x$, where R is revenue and x is the number of units sold per week (both in thousands). Find the break-even points for the company—that is, the output x at which revenue equals cost.

★ 20. Repeat Problem 19, with the company selling each unit for $6 each.

21. Quantity discounts can lead to quadratic cost equations for purchasers. Suppose the cost of purchasing x units of a resource is given by $C = 120 - 0.05x^2$. If the purchase cost is $100, find the quantity purchased.

22. Repeat Problem 21 for a purchase cost of $75.

AGRICULTURE

★ 23. An apple grower is preparing a new section of his orchard for planting. For the particular variety of dwarf tree to be planted, he can expect a yield of 40 pounds per tree if he plants 20 trees. For each additional tree planted, the crowding will reduce the yield per tree by 1 pound. Thus, the total yield when x trees are added is $(20 + x)(40 - x) = 800 + 20x - x^2$. How many trees should the grower plant to get a total yield of 875 pounds of apples?

★ 24. The grower in Problem 23 also plans a blackberry patch. In the space available for his patch, he can plant 20 bushes and expect a yield of 14 pounds per bush. Each additional bush planted will reduce the yield by $\frac{1}{5}$ pound. Can the grower obtain a total yield of 420 pounds of blackberries?

★ 25. An amateur California wine maker has room to plant 40 grapevines spaced so as ultimately to yield 8 liters of wine for each vine planted. He can plant the vines closer together but will lose 0.1 liter of wine per vine for each extra vine planted. Can the total amount of wine obtained be increased to 380 liters by planting more vines?

★ 26. A farmer plans a small strawberry patch. She can plant 60 plants that will yield $\frac{1}{2}$ pound of strawberries per plant. Each additional plant included will decrease the yield per plant by $\frac{1}{200}$ of a pound. How many plants will provide a total yield of 32 pounds of strawberries?

COMMUNICATIONS

★ 27. The number of telephone connections c possible through a switchboard to which n telephones are connected is given by the formula $c = n(n - 1)/2$. How many telephones n could be handled by a switchboard that had the capacity of 190 connections? [*Hint:* Find n when $c = 190$.]

★ 28. Repeat Problem 27 for a switchboard with a capacity of 435 connections.

GEOMETRY

Use the Pythagorean theorem where needed: a triangle is a right triangle if and only if the square of the longest side is the sum of the squares of the two shorter sides. See the figure. The triangle is a right triangle exactly when $a^2 + b^2 = c^2$.

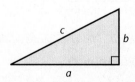

29. Find the length of each side of a right triangle if the second-longest side is 1 meter longer than the shortest side and the longest side is 2 meters longer than the shortest side.

30. Find the length of each side of a right triangle if the two shorter sides are 2 and 4 centimeters shorter than the longest side.

★ **31.** Find r in the accompanying figure. Express the answer in simplest radical form. (The radius of the smaller circle is 1 inch.)

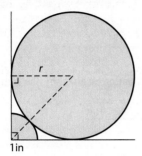

1 in

★ **32.** Approximately how far along the earth's surface would a person be able to see from the top of a mountain 2 miles high (see the figure)? Use a calculator to estimate the answer to the nearest mile.

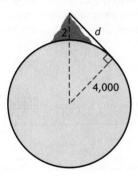

POLICE SCIENCE

33. Skid marks are often used to estimate the speed of a car in an accident. It is common practice for an officer to drive the car in question (if it is still running) at a speed of 20 to 30 miles per hour and skid it to a stop near the original skid marks. It is known (from physics) that the speed of the car and the length of the skid marks are related by the formula

$$\frac{d_a}{v_a^2} = \frac{d_t}{v_t^2}$$

where

d_a = Length of accident car's skid marks

d_t = Length of car's test skid marks

v_a = Speed of car during accident (to be found)

v_t = Speed of car during test

Estimate the speed of an accident vehicle if its skid marks are 100 feet and if during the test it is driven at 30 miles per hour and produces skid marks of 36 feet.

34. Repeat Problem 33 for an accident vehicle that left skid marks 196 feet long.

RATE-TIME PROBLEMS

★★ **35.** Two boats travel at right angles to each other after leaving the same dock at the same time; 1 hour later they are 25 miles apart. If one travels 17 miles per hour faster than the other, what is the rate of each? [*Hint:* See the theorem preceding Problem 29.]

★★ **36.** Repeat Problem 35 with one boat traveling 1 mile per hour faster than the other; after 1 hour they are 5 miles apart.

★★ **37.** A fishing boat takes 4 hours longer to go 30 miles up a stream than to return. If the boat cruises at 10 miles per hour in still water, what is the rate of the current?

★★ **38.** A speedboat takes 1 hour longer to go 60 miles up a river than to return. If the boat can cruise at 25 miles per hour on still water, what is the rate of the current?

HEALTH CARE

39. The body mass index, or Quetelet's index, is sometimes used by insurance companies to evaluate the weight of a person. The index is defined to be the ratio

$$\frac{\text{Weight in kilograms}}{(\text{Height in meters})^2}$$

How tall is a person who weighs 81 kilograms and has an index of 25?

40. Repeat Problem 39 for a person weighing 73.5 kilograms and having an index of 24.

CHAPTER SUMMARY

9-1   SOLUTION BY FACTORING AND BY SQUARE ROOT

The **standard form** of a **quadratic equation** in the variable x is $ax^2 + bx + c = 0$, where a, b and c are constants with $a \neq 0$. Some quadratic equations can be solved by factoring. A quadratic equation with no first-degree term can be solved by the **square root method:** if $x^2 = N$ with $N > 0$, then $x = \pm\sqrt{N}$.

9-2   SOLUTION BY COMPLETING THE SQUARES

To **complete the square** of $x^2 + bx$, add $(b/2)^2$ to obtain the perfect square $x^2 + bx + (b/2)^2 = (x + b/2)^2$. Quadratic equations can be solved by *completing the square*:

1. Write the equation in standard form.
2. If $a \neq 1$, divide both sides by a.
3. Shift the constant term to the other side of the equation.
4. Complete the square.
5. Solve by the method of square roots.

9-3   THE QUADRATIC FORMULA

The solution of the standard-form quadratic equation is given by the **quadratic formula:**

$$x = \frac{-b \pm \sqrt{b^2 - 4ac}}{2a}$$

The expression $b^2 - 4ac$ is called the **discriminant.** The equation has two real roots if the discriminant is positive, one real root if it is zero, and no real roots if it is negative.

9-4   GRAPHING QUADRATIC EQUATIONS

The graph of $y = ax^2 + bx + c$ is similar to the graph of $y = x^2$ for $a > 0$; for $a < 0$, the graph is similar to $y = -x^2$. The graph is called a **parabola.** The highest or lowest point of the parabola is called the **vertex** of the parabola. The vertex occurs at

$$x = -\frac{b}{2a}$$

Plotting the vertex and a few points corresponding to x values to either side of the vertex, and using the symmetry, will give the graph of $y = ax^2 + bx + c$ fairly quickly.

CHAPTER REVIEW EXERCISE

Work through all the problems in this chapter review and check the answers in the back of the book. Answers to all review problems are there, and following each answer is a number in italics indicating the section in which that type of problem is discussed. Where weaknesses show up, review appropriate sections in the text.

A *Write in standard form $ax^2 + bx + c = 0$ and identify a, b, and c:*

1. $4x = 2 - 3x^2$ **2.** $2x^2 = 9 - 5x$

Find all real solutions by factoring or square root methods:

3. $x^2 = 25$ **4.** $x^2 - 3x = 0$

5. $(2x - 1)(x + 3) = 0$ **6.** $x^2 - 5x + 6 = 0$

7. $x^2 - 2x - 15 = 0$ **8.** $x^2 - 4 = 0$

9. $1 - x^2 = 0$ **10.** $x^2 = \dfrac{4}{25}$

Complete the square and factor:

11. $x^2 + 12x$ **12.** $x^2 - 14x$

Solve by the method of completing the square:

13. $x^2 + 12x = 13$ **14.** $x^2 - 14x = -13$

Apply the discriminant test to determine the number of real roots and, if real roots exist, solve by use of the quadratic formula:

15. $x^2 + 3x + 1 = 0$ **16.** $x^2 + 3x - 10 = 0$

17. $2x^2 - 3x + 1 = 0$ **18.** $3x^2 + 7x - 4 = 0$

In Problems 19–21, graph the equation:

19. $y = -2x^2$ **20.** $y = x^2 - 3$

21. $y = -4x^2 + 1$

22. Find two positive numbers whose product is 27 if one is 6 more than the other.

23. Divide 18 into two parts so that their product is 72.

B *Find all real solutions by factoring or square root methods:*

24. $3x^2 = 36$

25. $10x^2 = 20x$

26. $(x - 2)^2 = 16$

27. $3t^2 - 8t - 3 = 0$

28. $2x = \dfrac{3}{x} - 5$

29. $x^2 - 5x - 6 = 0$

30. $x^2 = x + 20$

31. $x^2 + 8x - 20 = 0$

Complete the square and factor:

32. $x^2 + 13x$

33. $x^2 - 15x$

Solve by the method of completing the square:

34. $x^2 + 13x = -36$

35. $x^2 - 15x = 100$

Solve:

36. $x^2 - 6x - 3 = 0$

37. $3x^2 = 2(x + 1)$

38. $2x^2 - 2x = 40$

39. $x^2 + 2x + 3 = 0$

40. $x^2 + 2x - 3 = 0$

41. $5x^2 - 3x - 1 = 0$

42. $3x^2 - 5x + 3 = 0$

43. $8x^2 + 9x + 2 = 0$

Graph:

44. $y = -x^2 + 4$

45. $y = x^2 + 2x + 1$

46. $y = 2x^2 + 4x - 3$

47. $y = 2x^2 - 3x + 4$

In Problems 48–49, solve for the indicated variable:

48. $p = 0.003v^2$; v (v positive)

49. $F = \dfrac{GmM}{d^2}$; d (d positive)

50. The perimeter of a rectangle is 22 inches. If its area is 30 square inches, find the length of each side.

51. The illumination from a light source is proportional to the reciprocal of the square of the distance to the source, that is,

$$\frac{I}{1/d^2}$$

is a constant. If the illumination is 100 lumens at a distance of 5 feet, at what distance will the illumination be 625 lumens?

52. If the cost in dollars of producing x items is given by $C = 1{,}200 - 10x - x^2$, what number of items produced will result in a cost of $936?

C *Find all real solutions by factoring or square root methods:*

53. $x^2 - 2x - 48 = 0$

54. $2x^2 - 3x - 9 = 0$

55. $(x - \frac{3}{5})^2 - \frac{2}{5} = 0$

56. $5x^2 + 5x = 5(x + 20)$

57. $2x^2 + 27 = 0$

Solve:

58. $(t - \frac{3}{2})^2 = \frac{3}{2}$

59. $\dfrac{8m^2 + 15}{2m} = 13$

60. $2x^2 - 2x - 3 = 0$

61. $3x - 1 = \dfrac{2(x + 1)}{x + 2}$

62. $4x^2 + 4x + 1 = 0$

63. $x^2 - 3x + 1 = 0$

64. $x^2 - 11x + 9 = 0$

65. $2x^2 - 9x + 10 = 0$

66. $x^2 + 12x - 28 = 0$

67. $4x^2 + 12x + 9 = 0$

68. $x = \dfrac{7x - 10}{x}$

Graph:

69. $y = (x + 4)^2$

70. $y = x^2 - 3x - 4$

71. $y = 4x^2 + 4x + 1$

In Problems 72–73, solve for the indicated variable:

72. $-16t^2 + v_0t + s_0 = 0$; t

73. $A = \pi(R^2 - r^2)$; r (r positive)

74. A forest service airplane flies directly into a head wind for 140 miles before turning around and returning to its base. The airspeed of the plane is 120 miles per hour without any wind. If the return trip takes 24 minutes less than the trip out, what is the wind speed?

75. A **golden rectangle** is one that has the property that when a square with side equal to the short side of the rectangle is removed from one end, the ratio of the sides of the remaining rectangle is the same as the ratio of the sides of the original rectangle. If the shorter side of the original rectangle is 1, find the shorter side of the remaining rectangle (see the figure). This number is called the **golden ratio,** and it turns up frequently in the history of mathematics.

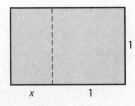

CHAPTER PRACTICE TEST

The following practice test is provided for you to test your knowledge of the material in this chapter. You should try to complete it in 50 minutes or less. The answers in the back of the book indicate the section in the text that covers the material in the question. Actual tests in your class may vary from the practice test in difficulty, length, or emphasis, depending on the goals of your course or instructor.

Solve by factoring or square root methods:

1. $x^2 - 16 = 0$

2. $x^2 - 4x - 12 = 0$

3. $x^2 + 3x - 28 = 0$

4. $\dfrac{x^2}{8} = \dfrac{9}{2}$

Complete the square and factor:

5. $x^2 + 6x$

6. $x^2 + 7x$

Solve by completing the square:

7. $x^2 + 6x - 5 = 0$

8. $x^2 + 7x = -6$

Apply the discriminant test to determine the number of real roots. If real roots exist, solve by use of the quadratic formula:

9. $x^2 + 7x - 2 = 0$

10. $x^2 + 7x + 14 = 0$

Solve:

11. $x^2 + 10x + 25 = 0$

12. $x^2 - x - 1 = 0$

13. $x^2 + 2x + 3 = 0$

14. $2x^2 - 5x - 8 = 0$

15. $5x^2 + 13x + 6 = 0$

In Problems 16–18, graph the equation:

16. $y = x^2 + 3$

17. $y = -x^2 + 4x - 3$

18. $y = x^2 + 2x + 3$

19. Find two positive numbers whose product is 238 if one number is 3 less than the other.

20. A gardener wants to build a rectangular garden pool surrounded by brick edging of uniform width (see the figure). The area for the pool and edging together is 6 feet by 9 feet. The pool is to cover an area of 28 square feet. How wide should the brick edging be?

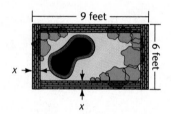

CUMULATIVE REVIEW EXERCISE—CHAPTERS 1–9

This set of exercises reviews the major concepts and techniques of Chapters 1–9. Work through all the problems, and check the answers in the back of the book. Answers to all review problems are there, and following each answer is a number in italics indicating the section in which that type of problem is discussed. Where weaknesses show up, review appropriate sections in the text.

A *Find, evaluate, or calculate the quantity requested:*

1. The LCM of 18, 30, and 40

2. The LCM of x^2, $x^2 + x$, and $x^2 - x$

3. $-2^3 - 3 \cdot 5 + 1$

4. The greater number of $\frac{3}{11}$ and $\frac{4}{15}$

5. The sum of $x^2 + 2x + 3$ and $2x^2 - 3x + 4$

6. The product of $x^2 + x + 1$ and $x - 1$

7. The product of $x + 2$ and $x - 5$

8. The sum of $\dfrac{x + 1}{3}$ and $\dfrac{x^2 - 1}{3}$

9. The product of $\dfrac{x + 1}{3x}$ and $\dfrac{3x + 6}{x + 1}$

10. The quotient of $\dfrac{x + 1}{x}$ divided by $\dfrac{x + 2}{x^2}$

11. $\dfrac{\dfrac{x}{2}}{\dfrac{x^2}{3}}$ written as a simple fraction

12. The slope of the line through the points $(3, -7)$ and $(-2, 8)$

13. The equation in the form $y = mx + b$ for the line with slope 4 and passing through the point $(-2, 7)$

Rewrite the given expressions or quantity in the form requested:

14. $\dfrac{x^{-1}y^2}{z^{-3}}$ using only positive exponents

15. $\sqrt{4x^3}$ in simplest radical form, for $x > 0$

16. 90 as a product of primes

17. $3x^2 + 2x - 4x^2 + 5x - 6x^2$ with like terms combined

18. $3x^3 + 6x^2 + 9x$ with common factors factored out

19. $x^2 + 6x + 9$ in factored form

20. $\dfrac{x^2y}{3z}$ as a fraction with denominator $6xz^2$

21. $\dfrac{x^3 - x}{x^2 + x}$ in lowest terms

22. $x^2 - 4y^2$ in factored form

23. $4x^2 + 9$ in factored form

24. $x^2 + 5x - 6$ in factored form

25. $x^2 + 2xy + 3x + 6y$ in factored form

26. 0.000 000 015 in scientific notation

27. 4.32×10^7 in decimal form

28. $x^2x^{-3}x^4$ as a power of x

29. $\dfrac{x^2}{x^5}$ as a power of x

30. $\dfrac{(xy)^3}{(xy^2)^2}$ as a power of x times a power of y

31. $\sqrt{28} + \sqrt{63}$ in simplest radical form

32. $2x - (3 - 4x^2)$ in standard quadratic form

Solve the equation, inequality, or system:

33. $2x - 3 = 4x + 7$

34. $\dfrac{1}{3}x + \dfrac{1}{4} = \dfrac{1}{5}x + \dfrac{1}{6}$

35. $\dfrac{1}{x} + \dfrac{1}{2x} = \dfrac{1}{3x^2}$

36. $2x - 3 > 4x - 5$

37. $x^2 - 5 = 0$

38. $x^2 - 2x + 1 = 0$

39. $\begin{array}{l} x + y = 5 \\ 2x - y = 12 \end{array}$

40. $2x + 3y = 4$ for y

41. $\sqrt{x + 1} = 2$

Graph:

42. $y \geq 2x + 3$

43. $y = -3x + 4$

44. The line through the point $(3, -5)$ and having slope -1

45. $y = x^2 + 3$

Set up the appropriate equation, or equations, and solve:

46. A football team has a ratio of points scored to points allowed of $5:3$. If the team has scored 220 points, how many points has it allowed?

47. Forty coins consisting only of dimes and quarters have a total value of $6.55. How many of each type of coin are there?

48. A piece of real estate that has increased by 30% in total value over the last 3 years is now worth $221,000. What was its value 3 years ago?

49. A package delivery service covers $\frac{1}{4}$ of the distance of its rural route at an average speed of 30 miles per hour and the other $\frac{3}{4}$ of the route at an average of 60 mph. If the route takes $12\frac{1}{2}$ hours to cover, how many miles long is it?

50. A traveler returning from an extended vacation has 36 rolls of slide and print film processed at a total cost of $215.20. Each roll of slide film costs $5.60, and each roll of print film costs $6.40 for processing. How many rolls of each type of film were processed?

B *Find, evaluate, or calculate the quantity requested:*

51. The sum $\dfrac{3}{x} + \dfrac{4x}{x + 1}$

52. The difference $\dfrac{5}{x} - \dfrac{6x + 7}{x + 2}$

53. The quotient $x^4 + 4x^3 + 7x^2 + 10x + 8$ divided by $x + 2$

54. The LCD of $\dfrac{1}{x}, \dfrac{x + 1}{x - 2},$ and $\dfrac{x + 3}{x^2 - 4}$

55. The equation in the form $y = mx + b$ for the line containing the points $(-2, 1)$ and $(4, 8)$

56. The product, in reduced form,
$$\dfrac{(x + 1)(x + 2)}{x + 3} \times \dfrac{(x + 2)(x + 3)}{x + 1}$$

57. The quotient, in reduced form,
$$\dfrac{(x + 1)(x + 2)}{x + 3} \div \dfrac{(x + 2)(x + 3)}{x + 1}$$

Rewrite the given expression or quantity in the form requested:

58. $\dfrac{1 + \dfrac{1}{x}}{\dfrac{x}{x + 1}}$ as a simple fraction

59. $x^2 + 8x - 48$ in factored form

60. $2x^2 + 5x + 2$ in factored form

61. $3x^3 - 24$ in factored form

62. $x^6 + 27$ in factored form

63. $3x + 2(x - y) + 1 = 4y - 5(x - y) + 6$ in the form $y = mx + b$

64. $\dfrac{3x}{\sqrt{x}}$ in simplest radical form

Solve the equation, inequality, or system:

65. $3x - 2(1 - x) = 4x - 5(1 - x)$

66. $\dfrac{x^2}{4} - \dfrac{y^2}{9} = 1$ for y

67. $4\sqrt{x - 3} = x - 8$

68. $x^2 - 10x - 56 = 0$

69. $x^2 - 10x + 20 = 0$

Graph:

70. The system $x + y \le 8$
$$x - y \ge 1$$
$$x \ge 0$$
$$y \ge 0$$

71. $y = x^2 + 4x$

Set up the appropriate equation, or equations, and solve:

72. How many liters of an 80% acid solution must be added to 5 liters of a 30% solution to obtain a 45% solution?

73. How many gallons each of a 30% acid solution and an 80% solution must be combined to obtain 20 gallons of a 45% solution?

74. A long-distance trucker leaves his terminal at 6:30 A.M. and drives at an average speed of 54 miles per hour. A second trucker leaves at 8:30 A.M. driving the same route at an average of 64 mph. How long will it take the second trucker to catch up to the first?

75. A theater production company determines that it can sell 800 tickets to a series of plays if it charges $9 per ticket. For each $1 increase in price it will lose 40 ticket sales. If the company has total revenue from ticket sales of $8,160, what price did the company charge per ticket? How many tickets were sold? [*Caution:* there may be more than one solution.]

76. On a college campus, all students take a particular freshman English course. The ratio of the number of students who pass the course to the number who fail is 23:3. How many students were enrolled in the course if the number who failed in one fall semester was 57?

77. A minor league hockey team determines that it can sell an average of 800 tickets at $9 per game and 1,160 at $7.50 per game. If the relationship between average ticket sales and price is linear, how many tickets would be sold if the price is set at $10 per ticket?

C *Translate each of the following statements into an algebraic equation with x as the variable. Solve the resulting equation.*

78. The product of a quantity that is 2 less than a given number and a quantity that is 5 more than the number is 6 times the number.

79. The sum of 3 times a given number and 4 times the number's reciprocal is 8 more than 15 times the number.

80. The sum of the squares of three consecutive positive integers is 194.

Rewrite the given expression or quantity in the form requested:

81. $6x^2 + 2x - 20$ in factored form

82. $4x^2 + 21x - 18$ in factored form

83. $4x^2 - 10x - 6$ in factored form

84. $12x^2 - 25x + 12$ in factored form

85. $\dfrac{x - \dfrac{x}{x + 1}}{1 + \dfrac{1}{x - 1}}$ as a simple fraction

86. $\dfrac{1}{\sqrt{x - 1}}$ in simplest radical form

87. $\dfrac{1 + \dfrac{1}{\sqrt{x}}}{1 - \dfrac{1}{\sqrt{x}}}$ as a simple fraction in simplest radical form

Solve:

88. $\dfrac{x - \dfrac{1}{x}}{x - 1} = \dfrac{3}{2}$

89. $\dfrac{x - \dfrac{1}{x}}{x + 1} = \dfrac{1}{x}$

90. $\sqrt{2 - x} = x$

91. $(x - 2)(x + 2) = 5$

92. $6x^2 - 11x - 340 = 0$

Set up the appropriate equation, or equations, and solve:

93. How many gallons of a 70% antifreeze solution must be replaced by pure water to obtain 35 gallons of a 40% solution?

94. A rental truck averages 9.2 miles per gallon of gasoline when fully loaded and 10.6 miles per gallon when empty. The truck was fully loaded, driven to its destination, emptied, and then driven to a return center. The total trip was 116 miles and used 12 gallons of gasoline. How long was each segment of the trip?

95. A riverboat used for sight-seeing cruises can travel at 16 miles per hour in still water. A cruise that goes 8 miles up the river and then returns takes 64 minutes. How fast is the river current?

96. The profit obtained from the sales of x units of a product is equal to the total cost of the x units subtracted from the product of the price per unit times the number of units x. The total cost of x units is $40x + 200$ dollars, and for x units the price is set at $500 - x$ dollars. How many units are sold if the profit is $52,700?

97. A teacher with a starting salary of $20,000 receives a certain percentage raise after the first year and twice that percentage raise after the second. If his salary is then $22,306.25, what were the percentages of the raises?

98. The number of diagonals in a regular polygon with n sides is given by $[n(n - 1)]/2$. If such a polygon has 9,180 diagonals, how many sides does it have?

99. Find the dimensions of a rectangle that has perimeter 78 yards and area 350 square yards.

100. A farmer with 100 acres to plant will divide his planting between sweet corn and asparagus. Sweet corn will require 2 hours of irrigation per acre and asparagus will require 3 hours per acre. How many acres of each crop should be planted if the farmer is to use 238 hours of irrigation?

Appendix A
SETS

This appendix introduces basic set terminology, relationships, and operations. The mathematical use of the word *set* does not differ much from the way it is used in everyday language. Words such as "set," "collection," "bunch," and "flock" all convey the same idea. Thus, we think of a set as any collection of objects with the important property that, given any object, it is either a member of the set or it is not. For example, the letter p belongs to our alphabet, but the Greek letter π does not.

♦ **SET MEMBERSHIP AND EQUALITY**

If an object a is in set A, we say that a **is an element of** or **is a member of** set A and write

$$a \in A$$

If an object **is not an element of** set A, we write

$$a \notin A$$

Sets are often specified by **listing** their elements between braces { }. For example,

$$\{2, 3, 5, 7\}$$

represents the set with elements 2, 3, 5, and 7. For this set, $3 \in \{2, 3, 5, 7\}$, and $4 \notin \{2, 3, 5, 7\}$. If two sets A and B have exactly the same elements, they are said to be **equal,** and we write

$$A = B$$

We write

$$A \neq B$$

if sets A and B are **not equal.**

The order of listing the elements in a set does not matter; thus,

$$\{3, 4, 5\} = \{4, 3, 5\} = \{5, 4, 3\}$$

Also, elements in a set are not listed more than once. For example, the set of letters in the word "letter" is

$$\{e, l, t, r\}$$

Example 1
Set Membership
and Equality

If $A = \{2, 4, 6\}$, $B = \{3, 5, 7\}$, and $C = \{4, 6, 2\}$, replace each question mark with $\in$, $\notin$, $=$, or $\neq$, as appropriate:

(A) $4 \,?\, A$ (B) $3 \,?\, A$ (C) $7 \,?\, B$
(D) $2 \,?\, B$ (E) $A \,?\, C$ (F) $A \,?\, B$

Solution

(A) $4 \in A$ (B) $3 \notin A$ (C) $7 \in B$
(D) $2 \notin B$ (E) $A = C$ (F) $A \neq B$

Matched Problem 1

If $P = \{1, 3, 5\}$, $Q = \{2, 3, 4\}$, and $R = \{3, 4, 2\}$, replace each question mark with $\in$, $\notin$, $=$, or $\neq$, as appropriate:

(A) $1 \,?\, P$ (B) $3 \,?\, Q$ (C) $5 \,?\, R$
(D) $4 \,?\, P$ (E) $P \,?\, Q$ (F) $Q \,?\, R$

♦ **SUBSETS**

We may be interested in sets within sets, called subsets. We say that a set A is a **subset** of set B if every element in set A is in set B. For example, the set of all women in a mathematics class would form a subset of all students in the class. The notation

$$A \subset B$$

is used to indicate that A is a subset of B.

A set with no elements is called the **empty** or **null** set. It is symbolized by

$$\varnothing$$

For example, the set of all months of the year beginning with the letter B is an empty, or null, set and would be designated by $\varnothing$. For any set A, $\varnothing \subset A$ and $A \subset A$.

<table>
<tr><td>

Example 2
Subsets

</td><td>

Which of the following are subsets of the set $\{2, 3, 5, 7\}$?

$A = \{2, 3, 5\}$ $B = \{2\}$ $C = \{2, 3, 5, 7\}$ $D = \varnothing$ $E = \{1, 2, 3\}$
$F = 3$

</td></tr>
</table>

Solution   The sets A, B, C, and D are all subsets of $\{2, 3, 5, 7\}$. In each case every element in the subset is in $\{2, 3, 5, 7\}$. The set $\{1, 2, 3\}$ is not a subset of $\{2, 3, 5, 7\}$, since $1 \notin \{2, 3, 5, 7\}$. The object 3 is different from the set $\{3\}$. We have $3 \in \{2, 3, 5, 7\}$ and $\{3\} \subset \{2, 3, 5, 7\}$. However, 3, and therefore F, is not a subset of $\{2, 3, 5, 7\}$.

Matched Problem 2   Which of the following are subsets of the set $\{2, 4, 6, 8\}$?

$A = \{2, 4\}$ $B = \{2, 4, 6, 8\}$ $C = \{4\}$ $D = 4$ $E = \varnothing$
$F = \{2, 4, 6, 8, 10\}$

♦ SPECIFYING SETS

The method of specifying a set by listing the elements, as in Example 1, is clear and convenient for small sets. However, if we are interested in specifying a set with a large number of elements, say, the set of all whole numbers from 10 to 10,000, then listing these elements would be impractical. The **rule method** for specifying sets takes care of situations of this type, as well as others. Using the rule method we would write

$$\{x \mid x \text{ is a whole number from 10 to 10,000}\}$$

which is read "the set of all elements x such that x is a whole number from 10 to 10,000." The vertical bar represents "such that."

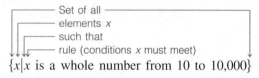

<table>
<tr><td>

Example 3
Describing Sets

</td><td>

Let $M = \{3, 4, 5, 6\}$ and $N = \{4, 5, 6, 7, 8\}$. Describe the set of all elements in M that are also in N using **(A)** the rule method and **(B)** the listing method.

</td></tr>
</table>

Solution   **(A)** Rule method: $\{x \mid x \in M \text{ and } x \in N\}$
         (B) Listing method: $\{4, 5, 6\}$

Matched Problem 3   Using the sets M and N in Example 3, describe the set of all elements that are in either M or N or both, using (A) the rule method and (B) the listing method.

♦ UNION AND INTERSECTION

Sets determined by "and" such as in Example 3 occur frequently in practice; the same is true for those determined by "or" as in Matched Problem 3. For this reason, we make the following definitions:

$$A \cap B = \textbf{Intersection of the sets } A \text{ and } B$$
$$= \text{Set of all elements belonging to both sets } A \text{ and } B$$
$$= \{x \mid x \in A \text{ and } x \in B\}$$

$$A \cup B = \textbf{Union of the sets } A \text{ and } B$$
$$= \text{Set of elements belonging to set } A \text{ or to set } B$$
$$= \{x \mid x \in A \text{ or } x \in B\}$$

The word ''or'' here is used in an inclusive sense meaning one or the other or both.

Example 4
Unions and
Intersections

Let $A = \{2, 3, 5, 7\}$ and $B = \{1, 2, 3, 4\}$. Find $A \cap B$ and $A \cup B$.

Solution

$$A \cap B = \{2, 3\}$$

The set of all elements belonging to set A and to set B

$$A \cup B = \{1, 2, 3, 4, 5, 7\}$$

The set of all elements belonging to set A or to set B or to both.

Matched Problem 4   Let $A = \{2, 4, 6, 8\}$ and $B = \{1, 2, 3, 4\}$. Find $A \cap B$ and $A \cup B$.

♦   **VENN DIAGRAMS**

Set relations and operations may be visualized in diagrams called **Venn diagrams.**

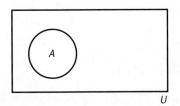

Here the rectangle represents all objects under consideration; this set is called the **universal set** and usually is denoted by U. It depends on the context in which one is working. If, for instance, you were interested in sets of students at a particular college, the universal set could be all students at that school. In the diagram, A represents a subset of U. To represent $B \subset A$, draw

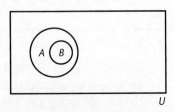

The intersection and union of two sets A and B are represented by the shaded regions in the following diagrams:

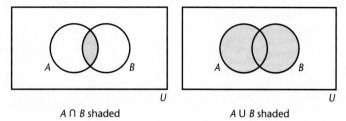

$A \cap B$ shaded $A \cup B$ shaded

♦ COMPLEMENTS

In the Venn diagram

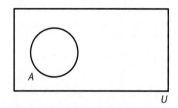

the region outside the circle represents all those elements in the universal set that are not in A, that is, the set

$$\{x \mid x \in U \text{ but } x \notin A\}$$

This set is called the **complement** of A and is usually denoted by A'.

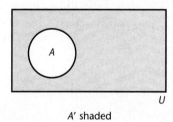

A' shaded

Example 5
Set Complements

Let $U = \{1, 2, 3, 4, 5, 6, 7, 8, 9, 10\}$ and $A = \{2, 3, 5, 7\}$. Give the complement of A by the listing method.

Solution

$$A' = \{1, 4, 6, 8, 9, 10\}$$

Matched Problem 5   Let U be as in Example 5 and $B = \{2, 4, 6, 8, 10\}$. Give B' by the listing method.

Answers to
Matched Problems

1. **(A)** $1 \in P$ **(B)** $3 \in Q$ **(C)** $5 \notin R$ **(D)** $4 \notin P$ **(E)** $P \neq Q$ **(F)** $Q = R$
2. A, B, C, E
3. Rule method: $\{x \mid x \in M \text{ or } x \in N\}$
   Listing method: $\{3, 4, 5, 6, 7, 8\}$
4. $A \cap B = \{2, 4\}$, $A \cup B = \{1, 2, 3, 4, 6, 8\}$
5. $B' = \{1, 3, 5, 7, 9\}$

EXERCISE A

A *In Problems 1–10, indicate which statements are true (T) and which are false (F):*

1. $4 \in \{2, 3, 4\}$
2. $7 \notin \{2, 3, 4\}$
3. $6 \notin \{2, 3, 4\}$
4. $7 \in \{2, 3, 4\}$
5. $\{3, 4, 5\} = \{5, 3, 4\}$
6. $\{1, 2, 3, 4\} = \{4, 3, 2, 1\}$
7. $\{3, 5, 7\} \neq \{4, 7, 5, 3\}$
8. $\{4, 6, 3\} \neq \{6, 3, 4\}$
9. $\{2, 3\} \subset \{2, 3, 4\}$
10. $\{4, 5\} \subset \{2, 3, 4\}$

Given sets

$$P = \{1, 3, 5, 7\} \qquad Q = \{2, 4, 6, 8\} \qquad R = \{5, 1, 7, 3\}$$

replace each question mark with $\in$, $\notin$, $=$, or $\neq$, as appropriate:

11. $5 \ ? \ P$
12. $6 \ ? \ Q$
13. $6 \ ? \ R$
14. $4 \ ? \ P$
15. $P \ ? \ R$
16. $Q \ ? \ R$
17. $P \ ? \ Q$
18. $R \ ? \ P$

B *Indicate the following sets by using the listing method. If the set is empty, write $\varnothing$.*

19. $\{x | x \text{ is a counting number between 5 and 10}\}$
20. $\{x | x \text{ is a counting number between 10 and 15}\}$
21. $\{x | x \text{ is a counting number between 7 and 8}\}$
22. $\{x | x \text{ is a counting number between 10 and 11}\}$
23. $\{x | x \text{ is a day of the week}\}$
24. $\{x | x \text{ is a month of the year}\}$
25. $\{x | x \text{ is a letter in ``alababa''}\}$
26. $\{x | x \text{ is a letter in ``millimeter''}\}$

27. $\{u | u \text{ is a state in the United States smaller than Rhode Island}\}$
28. $\{u | u \text{ is a day of the week starting with the letter } k\}$

If $U = \{1, 2, 3, 4, 5, 6, 7, 8\}$ and
$$A = \{1, 2, 3, 4\} \qquad B = \{2, 4, 6, 8\} \qquad C = \{1, 3, 5, 7\}$$

indicate each set by using the listing method:

29. $A \cap B$
30. $A \cap C$
31. $B \cap C$
32. $A \cup B$
33. $B \cup C$
34. $A \cup C$
35. A'
36. B'
37. C'

C 38. $\varnothing'$
39. $(A \cup B)'$
40. $(A \cap B)'$
41. $A' \cap B'$
42. $A' \cup B'$

43. List all the subsets of $\{1, 2\}$. (There are a total of four.)
44. List all the subsets of $\{1, 2, 3\}$. (There are a total of eight.)

A general Venn diagram for three sets A, B, and C is

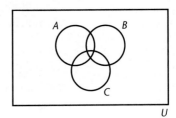

Use such a Venn diagram and shade the indicated set:

45. $(A \cup B) \cap C$
46. $(A \cap B) \cup C$
47. $(A \cap B) \cap C$
48. $(A \cap B) \cap C'$
49. $(A \cup B)' \cap C'$
50. $A \cap (B \cup C)'$

Appendix B
FUNCTIONS

The relationships

$$y = mx + b \qquad \text{(see Chapter 4)}$$
$$y = ax^2 + bx + c \qquad \text{(see Chapter 9)}$$

share the important characteristic that for each value of x there is just one value of y determined. These are examples of a relationship or correspondence between variables that will be called a *function*. The function concept is among the most important concepts in mathematics, and it will be studied extensively in subsequent courses. The concept is introduced briefly in this appendix.

♦ FUNCTIONS

One of the most important mathematical aspects of the sciences is establishing relationships between various phenomena. Once a relation is known, predictions can be made. An engineer can use a formula to predict pressures on a bridge for various wind speeds; an economist would like to predict unemployment rates given various levels of government spending; a chemist can use a formula to predict the pressure of an enclosed gas given its temperature; and so on. Establishing and working with such relationships is so fundamental to both pure and applied science that people have found it desirable to describe them in the precise language of mathematics.

We build a basis for such a mathematical description by looking at the variables connected with phenomena such as those mentioned just above. In each case, the values of one variable provide information about the related second variable:

Variable	Related Variable
Wind speed	Pressure on bridge
Government spending	Unemployment rate
Temperature	Pressure of gas

The key feature of these relationships is that for any value of the given variable, the related variable should be completely determined. For example, once the wind speed is known, the pressure on the bridge should be completely determined. That is, the pressure on the bridge is a unique number corresponding to the wind speed. The rule of correspondence that determines the pressure from the wind speed is called a *function,* and we say that pressure is a *function of* wind speed.

We can also look for functions in more commonplace settings. For example:

To each item on the shelf in a grocery store there corresponds a price.

To each square there corresponds an area.

To each student there corresponds a grade point average.

To each letter on a telephone dial there corresponds a number on the dial.

Once again, for each value of the first variable, there is a unique corresponding value of the related variable:

Variable	Related Variable
Grocery item	Price
Square	Area
Student	GPA
Letter on phone dial	Number on phone dial

If we denote the first variable by x and the related variable by y, we will say that y is a function of x if for each value of x there is a unique corresponding value for y. The actual rule of correspondence will be called a function. Here we allow the word "value" to represent an object—for example, a grocery item, student, or letter—as well as a number. We thus have the following informal definition of a function:

Informal Definition of a Function

A **function** is a rule that produces a correspondence between a first variable x and a second variable y such that to each value of x there corresponds *one and only one* value for y. In this case we say y is a function of x.

If a variable y is a function of a variable x, we will call the variable denoted by x the **independent variable** and the variable denoted by y the **dependent variable.** The value of the dependent variable is determined by, or dependent on, the value of the independent variable.

Let us take a closer look at the relationship between letters and numbers on a telephone dial as shown in the figure.

If we let the variable x represent a letter from among A, B, C, D, E, F, G, H, I, J, K, L, M, N, O, P, R, S, T, U, V, W, X, and Y and the variable y represent the corresponding number on the phone dial, then it is clear that every x corresponds to a unique y. That is, every letter uniquely determines the corresponding number. For example, corresponding to J is the number 5, and only that number. Also, corresponding to W is the number 9, and only that number, and so forth. Here the phone dial number is a function of the phone dial letter.

On the other hand, the phone dial letter is not a function of the phone dial number. For instance, corresponding to the number 5 there are *three* letters J, K, and L. Corresponding to the number 9 there are also three letters: W, X, and Y. The same is true for all the numbers 2, 3, 4, 5, 6, 7, 8, and 9, but corresponding to the numbers 0 and 1 there are no letters at all. Thus, this correspondence does not give a function.

The set of allowable values for the independent variable is called the **domain** of the function. The set of values for the dependent variable is called the **range** of the function. In the example where the number on the phone dial is determined by the letter, we would have domain and range as follows:

Domain: A, B, C, D, E, F, G, H, I, J, K, L, M, N, O, P, R, S, T, U, V, W, X, Y

Range: 2, 3, 4, 5, 6, 7, 8, 9

The numbers 0 and 1 are not in the range, since no letter will have 0 or 1 corresponding to it.

For most of the functions you will encounter in beginning college mathematics courses, the variables being related will represent numbers. That is, both the domain and range will be sets of numbers. The examples of the phone dial and the student-to-GPA correspondence indicate that this need not be the case. The function concept is more general than just relationships between numerical variables. This leads to a more formal, general definition of a function:

Formal Definition of a Function

A **function** is a rule that produces a correspondence between a first set, called the domain, and a second set, called the range, such that to each element in the domain there corresponds *one and only one* element in the range.

Example 1
Identifying Functions

The correspondence rules in this example are given by arrows. Read $1 \longrightarrow 5$ as "1 corresponds to 5." Indicate which rules are functions:

(A) Domain   Range
   $1 \longrightarrow 5$
   $2 \longrightarrow 7$
   $3 \longrightarrow 9$

(B) Domain   Range
   $-2 \longrightarrow -1$
   $0 \longrightarrow 0$
   2
   $4 \longrightarrow 1$

(C) Domain   Range
   $3 \longrightarrow 1$
   $\longrightarrow 3$
   $7 \longrightarrow 8$
   $9 \longrightarrow 9$

Solution
(A) *Function.* Exactly one range value corresponds to each domain value.
(B) *Function.* Exactly one range value corresponds to each domain value.
(C) *Not a function.* Two range values correspond to the domain value 3.

Matched Problem 1

Indicate which correspondence rules are functions:

(A) Domain   Range
   $-5 \longrightarrow 6$
   -3
   $3 \longrightarrow 0$

(B) Domain   Range
   $1 \longrightarrow 5$
   $2 \longrightarrow 6$
   $\longrightarrow 7$
   $3 \longrightarrow 8$

(C) Domain   Range
   -1
   0
   $1 \longrightarrow 5$
   3

◆ COMMON WAYS OF SPECIFYING FUNCTIONS

The arrow method of specifying functions illustrated in Example 1 above is convenient only when the domain has a small number of elements. Given the domain consisting of the numbers 1, 2, and 3, we can also specify the function from Example 1(A) in several ways:

1. By an equation such as $y = 2x + 3$
2. By a table:

x	y
1	5
2	7
3	9

3. By a graph:

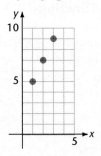

4. By a set of ordered pairs: $(1, 5), (2, 7), (3, 9)$

If a correspondence rule is specified by a set of ordered pairs of elements, then the set of first components forms the domain and the set of second components forms the range. Tables and ordered pairs are useful for specifying functions when the number of elements in the domain is small or for describing part of a function rule. Equations and graphs can be used whenever the domain and range are sets of real numbers.

Example 2
Identifying Functions

Given the set of ordered pairs:

$$(0, 0), (1, -1), (1, 1), (4, -2), (4, 2)$$

(A) Write this rule, using arrows as in Example 1. Indicate domain and range.
(B) Graph the set in a rectangular coordinate system.
(C) Is the rule a function? Explain.

Solution **(A)**

Domain	Range
0 ⟶	0
	−1
1 ⟹	1
	−2
4 ⟹	2

(B)

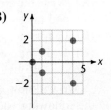

(C) The rule is not a function, since more than one range value corresponds to a given domain value.

For a rule given by ordered pairs to be a function, no two ordered pairs can have the same first coordinate. There is also an easy way to determine whether a rule is a function if you have the graph of an equation:

Vertical Line Test for a Function

A rule specified by an equation is a function if each vertical line in the coordinate system passes through *at most* one point on the graph of the equation. The same test may be applied to functions specified by a set of ordered pairs.

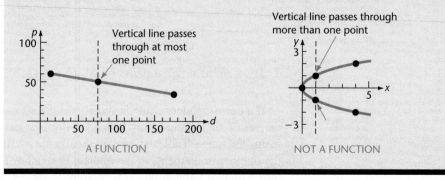

A FUNCTION   NOT A FUNCTION

If a vertical line passes through more than one point on the graph, then all points of intersection will have the same first coordinate, and thus the same domain value, and different second coordinates, and thus different range values. Since more than one range value is associated with a given domain value, the rule is not a function.

Matched Problem 2   Given the set of ordered pairs:

$$(-2, 4), (-1, 1), (0, 0), (1, 1), (2, 4)$$

(A) Write the correspondence rule, using arrows as in Example 1. Indicate domain and range.
(B) Graph the set of points F.
(C) Is the rule a function? Explain.

We will now concentrate on rules specified by equations in two variables. All the following equations give y as a function of x:

$$y = 2x + 3 \qquad y = \sqrt{x}$$

$$y = x^2 \qquad y = \frac{1}{x - 1}$$

$$y = |x|$$

In each case, for any allowable value x we obtain exactly one value for y. For example, using the function given by $y = 2x + 3$:

If $x = 2$, then $y = 2(2) + 3 = 7$.

If $x = -1$, then $y = 2(-1) + 3 = 1$.

Using the function given by $y = \sqrt{x}$:

If $x = 9$, then $y = \sqrt{9} = 3$.

> Recall that the symbol $\sqrt{x}$ represents only the *positive* square root of x, so that $\sqrt{x}$ gives only one number corresponding to x.

If $x = 36$, then $y = \sqrt{36} = 6$.

The allowable values of x are the values in the domain. If a domain is not specified, it is usually assumed to consist of all values of x for which the equation provides a value of y. In the first three equations above, any real number can be substituted for x, so the domain in each case is the set of all real numbers. In the case of $y = \sqrt{x}$, the domain is all $x \geq 0$. For $y = 1/(x - 1)$, the domain is all $x \neq 1$.

All equations in two variables specify correspondence rules, but when does an equation specify a function? Suppose x is to be the independent variable and y the dependent variable. If we can solve the equation uniquely for y in terms of x, then the equation does specify y as a function of x. On the other hand, if we can find a value for x that gives two or more corresponding values for y, then the equation does not specify y as a function of x. This will happen, for example, when solving for y yields something like $y = \pm\sqrt{x}$ and thus gives multiple values for a given x.

Example 3
Specifying Functions by Equations

Given the equation, with independent variable x and dependent variable y, decide whether or not the equation determines y as a function of x:

(A) $x^2 + y^2 = 4$ **(B)** $x^2 + y = 4$

Solution **(A)** Solve the equation for y:

$$x^2 + y^2 = 4$$
$$y^2 = 4 - x^2$$
$$y = \pm\sqrt{4 - x^2}$$

Is there a value of x that will produce more than one value of y? Yes. For example, if $x = 0$, then

$$y = \pm\sqrt{4} = \pm 2$$

Thus, two values of y result from one value of x. Therefore, this rule is not a function.

(B) Solve the equation for y:

$$x^2 + y = 4$$
$$y = 4 - x^2$$

Thus, since we can solve the equation for y uniquely, it does specify y as a function of x. For example, if we let $x = 1$, then $y = 3$, and no other value.

Matched Problem 3

Given the equation, with independent variable x and dependent variable y, decide whether or not the equation determines y as a function of x:

(A) $3 = x^2 - y$ **(B)** $y^2 = x - 3$

♦ FUNCTION NOTATION

We use different letters to denote names for numbers. In essentially the same way, we will now use different letters to denote names for functions. For example, f and g may be used to name the following two functions specified by equations:

$$f: \quad y = 2x + 1$$

$$g: \quad y = x^2 + 2x - 3$$

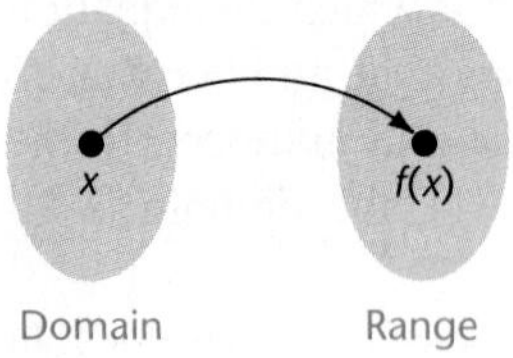

If x represents an element in the domain of a function f, then we may use the symbol

$$f(x)$$

Domain   Range

Figure 1

in place of y to designate the number in the range of the function f that corresponds to x (see Figure 1).

CAUTION

This new function symbol is not the product of f and x. The symbol $f(x)$ is read "f of x" or "the value of f at x." The variable x is an independent variable; y and $f(x)$ both name the dependent variable.

This new function notation is extremely useful. For example, in place of the more formal representation of the functions f and g above, we can now write

$$f(x) = 2x + 1 \qquad \text{and} \qquad g(x) = x^2 + 2x - 3$$

The function symbols $f(x)$ and $g(x)$ also allow us to describe the correspondence for particular values of x. For example, if we write $f(3)$ and $g(5)$, then each symbol indicates in a concise way that these are range values of particular functions associated with particular domain values. Let us find $f(3)$ and $g(5)$.

To find $f(3)$, we replace x by 3 where x occurs in

$$f(x) = 2x + 1$$

and evaluate the right side:

$$
\begin{aligned}
f(3) &= 2 \cdot 3 + 1 \\
&= 6 + 1 \\
&= 7
\end{aligned}
$$

Thus,

$$f(3) = 7$$

The function f assigns the range value 7 to the domain value 3; the ordered pair (3, 7) belongs to f.

To find $g(5)$, we replace x by 5 wherever x occurs in

$$g(x) = x^2 + 2x - 3$$

and evaluate the right side:

$$g(5) = 5^2 + 2 \cdot 5 - 3$$
$$= 25 + 10 - 3$$
$$= 32$$

Thus,

$$g(5) = 32$$

The function g assigns the range value 32 to the domain value 5; the ordered pair (5, 32) belongs to g.

It is very important to understand and remember the definition of $f(x)$:

The Function Symbol $f(x)$

For any element x in the domain of the function f, the function symbol

$$f(x)$$

represents the element in the range of f corresponding to x in the domain of f.

We may think of x as an input value into the function rule. Then $f(x)$ is the corresponding output value. The input-output way of looking at a function rule suggests viewing a function as a "machine" that converts inputs x into outputs y. Figure 2 illustrates this. The function notation can then be thought of as playing the role of the "function machine": $f(\ )$ converts x into $f(x)$.

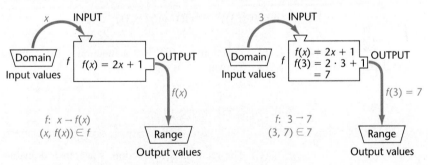

Figure 2

For the function $f(x) = 2x + 1$, the rule, or machine, takes each domain value, multiplies it by 2, then adds 1 to the result to produce the range value. Different rules inside the machine result in different functions.

Example 4
Using the $f(x)$ Notation

Let $f(x) = \dfrac{x}{2} + 1$ and $g(x) = 1 - x^2$. Find:

(A) $f(6)$ **(B)** $g(-2)$ **(C)** $f(4) + g(0)$

Solution   **(A)** $f(6) = \dfrac{6}{2} + 1 = 3 + 1 = 4$

(B) $g(-2) = 1 - (-2)^2 = 1 - 4 = -3$

$$\underset{\substack{\downarrow \\ f(4)}}{} \qquad \underset{\substack{\downarrow \\ g(0)}}{}$$

(C) $f(4) + g(0) = \left(\dfrac{4}{2} + 1\right) + (1 - 0^2) = 3 + 1 = 4$

Matched Problem 4 Let $f(x) = \dfrac{x}{3} - 2$ and $g(x) = 4 - x^2$. Find:

(A) $f(9)$ (B) $g(-2)$ (C) $f(0) + g(2)$

Example 5
Using the $f(x)$ Notation

Let $f(x) = \dfrac{x}{2} + 1$ and $g(x) = 1 - x^2$. Find:

(A) $f(g(3))$ (B) $g(f(-4))$

Solution (A) $f(g(3)) = f(1 - 3^2)$ Evaluate $g(3)$ first; then evaluate f for this value.

$$= f(-8) = \dfrac{-8}{2} + 1 = -3$$

(B) $g(f(-4)) = g\left(\dfrac{-4}{2} + 1\right)$

$$= g(-1) = 1 - (-1)^2 = 0$$

Matched Problem 5 Let $f(x) = \dfrac{x}{3} - 2$ and $g(x) = 3 - x^2$. Find:

(A) $f(g(3))$ (B) $g(f(-3))$

Answers to Matched Problems

1. (A) Function   (B) Not a function   (C) Function
2. (A) Domain   Range   (B)

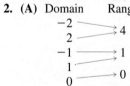
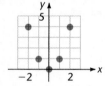

(C) The relation is a function, since each domain value corresponds to exactly one range value.
3. (A) Specifies a function: solving for y gives $y = x^2 - 3$
   (B) Does not specify a function: if $x = 4$, for example, then $y = \pm 1$
4. (A) 1   (B) 0   (C) -2
5. (A) -4 (B) -6

EXERCISE B

A *Indicate whether each correspondence rule is or is not a function:*

1. Domain   Range
   3 ⟶ 0
   5 ⟶ 1
   7 ⟶ 2

2. Domain   Range
   -1 ⟶ 5
   -2 ⟶ 7
   -3 ⟶ 9

3. Domain   Range
   3 ⟶ 5
      ⟶ 6
   4 ⟶ 7
   5 ⟶ 8

4. Domain   Range
   8 ⟶ 0
   9 ⟶ 1
      ⟶ 2
   10 ⟶ 3

5. Domain   Range

3 ⟶ 5
6
9 ⟶ 6
12

6. Domain   Range

−2
−1 ⟶ 6
0
1

Each rule is specified by a graph. Indicate whether it is a function with x as the independent variable:

7.

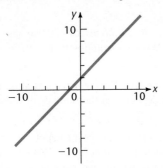

8.

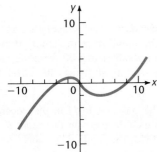

9.

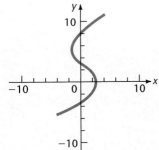

10.

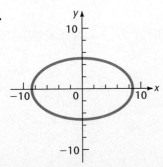

11.

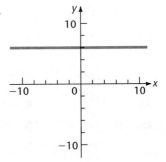

12.

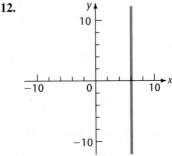

In Problems 13–18, let $f(x) = 3x - 2$. Find:

13. $f(2)$ **14.** $f(1)$ **15.** $f(-2)$

16. $f(-1)$ **17.** $f(0)$ **18.** $f(4)$

In Problems 19–24, let $g(x) = x - x^2$. Find:

19. $g(2)$ **20.** $g(1)$ **21.** $g(4)$

22. $g(5)$ **23.** $g(-2)$ **24.** $g(-1)$

B *Each equation specifies a relationship between x and y. Which specify functions, given that x is the independent variable?*

25. $y = 3x - 1$ **26.** $y = \dfrac{x}{2} - 1$

27. $y = x^2 - 3x + 1$ **28.** $y = x^3$

29. $y^2 = x$ **30.** $x^2 + y^2 = 25$

31. $x = y^2 - y$ **32.** $x = (y - 1)(y + 2)$

33. $y = x^4 - 3x^2$ **34.** $2x - 3y = 5$

35. $y = \dfrac{x + 1}{x - 1}$ **36.** $y = \dfrac{x^2}{1 - x}$

Graph the correspondence rule given by the listed ordered pairs. State its domain and range, and indicate which are functions. The variable x is independent.

37. $F = \{(1, 1), (2, 1), (3, 2), (3, 3)\}$

38. $f = \{(2, 4), (4, 2), (2, 0), (4, -2)\}$

39. $G = \{(-1, -2), (0, -1), (1, 0), (2, 1), (3, 2), (4, 1)\}$

40. $g = \{(-2, 0), (0, 2), (2, 0)\}$

In Problems 41–58 refer to the functions

$$f(x) = 10x - 7 \qquad g(t) = 6 - 2t \qquad F(u) = 3u^2$$
$$G(v) = v - v^2$$

Evaluate as indicated:

41. $f(-2)$ **42.** $F(-1)$ **43.** $g(2)$

44. $G(-3)$ **45.** $g(0)$ **46.** $G(0)$

47. $f(3) + g(2)$ **48.** $F(2) + G(3)$

49. $2g(-1) - 3G(-1)$ **50.** $4G(-2) - g(-3)$

51. $\dfrac{f(2) \cdot g(-4)}{G(-1)}$ **52.** $\dfrac{F(-1) \cdot G(2)}{g(-1)}$

53. $F(g(1))$ **54.** $G(F(1))$

55. $g(f(1))$ **56.** $g(G(0))$

57. $f(G(1))$ **58.** $G(g(2))$

59. If $A(w) = \dfrac{w - 3}{w + 5}$, find $A(5)$, $A(0)$, and $A(-5)$.

60. If $h(s) = \dfrac{s}{s - 2}$, find $h(3)$, $h(0)$, and $h(2)$.

C *Graph the correspondence rule given by the equation and for the given values of x. Indicate which are functions. The variable x is the independent variable.*

61. $y = 5 - 2x$ for $x = 0, 1, 2, 3, 4$

62. $y = \dfrac{x}{2} - 4$ for $x = 0, 2, 4$

63. $y^2 = x$ for $x = 0, 1, 4$

64. $y = x^2$ for $x = -2, 0, 2$

65. $x^2 + y^2 = 4$ for $x = -2, 0, 2$

66. $x^2 + y^2 = 9$ for $x = -3, 0, 3$

APPLICATIONS

Each of the statements in Problems 67–70 can be described by a function. Write an equation that specifies the function.

67. Cost Function The cost $C(x)$ of x records at \$5 per record. (The cost depends on the number of records purchased.)

68. Cost Function The total daily cost $C(x)$ of manufacturing x pairs of skis if fixed costs are \$800 per day and the variable costs are \$60 per pairs of skis. (The total daily cost depends on the number of skis manufactured per day.)

69. Temperature Conversion The temperature in degrees Celsius can be found from the temperature in degrees Fahrenheit by subtracting 32 from the Fahrenheit temperature and multiplying the difference by $\frac{5}{9}$.

70. Earth Science The pressure $P(d)$ in the ocean in pounds per square inch depends on the depth d. To find the pressure, divide the depth by 33, add 1 to the quotient, and then multiply the result by 15.

Answers to Selected Problems

CHAPTER 1

EXERCISE 1-1

1. 6, 13   **3.** 67, 402   **5.** Even: 14, 28; odd: 9, 33   **7.** Even: 426; odd: 23, 105, 77
9. Composite: 6, 9; prime: 2, 11   **11.** Composite: 12, 27; prime: 17, 23   **13.** Composite: 21, 34; prime: 5, 13, 29
15. Composite: 57; prime: 37, 47, 67   **17.** 20, 22, 24, 26, 28, 30   **19.** 21, 23, 25, 27, 29
21. 20, 21, 22, 24, 25, 26, 27, 28, 30   **23.** 23, 29   **25.** $2 \cdot 5$ **27.** $2 \cdot 3 \cdot 5$ **29.** $2 \cdot 2 \cdot 2 \cdot 5$ **31.** $2 \cdot 2 \cdot 2 \cdot 7$
33. $2 \cdot 2 \cdot 3 \cdot 7$ **35.** $2 \cdot 2 \cdot 3 \cdot 5$ **37.** $2 \cdot 2 \cdot 3 \cdot 3 \cdot 3$ **39.** $2 \cdot 3 \cdot 5 \cdot 7$ **41.** 36   **43.** 48   **45.** 24
47. 60   **49.** 90   **51.** 5,390   **53.** 630   **55.** $3 \cdot 7 \cdot 13$ **57.** $2 \cdot 2 \cdot 2 \cdot 2 \cdot 5 \cdot 7$ **59.** $2 \cdot 2 \cdot 3 \cdot 3 \cdot 5$
61. $2 \cdot 2 \cdot 3 \cdot 5 \cdot 5$ **63.** $2 \cdot 3 \cdot 3 \cdot 5 \cdot 7$ **65.** $2 \cdot 2 \cdot 2 \cdot 5 \cdot 5 \cdot 11$ **67.** $2 \cdot 3 \cdot 3 \cdot 7 \cdot 11$ **69.** No, no, no, yes
71. Finite   **73.** Infinite   **75.** Finite   **77.** 1,400   **79.** 2,310   **81.** 4,620   **83.** 7   **85.** 12   **87.** 1
89. 13   **91.** 3   **93.** Product 192; LCM 48; GCF 4; LCM × GCF 192   **95.** 2,310; 2,310; 1; 2,310
97. 5,880; 420; 14; 5,880   **99.** 1,080; 180; 1; 180; for two numbers the product is equal to LCM · GCF, but this is not true for three numbers.

EXERCISE 1-2

1. 13   **3.** 2   **5.** 17   **7.** 2   **9.** 24   **11.** 80   **13.** 1   **15.** 21   **17.** 8   **19.** 10   **21.** 5
23. 2   **25.** 18   **27.** 12   **29.** 4   **31.** $5x$ **33.** $x + 5$ **35.** $x - 5$ **37.** $5 - x$
39. Constant: $\frac{1}{2}$; variables: A, b, h **41.** Constants: none; variables: d, r, t **43.** Constants: none; variables: I, p, r, t
45. Constants: 2, 3; variables: x, y **47.** Constants: 2, 3; variables: u, v **49.** $A = 18$ cm^2, $P = 18$ cm
51. $A = 80$ km^2, $P = 36$ km   **53.** 20   **55.** 33   **57.** 14   **59.** 8   **61.** 5   **63.** 3   **65.** 22
67. 684 km   **69.** 600 words   **71.** $2x + 3$ **73.** $12x - 3$ **75.** $3(x - 8)$ **77.** $2(x + 4)$ **79.** 105
81. 16   **83.** 3   **85.** 4   **87.** 16   **89.** 47   **91.** 420   **93.** $3(t + 2)$ **95.** $t + (t + 1) + (t + 2)$
97. $t + (t + 2) + (t + 4)$ **99.** $(t + 1)(t + 3)(t + 5)$

EXERCISE 1-3

1. T   **3.** F   **5.** T   **7.** T   **9.** T   **11.** T   **13.** F   **15.** T   **17.** T   **19.** T   **21.** F
23. $5 = x + 3$ **25.** $8 = x - 3$ **27.** $18 = 3x$ **29.** $49 = 2x + 7$ **31.** $52 = 5x - 8$ **33.** $4x = 3x + 3$
35. $3x = 44 + 2$ **37.** $2x = 3x - 108$ **39.** $2x + 2 = (x + 41) - 6$ **41.** $2x = (x + 38) + 5$
43. $2x + 4 = 5x - 16$ **45.** $4x + 7 = 5x - 5$ **47.** $11x = 8x + 110$ **49.** $2x + 4 = 3x$ **51.** $x + 5 = 3(x - 4)$
53. $x + (x + 1) + (x + 2) = 90$ **55.** $x + (x + 2) = 54$ **57.** $5x + 5(x + 1) + 5(x + 2) = 75$ **59.** Expression
61. $x + 2x = 54$ **63.** $50 = x(x + 10)$ **65.** $1{,}440 = (2w)w$ **67.** $864 = (2w)w(2w)$ **69.** $108 = l + 2(w + h)$

EXERCISE 1-4

1. $x + 10$ **3.** $9 + x$ **5.** $15x$ **7.** $24x$ **9.** $28ab$ **11.** $16 + a + b$ **13.** $13 + x + x$
15. $11 + x + x$ **17.** $7 + 5x$ **19.** $60x$ **21.** $60xy$ **23.** $60xy$ **25.** $x + x + 25$ **27.** $x + x + 19$
29. $30abc$ **31.** $60uvw$ **33.** $x + y + z + 14$ **35.** $u + v + w + 19$ **37.** $x + y + z + w + 6$ **39.** $24abcd$
41. $x + y + z + w + 20$ **43.** Commutative $+$ **45.** Associative $\times$ **47.** Commutative $\times$
49. Commutative $+$ **51.** Commutative $+$ **53.** Commutative $+$ **55.** Associative $+$ **57.** $x + y + 4$
59. $a + b + 13$ **61.** $x + y + 9$ **63.** $a + b + 17$ **65.** $x + y + z + 17$ **67.** $a + 20b + 2c + 5$
69. **(C)** is false, since $9 - 7 \neq 7 - 9$.   **(D)** is false, since $4 \div 8 \neq 8 \div 4$.

EXERCISE 1-5

1. xxx **3.** $2xxxyy$ **5.** $3wwxyyy$ **7.** x^3 **9.** $2x^3y^2$ **11.** $3xy^2z^3$ **13.** u^{14} **15.** a^6 **17.** w^{19}
19. y^{16} **21.** 3^{30} (not 9^{30})   **23.** 9^{11} (not 81^{11})   **25.** 64   **27.** 225   **29.** 53   **31.** 196   **33.** x^7

35. y^{10} **37.** $24x^9$ **39.** a^3b^3 **41.** $12x^2y^2$ **43.** $6x^4y^2$ **45.** $12x^3y^5$ **47.** $x^2y^2z^2$ **49.** $x^3y^4z^7$
51. $24x^4y^4z^5$ **53.** 5,832   **55.** 24   **57.** 32,768   **59.** 128   **61.** 132,651   **63.** 27   **65.** 5,832
67. 2   **69.** 1,272   **71.** 30   **73.** 4   **75.** 10   **77.** 16,000   **79.** 795   **81.** 30   **83.** 1
85. **(A)** $s = 16t^2$ **(B)** Constants: 16, 2; variables: s, t **(C)** 1,024 ft

EXERCISE 1-6

1. Both 12   **3.** Both 45   **5.** $4x + 4y$ **7.** $7m + 7n$ **9.** $6x + 12$ **11.** $10 + 5m$ **13.** $12x + 48$
15. $3x + 3y$ **17.** $3(x + y)$ **19.** $5(m + n)$ **21.** $a(x + y)$ **23.** $2(x + 2)$ **25.** $3(2x + 5)$
27. $3(3a + 8)$ **29.** $6(2x + 7)$ **31.** $2x + 2y + 2z$ **33.** $3x + 3y + 3z$ **35.** $7(x + y + z)$
37. $2(m + n + 3)$ **39.** $x + x^2$ **41.** $y + y^3$ **43.** $6x^2 + 15x$ **45.** $2m^4 + 6m^3$ **47.** $6x^3 + 9x^2 + 3x$
49. $10x^3 + 15x^2 + 5x + 10$ **51.** $6x^5 + 9x^4 + 3x^3 + 6x^2$ **53.** $10x$ **55.** $11u$ **57.** $5xy$ **59.** $10x^2y$
61. $(7 + 2 + 5)x = 14x$ **63.** $18(2x + 5)$ **65.** $6(3a + 7)$ **67.** $x(x + 2)$ **69.** $u(u + 1)$ **71.** $2x(x^2 + 2)$
73. $xy(x + y)$ **75.** $abc(a + b)$ **77.** $x(x + y + z)$ **79.** $3m(m^2 + 2m + 3)$ **81.** $5uv(u + 3v)$
83. $8m^5n^4 + 4m^3n^5$ **85.** $6x^3y^4 + 12x^3y + 3x^2y^3$ **87.** $12x^4yz^4 + 4x^2y^2z^4$ **89.** $abc(a + b + c)$
91. $4xyz(4x^2z + xy + 3yz^2)$ **93.** $x^2 + 7x + 10$ **95.** $a^2 + 8a + 12$ **97.** $x^2 + 5x + 6$ **99.** $uc + vc + ud + vd$

EXERCISE 1-7

1. 4   **3.** 8   **5.** 1   **7.** 1   **9.** 2   **11.** 5   **13.** $8x, x; 5y$ **15.** $3x, 4x; 2y, 5y$
17. $6x^2, 3x^2, x^2; x^3, 4x^3$ **19.** $2u^2v, u^2v; 3uv^2, 5uv^2$ **21.** $9x$ **23.** $4u$ **25.** $9x^2$ **27.** $13xy$ **29.** $10x$
31. $16a$ **33.** $7x + 4y$ **35.** $3x + 5y + 6$ **37.** $m^2n, 5m^2n; 4mn^2, mn^2; 2mn, 3mn$ **39.** $6t^2$ **41.** $6a^4$
43. $16xy^2$ **45.** $4x + 7y + 7z$ **47.** $11x^3 + 4x^2 + 4x$ **49.** $4x^2 + 3xy + 2y^2$ **51.** $6x + 9$
53. $4t^2 + 8t + 10$ **55.** $3x^3 + 3x^2y + 4xy^2 + 2y^3$ **57.** $8x + 31$ **59.** $3x^2 + 4x$ **61.** $11t^2 + 13t + 17$
63. $3y^3 + 3y^2 + 4y$ **65.** $6x^2 + 5xy + 6y^2$ **67.** $4x^4 + 3x^2y^2 + 3y^4$ **69.** $4m^6 + 9m^5 + 4m^4$
71. $9x^2y^2 + 5x^3y^3$ **73.** $8u^3v^3 + 7u^4v^2$ **75.** $x^2 + 3x + 2$ **77.** $x^2 + 8x + 12$ **79.** $6x^2 + 13x + 6$
81. $2x^2 + 5xy + 2y^2$ **83.** $x^3 + 5x^2 + 11x + 15$ **85.** $y(y + 2); y^2 + 2y$ **87.** $x + x(x + 2) = 180; x^2 + 3x = 180$
89. $x + (x + 1) + (x + 2) + (x + 3); 4x + 6$

CHAPTER 1 REVIEW EXERCISE

The italicized number or numbers in parentheses following each answer indicates the section or sections in which that type of problem is discussed.

1. **(A)** $\{11, 13, 15\}$ **(B)** $\{11, 13\}$ *(1-1)*   **2.** **(A)** 3   **(B)** 1   *(1-7)*   **3.** 30   *(1-1)*   **4.** 90   *(1-1)*
5. 84   *(1-1)*   **6.** 140   *(1-1)*   **7.** $2 \cdot 3 \cdot 7$ *(1-1)*   **8.** $2 \cdot 2 \cdot 2 \cdot 7$ *(1-1)*   **9.** $2 \cdot 2 \cdot 3 \cdot 7$ *(1-1)*
10. $2 \cdot 7 \cdot 7$ *(1-1)*   **11.** 2   *(1-2)*   **12.** 11   *(1-2)*   **13.** 1   *(1-2)*   **14.** 20   *(1-2)*   **15.** x^{25} *(1-4)*
16. $6x^8$ *(1-4)*   **17.** 2^{25} (not 4^{25})   *(1-4)*   **18.** $x^2 + x$ *(1-6)*   **19.** $10x + 15y + 5z$ *(1-6)*
20. $6u^3 + 3u^2$ *(1-6)*   **21.** $9y$ *(1-7)*   **22.** $5m + 5n$ *(1-7)*   **23.** $7x^2 + 3x$ *(1-7)*   **24.** $8x^2y + 2xy^2$ *(1-7)*
25. $3(m + n)$ *(1-6)*   **26.** $8(u + v + w)$ *(1-6)*   **27.** $x(y + w)$ *(1-6)*   **28.** $4(x + 2w)$ *(1-6)*
29. $(x + y)z$ *(1-6)*   **30.** $x(x + 1)$ *(1-6)*   **31.** $ac(b + d)$ *(1-6)*   **32.** $3a(b + 4c)$ *(1-6)*   **33.** $12x$ *(1-2)*
34. $3x + 3$ *(1-2)*   **35.** $2x - 5$ *(1-2)*   **36.** $3(x - 4)$ *(1-2)*   **37.** **(A)** 23, 29, 31   **(B)** Finite   *(1-1)*
38. **(A)** 3   **(B)** 1   **(C)** 1   *(1-5, 1-7)*   **39.** $2 \cdot 2 \cdot 2 \cdot 3 \cdot 5$ *(1-1)*   **40.** $2 \cdot 2 \cdot 2 \cdot 2 \cdot 2 \cdot 3$ *(1-1)*
41. $2 \cdot 5 \cdot 11$ *(1-1)*   **42.** $2 \cdot 2 \cdot 5 \cdot 7$ *(1-1)*   **43.** 36   *(1-1)*   **44.** 90   *(1-1)*   **45.** 90   *(1-1)*
46. 180   *(1-1)*   **47.** 420   *(1-1)*   **48.** 11   *(1-5)*   **49.** 12   *(1-5)*   **50.** 17   *(1-5)*   **51.** 6   *(1-2)*
52. 6   *(1-2)*   **53.** 12   *(1-2)*   **54.** 20   *(1-2)*   **55.** 24   *(1-2)*   **56.** 36   *(1-2)*   **57.** 37   *(1-5)*
58. 120   *(1-5)*   **59.** $18x^8$ *(1-5)*   **60.** $12x^3y^5z^4$ *(1-5)*   **61.** $6y^5 + 3y^4 + 15y^3$ *(1-6)*
62. $19u^2 + 7u + 17$ *(1-6, 1-7)*   **63.** $8x^2 + 22x$ *(1-6, 1-7)*   **64.** $3x^2 + 7xy + 2y^2$ *(1-6, 1-7)*
65. $u(u^2 + u + 1)$ *(1-6)*   **66.** $3xy(2x + y)$ *(1-6)*   **67.** $3m^2(m^3 + 2m^2 + 5)$ *(1-6)*   **68.** $24 = 2x - 6$ *(1-3)*
69. $3x = x + 12$ *(1-3)*   **70.** $x + (x + 1) + (x + 2) + (x + 3) = 138$ *(1-3)*   **71.** $x + (x + 2) + (x + 4) = 78$ *(1-3)*
72. 5,940   *(1-1)*   **73.** 5,940   *(1-1)*   **74.** 1,200   *(1-1)*   **75.** $13 \cdot 13$ *(1-1)*   **76.** $2 \cdot 2 \cdot 3 \cdot 3 \cdot 3 \cdot 5$ *(1-1)*
77. $3 \cdot 3 \cdot 3 \cdot 5 \cdot 11$ *(1-1)*   **78.** $2 \cdot 2 \cdot 2 \cdot 3 \cdot 3 \cdot 5 \cdot 7$ *(1-1)*   **79.** 48   *(1-2)*   **80.** 39   *(1-2)*   **81.** 78   *(1-5)*
82. 320   *(1-5)*   **83.** $10u^5v^4 + 5u^4v^3 + 10u^3v^2$ *(1-5, 1-6)*   **84.** $7x^5 + 11x^3 + 6x^2$ *(1-6, 1-7)*
85. $8x^2 + 10x + 3$ *(1-6, 1-7)*   **86.** $3x^2yz(4xz + 3)$ *(1-6)*   **87.** $5x^2y^2(4x + y + 3)$ *(1-6)*
88. $3abc(1 + 3a + 6a^2b)$ *(1-6)*   **89.** $4(2x^4y + yz^2 + 3x^2z^4)$ *(1-6)*   **90.** $3y(xz + 5xy + 6z)$ *(1-6)*
91. No common factor   *(1-6)*   **92.** Commutative $\times$ *(1-4)*   **93.** Associative $+$ *(1-4)*
94. Commutative $+$ *(1-4)*   **95.** Associative $+$ *(1-4)*   **96.** Commutative $\times$ *(1-4)*
97. Distributive $\times$ over $+$ *(1-6)*   **98.** Commutative $+$ *(1-4)*   **99.** $x^2 = (x - 2)(x - 1)$ *(1-2)*
100. $4x = (x + 2) + (x + 4)$ *(1-3)*

CHAPTER 1 PRACTICE TEST

1. 52 *(1-1)* **2.** 192 *(1-1)* **3.** 3 *(1-5)* **4.** 1 *(1-5)* **5.** 25 *(1-5)* **6.** 34 *(1-5)* **7.** 40 *(1-5)*
8. $2 \cdot 3 \cdot 5 \cdot 5$ *(1-1)* **9.** $x^3y^4z^2$ *(1-5)* **10.** x^{13} *(1-5)* **11.** $a(b + c + d)$ *(1-6)* **12.** $xy + xz + xw$ *(1-6)*
13. $6x^2 + 10xy + 16xz$ *(1-6)* **14.** $3xy(y + 3xy + 5x)$ *(1-6)* **15.** $23x^2$ *(1-6)* **16.** $18ab + 10c$ *(1-6)*
17. $x^2 + 10x + 21$ *(1-6)* **18.** $15x^{11}$ *(1-5)* **19.** $3x(6 + 5y)$ *(1-6)* **20.** $x^3 + 3x^2 + 2x$ *(1-6)*
21. Distributive $\times$ over $+$, Associative $+$ *(1-6)* **22.** Distributive $\times$ over $+$, Distributive $\times$ over $+$ *(1-4)*
23. Associative $\times$, Distributive $\times$ over $+$ *(1-4)* **24.** $x + 11 = 3x - 7$ *(1-3)* **25.** $2x^2 = (x + 2)^2 + 1$ *(1-3)*

CHAPTER 2

EXERCISE 2-1

1. $-8, -2, +3, +9$ **3.** $a = -13, b = -7, c = -1, d = +6, e = +15$ **5.**

7. **9.** $+4$ **11.** -10 **13.** -3 **15.** $+9$ **17.** -12 **19.** $+10$
21. $1, 4, 17, 6{,}035$ **23.** $-21, -2$ **25.** $-21, -2, 0, 1, 4, 17, 6{,}035$ **27.** $-\frac{3}{8}, 3.14, \sqrt{13}, \frac{2}{9}$ **29.** $+20{,}270$
31. -280 **33.** $+6{,}960$ **35.** $-4{,}020$ **37.** -5 **39.** $+27$ **41.** -3 **43.** $+25$ **45.** -10
47. -9 **49.** $+12$ **51.** -200 **53.** -110 **55.** $+6$ **57.** -6 **59.** 0 **63.** $+1$ **65.** -1
67. $+2$ **69.** $+17$ **71.** -3 **73.** $+6$ **75.** $+6$ **77.** $+10$ **79.** $+4$

EXERCISE 2-2

1. -9 **3.** $+2$ **5.** $+4$ **7.** $+6$ **9.** 0 **11.** Sometimes **13.** Never **15.** -11 **17.** -5
19. $+13$ **21.** $+11$ **23.** -10 **25.** $+2$ or -2 **27.** No solution **29.** $+6$ **31.** $+5$ **33.** -5
35. -15 **37.** $+16$ **39.** $+7$ **41.** $+5$ **43.** -8 **45.** -13 **47.** -4 **49.** -7 **51.** $+5$
53. -7 **55.** -5 **57.** $+5$ **59.** $+2$ **61.** $\{+5\}$ **63.** $\{+3\}$ **65.** $\{-6, +6\}$ **67.** No solution
69. No solution **71.** $\{-15, +15\}$ **73.** $\{-21, +21\}$ **75.** $\{+18\}$ **77.** $\{+13\}$ **79.** $\{0\}$
81. Set of all integers less than or equal to 0 **83.** Set of all integers Z
85. Set of all integers greater than or equal to 0 **87.** $\{0\}$

EXERCISE 2-3

1. $+11$ **3.** -3 **5.** -2 **7.** -8 **9.** $+3$ **11.** $+9$ **13.** -6 **15.** -9 **17.** -9 **19.** -2
21. -4 **23.** -5 **25.** -4 **27.** -12 **29.** -622 **31.** -38 **33.** -668 **35.** -36 **37.** -4
39. -4 **41.** -5 **43.** $+5$ **45.** -77 **47.** 0 **49.** -8 **51.** $+12$ **53.** 0 **55.** $-10{,}143$
57. $-3{,}020$ **59.** $+14$ **61.** -6 **63.** $+4$ **65.** 0 **67.** -2 **69.** $+3$ **71.** -8 **73.** -2
75. -1 **77.** $+27$ **79.** -1 **81.** \$23 **83.** $-1{,}493$ ft **85.** \$92 owed **87.** $+17$ **89.** $+18$
91. 0 **93.** $-m$ **95.** Commutative property, associative property, addition of opposites, definition of addition

EXERCISE 2-4

1. $+5$ **3.** $+13$ **5.** -5 **7.** -5 **9.** $+5$ **11.** $7 > 5$ **13.** $5 < 7$ **15.** $-7 < -5$
17. $-5 > -7$ **19.** $0 < 8$ **21.** $0 > -8$ **23.** $-7 < 5$ **25.** $-842 < 0$ **27.** $900 > -1{,}000$ **29.** $+14$
31. -4 **33.** -6 **35.** -5 **37.** $+15$ **39.** $+87$ **41.** -315 **43.** -245 **45.** $+17{,}873$
47. $-5{,}230$ **49.** $+819$ **51.** $-1{,}705$ **53.** $+1$ **55.** -3 **57.** 0 **59.** 0 **61.** $+8$ **63.** $+1$
65. 0 **67.** $+2$ **69.** $+7$ **71.** $+4$ **73.** $+2$ **75.** $+3$ **77.** 0 **79.** $+3$ **81.** 0 **83.** $+10$
85. $(+29{,}141) - (-35{,}800) = +64{,}941$ ft **87.** $(-245) - (-280) = +35$ ft **89.** $+7$ **91.** -10 **93.** True
95. False; $(+7) - (-3) = +10, (-3) - (+7) = -10$ **97.** True **99.** False; $|(+9) + (-3)| = +6, |+9| + |-3| = +12$

EXERCISE 2-5

1. $+32$ **3.** -32 **5.** 0 **7.** $+2$ **9.** -3 **11.** Not defined **13.** -14 **15.** -14 **17.** 0
19. $+3$ **21.** -3 **23.** 0 **25.** -5 **27.** -7 **29.** $+2$ **31.** $+4$ **33.** -4 **35.** $+30$
37. -10 **39.** -8 **41.** -51 **43.** $+17$ **45.** -6 **47.** $+8$ **49.** 0 **51.** 0 **53.** Not defined
55. $+4$ **57.** $+6$ **59.** -20 **61.** 0 **63.** (A) $+12$ (B) $+12$ (C) $+12$
65. (A) -12 (B) -12 (C) -12 **67.** (A) $+2$ (B) $+2$ (C) $+2$ **69.** (A) -2 (B) -2 (C) -2
71. (A) -12 (B) -12 **73.** (A) $+12$ (B) $+12$ **75.** (A) -2 (B) -2 **77.** (A) $+2$ (B) $+2$
79. (A) $+3$ (B) $+3$ **81.** $+8$ **83.** $+8$ **85.** 0 **87.** No solution **89.** $+53{,}116$ **91.** $-9{,}728$
93. -27 **95.** $+12$ **97.** $+1{,}469$

EXERCISE 2-6

1. 5   **3.** -13 **5.** -3 **7.** -2 **9.** -3 **11.** 1   **13.** -2 **15.** $4x$ **17.** $-4x$ **19.** $-12y$
21. $-3x - 3y$ **23.** $-5x + 3y$ **25.** $6m - 2n$ **27.** $-x + 4y$ **29.** $3x - 2y$ **31.** $-x + y$
33. $-x + 8y$ **35.** $4x - 8y$ **37.** $6xy$ **39.** $-3x^2y$ **41.** $2x^2 + 2x - 3$ **43.** $2x^2y + 5xy^2 - 6xy$
45. $-2x - y$ **47.** $2t - 20$ **49.** $-3y + 4$ **51.** $-10x$ **53.** $x - 14$ **55.** $6t^2 - 16t$ **57.** $-9,683$
59. $34x - 258y$ **61.** $-48u - 59v$ **63.** $3x - y$ **65.** $-3x + y$ **67.** $y + 2z$ **69.** $x - y + z$
71. $2x + 3y$ **73.** $-2x - 3y$ **75.** $a - b + c$ **77.** $-a + b - c$ **79.** $x + 2y, -z - 2w$
81. $-x - 2y, z + 2w$ **83.** $P = 2x + 2(x - 5) = 4x - 10$ **85.** 1   **87.** $-8x^2 - 16x$ **89.** $13x^2 - 26x + 10$
91. Value in cents $= 25x + 10(x + 4) = 35x + 40$

EXERCISE 2-7

1. 3   **3.** -3 **5.** -12 **7.** 5   **9.** -3 **11.** -13 **13.** 8   **15.** -4 **17.** -4 **19.** 3
21. 0   **23.** 3   **25.** 2   **27.** -3 **29.** 7   **31.** 4   **33.** 2   **35.** -4 **37.** 8   **39.** 7   **41.** 5
43. No solution   **45.** 16   **47.** 16   **49.** -15 **51.** 4   **53.** 6   **55.** No solution   **57.** All numbers
59. No solution   **61.** No solution   **63.** All numbers   **65.** -3 **67.** All numbers   **69.** $b = \dfrac{A}{h}$

71. $r = \dfrac{d}{t}$ **73.** $r = \dfrac{I}{pt}$ **75.** $P = \dfrac{A}{1 + rt}$ **77.** $x = \dfrac{y - b}{m}$ **79.** $a = P - b - c$ **81.** -7 **83.** -5

85. 2   **87.** 1   **89.** -6 **91.** $3x = 12, x = 4$ **93.** $-2x = 6, 2x = -6, x = -3$

EXERCISE 2-8

1. 25, 26, 27   **3.** 16, 18, 20   **5.** 8 hr   **7.** 15 mph   **9.** 13 ft above and 104 ft below   **11.** 239,000 miles
13. 3, 3, and 12 ft   **15.** 7, 9, 11   **17.** 10 ft by 23 ft   **19.** 23, 69, and 69 in   **21.** 7 quarters and 10 dimes
23. 9 three-point baskets, 20 two-point baskets   **25.** 5 sec   **27.** 22,000 ft   **29.** Assistant 3 hr, mechanic 5 hr
31. 8 miles   **33.** 2 hr   **35.** 70 hr (or 2 days and 22 hr); 1,750 miles   **37.** Noon   **39.** 25 min
41. Any four consecutive integers

CHAPTER 2 REVIEW EXERCISE

1. $-12, -9, -3, 0, 4, 11$ *(2-1)*   **2.**   *(2-1)*   **3.** -4 *(2-2)*   **4.** $+3$ *(2-2)*
5. $+8$ *(2-2)*   **6.** -15 *(2-2)*   **7.** -5 *(2-3)*   **8.** -13 *(2-3)*   **9.** $+6$ *(2-4)*   **10.** -3 *(2-4)*
11. -17 *(2-3)*   **12.** $+19$ *(2-4)*   **13.** $+28$ *(2-5)*   **14.** -18 *(2-5)*   **15.** -72 *(2-5)*   **16.** -6 *(2-5)*
17. -4 *(2-5)*   **18.** $+6$ *(2-5)*   **19.** 0   *(2-5)*   **20.** Not defined   *(2-5)*   **21.** -2 *(2-3)*   **22.** -16 *(2-5)*
23. -8 *(2-5)*   **24.** $+8$ *(2-5)*   **25.** 0   *(2-5)*   **26.** -6 *(2-5)*   **27.** $+4$ *(2-6)*   **28.** -13 *(2-6)*
29. $+2$ *(2-6)*   **30.** $+12$ *(2-6)*   **31.** (A) $+7$ (B) -18 *(2-6)*   **32.** $+44$ *(2-6)*   **33.** $2x - 8$ *(2-6)*
34. $5x - 2$ *(2-6)*   **35.** $2m + 9n$ *(2-6)*   **36.** $-6x - 18y$ *(2-6)*   **37.** $x = +5$ *(2-7)*   **38.** $x = -3$ *(2-7)*
39. $x = -2$ *(2-7)*   **40.** $x = -7$ *(2-7)*   **41.** $x = +18$ *(2-7)*   **42.** $x = +6$ *(2-7)*
43. (A) -245 (B) $+14,495$ *(2-1)*   **44.** 52, 53, 54   *(2-8)*   **45.** 60 yds   *(2-8)*   **46.** $30°, 60°, 90°$ *(2-8)*
47. $+12$ *(2-2)*   **48.** $+3$ *(2-2)*   **49.** -2 *(2-2)*   **50.** -10 *(2-4)*   **51.** $+24$ *(2-5)*   **52.** -14 *(2-3)*
53. -4 *(2-5)*   **54.** -15 *(2-4)*   **55.** $+3$ *(2-2)*   **56.** $+9$ *(2-3)*   **57.** $+17$ *(2-4)*   **58.** -6 *(2-5)*
59. $+12$ *(2-5)*   **60.** 0   *(2-5)*   **61.** $+34$ *(2-5)*   **62.** 0   *(2-5)*   **63.** $-2x^2y^2 - 5xy$ *(2-6)*
64. $4y^3 - 7y^2 + 18y$ *(2-6)*   **65.** $10x - 24y$ *(2-6)*   **66.** $-6x^3y^2 - 10x^2y + 3xy^2$ *(2-6)*   **67.** $b + 2c$ *(2-6)*
68. $2y - 3$ *(2-6)*   **69.** $x - 2y$ *(2-6)*   **70.** $m = 9$ *(2-7)*   **71.** $x = 2$ *(2-7)*   **72.** No solution   *(2-7)*
73. All numbers   *(2-7)*   **74.** $x = -1$ *(2-7)*   **75.** $x = +1$ *(2-7)*   **76.** 44, 46, 48, 50   *(2-8)*
77. 4 nickels and 5 quarters   *(2-8)*   **78.** -45 *(2-1)*   **79.** (A) $+15$ (B) $+5$ *(2-4)*
80. (A) $+1$ (B) $+4$ *(2-5)*   **81.** $-11x - 8$ *(2-6)*   **82.** All numbers   *(2-7)*   **83.** All numbers   *(2-7)*
84. No solution   *(2-7)*   **85.** $+1$ *(2-7)*   **86.** $x = \dfrac{c - by}{a}$ *(2-7)*   **87.** $m = \dfrac{E}{c^2}$ *(2-7)*
88. $z = 1 - x - y$ *(2-7)*   **89.** $x = y - z$ *(2-7)*   **90.** 40, 50 laps   *(2-8)*   **91.** 37, 43 pounds   *(2-8)*
92. 17 hr   *(2-8)*   **93.** Addition prop. of $=$ *(2-7)*   **94.** Commutative $+$ *(2-3)*   **95.** Distributive prop.   *(2-5)*
96. Division prop. $=$ *(2-7)*   **97.** Commutative $\times$ *(2-5)*   **98.** Addition prop. $=$ *(2-7)*
99. Associative $\times$ *(2-5)*   **100.** Division prop. $=$ *(2-7)*

CHAPTER 2 PRACTICE TEST

1. $+12$ *(2-4)*   **2.** $+16$ *(2-4)*   **3.** -30 *(2-3, 2-5)*   **4.** $+5$ *(2-2)*   **5.** $+23$ *(2-4, 2-5)*
6. -18 *(2-4)*   **7.** $+3$ *(2-6)*   **8.** $-x$ *(2-2)*   **9.** $-x + 3$ *(2-6)*   **10.** $-x^2 - 3x - 2$ *(2-6)*

11. $-a - b + c$ *(2-6)*　　**12.** $-x^4 - 4x^3 + 5x^2 + 4x$ *(2-6)*　　**13.** $x - 8$ *(2-6)*　　**14.** $4x - 3y + 2z$ *(2-6)*
15. 7 *(2-7)*　　**16.** 8 *(2-7)*　　**17.** All numbers *(2-7)*　　**18.** All numbers *(2-7)*　　**19.** No solution *(2-7)*
20. 0 *(2-7)*　　**21.** $\dfrac{M - l - 2h}{2}$ *(2-7)*　　**22.** 0 *(2-4)*　　**23.** 103 cm by 209 cm *(2-8)*　　**24.** 58, 59, 60 *(2-8)*
25. 3 hr *(2-8)*

CHAPTER 3

EXERCISE 3-1

1. 24　　**3.** 11　　**5.** 7　　**7.** 18　　**9.** 54　　**11.** 20　　**13.** 16　　**15.** 6　　**17.** $\frac{4}{7}$　　**19.** $\frac{5}{8}$　　**21.** $\frac{3}{8}$
23. $\frac{3}{5}$　　**25.** $\frac{8}{15}$　　**27.** $\frac{3}{4}$　　**29.** $\frac{1}{9}$　　**31.** $\frac{5}{2}$　　**33.** $\frac{2}{3}$　　**35.** $\frac{2}{3}$　　**37.** $\frac{11}{12}$　　**39.** $\frac{1}{12}$　　**41.** $\frac{19}{24}$　　**43.** $\frac{13}{36}$
45. $\frac{2}{3}$　　**47.** $\frac{2}{27}$　　**49.** 5.8　　**51.** 31.662　　**53.** 17.73　　**55.** 79.64　　**57.** 32.5　　**59.** 0.00852　　**61.** 0.83
63. 0.74　　**65.** 38.2　　**67.** 7.9　　**69.** 0.67　　**71.** 0.09　　**73.** 2.16　　**75.** 0.006　　**77.** 0.074　　**79.** 0.231
81. 12%　　**83.** 8%　　**85.** 325%　　**87.** 0.7%　　**89.** 7.2%　　**91.** 40.5%　　**93.** 48.36　　**95.** 240
97. 250　　**99.** 1.56　　**101.** 0.08

EXERCISE 3-2

1. $a: -\frac{9}{4}, b: -\frac{3}{4}, c: \frac{7}{4}$　　**3.** $a: -\frac{3}{2}, b: -\frac{1}{4}, c: \frac{1}{2}, d: \frac{11}{4}$　　**5.**　　**7.**
9. 7　　**11.** Already in lowest terms　　**13.** 9　　**15.** 6　　**17.** 3/8　　**19.** 21/40　　**21.** 2/5　　**23.** 2/5
25. 2　　**27.** 15　　**29.** 3　　**31.** $-24a$　　**33.** -11　　**35.** -7　　**37.** $-18k$　　**39.** $9x^2$　　**41.** $3x^2$
43. $3b$　　**45.** $2ab$　　**47.** $-24xy$　　**49.** $-20xy^3$　　**51.** -16　　**53.** $-6ab$　　**55.** 3/2　　**57.** $-1/4$
59. $\dfrac{1}{4y}$　　**61.** $\dfrac{4a}{b}$　　**63.** $-\dfrac{5xy}{8}$　　**65.** $-\dfrac{3ab}{8}$　　**67.** $-\dfrac{3x^2y}{5}$　　**69.** $-\dfrac{4xz^2}{7y}$　　**71.** $\dfrac{-y^2}{4x}$　　**73.** $\dfrac{12y}{5x}$
75. $\dfrac{5xy}{8}$　　**77.** $\dfrac{2ab^2}{5}$　　**79.** $-\dfrac{10}{11ab^2}$　　**81.** $-\dfrac{7}{13xy^2}$　　**83.** $\frac{2}{3}$　　**85.** $\dfrac{x^2 + y}{3(x + y^2)}$　　**87.** $\frac{3}{5}$　　**89.** $-3/8$
91. 2/11　　**93.** 1/5　　**95.** $\dfrac{1}{6(x + y)}$

EXERCISE 3-3

1. $\frac{6}{35}$　　**3.** $-7/3$　　**5.** $\dfrac{28x}{15y}$　　**7.** $\dfrac{4ab}{15}$　　**9.** $\dfrac{3x^2}{2y^3}$　　**11.** $-\dfrac{3x^2}{10y^2}$　　**13.** $\dfrac{-6}{77}$ or $-\frac{6}{77}$　　**15.** $\frac{10}{21}$　　**17.** $\dfrac{2ab}{3c}$
19. $\dfrac{6a^2b^2}{35}$　　**21.** $\frac{21}{25}$　　**23.** $-\frac{15}{16}$　　**25.** $\dfrac{14xy}{15}$　　**27.** $-\dfrac{9x^2}{2y}$　　**29.** $\dfrac{-9}{14}$　　**31.** $\dfrac{4a}{9b^3}$　　**33.** $\dfrac{5x^2}{8y^2}$　　**35.** $\frac{2}{3}$
37. $\frac{4}{375}$　　**39.** 5　　**41.** $\frac{3}{2}$　　**43.** $\dfrac{-3}{4}$　　**45.** 27/28　　**47.** 7/48　　**49.** 9/2　　**51.** $\dfrac{1}{z}$　　**53.** 4　　**55.** $2y^2$
57. y　　**59.** $\dfrac{2}{3x}$　　**61.** $\dfrac{3x}{2y}$　　**63.** y　　**65.** $\dfrac{3ad}{2c}$　　**67.** $\dfrac{3v}{2u}$　　**69.** $\dfrac{-2x^2}{3y}$　　**71.** $\frac{81}{100}$　　**73.** 1　　**75.** -2
77. $\frac{3}{25}$　　**79.** $\frac{3}{25}$　　**81.** $-\dfrac{x^3}{2yz}$　　**83.** $-\dfrac{x^3}{2yz}$　　**85.** $\dfrac{adf}{bce}$　　**87.** $\dfrac{abc}{d^3}$　　**89.** $\dfrac{abc}{d^3}$

EXERCISE 3-4

1. 2　　**3.** $\frac{4}{5}$　　**5.** $\frac{7}{8}$　　**7.** $\frac{19}{15}$　　**9.** $\frac{4}{11}$　　**11.** $\frac{10}{11}$　　**13.** $\frac{1}{8}$　　**15.** $-\frac{1}{15}$　　**17.** $\dfrac{-3}{5xy}$　　**19.** $\dfrac{5y}{x}$　　**21.** $\dfrac{6}{7y}$
23. $\dfrac{7}{6x}$　　**25.** $\dfrac{13x}{6}$　　**27.** $\dfrac{9 - 10x}{15x}$　　**29.** $\frac{65}{84}$　　**31.** $\frac{130}{63}$　　**33.** $\frac{1}{20}$　　**35.** 4　　**37.** $\frac{81}{200}$　　**39.** $6x^2$
41. $24m^3$　　**43.** $24xy$　　**45.** $6y^3$　　**47.** 252　　**49.** 180　　**51.** $\dfrac{x^2 - y^2}{xy}$　　**53.** $\dfrac{x - 2y}{y}$　　**55.** $\dfrac{5x + 3}{x}$
57. $\dfrac{1 - 3x}{xy}$　　**59.** $\dfrac{9 + 8x}{6x^2}$　　**61.** $\dfrac{15 - 2m^2}{24m^3}$　　**63.** $\frac{5}{3}$　　**65.** $\dfrac{3x^2 - 4x - 6}{12}$　　**67.** $\dfrac{18y - 16x + 3}{24xy}$
69. $\dfrac{18 + 4y + 3y^2 - 18y^3}{6y^3}$　　**71.** $\dfrac{22y + 9}{252}$　　**73.** $\dfrac{15x^2 + 10x - 6}{180}$　　**75.** $\dfrac{x - 2x^2}{12}$　　**77.** $\dfrac{3x + x^2}{6}$　　**79.** $\frac{2}{7}$

81. $\frac{8}{15}$ **83.** 3/7 **85.** $-10/27$ **87.** $-6/11$ **89.** $69\frac{2}{7}$ **91.** 60.5, 60, 63.5, 66.5, 67, 69, 65.5 **93.** 11

95. 8 **97.** $\dfrac{23x + 4y}{7}$

EXERCISE 3-5

1. -35 **3.** 6 **5.** $\frac{15}{4}$ **7.** $-5/3$ **9.** $-7/3$ **11.** 8 **13.** 12 **15.** 6 **17.** -6 **19.** 36
21. $-\frac{4}{3}$ **23.** $-1/4$ **25.** $-31/10$ **27.** 20 **29.** 30 **31.** -20 **33.** 2.3 **35.** 15 **37.** $-\frac{5}{6}$
39. $\frac{27}{5}$ **41.** 9 **43.** 19/16 **45.** 1/14 **47.** 0 **49.** No solution **51.** All numbers **53.** -1
55. 150 **57.** 11 **59.** -1.5 **61.** 1.5 **63.** 3 **65.** -12 **67.** $\frac{11}{5}$ **69.** 61/4 **71.** 35/13
73. 13/77 **75.** No solution **77.** 7.2368 **79.** -22.1393 **81.** -2.0355 **83.** 3.6142 **85.** 0.2031
87. 80 **89.** 30

EXERCISE 3-6

1. $\frac{1}{2}x$ or $\dfrac{x}{2}$ **3.** $\frac{2}{3}x$ or $\dfrac{2x}{3}$ **5.** $\dfrac{x}{3} + 2$ **7.** $\dfrac{2x}{3} - 8$ **9.** $0.8x$ **11.** $0.6x - 3$ **13.** $0.2(x + 5)$

15. $\frac{1}{2}(2x - 3)$ or $\dfrac{2x - 3}{2}$ **17.** $0.8(0.5x - 2)$ **19.** (A) $\dfrac{x}{4} + 2 = \frac{1}{2}$ (B) -6 **21.** (A) $\dfrac{x}{2} - 2 = \dfrac{x}{3}$ (B) 12

23. (A) $58x = 20.3$ (B) $0.35 = 35\%$ **25.** (A) $0.8x = 45$ (B) 56.25
27. (A) $x + 3 = 1.4x$ (B) 7.5 **29.** (A) $x - 14 = 0.3x$ (B) 20
31. (A) $0.7x - 5 = 16$ (B) 30 **33.** (A) $0.6x - 70 = 0.25x$ (B) 200

35. (A) $\dfrac{x}{2} - 5 = \dfrac{x}{3} + 3$ (B) 48 **37.** (A) $\dfrac{2x}{3} + 5 = \dfrac{x}{4} - 10$ (B) -36 **39.** 7.2 meters **41.** 75 meters

43. 9 cm by 27 cm **45.** 84 meters by 24 meters **47.** $210 **49.** $197.60 **51.** $1.98 **53.** $33
55. 7,840,000 **57.** 1,166,000 **59.** 879,000 **61.** 39 cents (approximately) **63.** $755,000 **65.** $400
67. $29,200 **69.** 23.7 million **71.** $62.16; no **73.** $52 **75.** $184.60 **77.** 45 cm by 11 cm
79. 12, 12, and 18 feet **81.** $h = w = 15$, $l = 45$ inches **83.** 15.25 centimeters

EXERCISE 3-7

1. $\frac{1}{4}$ **3.** $\frac{5}{1}$ **5.** $\frac{1}{3}$ **7.** 300/14 **9.** 6000/2.4 **11.** 8 **13.** 18 **15.** 4 **17.** 460 **19.** 12,600
21. 600 men **23.** 36 cm **25.** 210 **27.** 125 km **29.** 300 gal. **31.** $15,600 **33.** 700 goals
35. 350 miles **37.** $30.82 **39.** $37.50 **41.** 47 cents **43.** 300,000 pesos **45.** 234 marks
47. 2.4 grams **49.** $5\frac{1}{3}$ quarts **51.** 4 in. **53.** $1\frac{5}{8}$ inches **55.** $90 per share **57.** $5.95 **59.** 40 kg
61. 4.5 **63.** 42 **65.** 70 **67.** 70 **69.** 5.44 kg **71.** 24.86 miles **73.** 35 oz **75.** 109.36 yards
77. 155.925 grams **79.** 353.98 km **81.** 620 **83.** 0.93 miles **85.** 300.5 marks **87.** 21,583,000
89. 657

EXERCISE 3-8

1. $16\frac{2}{3}$ min **3.** 40% **5.** 19 **7.** (A) 400 mi (B) 405 mi **9.** 1.5 kg **11.** 37.5 cm **13.** 62 in
15. 463 grams **17.** 70,000 per year **19.** 180 m **21.** 407 feet

23. $T = 80 - 5.5\left(\dfrac{h}{1,000}\right)$ or $T = 80 - 0.0055h$; 10,000 ft **25.** 2/15

27. (A) 15 in (B) 20 in (C) 22.5 in (D) 24 in **29.** 225 kg **31.** 170 cm **33.** 1.6 cc **35.** 87
37. $37,114,547 **39.** 79 **41.** 1.15 **43.** 10 **45.** 0.459 **47.** 2.25, 2.25, 4.5, and 9 inches
49. 45 mph **51.** 900 **53.** 110 **55.** 31,700

CHAPTER 3 REVIEW EXERCISE

1. (3-2) **2.** $-1.2, -0.9, 0.2, 0.9$ (3-2) **3.** 15 (3-2) **4.** 4 (3-2) **5.** 70 (3-2)

6. 12 (3-2) **7.** 2/5 (3-2) **8.** 12/7 (3-2) **9.** $\dfrac{5y}{2}$ (3-2) **10.** $\dfrac{15x}{8y}$ (3-2) **11.** $\dfrac{6}{5xy}$ (3-2)

12. $\dfrac{5x}{7}$ (3-4) **13.** $-\dfrac{3a}{4}$ (3-4) **14.** $\dfrac{5y}{6}$ (3-4) **15.** $\dfrac{6 - 5xy}{4y}$ (3-4) **16.** $-\dfrac{7x}{12}$ (3-4)

17. $-\dfrac{3a}{4}$ (3-3) **18.** $-\dfrac{13b}{20}$ (3-4) **19.** $\dfrac{8y}{25}$ (3-3) **20.** $\frac{5}{6}$ (3-5) **21.** $\frac{3}{2}$ (3-5) **22.** 6 (3-5)

23. 6 (3-5) **24.** 1 (3-5) **25.** 2 (3-5) **26.** $-40/7$ (3-5) **27.** -70 (3-5)

28. $x = -11/12$ *(3-5)*   **29.** 16.8 *(3-1, 3-6)*   **30.** 40 *(3-1, 3-6)*   **31.** **(A)** $0.45x = 60$ **(B)** $133\frac{1}{3}$ *(3-6)*
32. **(A)** $\frac{3}{10}x = \frac{2}{5}$ **(B)** $x = \frac{4}{3}$ *(3-6)*   **33.** 340 *(3-7)*   **34.** 7,500 *(3-7)*   **35.** 20 *(3-6)*

36. 15 cm by 25 cm *(3-6)*   **37.** 1,380 *(3-7)*   **38.** 21/11 *(3-2)*   **39.** 5/12 *(3-2)*   **40.** $\dfrac{14y}{11}$ *(3-2)*

41. $\dfrac{2}{5x^2}$ *(3-2)*   **42.** $\dfrac{9y}{10z}$ *(3-2)*   **43.** $\dfrac{9y^2 + 10z^2}{15xyz}$ *(3-4)*   **44.** $\dfrac{9y^2 - 10z^2}{15xyz}$ *(3-4)*   **45.** $\dfrac{9x}{4y}$ *(3-3)*

46. $\dfrac{9x - 4y}{12x^2y^2}$ *(3-4)*   **47.** $\dfrac{3 - 2x + x^2}{x^2}$ *(3-4)*   **48.** $\dfrac{3xz + 18xy - 4yz - 24xyz}{12xyz}$ *(3-4)*   **49.** $\frac{17}{18}$ *(3-4)*

50. $-\frac{9}{10}$ *(3-1, 3-4)*   **51.** $\dfrac{167x}{60}$ *(3-4)*   **52.** $\dfrac{2x + 9x^2}{9}$ *(3-4)*   **53.** $-x^2$ *(3-4)*   **54.** $2xy - 3$ *(3-5)*

55. -12 *(3-5)*   **56.** 41 *(3-5)*   **57.** 0.6 *(3-5)*   **58.** 245 *(3-5)*   **59.** 22/3 *(3-5)*   **60.** 11 *(3-5)*

61. $\dfrac{x}{450} = \dfrac{20}{3}$; $x = 3,000$ *(3-7)*   **62.** $\dfrac{x}{40} = \dfrac{1}{2.54}$; $x = 15.75$ in *(3-7)*   **63.** Approx. \$24.56 billion *(3-6)*

64. $22\frac{2}{9}$ *(3-7)*   **65.** $A = 36°$, $B = 144°$ *(3-6)*   **66.** 30/7, 90/7, 90/7 cm *(3-6)*   **67.** 44 *(3-6)*

68. \$4,128 *(3-7)*   **69.** \$22.50 *(3-7)*   **70.** \$37.7 billion *(3-7)*   **71.** $\dfrac{-4}{9}$ *(3-3)*

72. $\dfrac{27y^2 - 12xy + 25x^2}{90x^2y^2}$ *(3-4)*   **73.** $\dfrac{20xz + 135xy^2 - 54y^3}{18yz}$ *(3-4)*   **74.** $\dfrac{65x}{18y}$ *(3-4)*   **75.** $-\dfrac{25x^2y^2}{z^2}$ *(3-3)*

76. $-\dfrac{25x^2}{9y^2}$ *(3-3)*   **77.** $\dfrac{135xy^2 - 20xz}{54y^3}$ *(3-4)*   **78.** $\dfrac{20xz - 135xy^2}{54y^3}$ *(3-4)*   **79.** $\dfrac{27y^2 - 12xy - 25x^2}{90x^2y^2}$ *(3-4)*

80. $\dfrac{12xy - 25x^2}{27y^2}$ *(3-4)*   **81.** $\dfrac{3}{4yz}$ *(3-2)*   **82.** $\dfrac{15ab}{28c}$ *(3-2)*   **83.** $-14/3$ *(3-5)*   **84.** No solution *(3-5)*

85. No solution *(3-5)*   **86.** All numbers *(3-5)*   **87.** No solution *(3-5)*   **88.** -3 *(3-5)*   **89.** $-3/8$ *(3-5)*
90. No solution *(3-5)*   **91.** 490 *(3-7)*   **92.** 6% *(3-6)*   **93.** 5 *(3-6)*   **94.** 30°, 90°, 120°, 150°, 150° *(3-6)*
95. 92 million *(3-6)*   **96.** Approx. 3,400,000 *(3-7)*   **97.** 20 *(3-7)*   **98.** Approx. 16.71 *(3-4, 3-5)*

CHAPTER 3 PRACTICE TEST

1. $120x^2$ *(3-4)*   **2.** 12/5 *(3-2)*   **3.** $\dfrac{3yz}{5}$ *(3-2)*   **4.** $\dfrac{15 + 4x}{20}$ *(3-4)*   **5.** $-1/24$ *(3-4)*   **6.** 8/5 *(3-3)*

7. $\dfrac{5x}{12}$ *(3-3)*   **8.** $\dfrac{7x + 1}{56}$ *(3-4)*   **9.** $\dfrac{14x}{15}$ *(3-4)*   **10.** $\dfrac{2 - 5x}{10}$ *(3-4)*   **11.** $\dfrac{9ab}{4}$ *(3-3)*   **12.** 18 *(3-5)*

13. 4.2 *(3-5)*   **14.** -5.2 *(3-5)*   **15.** 1/10 *(3-5)*   **16.** 32 *(3-5)*   **17.** 1 *(3-4, 3-6)*   **18.** 1,400 *(3-7)*
19. 2.2 *(3-7)*   **20.** $5\frac{1}{3}$ *(3-5)*   **21.** 180 *(3-6)*

CUMULATIVE REVIEW EXERCISES—CHAPTERS 1–3

1. *(3-2)*   **2.** $-9/5$, $-6/5$, 0, 1, 8/5 *(3-2)*

3. **(A)** 3 **(B)** $-5, -2, 3$ **(C)** All *(3-2)*   **4.** **(A)** 0, 1 **(B)** 0, 1 *(3-2)*
5. **(A)** x **(B)** 2 **(C)** 3 *(1-5, 1-7)*   **6.** $9\frac{3}{4}$ *(3-4)*   **7.** $2 \cdot 2 \cdot 2 \cdot 2 \cdot 5$ *(1-1)*
8. **(A)** -12 **(B)** 12 *(2-2, 3-2)*   **9.** **(A)** 3.4 **(B)** 3.4 *(2-2, 3-2)*   **10.** **(A)** $-2/3$ **(B)** 2/3 *(2-2, 3-2)*
11. **(A)** 5/8 **(B)** 5/8 *(2-2, 3-2)*   **12.** **(A)** -7.6 **(B)** 7.6 *(2-2, 3-2)*   **13.** **(A)** 5 **(B)** 5 *(2-2, 3-2)*
14. 84 *(1-1)*   **15.** 210 *(1-1)*   **16.** $12xy$ *(3-4)*   **17.** **(A)** 1 **(B)** 145 *(1-4)*
18. **(A)** 13 **(B)** 17 *(1-4)*   **19.** **(A)** 8.7 **(B)** 6.82 *(1-4)*   **20.** **(A)** 7/3 **(B)** 1 *(1-4)*
21. $x + 5y$ *(1-6)*   **22.** $-20y$ *(1-6)*   **23.** ab *(1-4)*   **24.** 11 *(1-5)*   **25.** $x + 5x^2$ *(1-6)*
26. $10x - 2xy$ *(1-7)*   **27.** $a - 11b$ *(1-7)*   **28.** $\frac{7}{10}x + \frac{5}{8}y$ *(3-4)*   **29.** $x^3 + 3x^4$ *(1-7)*   **30.** $10 - 2x$ *(2-6)*
31. $-\frac{3}{2}x^3$ *(3-3)*   **32.** $-\dfrac{7a}{20}$ *(3-4)*   **33.** 4/9 *(3-3)*   **34.** $-\dfrac{b^2}{10}$ *(3-3)*   **35.** 1 *(2-7)*   **36.** -1 *(2-7)*
37. 1 *(3-7)*   **38.** $-25/6$ *(3-5)*   **39.** 31.5 *(3-5)*   **40.** 9/4 *(3-5)*   **41.** 200 *(3-6)*
42. $27 + x = 3x - 3$, $x = 15$ *(2-7)*   **43.** 5,962 *(3-7)*   **44.** 42×14 cm *(3-6)*   **45.** 47.7 million *(3-7)*
46. $120ab^2c^3$ *(3-4)*   **47.** $3 \cdot 5 \cdot 5 \cdot 7$ *(1-1)*   **48.** 6 *(1-5)*   **49.** 18 *(1-5)*   **50.** 8 *(1-5)*
51. 25.8 *(1-5)*   **52.** 5 *(2-2)*   **53.** 15 *(2-2)*   **54.** 3/4 *(3-2)*   **55.** 10.2 *(3-2)*   **56.** 45 *(1-5)*
57. -21.417 *(1-5, 3-1)*   **58.** $-1/3$ *(1-5, 3-1)*   **59.** $-35/64$ *(1-5, 3-1)*   **60.** $6x^5y^4$ *(1-5)*
61. $-\frac{1}{2}x^3y^2$ *(3-3)*   **62.** $2x^2 - x - 1$ *(2-6)*   **63.** $\dfrac{6x^2 + 4x - 3}{12}$ *(3-4)*   **64.** $2a^3b^3 + a^2b^4$ *(1-7)*
65. $5x^2 - 5y^2$ *(2-6)*   **66.** $3xy(2 + 5y - 3x)$ *(1-7)*   **67.** $3a^2(4a - 6 + 9a^2)$ *(1-7)*   **68.** All numbers *(2-7)*

69. No solution *(2-7)*   **70.** 1 *(2-7)*   **71.** $-23/5$ *(3-5)*   **72.** No solution *(3-5)*   **73.** All numbers *(3-5)*
74. 29 *(3-6)*   **75.** 36, 37, 38, 39 *(3-6)*   **76.** 8.57 mph *(3-7)*   **77.** \$4.14 *(3-6)*   **78.** All numbers *(3-5)*

79. No solution *(3-5)*   **80.** $2\frac{7}{8}$ *(3-5)*   **81.** -2 *(3-5)*   **82.** All numbers *(3-5)*   **83.** $\dfrac{w+220}{5.5}$ *(3-5)*

84. $\dfrac{11p-165}{5}$ *(3-5)*   **85.** $\dfrac{P_2V_2}{V_1}$ *(3-7)*   **86.** $\dfrac{13x-4}{30}$ *(3-4)*   **87.** $\dfrac{7x-1}{15}$ *(3-4)*

88. $\dfrac{22x-3x^2}{6}$ or $\frac{11}{3}x - \frac{1}{2}x^2$ *(3-4)*   **89.** e *(1-6)*   **90.** 1 *(2-7)*   **91.** h *(1-4)*   **92.** b *(1-4)*

93. d *(1-4)*   **94.** j *(2-7)*   **95.** a *(1-4)*   **96.** k *(2-7)*   **97.** 370 *(3-6)*
98. Yes, in 45 seconds the team A runner runs 375 m, so the team B runner finishes 1 m ahead *(3-7)*
99. 209,205.02 *(3-7)*   **100.** $0.52359\cdots$ *(3-7)*

CHAPTER 4

EXERCISE 4-1

1. T   **3.** T   **5.** T   **7.** T   **9.** T   **11.** T   **13.** 3.162   **15.** 10.583   **17.** 45.978
19. $A(2,1), B(-2,4), C(-4,-2), D(4,-4)$ **21.** $J(1,4), K(-3,1), L(-2,-3), M(2,-1)$
23. $A(3,0), B(-3,9), C(-8,-3), D(1,-8)$ **25.** $J(8,5), K(0,4), L(-4,-5), M(8,-9)$ **27.**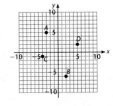

29. 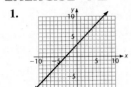   **31.** $A(2.5,-1.5), B(0,2.5), C(-3.5,3.5), D(-2.5,-3)$

33. $J(1.5,-4.5), K(3.5,3), L(-2.5,0), M(-1.5,-1.5)$ **35.** 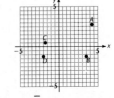   **37.**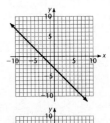

39. (A) II   (B) IV   (C) III   **41.** (A) II   (B) III   (C) IV   **43.** 0.2500   **45.** $2.5\overline{5}$
47. $0.142857\overline{142857}$ **49.** $0.5\overline{5}$ **51.** $0.32\overline{32}$ **53.** $0.571428\overline{571428}$ **55.** $0.63\overline{63}$ **57.** 0.015625
59. I and III   **61.** 14/33   **63.** 3/11   **65.** 41/333   **67.** 82/333

EXERCISE 4-2

1. 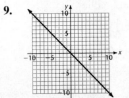   **3.** 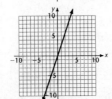   **5.** 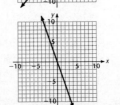   **7.**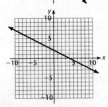

9. 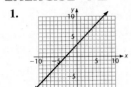   **11.** 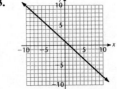   **13.** 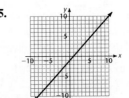   **15.**

17.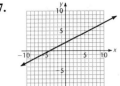

19.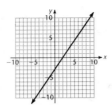

21.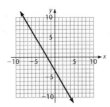

23.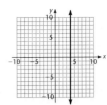

25.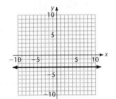

27.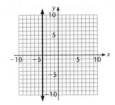

29.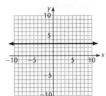

31.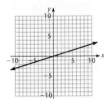

33.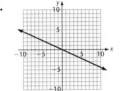

35.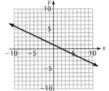

37.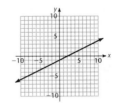

39.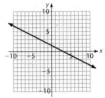

41.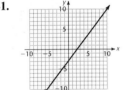

43.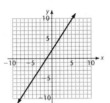

45.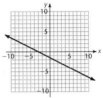

47.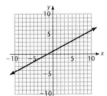

49.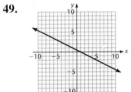

51.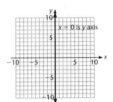

53.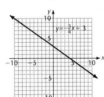

55.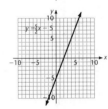

57.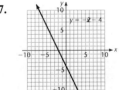

59.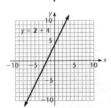

61.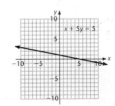

63.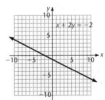

65.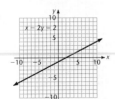

67.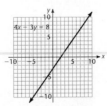

69.

71.

73.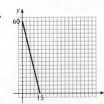

75.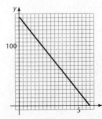

77.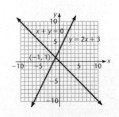

79.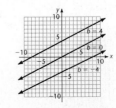

81.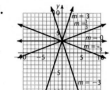

83.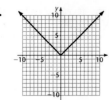

85.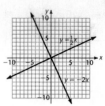

87. (a) and (b)   **89.** 25

91. **(A)** $55,000 **(B)** 8,000 **(C)** Start up costs   **93.** **(A)** In the first weeks **(B)** 4 weeks   **95.**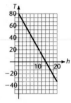

97. 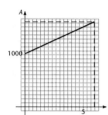   **99.** 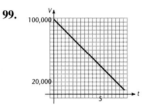   **101.** 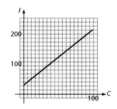   **103.**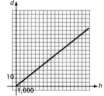

105.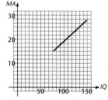

EXERCISE 4-3

1. 1   **3.** 2   **5.** 2   **7.** 4/9   **9.** slope 3, y intercept 5   **11.** slope -2, y intercept 4

13. slope 1/5, y intercept $-2/5$ **15.** slope -0.35, y intercept 0.8   **17.** $y = 3x - 1$ **19.** $y = -4x + 1$

21. $y = \frac{2}{3}x + \frac{1}{2}$ **23.** $y = -3.5x + 6.25$ **25.** $y = 2x$ **27.** $y = -2x + 4$ **29.** $y = \frac{1}{2}x + \frac{3}{40}$

31. $y = 2.5x + 0.4$ **33.** $y = x + 1$ **35.** $y = 2x + 3$ **37.** $y = 2x - \frac{1}{6}$ **39.** $y = \frac{4}{9}x + \frac{1}{9}$

41. $-\frac{3}{4}$ **43.** -1 **45.** -8 **47.** -0.6 **49.** slope $\frac{2}{3}$, y intercept $-\frac{4}{3}$ **51.** slope $\frac{3}{5}$, y intercept $\frac{4}{5}$

53. slope $\frac{3}{2}$, y intercept -3 **55.** slope $-\frac{2}{3}$, y intercept $\frac{1}{2}$ **57.** $y = \frac{1}{5}x + \frac{14}{5}$ **59.** $y = -\frac{1}{2}x + \frac{7}{6}$

61. $y = 0.8x - 1$ **63.** $y = -4x - 21$ **65.** $y = -\frac{3}{4}x + \frac{11}{4}$ **67.** $y = -x$ **69.** $y = -8x + 3$

71. $y = -0.6x + 0.24$ **73.**    **75.**   **77.**

79. 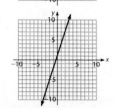   **81.** 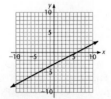   **83.** 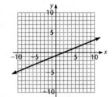   **85.**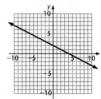

87. 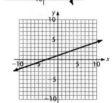   **89.** 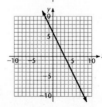   **91.**   **93.**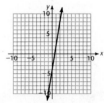

95. Slope $\frac{5}{11}$ represents increase of pressure in pounds per square inch per foot of depth.

97. **(A)** $R = 1.5C + 1$ **(B)** slope 1.5, y intercept 1 **(C)** The retail price is a fixed cost of \$1 plus a 50% markup on the variable cost. The y intercept 1 is the fixed cost. The slope 1.5 represents 150%, 100% of the variable cost plus a 50% markup.

99. **(A)** $F = \frac{9}{5}C + 32$ **(B)** $C = \frac{5}{9}F - \frac{160}{9}$

101. 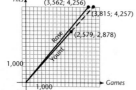 The slope represents the ratio hits per game; 3,815 games.

EXERCISE 4-4

1. $(3, 2)$ **3.** $(3, 2)$ **5.** $(9, 2)$ **7.** No solution **9.** Solution is the line $y = x - 7$ **11.** $(1, 3)$
13. $(0, 3)$ **15.** $(3, 0)$ **17.** No solution **19.** $(2, 4)$ **21.** $(6, 8)$ **23.** $(-4, -3)$ **25.** No solution
27. An infinite number of solutions. Any solution of one is a solution of the other. **29.** $(1, 4)$ **31.** $(25, 11)$
33. $(-4, -1)$ **35.** The solution is the line $y = -3x + 2$ **37.** The solution is the line $y = \frac{1}{2}x + 1$ **39.** $(33, -15)$
41. The solution is the line $y = -\frac{3}{2}x$ **43.** $(9, -2)$ **45.** $(3, -10)$ **47.** $(3, -4)$ **49.** Parallel
51. Same line **53.** Intersecting

EXERCISE 4-5

1. $x = 2, y = 3$ **3.** $x = -2, y = 3$ **5.** $x = 5, y = 10$ **7.** The solution is the line $y = 2x - 3$
9. The solution is the line $y = -2x - 1$ **11.** $x = 3, y = -1$ **13.** $x = 7, y = 19$ **15.** $x = -9, y = 14$
17. $x = -1, y = -6$ **19.** $x = 1, y = 4$ **21.** $x = 2, y = 4$ **23.** $m = 8, n = 6$ **25.** $u = -2, v = -3$
27. $x = 7, y = 1$ **29.** No solution; lines are parallel **31.** $x = 10, y = -4$ **33.** $x = -\dfrac{1}{2}, y = -\dfrac{3}{2}$
35. $x = 250, y = 100$ **37.** $x = 4,000, y = 280$ **39.** $x = 8, y = -1$ **41.** $x = 15, y = 2$ **43.** $x = 2, y = \frac{7}{4}$
45. $x = \frac{5}{3}, y = 0$ **47.** $x = 5, y = 1$ **49.** The solution is the line $y = \frac{1}{6}x - \frac{1}{2}$ **51.** No solution; lines are parallel

EXERCISE 4-6

1. $x = 3, y = 2$ **3.** $x = 1, y = 4$ **5.** $x = 2, y = -4$ **7.** $x = 2, y = -1$ **9.** $x = 3, y = 2$
11. $x = -1, y = 2$ **13.** $x = -2, y = 2$ **15.** $x = -1, y = 2$ **17.** $x = 1, y = -5$ **19.** $p = -\frac{4}{3}, q = 1$
21. $m = \frac{3}{2}, n = -\frac{2}{3}$ **23.** No solution **25.** Solution is the line $y = \frac{3}{5}x - 3$ **27.** $x = \frac{1}{3}, y = -2$
29. $x = 1, y = 4$ **31.** $m = -2, n = 2$ **33.** $x = 0, y = 0$ **35.** The solution is the line $y = -x$
37. No solution; lines are parallel **39.** $x = 1, y = 0.2$ **41.** $x = 3, y = -1$ **43.** $x = 6, y = 4$
45. $x = -6, y = 12$ **47.** No solution; lines are parallel **49.** Equivalent—multiply the second equation by -1
51. Equivalent—replace the second equation by the sum of the equations
53. Not equivalent—original system has solution $x = \frac{38}{7}, y = -\frac{16}{7}$; this system has solution $x = \frac{50}{7}, y = -\frac{24}{7}$
55. Limes: 11 cents each; lemons: 4 cents each **57.** 27 two-point shots, 7 three-point shots

EXERCISE 4-7

1. $48t = 156, t = 3.25$ hr **3.** $20t = 12,000$; 600 minutes or 10 hours **5.** $12t = 30, t = 2.5$ min
7. $7.5r = 517.5$; \$69 per hour **9.** $r \cdot 40 = 220, r = \$5.50$ per hr **11.** $50r = 10,500$; 210 items per hour
13. $r(5.5) = 550, r = 100$ km/hr **15.** $2.5r = 15$; 6 essays per hour **17.** $55t + 50t = 630, t = 6$ hr
19. $600r = 180$; .300 **21.** $20t + 30t = 30,000, t = 600$ min (10 hr)
23. If t = Time to catch up, then $50t = 45(t + 1)$ and $t = 9$ hr.
25. If t = Time to complete the job, then $20t + 30(t - 60) = 30,000$; $t = 636$ min (10.6 hr).
27. If t = Time that assistant worked, then $21t + 35(t - 5) = 1,505, t = 30$ hr (assistant), $t - 5 = 25$ hr (chemist).
29. $60t + 12(t - 120) = 30,960$; 450 minutes or $7\frac{1}{2}$ hours **31.** $\dfrac{d}{3} - \dfrac{d}{5} = 12, d = 90$ miles **33.** 21,120 feet
35. 180 **37.** 620 games on grass, 1,000 games on turf **39.** 244.8 miles **41.** 18 hours

EXERCISE 4-8

1. 55 dimes, 82 quarters **3.** 324 student tickets, 410 general admission tickets
5. 192 two-point shots, 88 three-point shots **7.** 28 **9.** 44 twenty-pound bags, 16 fifty-pound bags

11. 6 **13.** 310 acres of corn, 230 acres of soybeans **15.** 37.5 dl **17.** 300 dl
19. 80 cl of 20% solution, 120 cl of 60% solution
21. 60 lb of $3.50-per-pound coffee and 40 lb of $4.75-per-pound coffee **23.** $24,000 in 8% fund, $16,000 in 13% fund
25. 8.8 gallons of 87 octane, 13.2 gallons of 92 octane **27.** 5 liters **29.** 50 twenty-pound bags, 24 fifty-pound bags
31. $70\frac{1}{2}$-lb packages and $30\frac{1}{3}$-lb packages **33.** 60 pounds of the first, 50 pounds of the second
35. $\frac{2}{5}$ ounces of 8-karat gold, $\frac{3}{5}$ ounces of 18-karat gold

EXERCISE 4-9

1. 11 ft and 7 ft **3.** 100 m by 200 m **5.** 35°, 60°, 85° **7.** 45°, 45°, 135°, 135° **9.** 48 and 33
11. 3,800 $7 tickets and 4,400 $11 tickets **13.** $6,000 at 6% and $2,000 at 8%
15. 40 lb of $3.70-per-pound coffee and 60 lb of $5.20-per-pound coffee **17.** 70 cl water, 50 cl alcohol
19. 25 ml of 30% solution and 75 ml of 70% solution **21.** 500 grams and 300 grams
23. 80 grams of mix 1 and 60 grams of mix 2 **25.** 3 ft from the 42-lb end (9 ft from the 14-lb end)
27. 16 in. and 20 in. **29.** 10 test tubes for 1 flask; 3 test tubes for 1 mixing dish
31. 92 quarter-pound packages, 58 half-pound packages **33.** canoeist 5 mph, current 3 mph
35. $a = 5, d = 7$ **37.** 50 and 26 years old **39.** 350 miles **41.** $3\frac{1}{2}$ hours **43.** $2\frac{1}{4}$ hours

CHAPTER 4 REVIEW EXERCISE

1. $A(4, 7), B(-9, 0), C(-4, -2)$ *(4-1)* **2.** $D(2, -4), E(0, 2), F(-4, 6)$ *(4-1)*
3. *(4-1)* **4.** *(4-1)* **5.** *(4-2)*

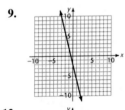

6. *(4-2)* **7.** *(4-2)* **8.** *(4-2)*

9. *(4-2)* **10.** *(4-2)* **11.** *(4-2)*

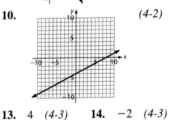 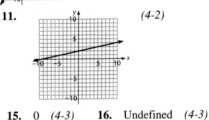 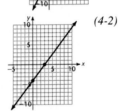

12. *(4-2)* **13.** 4 *(4-3)* **14.** −2 *(4-3)* **15.** 0 *(4-3)* **16.** Undefined *(4-3)*

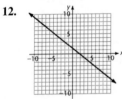

17. $-\frac{3}{5}$ *(4-3)* **18.** 19 *(4-3)* **19.** 13 *(4-3)* **20.** $y = -3x + 5$ *(4-3)* **21.** $y = -2x + 5$ *(4-3)*
22. $y = 3x - 1$ *(4-3)* **23.** $y = -14$ *(4-2)* **24.** $x = 5$ *(4-2)* **25.** $y = 3x - 9$ *(4-3)*
26. $y = -2x + 1$ *(4-3)* **27.** (6, 1) *(4-4)* **28.** (5, 0) *(4-4)* **29.** No solution *(4-4)*
30. Solution is line $y = 3x + 5$ *(4-4)* **31.** $x = 5, y = 3$ *(4-5, 4-6)* **32.** $x = 7, y = 31$ *(4-5, 4-6)*
33. $x = 2, y = 1$ *(4-6)* **34.** $x = -1, y = 2$ *(4-4 or 4-6)* **35.** Solution is line $y = 5x - 2$ *(4-5, 4-6)*
36. No solution *(4-5, 4-6)* **37.** 28 mpg *(4-7)* **38.** 72,000 acres *(4-7)*
39. $2x + 2y = 480; y = \frac{3}{2}x$; 96 by 144 yards *(4-5, 4-6)* **40.** $x + y = 30, 5x + 10y = 230$; 16 dimes, 14 nickels *(4-9)*
41. All are true *(4-1)* **42.** All are true *(4-1)* **43.** 4.472 *(4-1)* **44.** −13/2 *(4-3)* **45.** −16/15 *(4-3)*
46. 5 *(4-3)* **47.** $y = \frac{5}{3}x - \frac{7}{3}$ *(4-3)* **48.** $y = 4x - 7$ *(4-3)* **49.** $y = \frac{2}{7}x - 3$ *(4-3)*

50. $y = -\frac{1}{5}x - 3$ *(4-3)* **51.** 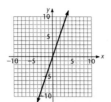 *(4-3)* **52.** 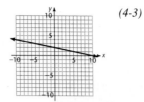 *(4-3)*

53. 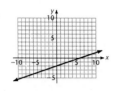 *(4-2)* **54.** 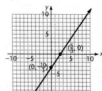 *(4-2)* **55.** 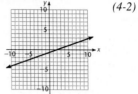 *(4-2)*

56. 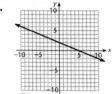 *(4-2)* **57.** 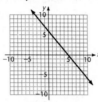 *(4-2)* **58.** 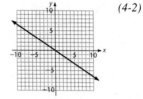 *(4-2)*

59. 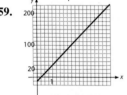 *(4-2)* **60.** 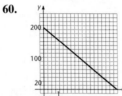 *(4-2)* **61.** $(3, 3)$ *(4-4)*

62. $(-18, 4)$ *(4-4)* **63.** $(1, 7)$ *(4-4)* **64.** $(2, 2)$ *(4-4)* **65.** Solution is the line $y = \frac{1}{3}x + \frac{1}{2}$ *(4-4)*

66. $x = 8, y = 6$ *(4-4)* **67.** $u = 1, v = -2$ *(4-6)* **68.** $m = -1, n = -3$ *(4-6)* **69.** $x = 0, y = 2$ *(4-5, 4-6)*

70. $x = 5, y = 4$ *(4-5, 4-6)* **71.** $x = 1.26, y = 0.07$ *(4-5, 4-6)* **72.** $x = 3, y = 4$ *(4-5, 4-6)*

73. $y = \frac{3}{5}x - \frac{7}{5}$ *(4-2)* **74.** $y = \frac{2}{9}x + \frac{5}{3}$ *(4-2)* **75.** $y = -\frac{1}{2}x + \frac{2}{5}$ *(4-2)* **76.** $y = \frac{1}{6}x - \frac{25}{12}$ *(4-2)*

77. $5x - y = 11$ *(4-2)* **78.** $x - 4y = 6$ *(4-2)* **79.** $-3.2x + 1.5y = 3.3$ *(4-2)* **80.** $x - 4y = 0$ *(4-2)*

81. 6.66 *(4-7)* **82.** 75 *(4-7)* **83.** 170 brownies, 140 cupcakes *(4-8)* **84.** $22t + 13t = 2,800; t = 80$ hr *(4-7)*

85. $54t = 48(t + 1); t = 8$ hr *(4-7)* **86.** $x + y = 6,000, 0.1x + 0.06y = 440;$ \$2,000 at 10% and \$4,000 at 6% *(4-9)*

87. $x + y = 100, 0.5x + 0.7y = 66;$ 20 ml of 50% solution and 80 ml of 70% solution *(4-9)*

88. No solution *(4-5, 4-6)* **89.** $x = 13/9, y = 10/9$ *(4-5, 4-6)* **90.** Unique solution *(4-4)*

91. No solution *(4-4)* **92.** Infinite number of solutions *(4-4)* **93.** Infinite number of solutions *(4-4)*

94. 66 and 24 *(4-5, 4-6)* **95.** 60, 40, and 30 cm *(4-5, 4-6)*

96. 40 games in first half season, 50 in second *(4-7)*

97. The second car is going 1.2 times as fast as the first, but neither speed can be determined from the information given. *(4-7)*

98. $90t + 110(t - 20) = 6,000; t = 41$ min *(4-7)* **99.** $0.4(12 - x) + x = 0.5(12); x = 2$ liters *(4-8)*

100. **(A)** 1.25 hr, 0.75 hr **(B)** 112.5 miles *(4-5, 4-6)*

CHAPTER 4 PRACTICE TEST

1. 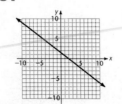 *(4-1)* **2.**   *(4-2)* **3.** 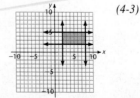 *(4-3)*

4. $y = 4x - 1$ *(4-3)* **5.** $y = -\frac{3}{5}x + 3$ *(4-3)* **6.** 2 *(4-3)* **7.** slope 1/2, y intercept $-7/2$ *(4-3)*

8. $(0, 2)$ *(4-4)* **9.** $(3, 7)$ *(4-5)* **10.** $(\frac{11}{3}, -2)$ *(4-6)* **11.** \$210,000 *(4-7)*

12. 150 by 375 m *(4-5, 4-6)* **13.** 6 lbs of 30% mix, 4 lbs of 55% mix *(4-8)* **14.** 3 hours *(4-7)*

CHAPTER 5

EXERCISE 5-1

1. -8 **3.** -6 **5.** $\frac{4}{9}$ **7.** $\frac{2}{3}$ **9.** $\frac{2}{13}$ **11.** $\frac{22}{7}$ **13.** 2π **15.** $\sqrt{9.9}$ **17.** $\sqrt{\dfrac{23}{9}}$ **19.** $a < d$

21. $b > a$ **23.** $e < f$ **25.** (number line)   **27.** (number line)   **29.** (number line)

31. (number line)   **33.** $-\frac{8}{15}$ **35.** -3.14 **37.** $5\sqrt{3}$ **39.** $\sqrt{3} \div 2$ **41.** (number line)

43. (number line)   **45.** (number line)   **47.** (number line)   **49.** (number line)   **51.** $-1 < x \le 6$

53. $2 < x \le 6$ **55.** $-\sqrt{2}$ **57.** $-\sqrt{2} \div 2$ **59.** $\sqrt{21}$ **61.** Right

63. Tropical storm: $39 \le w \le 73$; Hurricane: $w > 73$ **65.** $-170 \le t \le 134$ **67.** $-3 \le T_c \le 18$

69. $T_h \ge 10$ and $T_c \le -3$ **71.** $T_c \ge 18$ **73.** $T_c \ge 18$ and $10 - \dfrac{R}{25} \le R_d < 6$

EXERCISE 5-2

1. (number line)   **3.** (number line)

5. (number line)   **7.** (number line)

9. $-4 \le x < 7$ **11.** $-1 \le x \le 8$ **13.** $-3 < x \le 9$ **15.** $-3 < x < 6$ **17.** $x > 7$

19. $x < -7$ **21.** $x > 4$ **23.** $x \le -4$

25. $x < -21$ **27.** $x \ge 21$ **29.** $x < 2$

31. $x > 2$ **33.** $x \le 5$ **35.** $y \le -2$

37. $x < 10$ **39.** $x \ge 3$ **41.** $u \le -11$

43. $m < -\frac{7}{5}$ **45.** $x > -4$ **47.** $-1 < x < 2$

49. $-2 \le x \le 3$ **51.** $-20 \le C \le 20$ **53.** $14 \le F \le 77$

55. $-2 < x \le 3$ **57.** $-\frac{1}{3} < x < \frac{1}{6}$ **59.** $-\frac{6}{5} \le x < \frac{3}{5}$

61. $+\frac{13}{6} \le x \le \frac{19}{6}$ **63.** $+0.1 \le x \le 1.7$ **65.** $-\frac{2}{3} \le x < \frac{2}{3}$

67. $-2 < x \le 10$ **69.** $2x - 3 \ge -6;\ x \ge -\frac{3}{2}$ **71.** $2 \cdot 10 + 2w < 30;\ w < 5$ cm

73. $-62 \le \frac{5}{9}(F - 32) \le 57;\ -79.6 \le F \le 134.6$

75. $20 \le 70 - 5.5A \le 50;\ 3.6\overline{36} \le A \le 9.09\overline{09}$ or $3636\frac{7}{11} \le Altitude \le 9090\frac{1}{11}$

77. $45 \le 15 + \frac{15}{33}d \le 90;\ 66 \le d \le 165$ **79.** $2 \le 5 - \dfrac{v}{32} \le 5;\ 0 \le v \le 96$

81. $20 \le \dfrac{w}{1.78^2} \le 25;\ 63.368 \le w \le 79.21$ **83.** $100{,}000 \le 240{,}000 - 12{,}000t \le 150{,}000;\ 7\frac{1}{2} \le t \le 11\frac{2}{3}$

85. $40 \le \frac{20}{3}d \le 200;\ 6 \le d \le 30$ **87.** $100 \le \dfrac{m}{0.9144} \le 130;\ 91.44 \le m \le 118.872$

89. $0.35(60) + x \ge 0.6(60 + x);\ x \ge 37.5$ dl   **91.** $0.35(400) \le 0.2(400 + x);\ x \ge 300$ dl

EXERCISE 5-3

1.   **3.**   **5.**   **7.**

9. 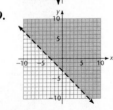   **11.** 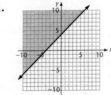   **13.** 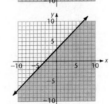   **15.**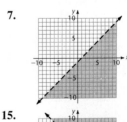

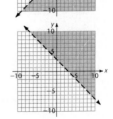

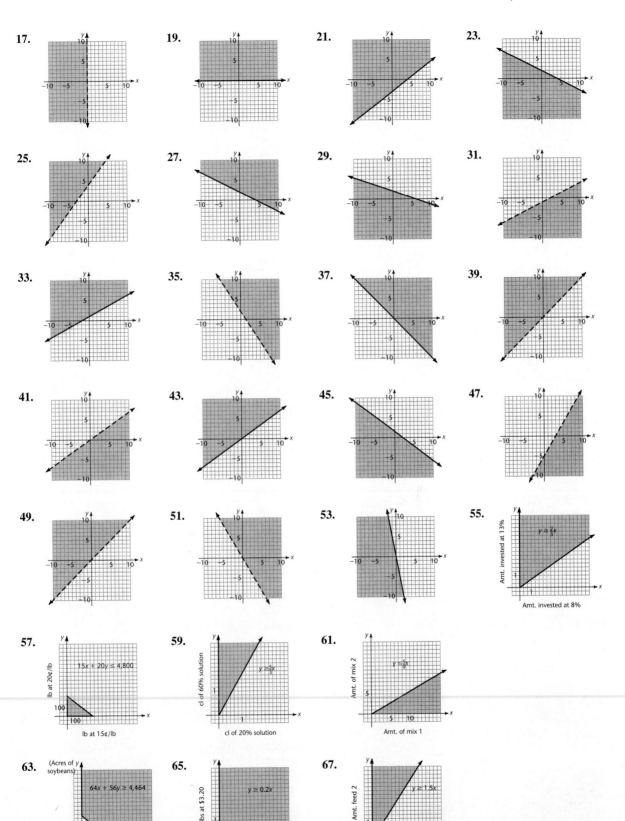

17.

19.

21.

23.

25.

27.

29.

31.

33.

35.

37.

39.

41.

43.

45.

47.

49.

51.

53.

55. Amt. invested at 13% ; $y \geq \frac{2}{3}x$; Amt. invested at 8%

57. lb at 20¢/lb ; $15x + 20y \leq 4,800$; 100 ; 100 ; lb at 15¢/lb

59. cl of 60% solution ; $y = \frac{5}{3}x$; 1 ; 1 ; cl of 20% solution

61. Amt. of mix 2 ; $y \leq \frac{5}{9}x$; 5 ; 5 ; 10 ; Amt. of mix 1

63. (Acres of soybeans) ; $64x + 56y \geq 4,464$; 10 ; 10 ; x (Acres of corn)

65. lbs at \$3.20 ; $y \geq 0.2x$; 1 ; 1 ; lbs at \$2.00

67. Amt. feed 2 ; $y \geq 1.5x$; 1 ; 1 ; Amt. feed 1

EXERCISE 5-4

1.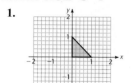

3.

5.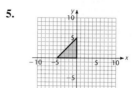

7.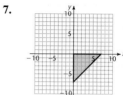

9.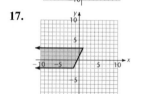

11.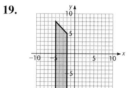

13.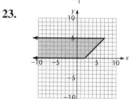

15.

17.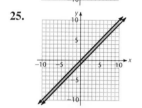

19.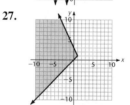

21.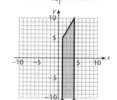

23.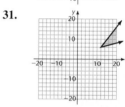

25.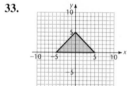

27.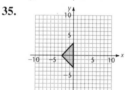

29.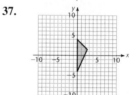

31.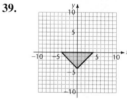

33.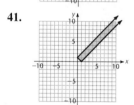

35.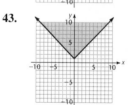

37.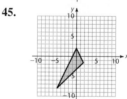

39.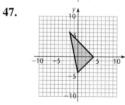

41.

43.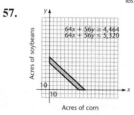

45.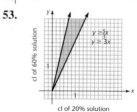

47.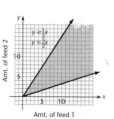

49.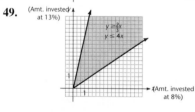

(Amt. invested at 13%)

$y \geq \frac{2}{3}x$
$y \leq 4x$

(Amt. invested at 8%)

51.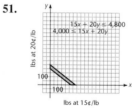

$15x + 20y \leq 4,800$
$4,000 \leq 15x + 20y$

lbs at 20¢/lb

lbs at 15¢/lb

53.

cl of 60% solution

$y \geq \frac{5}{3}x$
$y \geq 3x$

cl of 20% solution

55.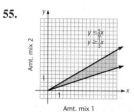

Amt. mix 2

$y \leq \frac{5}{3}x$
$y \geq \frac{1}{3}x$

Amt. mix 1

57.

Acres of soybeans

$64x + 56y \geq 4,464$
$64x + 56y \leq 5,320$

Acres of corn

59.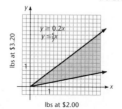

lbs at $3.20

$y \geq 0.2x$
$y \leq \frac{2}{3}x$

lbs at $2.00

61.

Amt. of feed 2

$y \geq \frac{1}{3}x$
$y \leq \frac{3}{2}x$

Amt. of feed 1

CHAPTER 5 REVIEW EXERCISE

1. $\frac{5}{18}$ *(5-1)*   **2.** $\frac{19}{24}$ *(5-1)*   **3.** $\sqrt{7}$ *(5-1)*   **4.** 2π *(5-1)*   **5.** 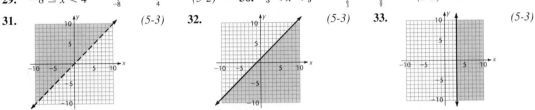 *(5-1)*

6. *(5-1)*   **7.** *(5-1)*   **8.** *(5-1)*

9. *(5-1)*   **10.** *(5-1)*   **11.** $x \geq 4$ *(5-2)*

12. $x < 16$ *(5-2)*   **13.** $x \leq 25$ *(5-2)*   **14.** $x > -10$ *(5-2)*   **15.** $x < -1$ *(5-2)*   **16.** $x \geq 0$ *(5-2)*

17. $x < 6$ *(5-2)*   **18.** $x \geq -3$ *(5-2)*   **19.** $x \leq -3$ *(5-2)*   **20.** $-4 \leq x \leq 3$ *(5-2)*   **21.** $-\frac{2}{3} \leq x \leq \frac{1}{3}$ *(5-2)*

22. $-189 \leq T \leq 39$ *(5-2)*   **23.** $5x - 5 \leq 10; x \leq 3$ *(5-2)*   **24.** $0 < w \leq 27$ *(5-2)*

25. $x > -1$ 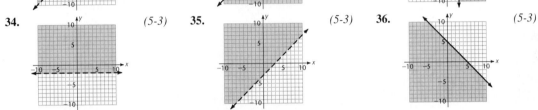 *(5-2)*   **26.** $x < -\frac{1}{5}$ *(5-2)*

27. $x \geq -8$ *(5-2)*   **28.** $-6 \leq x < 18$ *(5-2)*

29. $-8 \leq x < 4$ *(5-2)*   **30.** $\frac{2}{3} < x < \frac{8}{9}$ *(5-2)*

31. 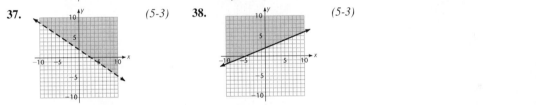 *(5-3)*   **32.** *(5-3)*   **33.** *(5-3)*

34. *(5-3)*   **35.** *(5-3)*   **36.** *(5-3)*

37. *(5-3)*   **38.** *(5-3)*

39. $140 \leq \frac{9}{5}C + 32 \leq 170; 60 \leq C \leq 76\frac{2}{3}$ *(5-2)*   **40.** $59 \leq \frac{9}{5}C + 32 \leq 86; 15 \leq C \leq 30$ *(5-2)*

41. 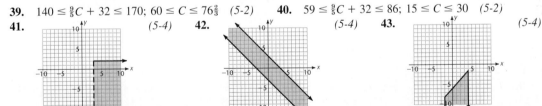 *(5-4)*   **42.** *(5-4)*   **43.** *(5-4)*

44. 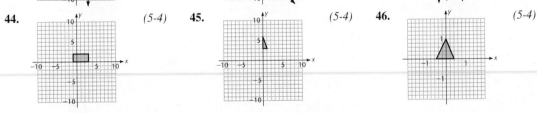 *(5-4)*   **45.** *(5-4)*   **46.** *(5-4)*

47. 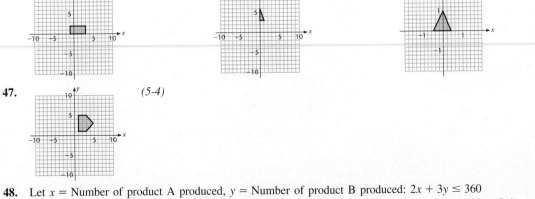 *(5-4)*

48. Let x = Number of product A produced, y = Number of product B produced: $2x + 3y \leq 360$
$$3x + 5y \leq 420 \quad (5\text{-}4)$$

49. Let x = Pounds of corn used, y = Pounds of commercial feed used: $15x + 40y \geq 150$
$$20x + 50y \geq 120$$

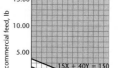

50. **(A)** True **(B)** False for any $z < 0$ **(C)** True **(D)** False for any $z < 0$

CHAPTER 5 PRACTICE TEST

1. $2\sqrt{3} + 3$ *(5-1)* **2.** 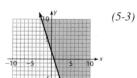 *(5-1)* **3.** $x \geq \frac{8}{5}$ *(5-2)*

4. $x \geq -\frac{1}{2}$ *(5-2)* **5.** $-4 < x \leq 2$ *(5-2)* **6.** *(5-3)*

7. 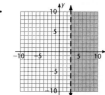 *(5-3)* **8.** 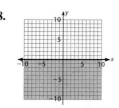 *(5-3)* **9.** 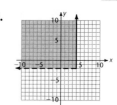 *(5-4)*

10. 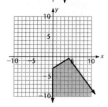 *(5-4)* **11.** $3 \leq l \leq 18$ *(5-2)* **12.** $2 \leq v < 4$ *(5-2)* **13.** $6 \leq x \leq 16$ *(5-2)*

CUMULATIVE REVIEW EXERCISES—CHAPTERS 1–5

1. 120 *(1-1)* **2.** 180 *(3-4)* **3.** $12x^2y$ *(3-4)* **4.** 26 *(1-5)* **5.** -9 *(2-5)* **6.** -8 *(2-5)*
7. -36 *(2-5)* **8.** 7 *(2-2, 2-3)* **9.** -2 *(2-2, 2-4)* **10.** 3/14 *(3-4)* **11.** 3.742 *(4-1)* **12.** 6 *(4-3)*
13. -4 *(4-3)* **14.** $y = -4x + 9$ *(4-3)* **15.** $x = -5$ *(4-2)* **16.** 3/2 *(3-4)* **17.** 120 *(3-6)*
18. $a = -6, b = 0, c = 8$ *(2-1)* **19.** $A(4, 2), B(-3, 0), C(-2, -5)$ *(4-1)* **20.** $2 \times 2 \times 2 \times 2 \times 2 \times 3 \times 3$ *(1-1)*
21. $12x - 15xy$ *(1-6, 2-6)* **22.** $7x - 2xy$ *(2-6)* **23.** 18/30 *(3-2)* **24.** 3/5 *(3-2)* **25.** 9/2 *(3-5)*
26. $x \geq \frac{3}{5}$ *(5-2)* **27.** $x = 5, y = -1$ *(2-7)* **28.** $x = 6, y = -3$ *(4-5, 4-6)* **29.** $-5 \leq x \leq 2$ *(5-2)*

30. $x = 15$ *(3-5)* **31.** $x = -\dfrac{y - 8}{3}$ *(2-7)* **32.** 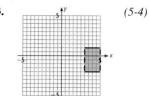 *(4-4)* **33.** *(5-4)*

34.  *(3-2)* **35.** 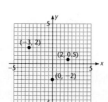 *(4-1)* **36.** *(5-1)*

37. 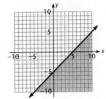 *(5-3)*   **38.** 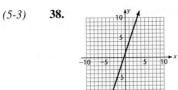 *(4-2)*   **39.** 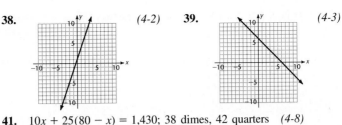 *(4-3)*

40. 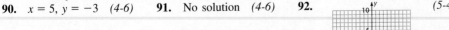 *(5-3)*   **41.** $10x + 25(80 - x) = 1,430$; 38 dimes, 42 quarters *(4-8)*

42. $\dfrac{9}{4} = \dfrac{432}{x}$; 192 *(3-7)*   **43.** $\dfrac{x + (x - 4) + x/3}{3} = 36$; 48, 44, 16 *(3-4, 3-5)*   **44.** $\dfrac{10^2}{172} = \dfrac{20^2}{A}$; 688 sq. cm *(3-7)*

45. 6 *(1-5)*   **46.** 3/4 *(1-5, 3-3)*   **47.** $\dfrac{7x}{12}$ *(3-4)*   **48.** 4/25 *(3-3)*   **49.** xy^2 *(3-4)*   **50.** $\dfrac{2x}{3}$ *(3-4)*

51. $y = -\tfrac{13}{3}x + \tfrac{28}{3}$ *(4-3)*   **52.** $3x + 2$ *(1-2)*   **53.** $x + 3 = 4x - 2$ *(1-3)*   **54.** $x^2 - 6 = 2x - 1$ *(1-3)*

55. $3x + 4 \geq 40$ *(1-3, 5-1)*   **56.** $x + (x + 1) + (x + 2) + (x + 3)$ *(1-2)*   **57.** $5 < (x - 1)^2$ *(5-1)*

58. $7x + 4$ *(1-2)*   **59.** $2x - 11 < 2x$ *(5-1)*   **60.** $\dfrac{x + (x + 1) + (x + 2)}{3}$ *(1-2, 3-4)*   **61.** $x(x + 3) = 154$ *(1-3)*

62. $\dfrac{15x^2y^3}{40xy^2}$ *(3-2)*   **63.** $2x - 2$ *(2-6)*   **64.** $a - (3b + 4c)$ *(2-6)*   **65.** $4x(3y + x - 9y^2)$ *(1-7, 2-6)*

66. $7x - y = 11$ *(3-2)*   **67.** $y = -\tfrac{3}{4}x + 3$ *(2-7)*   **68.** $-3ab + 12ac - 8bc$ *(2-6)*   **69.** $-x^2 - 3x - 4$ *(2-6)*

70. All numbers *(2-7)*   **71.** $x \geq -\tfrac{3}{2}$ *(5-2)*   **72.** $x = \tfrac{5}{2}, y = \tfrac{5}{2}$ *(4-5, 4-6)*   **73.** No solution *(4-5, 4-6)*

74. All numbers *(2-7)*   **75.** $x = 0, y = 5$ *(4-5, 4-6)*   **76.** $h = \dfrac{2A}{a + b}$ *(2-7)*   **77.** 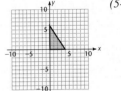 *(5-4)*

78. 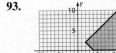 *(5-4)*   **79.** 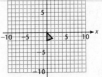 *(4-2)*   **80.** 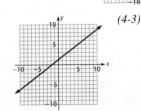 *(4-3)*

81. 55×35 yd *(3-7)*   **82.** 40°C *(3-5)*   **83.** $71\tfrac{3}{7}$ *(3-7)*

84. New process 22 hours, old process 28 hours *(4-5)*   **85.** 5:24 P.M. *(4-7)*   **86.** $1,830 *(3-6)*

87. $s = 100p - 2,000$; 4,000 *(4-3)*   **88.** x *(2-6)*   **89.** $-6x^2 + 26xy - 20y^2 + 2x + 8y$ *(2-6)*

90. $x = 5, y = -3$ *(4-6)*   **91.** No solution *(4-6)*   **92.** 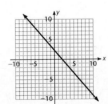 *(5-4)*

93. 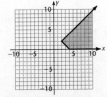 *(5-4)*   **94.** 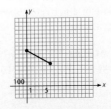 *(4-3)*   **95.** 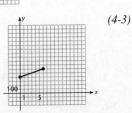 *(4-3)*

96. 1 liter *(4-8)*   **97.** 63 mph *(4-7)*   **98.** 9 pounds of \$2.45 coffee, 14 pounds of \$3.60 coffee *(4-8)*
99. 2 qts *(4-8)*   **100.** 240 on single-sided pages, 360 on double-sided pages *(4-8)*

CHAPTER 6

EXERCISE 6-1

1. 3   **3.** 0   **5.** 7   **7.** 2   **9.** 5   **11.** 7   **13.** $7x - 2$ **15.** $8x$ **17.** $x - 1$ **19.** $3x + 8$
21. $x^2 + 7x + 1$ **23.** $2x^2 - x + 1$ **25.** $3x + 4$ **27.** $x + 6$ **29.** $2x^2 + 3x + 1$ **31.** $3x^2 - 3x + 5$
33. 1   **35.** 3   **37.** 2   **39.** 6   **41.** 7   **43.** $4x^3 + 2x^2 + 3x$ **45.** $4x - y$ **47.** $x - y - z$
49. $-3x - xy + xy^2 + 3y^2$ **51.** $-x^3 + 4x^2 + 2x - 3$ **53.** $-a + 2b$ **55.** $-2a + b$ **57.** $-4xy + 5y^2$
59. $-x^2 + 5x - 1$ **61.** $-3x^2 - x - 1$ **63.** $-2x^2 + 3x$ **65.** $7x - 2$ **67.** $3x^2 - 6x + 3$ **69.** $-2x + 6$
71. $-2x^2 + 8x - 4$ **73.** $-2x^3 - x^2 - x + 11$ **75.** $b = -4, c = -6$ **77.** $-x^3 + 9x^2 - 8x - 13$
79. $b = 6, c = 6$

EXERCISE 6-2

1. $2x^2 + x - 6$ **3.** $2x^3 - 7x^2 + 13x - 5$ **5.** $x^3 - 6x^2y + 10xy^2 - 3y^3$ **7.** $a^3 + b^3$ **9.** $y^2 - 10y + 25$
11. $x^2 + 6x + 9$ **13.** $x^2 + 3x + 2$ **15.** $y^2 + 7y + 12$ **17.** $x^2 - 9x + 20$ **19.** $n^2 - 7n + 12$
21. $s^2 + 5s - 14$ **23.** $m^2 - 7m - 60$ **25.** $u^2 - 9$ **27.** $x^2 - 64$ **29.** $y^2 + 16y + 63$ **31.** $c^2 - 15c + 54$
33. $x^2 - 8x - 48$ **35.** $a^2 - b^2$ **37.** $x^2 + 4xy + 3y^2$ **39.** $3x^2 + 7x + 2$ **41.** $4t^2 - 11t + 6$
43. $3y^2 - 2y - 21$ **45.** $2x^2 + xy - 6y^2$ **47.** $6x^2 + x - 2$ **49.** $9y^2 - 4$ **51.** $5s^2 + 34s - 7$
53. $6m^2 - mn - 35n^2$ **55.** $12n^2 - 13n - 14$ **57.** $6x^2 - 13xy + 6y^2$ **59.** $x^3 + 6x^2y + 12xy^2 + 8y^3$
61. $2x^4 - 5x^3 + 5x^2 + 11x - 10$ **63.** $2x^4 + x^3y - 7x^2y^2 + 5xy^3 - y^4$ **65.** $4x^2 - 12x + 9$
67. $4x^2 - 20xy + 25y^2$ **69.** $16a^2 + 24ab + 9b^2$ **71.** $9u^2 + 12uv + 4v^2$ **73.** $-x^2 + 17x - 11$
75. $2x^3 - 13x^2 + 25x - 18$ **77.** $x^3 + x^2 - 3x + 1$ **79.** $6x^3 - 13x^2 + 14x - 12$ **81.** $x^4 - 2x^3 - 4x^2 + 11x - 6$
83. $x^4 + 2x^2 + 1$ **85.** $4a^6 - 4a^3b^2 + b^4$ **87.** $9x^6 + 24x^3y^2 + 16y^4$ **89.** $-2x^4 - x^3 - 4x^2 + x - 12$
91. $x^4 + 4x^3 + 10x^2 + 12x + 9$ **93.** $2x^4 + 3x^3 + 5x^2 - x + 7$ **95.** $2x^6 + 5x^5 + 4x^4 + 6x^3 + 6x^2 + 25x - 12$
97. $2x^3 + 2x^2 + 2x - 6$ **99.** $A^4 - B^4$

EXERCISE 6-3

1. $A(2x + 3)$ **3.** $5x(2x + 3)$ **5.** $2u(7u - 3)$ **7.** $2u(3u - 5v)$ **9.** $5mn(2m - 3n)$ **11.** $2x^2y(x - 3y)$
13. $(x + 2)(3x + 5)$ **15.** $(m - 4)(3m - 2)$ **17.** $(x + y)(x - y)$ **19.** $3x^2(2x^2 - 3x + 1)$
21. $2xy(4x^2 - 3xy + 2y^2)$ **23.** $4x^2(2x^2 - 3xy + y^2)$ **25.** $4xyz(x + 2y + 3z)$ **27.** $(2x + 3)(3x - 5)$
29. $(x + 1)(x - 1)$ **31.** $(2x - 3)(4x - 1)$ **33.** $x(y^2 + 2z)(2x - 3)$ **35.** $2x - 2$ **37.** $2x - 8$ **39.** $2u + 1$
41. $2y + 3xy$ **43.** $3x^3 - 3xy$ **45.** $3x(x - 1) + 2(x - 1) = (x - 1)(3x + 2)$
47. $3x(x - 4) - 2(x - 4) = (x - 4)(3x - 2)$ **49.** $4u(2u + 1) - (2u + 1) = (2u + 1)(4u - 1)$ **51.** $(x^2 - y)(2 + 3x)$
53. $(x^2 - y)(2 + 3x)$ **55.** $(x - 1)(3x + 2)$ **57.** $(x - 4)(3x - 2)$ **59.** $(2u + 1)(4u - 1)$
61. $(x^2 - y)(2 + 3x)$ **63.** $(x^2 - y)(2 + 3x)$ **65.** $2m(m - 4) + 5(m - 4) = (m - 4)(2m + 5)$
67. $3x(2x - 3) - 2(2x - 3) = (2x - 3)(3x - 2)$ **69.** $3u(u - 4) - (u - 4) = (u - 4)(3u - 1)$
71. $3u(2u + v) - 2v(2u + v) = (2u + v)(3u - 2v)$ **73.** $3x(2x + y) - 5y(2x + y) = (2x + y)(3x - 5y)$
75. $3u^2 - 12u - u + 4 = 3u(u - 4) - (u - 4)$ **77.** $6u^2 + 3uv - 4uv - 2v^2 = 3u(2u + v) - 2v(2u + v)$
        $= (u - 4)(3u - 1)$ $= (2u + v)(3u - 2v)$
79. $6x^2 + 3xy - 10xy - 5y^2 = 3x(2x + y) - 5y(2x + y)$ **81.** $3a^2 + 9ab + ab + 3b^2 = 3a(a + 3b) + b(a + 3b)$
                 $= (2x + y)(3x - 5y)$ $= (a + 3b)(3a + b)$
83. $uw - ux - vw + vx = u(w - x) - v(w - x)$
                      $= (w - x)(u - v)$

EXERCISE 6-4

1. Not a perfect square   **3.** $(x + 9)^2$ **5.** $(x + 2)^2$ **7.** Not a perfect square   **9.** Not a perfect square
11. Not a perfect square   **13.** Not a perfect square   **15.** Not a perfect square   **17.** $(x + 11)^2$
19. Not a perfect square   **21.** $(x - 4)(x + 4)$ **23.** Not factorable   **25.** $(x - 10)(x + 10)$ **27.** Not factorable
29. Not factorable   **31.** Not factorable   **33.** $3(x + 5)^2$ **35.** $4(x + 3)^2$ **37.** $6(x^2 + 6x + 10)$
39. $4(x - 3)^2$ **41.** $4(x^2 - 4x + 2)$ **43.** $6(x - 5)^2$ **45.** $4(x^2 + 4x + 16)$ **47.** $2(x^2 - 6x + 18)$
49. $5(x^2 + 2x + 4)$ **51.** $6(x^2 - 2)$ **53.** $8(x - 3)(x + 3)$ **55.** $3(x^2 + 16)$ **57.** $x(2x - 5)$
59. $6x(x + 6)$ **61.** $x(3x - 1)$ **63.** 49   **65.** 81   **67.** 100   **69.** 196   **71.** 16   **73.** 10   **75.** 26
77. 8   **79.** 28   **81.** 0

EXERCISE 6-5

1. $(x + 1)(x + 4)$ **3.** $(x + 2)(x + 3)$ **5.** $(x - 1)(x - 3)$ **7.** $(x - 2)(x - 5)$ **9.** Not factorable
11. Not factorable   **13.** $(x - 3)(x - 4)$ **15.** $(x + 6)(x + 2)$ **17.** $(x + 1)(x + 12)$ **19.** $(x + 3y)(x + 5y)$
21. $(x - 4)^2$ **23.** Not factorable   **25.** $(x + 3)(x - 4)$ **27.** $(x - 2)(x + 6)$ **29.** $(x - 12)(x + 1)$
31. $(x - 3y)(x - 7y)$ **33.** Not factorable   **35.** $(3x + 1)(x + 2)$ **37.** $(3x - 4)(x - 1)$ **39.** Not factorable
41. $(x + 1)^2$ **43.** $(x - 4)(3x - 2)$ **45.** $(3x - 2y)(x - 3y)$ **47.** $(n - 4)(n + 2)$ **49.** Not factorable
51. $(x - 1)(3x + 2)$ **53.** $(x + 6y)(x - 2y)$ **55.** $(u - 4)(3u + 1)$ **57.** $(3x + 5)(2x - 1)$ **59.** $(3s + 1)(s - 2)$
61. Not factorable   **63.** $(x - 2)(5x + 2)$ **65.** $(2u + v)(3u - 2v)$ **67.** $(4x - 3)(2x + 3)$
69. $(3u - 2v)(u + 3v)$ **71.** $(u - 4v)(4u - 3v)$ **73.** $(6x + y)(2x - 7y)$ **75.** $(12x - 5y)(x + 2y)$
77. $(x + y)(x - y)$ **79.** $(2x + y)(2x - y)$ **81.** Not factorable   **83.** Not factorable   **85.** $(3x - 2y)^2$
87. $(5x + y)^2$ **89.** Not factorable

EXERCISE 6-6

1. $(3x - 4)(x - 1)$ **3.** Not factorable   **5.** $(2x - 1)(x + 3)$ **7.** Not factorable   **9.** $(3x + 1)(2x + 1)$
11. $(2x + 1)^2$ **13.** $(4x - 1)(x + 1)$ **15.** $(4x - 1)(x - 1)$ **17.** $(2x - 1)(3x - 1)$ **19.** $(8x + 1)(x - 1)$
21. $(5x + 1)(x + 2)$ **23.** $(3x + 2)(2x + 1)$ **25.** $(7x + 2)(x + 1)$ **27.** $(4x + 1)(x + 2)$ **29.** $(3x + 2)(x + 1)$
31. $(x - 4)(3x - 2)$ **33.** $(3x + 5)(2x - 1)$ **35.** Not factorable   **37.** $(m - 4)(2m + 5)$ **39.** $(u - 4)(3u + 1)$
41. $(2u + v)(3u - 2v)$ **43.** Not factorable   **45.** $(4x - 3)(2x + 3)$ **47.** $2(2m - n)(m + 3n)$
49. Not factorable   **51.** $(u - 4v)(4u - 3v)$ **53.** $(4x + 1)(x - 2)$ **55.** $(3x - 2)(2x + 1)$ **57.** $2(3x + 1)(x - 1)$
59. $(5x + 2)(x + 3)$ **61.** $(3x + 2)(x + 1)$ **63.** $(8x + 3)(x + 2)$ **65.** $(5x - 3)(x + 2)$ **67.** $3(3x - 1)(x + 2)$
69. $(7x - 3)(x + 2)$ **71.** Not factorable   **73.** Not factorable   **75.** $(6x + y)(2x - 7y)$ **77.** $(6x + 5y)(3x - 4y)$
79. $(5x - 2)(2x - 3)$ **81.** $(8x - 5)(x + 3)$ **83.** $\pm 7, \pm 8, \pm 13$ **85.** $\pm 6, \pm 9$ **87.** $\pm 8, \pm 10, \pm 17$
89. 6, 10, 12   **91.** 2   **93.** 10

EXERCISE 6-7

1. $3x^2(2x + 3)$ **3.** $u^2(u + 2)(u + 4)$ **5.** $x(x - 2)(x - 3)$ **7.** $(x - 2)(x + 2)$ **9.** $(2x - 1)(2x + 1)$
11. Not factorable   **13.** $2(x - 2)(x + 2)$ **15.** $(3x - 4y)(3x + 4y)$ **17.** $3uv^2(2u - v)$ **19.** $xy(2x - y)(2x + y)$
21. $3x^2(x^2 + 9)$ **23.** Not factorable   **25.** $(x - 2y)(x^2 + 2xy + 4y^2)$ **27.** $8y(x - 3y)(x + 3y)$
29. $y^2(x - 3)(x + 3)$ **31.** $7(x + 1)^2$ **33.** $6(x - 5)^2$ **35.** $6(x + 2)(x + 4)$ **37.** $3x(x^2 - 2x + 5)$
39. $4x(3x - 2y)(x + 2y)$ **41.** $(x + 3)(x + y)$ **43.** $(x - 3)(x - y)$ **45.** $(2a + b)(c - 3d)$
47. $(m + n)(2u - v)$ **49.** $(x - 2)(x^2 + 2x + 4)$ **51.** $(x + 3)(x^2 - 3x + 9)$ **53.** $3m^2(m^2 + 4)$
55. $(x - 6)(x - 1)$ **57.** $(2y - 1)(2y + 1)$ **59.** $2x(3x + 2)(x - 2)$ **61.** $(u - 5)(u - v)$
63. $(a + 2b)(c - 3d)$ **65.** $(2x + 3y)(z + w)$ **67.** $2xy(2x + y)(x + 3y)$ **69.** $5y^2(6x + y)(2x - 7y)$
71. $3xy^2(2x - y + 1)$ **73.** $(x + 5)(x^2 + 1)$ **75.** $2x(2y + x)(4y^2 - 2yx + x^2)$ **77.** $2(2x - 1)(x + 3)$
79. $4(2x + 1)(x - 5)$ **81.** $(x^2 + 2)^2$ **83.** $(x^3 - 4)(x + 1)(x^2 - x + 1)$ **85.** $(x - 2)(x + 2)(x^2 + 4)$
87. $(x^2 + 3)(x^4 - 3x^2 + 9)$ **89.** $(x - 1)(x + 1)(x^2 + 1)$ **91.** $(x - y)(x + y)(x^2 + y^2)$ **93.** $(x^3 + 2)^2$
95. $(x^2 - 2)(x^2 - 3)$ **97.** $2(x^3 + 7)^2$

EXERCISE 6-8

1. 3, 4   **3.** $-6, 5$ **5.** $-4, \frac{2}{3}$ **7.** $-\frac{3}{4}, \frac{2}{5}$ **9.** $0, \frac{1}{4}$ **11.** 1, 5   **13.** 1, 3   **15.** $-2, 6$ **17.** 0, 3
19. 0, 2   **21.** $-5, 5$ **23.** -3 **25.** 4   **27.** $-4, -2$ **29.** $-4, 7$ **31.** $\frac{1}{2}, -3$ **33.** $\frac{2}{3}, 2$
35. $-4, 8$ **37.** $-\frac{2}{3}, 4$ **39.** Not factorable in the integers   **41.** 3, -4 **43.** $y = -3, 5$ **45.** $\frac{1}{2}, 2$
47. $-2, 6$ **49.** $-3, 8$ **51.** $-9, 8$ **53.** 2, 5   **55.** 2, -4 **57.** $x = -9, 7$ **59.** $-11, 2$ **61.** $-4, 6$
63. $-5, -4$ **65.** $3, \frac{5}{2}$ **67.** $0, -\frac{5}{4}$ **69.** $-3, 3$ **71.** $-1, 3$

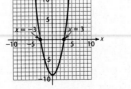

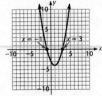

73. $-1, -5$ **75.** 3 in by 11 in   **77.** 4 cm by 24 cm   **79.** base 4 ft, height 5 ft

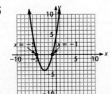

81. base 6 m, height 3 m   **83.** base 8 in, top 4 in, height 2 in   **85.** base 12 m, top 4 m, height 6 m   **87.** 1 in
89. 1 in   **91.** 30   **93.** Yes, for 60 vines

EXERCISE 6-9

1. $x + 2$ **3.** $2x - 3$ **5.** $2x + 3$, R $= 5$ **7.** $m - 2$ **9.** $2x + 3$ **11.** $2x + 5$, R $= -2$
13. $x + 5$, R $= -2$ **15.** $x - 7$ **17.** $x + 5$ **19.** $x + 6$ **21.** $x - 3$ **23.** $x + 6$ **25.** $x + 2$
27. $m + 3$, R $= 2$ **29.** $5c - 2$, R $= 8$ **31.** $3x + 2$, R $= -4$ **33.** $2y^2 + y - 3$ **35.** $x^2 + x + 1$
37. $x^3 - 2x^2 + 4x - 8$ **39.** $2y^2 - 5y + 13$, R $= -27$ **41.** $x - 1$, R $= 13$ **43.** $x + 5$ **45.** $x^2 + 4x + 16$
47. $2x - 3$, R $= 45$ **49.** $4x^2 + 1$ **51.** $2x^3 - 3x^2 - 5$, R $= 5$ **53.** $2x^2 - 3x + 2$, R $= 4$
55. $x^3 + 2x^2 - 12x + 33$, R $= -36$ **57.** $x^2 - 10$, R $= 36$ **59.** $x^3 - 3x^2 + 3x - 2$ **61.** $4, 3, 1$
63. $-3, -2, -1$ **65.** $-2, -7, 5$ **67.** $4, -2, 1$

CHAPTER 6 REVIEW EXERCISE

1. **(A)** 3 **(B)** 7 **(C)** 11 *(6-1)* **2.** **(A)** 4 **(B)** 8 **(C)** 3 *(6-1)* **3.** $5x^2 + 3x - 4$ *(6-1)*
4. $2x^2 - x + 7$ *(6-1)* **5.** $2x^4 + 5x^3 - 2x^2 + 10x - 3$ *(6-2)* **6.** $3x + 4$, R $= 2$ *(6-9)*
7. $6x^2 + 11x - 10$ *(6-2)* **8.** $6u^2 - uv - 12v^2$ *(6-2)* **9.** $3x^3 + x^2 - 3x + 6$ *(6-1)*
10. $-3x^3 + 3x^2 - 3x - 4$ *(6-1)* **11.** $-x^2 + 6x - 7$ *(6-1)* **12.** $9x^2 - 2x - 3$ *(6-1)*
13. $(x - 7)(x - 2)$ *(6-5)* **14.** $(3x - 4)(x - 2)$ *(6-5)* **15.** Not factorable *(6-5)* **16.** $2xy(2x - 3y)$ *(6-4)*
17. $x(x - 2)(x - 3)$ *(6-5)* **18.** $(2u - 3)(2u + 3)$ *(6-4)* **19.** $(x - 1)(x + 3)$ *(6-3)* **20.** Not factorable *(6-4)*
21. $(x + 7)(x - 3)$ *(6-5)* **22.** $(3x + 1)(x - 4)$ *(6-5, 6-6)* **23.** $(2x + 1)(x - 3)$ *(6-5, 6-6)*
24. $(x + 5)(x - 6)$ *(6-5)* **25.** $x(x + 2)(x + 3)$ *(6-5)* **26.** $(x - 4)^2$ *(6-4)* **27.** $(4x + 1)(x - 1)$ *(6-5, 6-6)*
28. $(x + 1)(x + 7)$ *(6-5)* **29.** $3(x + 2)(x + 4)$ *(6-5)* **30.** $(x + 11)^2$ *(6-4)* **31.** Not factorable *(6-4)*
32. $4(x + 5)(x - 5)$ *(6-4)* **33.** $(x + 4)(x + 6)$ *(6-5)* **34.** $(5x + 1)(x + 2)$ *(6-5, 6-6)*
35. $x^2(3x - 8)(3x + 8)$ *(6-4)* **36.** $(x + 20)^2$ *(6-4)* **37.** $x = -5, 2$ *(6-8)* **38.** $x = 0, 1/3$ *(6-8)*
39. $x = 3$ *(6-8)* **40.** $x = 0, -1, 9$ *(6-8)* **41.** $-2x^2 - 5x - 5$ *(6-1, 6-2)*
42. $4x^4 + 4x^3 + 3x^2 - x - 1$ *(6-2)* **43.** $7x^2 - 4x - 6$ *(6-1, 6-2)* **44.** $2x^4 - 9x^3 + 2x^2 + 5x$ *(6-1, 6-2)*
45. $x - 4$, R $= 3$ *(6-9)* **46.** $3x^2 + 2x - 2$, R $= -2$ *(6-9)* **47.** $x^2 + 3x + 8$ *(6-9)*
48. $x^3 + x^2 + 3x - 3$, R $= 3$ *(6-9)* **49.** $4x^3 - 12x^2 + 13x - 6$ *(6-2)* **50.** $27x^4 + 63x^3 - 66x^2 - 28x + 24$ *(6-2)*
51. $a^3 + b^3$ *(6-3)* **52.** $-2x + 20$ *(6-1, 6-2)* **53.** $x^4 + 2x^3 - 33x^2 + 74x - 35$ *(6-2)* **54.** $x^3 + x^2 - 3x$ *(6-1)*
55. $x^3 - 6x^2 + 11x - 6$ *(6-2)* **56.** $2x^2 - 10x + 14$ *(6-1, 6-2)* **57.** $3(u - 2)(u + 2)$ *(6-4)*
58. $(2x - 3y)(x + y)$ *(6-5, 6-6)* **59.** Not factorable *(6-5, 6-6)* **60.** $3y(2y - 5)(y + 3)$ *(6-5, 6-6)*
61. $2x(x^2 - 2xy - 5y^2)$ *(6-5, 6-6)* **62.** $3xy(4x^2 + 9y^2)$ *(6-4)* **63.** $(x - 1)(x - 3)(x + 3)$ *(6-3)*
64. Not factorable *(6-5, 6-6)* **65.** $(x - 5y)(x^2 + 5xy + 25y^2)$ *(6-7)* **66.** $(x^2 + 3)^2$ *(6-7)*
67. $(8x + 1)(x - 6)$ *(6-5, 6-6)* **68.** $(x^4 + y^2)(x^2 - y)(x^2 + y)$ *(6-7)* **69.** $(6x + 1)(x - 8)$ *(6-5, 6-6)*
70. $(x - z)(x^2 + xz + z^2)$ *(6-7)* **71.** $(x + y)(y + z)$ *(6-7)* **72.** $(3x + 2)(2x - 3)$ *(6-5, 6-6)*
73. $(x - y)(x + 4)$ *(6-7)* **74.** $(x + y)(x - 3)$ *(6-7)* **75.** $(2u - 3)(u + 3)$ *(6-7)* **76.** $(3x + 2)(2x - 1)$ *(6-7)*
77. $(x + z)(y + 1)$ *(6-7)* **78.** $x(x + 1)(x^2 + 1)$ *(6-7)* **79.** $-4, 1$ *(6-8)* **80.** $\frac{1}{2}, -2$ *(6-8)*
81. $x = 0, 11$ *(6-8)* **82.** $x = -2, -6$ *(6-8)* **83.** 0 *(6-1, 6-2)* **84.** $x^2 - 10$, R $= 20x + 64$ *(6-9)*
85. $6(2x - 1)(x + 2)$ *(6-5)* **86.** $(4x - 3)(3x + 4)$ *(6-5, 6-6)* **87.** $4(3x + 1)(x + 3)$ *(6-5, 6-6)*
88. $4(3x - 1)(x + 3)$ *(6-5, 6-6)* **89.** $(x + 6)(x^2 - 6x + 36)$ *(6-7)* **90.** $x^2(x^2 + 1)$ *(6-4)*
91. $x^2(y + z)(y - z)$ *(6-4)* **92.** $x(y - x)(x^2 + xy + y^2)$ *(6-7)* **93.** $3xy(6x - 5y)(2x + 3y)$ *(6-5, 6-6)*
94. $4u^3(3u - 3v - 5v^2)$ *(6-5, 6-6)* **95.** $2(3a + 2b)(c - 2d)$ *(6-7)* **96.** $(4u - 5v)(3x - 1)$ *(6-7)*
97. $(2x + 1)(4x^2 - 2x + 1)$ *(6-7)* **98.** $-2, 2, -7, 7$ *(6-5, 6-6)* **99.** $4, 2, -2$ *(6-8)*
100. 4 in by 12 in *(6-5, 6-6)*

CHAPTER 6 PRACTICE TEST

1. $3x^2 + 4x - 17$ *(6-1)* **2.** $3x^2 - x - 3$ *(6-1)* **3.** $3x^3 - 11x^2 - 12x + 32$ *(6-2)*
4. $-x^2 - 14x + 28$ *(6-1, 6-2)* **5.** $6x^4 - 37x^3 + 31x^2 + 124x - 160$ *(6-2)* **6.** $2x(3y + 5x)$ *(6-3)*
7. $(2x - 3)(2x + 3)$ *(6-4)* **8.** $(x - 6)^2$ *(6-4)* **9.** $(x - 5)(x + 3)$ *(6-5)* **10.** $(4x - 1)(16x^2 + 4x + 1)$ *(6-7)*
11. Not factorable *(6-4)* **12.** $(2x + 1)(3x - 4)$ *(6-5, 6-6)* **13.** $3(2x + 1)(x - 2)$ *(6-5, 6-6)*
14. $(a + b)(x + y)$ *(6-7)* **15.** Not factorable *(6-5, 6-6)* **16.** $x^2 - 3x + 18$ *(6-9)* **17.** $x = 1, -2$ *(6-8)*
18. $x = -9, 5$ *(6-8)*

CHAPTER 7

EXERCISE 7-1

1. $\dfrac{x}{3}$ **3.** $\dfrac{3}{5y^3}$ **5.** $\dfrac{1}{A}$ **7.** $\dfrac{2y^3}{7x}$ **9.** $\dfrac{1}{x + 3}$ **11.** $4(y - 5)$ **13.** $\dfrac{x}{3(x + 7)^2}$ **15.** $\dfrac{11a}{13(b + 3)}$

17. $\dfrac{x}{2}$ **19.** x **21.** $\dfrac{x^2}{2}$ **23.** $3 - y$ **25.** $\dfrac{2x^2}{x^2 - 3}$ **27.** $\dfrac{1}{n}$ **29.** $\dfrac{-1}{ab}$ **31.** $\dfrac{2x - 1}{3x}$ **33.** $\dfrac{2x + 1}{3x - 7}$

35. $\dfrac{x + 2}{x - 2}$ **37.** $\dfrac{x + 2}{2x}$ **39.** $\dfrac{x - 3}{x + 3}$ **41.** $\dfrac{x - 2}{x - 3}$ **43.** $\dfrac{x + 3}{2x + 1}$ **45.** $\dfrac{x + 1}{x}$ **47.** $\dfrac{y}{y - 3}$

49. $3x^2 - x + 2$ **51.** $\dfrac{2 + m - 3m^2}{m}$ **53.** $\dfrac{2x}{x^2 - 9}$ **55.** $2m^2 - mn + 3n^2$ **57.** $\dfrac{xy^2}{(x + y)^2}$ **59.** $x + 2$

61. $\dfrac{2x - 3y}{2xy}$ **63.** $x - 2y$ **65.** -1 **67.** $-\dfrac{x - 3}{x + 3}$ **69.** $-y$ **71.** $\dfrac{-1}{x + 2}$ **73.** $\dfrac{3 - x}{x}$ **75.** $\dfrac{x + 2}{x - 2}$

77. $\dfrac{x + 2}{x + y}$ **79.** $x - 5$ **81.** $\dfrac{x + 5}{2x}$ **83.** $\dfrac{x^2 + 2x + 4}{x + 2}$ **85.** $\dfrac{2x + 1}{x + 2}$ **87.** Already reduced   **89.** $\dfrac{x + 1}{x - 3}$

91. $\dfrac{2}{b + 1}$

EXERCISE 7-2

1. $\frac{5}{6}$ **3.** 2 **5.** $\dfrac{-9}{20}$ **7.** $\dfrac{-160}{891}$ **9.** $\dfrac{2y^3}{9u^2}$ **11.** $\dfrac{u^2w^2}{25y^2}$ **13.** $\dfrac{x^2}{5yz}$ **15.** $-\dfrac{10c^2}{b^3}$ **17.** $\dfrac{2}{x}$

19. $a + 1$ **21.** $\dfrac{3x(x + 3)}{2(x - 3)}$ **23.** $\dfrac{3(y + 1)}{y(y + 2)}$ **25.** $\dfrac{x - 2}{2x}$ **27.** $8d^6$ **29.** $-\dfrac{2x^2}{y^2}$ **31.** $-\dfrac{x^4}{4y^2z^3}$

33. $-\dfrac{25x^2}{8y^2z^2}$ **35.** $-\dfrac{25x^2}{9y^2z}$ **37.** $\dfrac{1}{y(x + 4)}$ **39.** $\dfrac{1}{2y - 1}$ **41.** $\dfrac{1}{x - 1}$ **43.** $\dfrac{(x - 5)^2}{x + 1}$ **45.** $\dfrac{x}{x + 5}$

47. $-3(x - 2)$ or $6 - 3x$ **49.** $a^2 - b^2$ **51.** $a + 1$ **53.** $\dfrac{m + 2}{m(m - 2)}$ **55.** $\dfrac{x^2(x + y)}{(x - y)^2}$ **57.** $\dfrac{x^5}{3(y + z)^2}$

59. $\dfrac{a + 2}{a - 2}$ **61.** $\dfrac{(x + 1)^2(x + 2)}{(x - 3)^2(x + 3)}$ **63.** All x except $x = 1$ **65.** All x except $x = -4$

67. Obtain

$$\dfrac{x^2 - (x + 2)^2}{(x + 1)(x + 5) - (x + 3)^2},$$

which simplifies to $x + 1$. Since $x = 108{,}641$, $x + 1 = 108{,}642$, the answer.

EXERCISE 7-3

1. $\dfrac{1}{x}$ **3.** $\dfrac{2}{x^2}$ **5.** $\dfrac{x}{y}$ **7.** $\dfrac{-x}{2y}$ **9.** 2 **11.** $\dfrac{1}{x - 1}$ **13.** 1 **15.** 1 **17.** $\dfrac{-1}{x(x + 1)}$

19. $\dfrac{x + 4}{x(x - 2)}$ **21.** $\dfrac{3(z + 1)}{z(z + 3)}$ **23.** $\dfrac{y + 4}{(y + 1)(y + 2)}$ **25.** $\dfrac{3(x - 3)}{(x - 1)(x + 1)}$ **27.** $\dfrac{2x + 1}{x(x + 1)}$ **29.** $\dfrac{1}{x(x - 1)}$

31. $\dfrac{-2x}{(x - 1)(x + 1)}$ **33.** $-\dfrac{1}{x}$ **35.** $-\dfrac{3x + 4}{x + 3}$ **37.** $\dfrac{-1}{y - 1}$ or $\dfrac{1}{1 - y}$ **39.** $\dfrac{3x - 7}{(x - 1)(x - 2)(x - 3)}$

41. $\dfrac{x - 7}{(x + 4)(x - 3)(x - 1)}$ **43.** $-\dfrac{x - 2}{x(x - 1)(x + 2)}$ **45.** $\dfrac{2}{(x + 3)(x + 4)}$ **47.** $\dfrac{1}{(x + 2)(x + 3)}$

49. $\dfrac{-3}{(x + 1)(x + 2)}$ **51.** $\dfrac{-x - 6}{(x - 2)^2(x + 2)}$ **53.** $\dfrac{5}{(x + 3)(x - 2)}$ **55.** $\dfrac{x^2 + 2x + 3}{x^2}$ **57.** $\dfrac{3x^2 - 2x + 1}{x^2}$

59. $-\dfrac{x + 4}{(x - 1)(x + 1)}$ **61.** $\dfrac{6x + 15}{(x + 1)(x + 4)}$ **63.** $\dfrac{6x + 10}{(x + 2)(x + 1)(x + 4)}$ **65.** $\dfrac{5}{x + 1}$ **67.** $\dfrac{4x - 1}{x(x - 1)(x + 1)}$

69. $\dfrac{6}{x - 3}$ **71.** $\dfrac{5}{x - 2}$ **73.** $\dfrac{1}{1 + y}$ **75.** $\dfrac{-3}{y - 1}$ or $\dfrac{3}{1 - y}$ **77.** $\dfrac{1}{xy}$ **79.** $\dfrac{x^2 + y^2}{x + y}$ **81.** $\dfrac{x - 2}{x - 1}$

83. $\dfrac{2(x + 8)(x + 2)}{(x - 2)(x + 3)(x + 4)}$

EXERCISE 7-4

1. $x = -9$ **3.** $y = 2$ **5.** $x = 4$ **7.** $x = 2$ **9.** $x = 1$ **11.** $x = 5$ **13.** $x = 7$ **15.** $x = 3$

17. $x = -4$ **19.** $t = 4$ **21.** $L = -4$ **23.** No solution   **25.** $N = -\frac{6}{5}$ **27.** No solution

29. $y = -3, 3$ **31.** $x = 9$ **33.** $x = 7$ **35.** $x = 0$ **37.** $x = 2$ **39.** $x = 8$ **41.** $n = \frac{53}{11}$ **43.** $x = \frac{3}{5}$

45. $x = 11$ **47.** $x = 5, -3$ **49.** $x = \frac{1}{2}, 2$ **51.** $x = 1$ **53.** $x = 2$ **55.** $x = 2$ **57.** $x = 5$
59. $x = -4$ **61.** $x = -5$ **63.** 336   **65.** 2,254   **67.** 8   **69.** 12

EXERCISE 7-5

1. $x = \dfrac{y + 8}{3}$ **3.** $x = -\dfrac{y - 2}{5}$ **5.** $x = \dfrac{y - 7}{2}$ **7.** $x = -\dfrac{y + 11}{3}$ **9.** $x = \dfrac{18y}{z}$ **11.** $x = \dfrac{-1}{yz}$

13. $x = \dfrac{y - 4z}{3}$ **15.** $x = yz - 1$ **17.** $I = A - P$ **19.** $r = \dfrac{d}{t}$ **21.** $t = \dfrac{I}{Pr}$ **23.** $\pi = \dfrac{C}{D}$

25. $x = -\dfrac{b}{a}$ **27.** $t = \dfrac{s + 5}{2}$ **29.** $y = \dfrac{3x - 12}{4}$ or $y = \frac{3}{4}x - 3$ **31.** $E = IR$ **33.** $R = \dfrac{E}{I}$

35. $l = \dfrac{A - 2b}{2}$ **37.** $b = \dfrac{2A - hB}{h}$ **39.** $D = \dfrac{2A}{d}$ **41.** $B = \dfrac{CL}{100}$ **43.** $L = \dfrac{100B}{C}$ **45.** $m = \dfrac{d^2 F}{GM}$

47. $d = \dfrac{M - P}{Mt}$ **49.** $F = \frac{9}{5}C + 32$ **51.** $c = \dfrac{b}{1 - 3d}$ **53.** $x = \dfrac{y}{z - 2}$ **55.** $b = \dfrac{a}{a + c - 1}$

57. $x = \dfrac{yz}{z - y}$ **59.** $x = \dfrac{2 - y}{y - 1}$ or $\dfrac{y - 2}{1 - y}$ **61.** $x = \dfrac{y + 1}{2y - 3}$ **63.** $b = \dfrac{a + 2}{2a - 1}$ **65.** $y = \dfrac{xz + 3z}{1 - 2x}$

67. $f = \dfrac{ab}{a + b}$ or $f = \dfrac{1}{\dfrac{1}{a} + \dfrac{1}{b}}$ **69.** $M = \dfrac{P}{1 - dt}$

EXERCISE 7-6

1. $\frac{5}{6}$ **3.** $\frac{7}{8}$ **5.** $-\frac{11}{9}$ **7.** $\dfrac{xz}{y}$ **9.** $a^2 c^2$ **11.** $\dfrac{a^2}{b^2}$ **13.** $-\dfrac{1}{w^2}$ **15.** $\dfrac{y(x + y)}{x}$ **17.** $\dfrac{1}{x(x + 1)}$

19. $\dfrac{x + 1}{x}$ **21.** $\dfrac{x + y}{y}$ **23.** $\dfrac{a - 1}{a}$ **25.** $\dfrac{xy}{x + y}$ **27.** $-\dfrac{b}{a}$ **29.** $\dfrac{1}{(x - 3)^2}$ **31.** $\dfrac{x - 1}{x + 1}$

33. $\dfrac{(x - 1)^2}{(x + 4)^2}$ **35.** $\dfrac{(x - 2)(x - 5)}{(x + 3)(x + 2)}$ **37.** $\dfrac{m - 3}{m(m - 2)}$ **39.** $\dfrac{(m + n)^2}{m^3}$ **41.** $x = 3$ **43.** No solution

45. No solution   **47.** 1   **49.** $\dfrac{x^2 + xy + y^2}{xy}$ **51.** $\dfrac{(x + y)^2}{(x - y)(x^2 - xy + y^2)}$ **53.** $-\frac{1}{2}$ **55.** $\dfrac{2x + 1}{x + 1}$

57. $\dfrac{x + 1}{x}$ **59.** No solution   **61.** $-\frac{1}{2}$ **63.** $t = \dfrac{rs}{r + s}$ **65.** $ERA = \dfrac{9R}{I}$

CHAPTER 7 REVIEW EXERCISE

1. $\dfrac{2x^2}{7}$ *(7-1)*   **2.** $\dfrac{x - 3}{14}$ *(7-1)*   **3.** $\frac{2}{5}$ *(7-1)*   **4.** q *(7-1)*   **5.** $\dfrac{x + 3}{x + 4}$ *(7-1)*   **6.** $x - 1$ *(7-1)*

7. $\dfrac{3x + 2}{3x}$ *(7-3)*   **8.** $\dfrac{2x + 11}{6x}$ *(7-3)*   **9.** $\dfrac{4x^2 y^2}{9(x - 3)}$ *(7-2)*   **10.** $\dfrac{(d - 2)^2}{d + 2}$ *(7-2)*   **11.** $\dfrac{x + 1}{2x(3x - 1)}$ *(7-3)*

12. $\dfrac{4}{x - 4}$ *(7-3)*   **13.** $\frac{3}{10}$ *(7-6)*   **14.** $\frac{28}{9}$ *(7-2)*   **15.** $\dfrac{x^2 - x + 1}{x^3}$ *(7-3)*   **16.** $\dfrac{2x - 1}{x(x - 1)}$ *(7-3)*

17. $\dfrac{x^2}{(x - 1)^2}$ *(7-2)*   **18.** $\dfrac{3x}{x + 1}$ *(7-2)*   **19.** $\dfrac{x}{2}$ *(7-6)*   **20.** $\dfrac{1}{x}$ *(7-6)*   **21.** $\dfrac{x - 1}{x}$ *(7-2)*

22. $\dfrac{-4}{x^2}$ *(7-3)*   **23.** $x = 9$ *(7-4)*   **24.** $x = 2$ *(7-4)*   **25.** $x = -3$ *(7-4)*   **26.** $m = 5$ *(7-4)*

27. No solution *(7-4)*   **28.** $b = \dfrac{2A}{h}$ *(7-5)*   **29.** $x = \dfrac{12y - 9}{4}$ *(7-5)*   **30.** $b = \dfrac{h}{A}$ *(7-5)*   **31.** $\dfrac{x + 7}{4}$ *(7-1)*

32. $\dfrac{x + 2}{x - 1}$ *(7-1)*   **33.** $\dfrac{x + 3}{x - 1}$ *(7-1)*   **34.** $\dfrac{a - b}{a + b}$ *(7-1)*   **35.** $-\dfrac{1}{z + 3}$ *(7-1)*   **36.** $-qr$ *(7-1)*

37. $\dfrac{y + 2}{y(y - 2)}$ *(7-2)*   **38.** 2 *(7-2)*   **39.** $\dfrac{-1}{(x + 2)(x + 3)}$ *(7-3)*   **40.** $\dfrac{5x - 3}{6x^2(x - 1)}$ *(7-3)*

41. $\dfrac{4m}{(m - 2)(m + 2)}$ *(7-3)*   **42.** $\dfrac{-2y}{(x - y)^2(x + y)}$ *(7-3)*   **43.** x *(7-6)*   **44.** $\dfrac{x - y}{x}$ *(7-6)*

45. $-\dfrac{8x + 15}{x(x + 5)}$ *(7-3)*   **46.** $\dfrac{x^2 + 4x - 4}{2(x - 2)(x + 2)}$ *(7-3)*   **47.** $\dfrac{a + 4}{a + 2}$ *(7-2)*   **48.** $\dfrac{b + 1}{b + 2}$ *(7-2)*

49. $\dfrac{x^3 - x^2 + x}{x^2 + 1}$ *(7-3)*   **50.** $\dfrac{3}{8x^2}$ *(7-2)*   **51.** $\dfrac{9x^3(x + 2)}{(x - 3)(x + 3)}$ *(7-6)*   **52.** $\dfrac{x - 1}{x(x + 1)^3}$ *(7-6)*

53. No solution *(7-4)*   **54.** $x = 2$ *(7-4)*   **55.** $x = 2$ *(7-4)*   **56.** $x = -\frac{1}{2}$ *(7-4)*   **57.** $x = -2$ *(7-4)*

58. $x = -5$ *(7-4)*   **59.** $L = \dfrac{2S - na}{n}$ *(7-5)*   **60.** $A = \dfrac{M}{x + 1}$ *(7-5)*   **61.** $a = \dfrac{S - 2bc}{2b + 2c}$ *(7-5)*

62. 7 *(7-4)*   **63.** $\dfrac{1}{x} + \dfrac{1}{2} = 2;\ \dfrac{2}{3}$ *(7-4)*   **64.** $\dfrac{x + 5}{5}$ *(7-1)*   **65.** $\dfrac{a^2 + 2a + 4}{a - 5}$ *(7-1)*   **66.** $\dfrac{p + 2}{p - 2}$ *(7-1)*

67. -1 *(7-2)*   **68.** $\dfrac{y + 4}{2x - y}$ *(7-3)*   **69.** $\dfrac{6 - 3x}{2x}$ *(7-6)*   **70.** $\dfrac{x - 2}{x - 1}$ *(7-6)*   **71.** 1 *(7-6)*

72. No solution *(7-4)*   **73.** $x = \dfrac{3y + 1}{2y - 3}$ *(7-5)*   **74.** $a = \dfrac{b}{b - 1}$ *(7-5)*   **75.** 10 *(7-4)*

CHAPTER 7 PRACTICE TEST

1. $\dfrac{3x^2}{5}$ *(7-1)*   **2.** $\frac{2}{3}$ *(7-1)*   **3.** $x + 2$ *(7-1)*   **4.** $\dfrac{1}{x - 1}$ *(7-1)*   **5.** $\dfrac{x + 1}{x - 1}$ *(7-1)*   **6.** $\dfrac{2(x - y)}{x + y}$ *(7-1)*

7. $\dfrac{x + 1}{x^2}$ *(7-3)*   **8.** $\dfrac{3x}{(x + 1)^2}$ *(7-3)*   **9.** $\dfrac{-x^3 + x - 1}{x(x + 1)}$ *(7-3)*   **10.** $\dfrac{x^2}{x - 1}$ *(7-2)*   **11.** $x^2 - 1$ *(7-2)*

12. $\dfrac{4y^2}{3x}$ *(7-6)*   **13.** $x^2 + x$ *(7-6)*   **14.** $x = 4$ *(7-4)*   **15.** $x = 29$ *(7-4)*   **16.** Any x except $x = 0$ *(7-4)*

17. No solution *(7-4)*   **18.** $x = \dfrac{2}{y - 2}$ *(7-5)*   **19.** 15 *(7-4)*   **20.** 238 *(7-4)*

CUMULATIVE REVIEW EXERCISES—CHAPTERS 1–7

1. 1,260 *(1-1)*   **2.** $3x(x^2 - 4)$ *(7-1)*   **3.** -1 *(1-5, 2-2, 2-5)*   **4.** $-\frac{5}{8}$ *(3-4)*   **5.** $5x^2 + x + 5$ *(6-1)*

6. $6x^3 - 6x^2 - 2x + 8$ *(6-1)*   **7.** $2x^3 + 13x^2 + 17x - 12$ *(6-2)*   **8.** $x^3 - 3x^2 + 3x - 1$ *(6-9)*

9. $8x^2 - 2xy - 15y^2$ *(6-2)*   **10.** $\dfrac{7x + 8}{12}$ *(7-3)*   **11.** $\dfrac{5x^2 - 2x + 8}{10}$ *(7-3)*   **12.** $4x^3$ *(7-2)*

13. $\dfrac{1}{4ac}$ *(7-2)*   **14.** $\dfrac{4}{x}$ *(7-6)*   **15.** Slope 3, y intercept -5 *(4-3)*   **16.** $y = -2x + 9$ *(4-3)*

17. 120 *(3-6)*   **18.** $2 \cdot 2 \cdot 2 \cdot 5 \cdot 7$ *(1-1)*   **19.** $8x^2y + 3xy - 8xy^2$ *(1-7)*   **20.** $4ab(a - 2 + 3b)$ *(6-3)*

21. $(x + 3)^2$ *(6-4)*   **22.** $\dfrac{9y^2}{15xy^2}$ *(7-1)*   **23.** $\dfrac{4x}{x - 2}$ *(7-1)*   **24.** $2(x - 3)(x + 3)$ *(6-4)*

25. $5(x^2 + 10)$ *(6-4)*   **26.** $(x - 4)(x - 5)$ *(6-5, 6-6)*   **27.** $(2x - y)(x + 5y)$ *(6-3)*   **28.** $\dfrac{x + 1}{x - 1}$ *(7-1)*

29. $x = \frac{9}{13}$ *(2-7)*   **30.** $x = \frac{31}{4}$ *(3-5)*   **31.** $x = -\frac{43}{21}$ *(7-4)*   **32.** $x = 1, 2$ *(6-8)*   **33.** $x = 2$ *(7-4)*

34. $x = 7, y = 3$ *(4-5, 4-6)*   **35.** $x = 7, y = -4$ *(4-6)*   **36.** $1 \le x < 9$ *(5-1)*   **37.** $x = \dfrac{y - 10}{3}$ *(7-5)*

38. $b = \dfrac{4}{2a + c}$ *(7-5)*   **39.**   *(5-3)*   **40.** $(9, 5)$ *(4-6)*

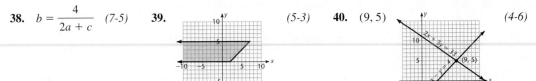

41.   *(4-2)*   **42.**   *(4-3)*   **43.**   *(5-3)*

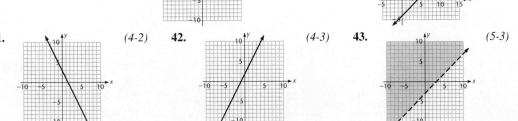

44. $\dfrac{x}{15{,}204 - x} = \dfrac{5}{7}$; 6,335 *(3-7)*   **45.** $\frac{12}{100} = 0.12$ *(3-7)*   **46.** \$22,600 *(3-6)*   **47.** 600 ml *(4-8)*

48. 30 min *(4-7)*   **49.** 80 boxes of oranges, 100 boxes of grapefruit *(4-8)*   **50.** $x^3 + 2x - 3$ *(6-9)*

51. $\dfrac{6x}{x^2 - 1}$ *(7-3)*   **52.** -191 *(2-2)*   **53.** $x^2(x + 3)$ *(7-3)*   **54.** $y = -\frac{4}{3}x - \frac{7}{3}$ or $4x + 3y = -7$ *(4-3)*

55. $x + 2$ *(7-2)*   **56.** 1 *(7-3)*   **57.** $\dfrac{x + 1}{x}$ *(7-6)*   **58.** $\dfrac{1}{x} + 6$ *(1-2)*   **59.** $\dfrac{1}{x + 6}$ *(1-2)*

60. $x + 80 > 8x$ *(5-1)*   **61.** $x - 3 \geq 3x$ *(5-1)*   **62.** $\dfrac{1}{x} + \dfrac{1}{x + 1} + \dfrac{1}{x + 2}$ *(1-2)*   **63.** $5x + \dfrac{4}{x} = \dfrac{11}{16x}$ *(1-2)*

64. $(2x - 3)(x + 5)$ *(6-5, 6-6)*   **65.** $(4x + 1)(x - 4)$ *(6-5, 6-6)*   **66.** $(x - 4)(x^2 + 4x + 16)$ *(6-7)*

67. $(x^2 + 1)(x - 1)(x + 1)$ *(6-7)*   **68.** $3(x + 2)(x^2 - 2x + 4)$ *(6-7)*   **69.** $-a^2 + 2a - 1$ *(2-6)*

70. $y = -\frac{11}{7}x + \frac{30}{7}$ *(4-3)*   **71.** -1 *(7-6)*   **72.** $(y - w)(3x - 4z)$ *(6-3)*   **73.** $x = \frac{9}{7}, y = -\frac{3}{7}$ *(4-6)*

74. $x \leq 5$ *(5-2)*   **75.** $x = 5$ *(7-4)*   **76.** $x = -1$ *(7-4)*   **77.** $x = \dfrac{y}{y - 1}$ *(7-5)*   **78.** $x = \dfrac{y}{z + 1}$ *(7-5)*

79. 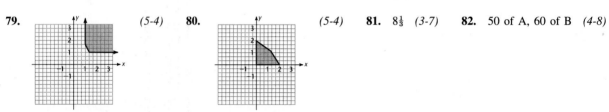 *(5-4)*   **80.** *(5-4)*   **81.** $8\frac{1}{3}$ *(3-7)*   **82.** 50 of A, 60 of B *(4-8)*

83. 2 hours after first worker begins *(4-7)*   **84.** \$2.85 *(3-6)*   **85.** \$45 *(4-5)*   **86.** $(3x - 2)(x + 4)$ *(6-5, 6-6)*

87. $(2x - 3)(x + 4)$ *(6-5, 6-6)*   **88.** $x^2 + 2x + 3$ *(6-5, 6-6)*   **89.** $(x - 1)(x + 1)(x^2 + x + 1)(x^2 - x + 1)$ *(6-7)*

90. $(x + 2y)(x + y)(x - y)$ *(6-7)*   **91.** $-\dfrac{1}{x}$ *(7-5)*   **92.** $-3, 1$ *(6-8)*   **93.** $-3, \frac{4}{3}$ *(6-8)*

94. All numbers except $x = 1, x = 2$ *(7-4)*   **95.** No solution *(7-4)*   **96.** $x = \dfrac{y}{2}$ *(7-5)*

97. No solution *(7-4)*   **98.** \$8,400 at 9%, 3,600 at 11% *(4-8)*   **99.** $m = -2, b = 240$ *(4-3)*

100. 1.25 liters *(4-8)*

CHAPTER 8

EXERCISE 8-1

1. x^{13} **3.** y^{15} **5.** x^6 **7.** x^{16} **9.** x^{15} **11.** y^{15} **13.** $x^8 y^8$ **15.** $(xy)^7$ **17.** $(xy)^4$ **19.** $x^5 y^5$

21. $\left(\dfrac{x}{y}\right)^4$ **23.** $\dfrac{x^2}{y^2}$ **25.** x^6 **27.** $\dfrac{1}{x^8}$ **29.** $\dfrac{1}{y^4}$ **31.** y^6 **33.** 10^{15} **35.** 10^{12} **37.** 6×10^{11}

39. 2×10^3 **41.** 1.5×10^6 **43.** -1 **45.** 1   **47.** -8 **49.** 9   **51.** $x^{20} y^8$ **53.** $27a^6 b^3$

55. $6x^7$ **57.** $\dfrac{1}{2a^4}$ **59.** $\dfrac{3a^3}{b}$ **61.** $a^8 b^{11}$ **63.** $5x^4 y^{12}$ **65.** $625 x^4 y^{12}$ **67.** $x^2 y^4$ **69.** $-x^2 y^4$

71. -16 **73.** -27 **75.** $-x^4$ **77.** $-x^5$ **79.** $16x^4 y^{12} z^{16}$ **81.** $6a^{15}$ **83.** $324 a^{20}$ **85.** $\dfrac{a^4 b^7}{2}$

87. $-a^6 b^9$ **89.** $-\dfrac{1}{a^2 b^5}$ **91.** a^5

EXERCISE 8-2

1. 1   **3.** 1   **5.** $\frac{1}{8}$ **7.** 27   **9.** $\frac{1}{4}$ **11.** $-\frac{1}{27}$ **13.** $\dfrac{1}{a^5}$ **15.** b^7 **17.** $\dfrac{1}{3^7}$ **19.** $\dfrac{1}{4^3}$ **21.** x^{11}

23. $\dfrac{1}{y^9}$ **25.** $\frac{1}{9}$ **27.** 3^4 **29.** 10^8 **31.** $\dfrac{1}{10^6}$ **33.** 10^4 **35.** $\dfrac{1}{x^6}$ **37.** y^8 **39.** $\dfrac{y^4}{x^2}$ **41.** $\dfrac{x^6}{y^3}$

43. $a^2 b$ **45.** $x^4 y^7$ **47.** 10^2 **49.** a^6 **51.** $\dfrac{1}{x^2}$ **53.** $\dfrac{b^2}{9a^4}$ **55.** $\dfrac{3y^2}{x^3}$ **57.** $3ab$ **59.** $\frac{12}{7}$

61. $\frac{36}{25}$ **63.** $\frac{1,000}{11}$ **65.** $\frac{3y^2}{2x^3}$ **67.** $\frac{4x^6y^4}{9}$ **69.** $\frac{b^4}{2a^4}$ **71.** $\frac{1}{10a^3}$ **73.** $\frac{ab}{a+b}$ **75.** $\frac{x^4y^4}{(y^2-x^2)^2}$

77. $\frac{p^2+q^2}{p^2q^2}$ **79.** $\frac{b-a}{ab}$ **81.** $\frac{(x+y)^2}{(xy)^2}$

EXERCISE 8-3

1. 2×10^3 **3.** 1.08×10^5 **5.** 4.12×10^2 **7.** 4.8×10^1 **9.** 6.3×10^{-1} **11.** 4.6×10^{-2}
13. 4.4×10^{-4} **15.** 9.5×10^{-5} **17.** 3×10^{-6} **19.** 1.8×10^{-6} **21.** 1×10^0 **23.** 6.35×10^9
25. 8.45×10^{-10} **27.** 4.805×10^{-7} **29.** $30,000$ **31.** 0.004 **33.** 0.08 **35.** $3,400,000$
37. $480,000,000$ **39.** $0.000\,24$ **41.** $0.000\,003\,2$ **43.** $52,500$ **45.** $0.004\,15$ **47.** 1.5×10^2
49. 2.4×10^{-1} **51.** 1.6×10^0 **53.** 1.1×10^4 **55.** 3.664×10^9 **57.** 9.814×10^{-3}
59. 1.61988×10^{11} **61.** 7.5276×10^{10} **63.** 2.2477075×10^{-1} **65.** 1×10^{-6} **67.** 5.6×10^{-9}
69. 1×10^{-5} **71.** 1.5×10^{15} **73.** 9.108×10^{-31} **75.** 2×10^{13} or $20,000,000,000,000$
77. 5×10^{-7} or $0.000\,000\,5$ **79.** 2×10^{-5} or $0.000\,02$ **81.** 1×10^{30} or $1,000,000,000,000,000,000,000,000,000,000$
83. 6.25×10^{22} or $62,500,000,000,000,000,000,000$ **85.** **(A)** 9×10^{-6} **(B)** $0.000\,009$ **(C)** 0.0009%
87. **(A)** 5×10^{-8} **(B)** $0.000\,000\,05$ **(C)** $0.000\,005\%$

EXERCISE 8-4

1. 4 **3.** -9 **5.** x **7.** $3m$ **9.** 9 **11.** 36 **13.** $1/2$ **15.** 10 **17.** $6\sqrt{2}$ **19.** $3\sqrt{7}$
21. $\frac{\sqrt{35}}{7}$ **23.** $\sqrt{3}$ **25.** a^2 **27.** x^2 **29.** $2x\sqrt{3}$ **31.** $\sqrt{2}$ **33.** $2\sqrt{2}$ **35.** $x\sqrt{x}$
37. $3y\sqrt{2y}$ **39.** $\frac{1}{2}$ **41.** $-\frac{2}{3}$ **43.** $\frac{1}{x}$ **45.** $\frac{\sqrt{3}}{3}$ **47.** $\frac{\sqrt{3}}{3}$ **49.** $\frac{\sqrt{x}}{x}$ **51.** $\frac{\sqrt{x}}{x}$ **53.** $5xy^2$
55. $2x^2y\sqrt{xy}$ **57.** $2x^3y^3\sqrt{2x}$ **59.** $\frac{\sqrt{3y}}{3y}$ **61.** $2x\sqrt{2y}$ **63.** $\frac{2}{3}x\sqrt{3xy}$ **65.** $\frac{\sqrt{6}}{3}$ **67.** $\frac{\sqrt{6mn}}{2n}$
69. $\frac{2a\sqrt{3ab}}{3b}$ **71.** $3\sqrt{2} \approx 4.24$ **73.** $\frac{\sqrt{5}}{5} \approx 0.45$ **75.** $\frac{\sqrt{66}}{2} \approx 4.06$ **77.** $\frac{\sqrt{2}}{2}$
79. In simplest radical form   **81.** $x\sqrt{x^2-2}$ **83.** $x+1$ **85.** Yes, yes
87. If $a^2 = b^2$, then a does not necessarily equal b. For example, let $a = 2$ and $b = -2$.   **89.** 62.85 miles
91. $1,518$ items

EXERCISE 8-5

1. $10\sqrt{3}$ **3.** $3\sqrt{x}$ **5.** $4\sqrt{7} - 3\sqrt{5}$ **7.** $4\sqrt{x}$ **9.** $9\sqrt{a}$ **11.** $-5\sqrt{x}$ **13.** $4\sqrt{5}$ **15.** $-\sqrt{a}$
17. $6\sqrt{2} - \sqrt{3}$ **19.** $2\sqrt{x} + 2\sqrt{y}$ **21.** $4\sqrt{7} + 5\sqrt{5}$ **23.** $5\sqrt{a} - \sqrt{b}$ **25.** $\sqrt{3}$ **27.** $4\sqrt{2}$
29. $40\sqrt{5}$ **31.** $2\sqrt{2} + 6\sqrt{3}$ **33.** $-\sqrt{x}$ **35.** $5\sqrt{2a}$ **37.** $-\sqrt{3x}$ **39.** $2\sqrt{6} + \sqrt{3}$ **41.** $6\sqrt{2x}$
43. $(1+x)\sqrt{x}$ **45.** $(4-2a)\sqrt{a}$ **47.** $5x^3\sqrt{5x}$ **49.** $3x\sqrt{3xy}$ **51.** $xy^2z^2\sqrt{xz}$ **53.** $4bc\sqrt{2ac}$
55. $-\sqrt{6}/6$ **57.** $5\sqrt{2xy}/2$ **59.** $2\sqrt{3} - \frac{\sqrt{2}}{2}$ **61.** $3\sqrt{2}$ **63.** $\frac{3\sqrt{x}}{4}$ **65.** $\frac{3}{25}\sqrt{2x}$

EXERCISE 8-6

1. $4\sqrt{5} + 8$ **3.** $10 - 2\sqrt{2}$ **5.** $2 + 3\sqrt{2}$ **7.** $5 - 4\sqrt{5}$ **9.** $3\sqrt{5} - 5$ **11.** $a + \sqrt{a}$ **13.** $2\sqrt{y} - y$
15. $2\sqrt{3} - \sqrt{6}$ **17.** $2\sqrt{3} - 2$ **19.** $-3\sqrt{3} - 1$ **21.** $2\sqrt{6} - 3$ **23.** $x + \sqrt{x} - 12$ **25.** $x + 5\sqrt{x} + 6$
27. $24 + 2\sqrt{x} - x$ **29.** $7 - 4\sqrt{3}$ **31.** $21 + 8\sqrt{5}$ **33.** $x - 2\sqrt{x} + 1$ **35.** $x + 4\sqrt{x} + 4$
37. $16 + 8\sqrt{x} + x$ **39.** $1 - 2\sqrt{x} + x$ **41.** -1 **43.** $x - 16$ **45.** $36 - x$
47. $15\sqrt{15} + 25\sqrt{3} + 9\sqrt{5} + 15$ **49.** $20x + 9\sqrt{x} - 18$ **51.** $2 - 11\sqrt{2}$ **53.** $6x - 13\sqrt{x} + 6$
55. $(3 - \sqrt{2})^2 - 6(3 - \sqrt{2}) + 7$ **57.** $(1 + \sqrt{3})^2 - 2(1 + \sqrt{3}) - 2$
$\quad = 9 - 6\sqrt{2} + 2 - 18 + 6\sqrt{2} + 7 = 0$ $\quad = 1 + 2\sqrt{3} + 3 - 2 - 2\sqrt{3} - 2 = 0$
59. $(4 - \sqrt{5})^2 - 8(4 - \sqrt{5}) + 11$ **61.** $(2 + \sqrt{6})^2 - 4(2 + \sqrt{6}) - 2$
$\quad = 16 - 8\sqrt{5} + 5 - 32 + 8\sqrt{5} + 11 = 0$ $\quad = 4 + 4\sqrt{6} + 6 - 8 - 4\sqrt{6} - 2 = 0$
63. $\frac{2 + \sqrt{2}}{3}$ **65.** $\frac{-1 - 2\sqrt{5}}{3}$ **67.** $2 - \sqrt{2}$ **69.** $1 - \sqrt{3}$ **71.** $\frac{\sqrt{11} - 3}{2}$ **73.** $\frac{\sqrt{5} - 1}{2}$
75. $\frac{y - 3\sqrt{y}}{y - 9}$ **77.** $\frac{7 + 4\sqrt{3}}{-1}$ or $-7 - 4\sqrt{3}$ **79.** $\frac{x + 5\sqrt{x} + 6}{x - 9}$ **81.** $4 + \sqrt{15}$ **83.** $3 + \sqrt{2} + \sqrt{3} + \sqrt{6}$

EXERCISE 8-7

1. 5 **3.** 5 **5.** No solution   **7.** 3 **9.** 1 **11.** -2 **13.** No solution   **15.** 7 **17.** 5
19. $0, 1$ **21.** $0, 9$ **23.** $0, 27$ **25.** $1, 25$ **27.** $1, 4$ **29.** 4 **31.** $1, 16$ **33.** $9, 25$ **35.** $4, 16$

37. 4 **39.** 9 **41.** 1 **43.** 9 **45.** 1 **47.** 30, 14 **49.** 1 **51.** 32 **53.** 5 **55.** 5 **57.** 7
59. 4 **61.** 12 **63.** No solution **65.** No solution **67.**

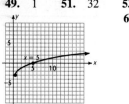

69.

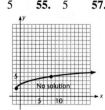

71.

CHAPTER 8 REVIEW EXERCISE

1. 81 *(8-1)* **2.** 1/9 *(8-2)* **3.** 1 *(8-2)* **4.** 9 *(8-2)* **5.** -5 *(8-4)* **6.** 0.000 03 *(8-3)*

7. 23,400,000 *(8-3)* **8.** 1/27 *(8-2)* **9.** 9 *(8-2)* **10.** 1/100 or 0.01 *(8-2)* **11.** 1 *(8-2)*

12. 6/100 or 0.06 *(8-3)* **13.** 1/400,000,000 or 0.000 000 002 5 *(8-3)* **14.** 1.5×10^{10} *(8-3)* **15.** $\dfrac{4x^4}{9y^6}$ *(8-1)*

16. $\dfrac{y^3}{x^2}$ *(8-2)* **17.** -5 *(8-4)* **18.** $2xy^2$ *(8-4)* **19.** $\dfrac{5\,'}{y}$ *(8-4)* **20.** $-3\sqrt{x}$ *(8-5)* **21.** $5 + 2\sqrt{5}$ *(8-6)*

22. x^{11} *(8-1)* **23.** x^{28} *(8-1)* **24.** $\dfrac{1}{x^3}$ *(8-1)* **25.** $x^8 y^{14}$ *(8-1)* **26.** $\dfrac{x^8}{y^{14}}$ *(8-1)* **27.** $\dfrac{1}{x^3}$ *(8-2)*

28. $\dfrac{1}{x^{28}}$ *(8-2)* **29.** x^{11} *(8-2)* **30.** $\dfrac{1}{x^8 y^{14}}$ *(8-2)* **31.** $\dfrac{x^8}{y^{14}}$ *(8-2)* **32.** y^7 *(8-2)* **33.** x^4 *(8-2)*

34. $\dfrac{\sqrt{3}}{3}$ *(8-6)* **35.** $\dfrac{2\sqrt{x}}{x}$ *(8-6)* **36.** $6x$ *(8-6)* **37.** 16 *(8-4)* **38.** No solution *(8-7)* **39.** 0 *(8-7)*

40. 0 *(8-7)* **41.** 7 *(8-7)* **42.** 12 *(8-7)* **43.** 6.543×10^{13} *(8-3)* **44.** 7.89×10^{-11} *(8-3)*

45. $\dfrac{4x^4}{y^6}$ *(8-2)* **46.** $\dfrac{m^2}{2n^5}$ *(8-2)* **47.** $6x^2 y^3 \sqrt{y}$ *(8-4)* **48.** $\dfrac{\sqrt{2y}}{2y}$ *(8-4)* **49.** $\dfrac{\sqrt{6xy}}{2y}$ *(8-4)*

50. $\dfrac{5\sqrt{6}}{6}$ *(8-5)* **51.** $1 + \sqrt{3}$ *(8-6)* **52.** $\dfrac{9a^2}{b^4}$ *(8-2)* **53.** 1 *(8-2)* **54.** $-\dfrac{y^3}{x^6}$ *(8-2)*

55. $3a^2 b^4$ *(8-2)* **56.** $7\sqrt{3}$ *(8-6)* **57.** $(3 + 2x)\sqrt{2x}$ *(8-5)* **58.** 2×10^{-3} *(8-3)* **59.** 1.5×10^{-9} *(8-3)*

60. 2.5×10^9 *(8-3)* **61.** 1 *(8-7)* **62.** 4, 0 *(8-7)* **63.** 1, 0 *(8-7)* **64.** 5 *(8-7)*

65. $(-1 + \sqrt{3})^2 + 2(-1 + \sqrt{3}) - 2 = 1 - 2\sqrt{3} + 3 - 2 + 2\sqrt{3} - 2 = 0$ *(8-6)* **66.** $\dfrac{n^{10}}{9m^{10}}$ *(8-2)*

67. $\dfrac{xy}{x + y}$ *(8-2)* **68.** $\dfrac{n^2 \sqrt{6m}}{3}$ *(8-4)* **69.** $3 + \sqrt{6}$ *(8-6)* **70.** $\dfrac{x - 4\sqrt{x} + 4}{x - 4}$ *(8-5)*

71. $2x\sqrt{x^2 + 4}$ *(8-4)* **72.** x^4 *(8-2)* **73.** $\dfrac{3b}{2ac}$ *(8-2)* **74.** $\dfrac{a^2 c^2 \sqrt{6}}{3b^2}$ *(8-6)* **75.** 9 *(8-7)*

76. 1/12 *(8-7)* **77.** 8 *(8-7)* **78.** 4 *(8-7)* **79.** 1.036 cc *(8-3)* **80.** $a^2 = b$ *(8-4)*

CHAPTER 8 PRACTICE TEST

1. 4/25 *(8-2)* **2.** 0.000 003 45 *(8-3)* **3.** 3,450,000 *(8-3)* **4.** -10 *(8-4)* **5.** x^{10} *(8-1)*

6. $\dfrac{1}{x^5 y^2}$ *(8-1)* **7.** $\dfrac{c^4}{a^2 b^3}$ *(8-2)* **8.** $\dfrac{xy}{x + y}$ *(8-2)* **9.** $5xy\sqrt{y}$ *(8-4)* **10.** $5\sqrt{5x}$ *(8-6)* **11.** $\dfrac{1}{y^4}$ *(8-2)*

12. $\dfrac{1}{x^3}$ *(8-2)* **13.** $\dfrac{\sqrt{5x}}{5x}$ *(8-6)* **14.** $x - 6\sqrt{x} + 5$ *(8-6)* **15.** $\dfrac{x - 6\sqrt{x} + 5}{x - 1}$ *(8-6)*

16. 7.654×10^{15} *(8-3)* **17.** 4.567×10^{-11} *(8-3)* **18.** 26 *(8-7)* **19.** 9 *(8-7)* **20.** 9 *(8-7)*

CHAPTER 9

EXERCISE 9-1

1. 3, 5   **3.** $-7, 4$ **5.** $-6, 7$ **7.** $-7, 3$ **9.** $-7, -4$ **11.** $-18, 2$ **13.** 12, 3   **15.** $-12, 2$
17. -6 **19.** 12, 2   **21.** $-9, 4$ **23.** ± 4 **25.** ± 8 **27.** $\pm\sqrt{3}$ **29.** $\pm\sqrt{5}$ **31.** $\pm\sqrt{18}$ or $\pm 3\sqrt{2}$

33. $\pm\sqrt{12}$ or $\pm 2\sqrt{3}$ **35.** $\pm\frac{2}{3}$ **37.** $\pm\frac{2}{3}$ **39.** $\pm\frac{2}{3}$ **41.** $\pm\frac{2}{5}$ **43.** $\pm\sqrt{\frac{3}{4}}$ or $\pm\dfrac{\sqrt{3}}{2}$

45. $\pm\sqrt{\frac{5}{2}}$ or $\pm\dfrac{\sqrt{10}}{2}$ **47.** $\pm\sqrt{\frac{1}{3}}$ or $\pm\dfrac{\sqrt{3}}{3}$ **49.** $-1, 5$ **51.** $3, -7$ **53.** $2\pm\sqrt{3}$ **55.** $-1, 2$

57. No real solution   **59.** $\frac{13}{6}, \frac{5}{6}$ **61.** $-\frac{1}{2}, 3$ **63.** $-2, \frac{1}{3}$ **65.** $-2, \frac{1}{4}$ **67.** $1, \frac{1}{2}$ **69.** $1, \frac{3}{2}$
71. $-1, -\frac{1}{2}$ **73.** $-3, \frac{1}{2}$ **75.** $-1, \frac{1}{2}$ **77.** $2, -\frac{1}{3}$ **79.** $-1, \frac{4}{3}$ **81.** $\pm\sqrt{c^2 - a^2}$

83. $\pm\sqrt{(h+R)^2 - R^2}$ or $\pm\sqrt{h^2 + 2hR}$ **85.** $\pm\sqrt{\dfrac{2DC}{H}}$ or $\dfrac{\sqrt{2DCH}}{H}$ **87.** 400 square meters

89. 110 yards by 330 yards   **91.** 30 mph

EXERCISE 9-2

1. $x^2 - 8x + 16 = (x-4)^2$ **3.** $x^2 + 10x + 25 = (x+5)^2$ **5.** $x^2 - 10x + 25 = (x-5)^2$
7. $x^2 - 6x + 9 = (x-3)^2$ **9.** $x^2 - 4x + 4 = (x-2)^2$ **11.** $4 \pm \sqrt{10}$ **13.** $-5 \pm \sqrt{26}$ **15.** $5 \pm \sqrt{10}$
17. $3 \pm \sqrt{3}$ **19.** $2 \pm \sqrt{2}$ **21.** $x^2 + 7x + \frac{49}{4} = (x+\frac{7}{2})^2$ **23.** $x^2 + 3x + \frac{9}{4} = (x+\frac{3}{2})^2$
25. $x^2 - 5x + \frac{25}{4} = (x-\frac{5}{2})^2$ **27.** $x^2 + 9x + \frac{81}{4} = (x+\frac{9}{2})^2$ **29.** $x^2 - 11x + \frac{121}{4} = (x-\frac{11}{2})^2$ **31.** $\dfrac{-7 \pm \sqrt{69}}{2}$
33. $\dfrac{-3 \pm \sqrt{5}}{2}$ **35.** $\dfrac{5 \pm \sqrt{37}}{2}$ **37.** $\dfrac{9 \pm \sqrt{33}}{2}$ **39.** $\dfrac{11 \pm \sqrt{77}}{2}$ **41.** $\dfrac{-2 \pm \sqrt{14}}{2}$ **43.** $\dfrac{-5 \pm \sqrt{57}}{4}$
45. $-6 \pm \sqrt{6}$ **47.** $\dfrac{3 \pm \sqrt{6}}{3}$ **49.** $1, -3$ **51.** $1, \frac{3}{2}$ **53.** $\frac{3}{2}, \frac{1}{2}$ **55.** $\frac{1}{2}, -\frac{1}{3}$ **57.** $\dfrac{-1 \pm \sqrt{6}}{2}$
59. $-1, \frac{1}{2}$ **61.** $\frac{3}{2}, -\frac{1}{2}$ **63.** $\frac{1}{2}, \frac{2}{3}$

EXERCISE 9-3

1. $a = 1, b = 4, c = 2$ **3.** $a = 1, b = -3, c = -2$ **5.** $a = 3, b = -2, c = 1$ **7.** $a = 2, b = 3, c = -1$
9. $a = 2, b = -5, c = 0$ **11.** $-2 \pm \sqrt{2}$ **13.** $\dfrac{3 \pm \sqrt{17}}{2}$ **15.** No solution   **17.** $\dfrac{-3 \pm \sqrt{17}}{4}$ **19.** $\frac{5}{2}, 0$

21. $3 \pm 2\sqrt{3}$ **23.** $\dfrac{-3 \pm \sqrt{13}}{2}$ **25.** $\dfrac{3 \pm \sqrt{3}}{2}$ **27.** $\dfrac{-1 \pm \sqrt{13}}{6}$ **29.** No real solution   **31.** $\dfrac{3 \pm \sqrt{33}}{4}$

33. $\dfrac{4 \pm \sqrt{11}}{5}$ **35.** No real roots   **37.** One, $x = -\frac{1}{6}$ **39.** Two, $x = \dfrac{7 \pm \sqrt{13}}{6}$ **41.** No real roots

43. One, $x = \frac{3}{4}$ **45.** Two, $\dfrac{-5 \pm \sqrt{105}}{2}$ **47.** Two, $4 \pm \sqrt{6}$ **49.** Two, $\dfrac{-15 \pm \sqrt{21}}{6}$ **51.** Two, $1, \frac{7}{5}$

53. No real roots   **55.** One, $\frac{5}{2}$ **57.** $-2, 3$ **59.** $0, -7$ **61.** -10 **63.** No real solution   **65.** $7, -2$

67. ± 4 **69.** $-1 \pm \sqrt{3}$ **71.** $0, 2$ **73.** $\dfrac{-1 \pm \sqrt{17}}{4}$ **75.** -4 **77.** $1 \pm \sqrt{2}$ **79.** $\dfrac{3 \pm \sqrt{3}}{2}$

81. $0, 1$ **83.** $\frac{1}{3}, -\frac{1}{2}$ **85.** $\pm 5\sqrt{2}$ **87.** $2 \pm \sqrt{3}$ **89.** No real solution   **91.** $-\frac{2}{3}, 3$ **93.** $5, -1$

95. $-2 \pm \sqrt{2}$ **97.** $\dfrac{2\sqrt{3}}{3}, -\dfrac{\sqrt{3}}{3}$ **99.** $-50, 2$

EXERCISE 9-4

1.   **3.**   **5.**   **7.**

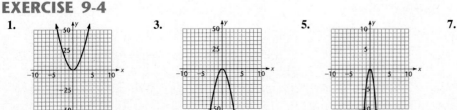
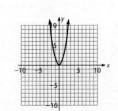

9.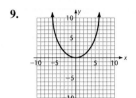

11.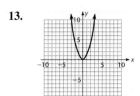

13.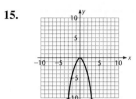

15.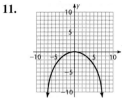

17.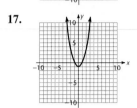

19.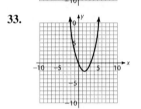

21.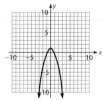

23.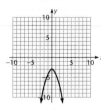

25.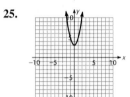

27.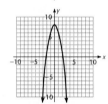

29.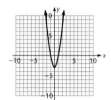

31.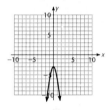

33.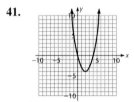

35.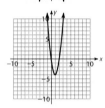

37.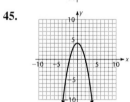

39.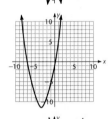

41.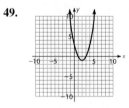

43.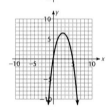

45.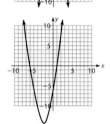

47.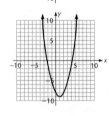

49.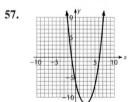

51.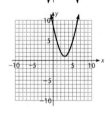

53.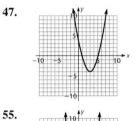

55.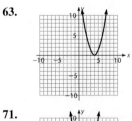

57.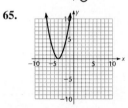

59.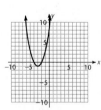

61.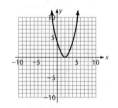

63.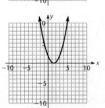

65.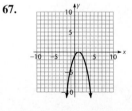

67.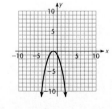

69.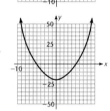

71.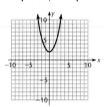

73.

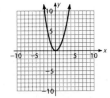

EXERCISE 9-5

1. $t = \sqrt{\dfrac{2d}{g}}$ **3.** $r = -1 + \sqrt{\dfrac{A}{P}}$ **5.** 8   **7.** 0, 2   **9.** $\frac{1}{4}$, 4   **11.** 15   **13.** 10   **15.** 20%

17. 2,000 and 8,000 units   **19.** Approx. 9,854 or 3,146 units   **21.** 20   **23.** 25 or 35   **25.** No   **27.** 20

29. 3 meters, 4 meters, 5 meters   **31.** $(1 + \sqrt{2})$ in ≈ 2.414 in   **33.** 50 mph   **35.** 7 mph and 24 mph

37. 5 mph   **39.** 1.8 m

CHAPTER 9 REVIEW EXERCISE

1. $3x^2 + 4x - 2 = 0$; $a = 3$, $b = 4$, $c = -2$ *(9-1)*   **2.** $2x^2 + 5x - 9 = 0$; $a = 2$, $b = 5$, $c = -9$ *(9-1)*

3. ± 5 *(9-1)*   **4.** 0, 3 *(9-1)*   **5.** $-3, \frac{1}{2}$ *(9-1)*   **6.** 2, 3 *(9-1)*   **7.** $-3, 5$ *(9-1)*   **8.** $-2, 2$ *(9-1)*

9. $-1, 1$ *(9-1)*   **10.** $-\frac{2}{5}, \frac{2}{5}$ *(9-1)*   **11.** $x^2 + 12x + 36 = (x + 6)^2$ *(9-2)*   **12.** $x^2 - 14x + 49 = (x - 7)^2$ *(9-2)*

13. $-13, 1$ *(9-2)*   **14.** Two, 13, 1 *(9-2)*   **15.** Two, $\dfrac{-3 \pm \sqrt{5}}{2}$ *(9-3)*   **16.** Two, $-5, 2$ *(9-1)*

17. Two, $1, \frac{1}{2}$ *(9-3)*   **18.** $\dfrac{-7 \pm \sqrt{97}}{6}$ *(9-3)*   **19.**   *(9-4)*   **20.**

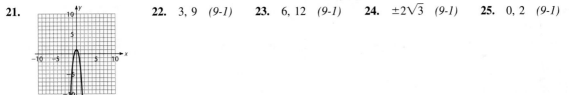

21.   **22.** 3, 9 *(9-1)*   **23.** 6, 12 *(9-1)*   **24.** $\pm 2\sqrt{3}$ *(9-1)*   **25.** 0, 2 *(9-1)*

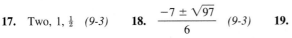

26. $-2, 6$ *(9-1)*   **27.** $-\frac{1}{3}, 3$ *(9-1)*   **28.** $\frac{1}{2}, -3$ *(9-1)*   **29.** 6, -1 *(9-1)*   **30.** 5, -4 *(9-1)*

31. $-10, 2$ *(9-1)*   **32.** $x^2 + 13x + \frac{169}{4} = (x + \frac{13}{2})^2$ *(9-2)*   **33.** $x^2 - 15x + \frac{225}{4} = (x - \frac{15}{2})^2$ *(9-2)*

34. $-9, -4$ *(9-2)*   **35.** 20, -5 *(9-2)*   **36.** $3 \pm 2\sqrt{3}$ *(9-2)*   **37.** $\dfrac{1 \pm \sqrt{7}}{3}$ *(9-3)*   **38.** 5, -4 *(9-1)*

39. No real rolution *(9-3)*   **40.** $-3, 1$ *(9-1, 9-3)*   **41.** $\dfrac{3 \pm \sqrt{29}}{10}$ *(9-3)*   **42.** No real solution *(9-3)*

43. $\dfrac{-9 \pm \sqrt{17}}{16}$ *(9-3)*   **44.**   *(9-4)*   **45.**   *(9-4)*

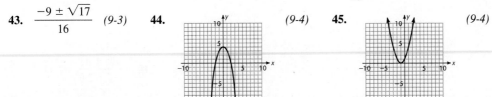

46.   *(9-4)*   **47.**   *(9-4)*   **48.** $\sqrt{\dfrac{p}{0.003}}$ *(9-3)*   **49.** $\sqrt{\dfrac{GmM}{F}}$ *(9-3)*

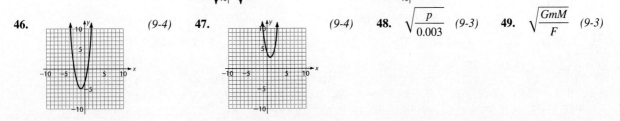

50. 6 in by 5 in *(9-5)*   **51.** 2 feet *(9-5)*   **52.** 12 *(9-5)*   **53.** 8, −6 *(9-1)*   **54.** 3, −$\frac{3}{2}$ *(9-1)*

55. $\dfrac{3 \pm \sqrt{10}}{5}$ *(9-1)*   **56.** ±2$\sqrt{5}$ *(9-1)*   **57.** No real solutions *(9-1)*   **58.** $\dfrac{3 \pm \sqrt{6}}{2}$ *(9-1)*   **59.** $\frac{3}{4}, \frac{5}{2}$ *(9-1)*

60. $\dfrac{1 \pm \sqrt{7}}{2}$ *(9-2)*   **61.** $\dfrac{-3 \pm \sqrt{57}}{6}$ *(9-3)*   **62.** −$\frac{1}{2}$ *(9-3)*   **63.** $\dfrac{3 \pm \sqrt{5}}{2}$ *(9-3)*   **64.** $\dfrac{11 \pm \sqrt{85}}{2}$ *(9-3)*

65. 2, $\frac{5}{2}$ *(9-3)*   **66.** −14, 2 *(9-1, 9-3)*   **67.** −$\frac{3}{2}$ *(9-1, 9-3)*   **68.** 2, 5 *(9-1, 9-3)*

69.   *(9-4)*   **70.**   *(9-4)*   **71.**   *(9-4)*

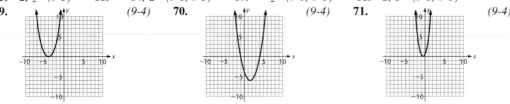

72. $\dfrac{-v_0 \pm \sqrt{v_0^2 + 64 s_0}}{-32}$ *(9-3)*   **73.** $\sqrt{\dfrac{\pi R^2 - A}{\pi}}$ *(9-3)*   **74.** 20 mph *(9-5)*   **75.** $\dfrac{-1 + \sqrt{5}}{2}$ *(9-5)*

CHAPTER 9 PRACTICE TEST

1. −4, 4 *(9-1)*   **2.** 6, −2 *(9-1)*   **3.** −7, 4 *(9-1)*   **4.** −6, 6 *(9-1)*   **5.** $x^2 + 6x + 9 = (x + 3)^2$ *(9-2)*

6. $x^2 + 7x + \frac{49}{2} = (x + \frac{7}{4})^2$ *(9-2)*   **7.** −3 ± $\sqrt{14}$ *(9-2)*   **8.** −6, −1 *(9-2)*   **9.** Two, $\dfrac{-7 \pm \sqrt{57}}{2}$ *(9-3)*

10. No real roots *(9-3)*   **11.** −5 *(9-1, 9-3)*   **12.** $\dfrac{1 \pm \sqrt{5}}{2}$ *(9-3)*   **13.** No real solution *(9-3)*

14. $\dfrac{5 \pm \sqrt{89}}{4}$ *(9-3)*   **15.** −2, −$\frac{3}{5}$ *(9-3)*   **16.**   *(9-4)*   **17.**   *(9-4)*

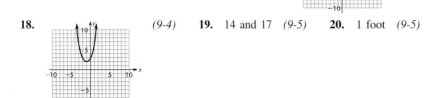

18.   *(9-4)*   **19.** 14 and 17 *(9-5)*   **20.** 1 foot *(9-5)*

CUMULATIVE REVIEW EXERCISES—CHAPTERS 1–9

1. 360 *(1-1)*   **2.** $x^4 - x^2$ *(6-1)*   **3.** −22 *(1-5)*   **4.** 3/11 *(3-1)*   **5.** $3x^2 - x + 7$ *(6-1)*

6. $x^3 - 1$ *(6-2)*   **7.** $x^2 - 3x - 10$ *(6-2)*   **8.** $\dfrac{x^2 + x}{3}$ *(7-3)*   **9.** $\dfrac{x + 2}{x}$ *(7-2)*   **10.** $\dfrac{x(x + 1)}{x + 2}$ *(1-1)*

11. $\dfrac{3}{2x}$ *(7-6)*   **12.** −3 *(4-3)*   **13.** $y = 4x + 15$ *(4-3)*   **14.** $\dfrac{y^2 z^3}{x}$ *(8-2)*   **15.** $2x\sqrt{x}$ *(8-4)*

16. $2 \cdot 3^2 \cdot 5$ *(1-1)*   **17.** $-7x^2 + 7x$ *(1-7)*   **18.** $3x(x^2 + 2x + 3)$ *(6-3)*   **19.** $(x + 3)^2$ *(6-4)*

20. $\dfrac{2x^3 yz}{6xz^2}$ *(3-2)*   **21.** $x - 1$ *(7-1)*   **22.** $(x - 2y)(x + 2y)$ *(6-4)*   **23.** Not factorable *(6-4)*

24. $(x + 6)(x - 1)$ *(6-5)*   **25.** $(x + 3)(x + 2)$ *(6-3)*   **26.** 1.5×10^{-8} *(8-3)*   **27.** 43,200,000 *(8-3)*

28. x^3 *(8-2)*   **29.** x^{-3} *(8-2)*   **30.** xy^{-1} *(8-2)*   **31.** $5\sqrt{7}$ *(8-5)*   **32.** $4x^2 + 2x - 3$ *(9-1)*

33. −5 *(2-7)*   **34.** −$\frac{5}{8}$ *(3-5)*   **35.** $\frac{2}{9}$ *(7-4)*   **36.** $1 > x$ *(5-2)*   **37.** −$\sqrt{5}$, $\sqrt{5}$ *(9-1)*

38. 1 *(9-1)*   **39.** $x = \frac{17}{3}$, $y = -\frac{2}{3}$ *(4-6)*   **40.** $y = -\frac{2}{3}x + \frac{4}{3}$ *(7-5)*   **41.** 3 *(8-7)*

42. 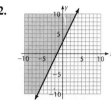 *(5-3)*   **43.** 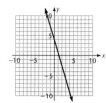 *(4-2)*   **44.** 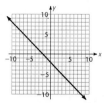 *(4-3)*

45. *(9-4)*   **46.** 132 *(3-7)*   **47.** 23 dimes, 17 quarters *(4-8)*   **48.** $170,000 *(3-6)*

49. 600 miles *(4-7)*   **50.** 19 rolls slide film, 17 rolls print film *(4-8)*   **51.** $\dfrac{4x^2 + 3x + 3}{x^2 + x}$ *(7-3)*

52. $\dfrac{-6x^2 - 2x + 10}{x^2 + 2x}$ *(7-3)*   **53.** $x^3 + 2x^2 + 3x + 4$ *(6-9)*   **54.** $x^3 - 4x$ *(7-3)*   **55.** $y = \frac{7}{6}x + \frac{20}{6}$ *(4-3)*

56. $(x + 2)^2$ *(7-2)*   **57.** $\dfrac{(x + 1)^2}{(x + 3)^2}$ *(7-2)*   **58.** $\dfrac{(x + 1)^2}{x^2}$ *(7-6)*   **59.** $(x + 12)(x - 4)$ *(6-5)*

60. $(2x + 1)(x + 2)$ *(6-5, 6-6)*   **61.** $3(x - 2)(x^2 + 2x + 4)$ *(6-7)*   **62.** $(x^2 + 3)(x^4 - 3x^2 + 9)$ *(6-7)*

63. $y = \frac{10}{11}x - \frac{5}{11}$ *(3-5)*   **64.** $3\sqrt{x}$ *(8-4)*   **65.** 3/4 *(3-5)*   **66.** $\pm\frac{3}{2}\sqrt{x^2 - 4}$ *(9-5)*   **67.** 28 *(8-7)*

68. $x = 14, -4$ *(6-5)*   **69.** $x = 5 \pm \sqrt{5}$ *(9-3)*   **70.** 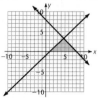 *(5-4)*

71. 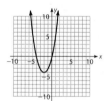 *(9-4)*   **72.** $2\frac{1}{7}$ *(3-6)*   **73.** 14 gal 30% solution, 6 gal 80% solution *(4-8)*

74. 10.8 hr *(4-7)*   **75.** 680 tickets at $12 or 480 tickets at $17 *(9-5)*   **76.** 494 *(3-7)*   **77.** 560 *(4-3)*

78. $(x - 2)(x + 5) = 6x; \; x = -2$ or 5 *(9-3)*   **79.** $3x + \dfrac{4}{x} = 15x + 8; \; x = \frac{1}{3}$ or -1 *(9-3)*

80. $x^2 + (x + 1)^2 + (x + 2)^2 = 194$; The integers are 7, 8, and 9 *(9-3)*   **81.** $2(3x - 5)(x + 2)$ *(6-5, 6-6)*

82. $(4x - 3)(x + 6)$ *(6-5, 6-6)*   **83.** $2(2x + 1)(x - 3)$ *(6-5, 6-6)*   **84.** $(4x - 3)(3x - 4)$ *(6-5, 6-6)*

85. $\dfrac{x^2 - x}{x + 1}$ *(7-6)*   **86.** $\dfrac{\sqrt{x} + 1}{x - 1}$ *(8-6)*   **87.** $\dfrac{(\sqrt{x} + 1)^2}{x - 1}$ *(8-6)*   **88.** 2 *(7-6)*   **89.** 2 *(7-6)*

90. 1 *(9-3)*   **91.** 3, -3 *(9-3)*   **92.** $\frac{17}{2}$, $-\frac{20}{3}$ *(3-6)*   **93.** 15 gal *(4-7)*

94. 73.6 miles to destination, 42.4 miles to return center *(4-7)*   **95.** 4 mph *(4-7)*   **96.** 230 *(9-5)*

97. 3.75% and 7.5% *(9-5)*   **98.** 136 *(9-5)*   **99.** 14 yd by 25 yd *(9-5)*

100. 62 acres of corn, 38 acres of asparagus *(4-8)*

APPENDIXES

EXERCISE A

1. T   **3.** T   **5.** T   **7.** T   **9.** T   **11.** $5 \in P$ **13.** $6 \notin R$ **15.** $P = R$ **17.** $P \neq Q$

19. $\{6, 7, 8, 9\}$ **21.** $\varnothing$ **23.** $\{$Su, M, T, W, Th, F, S$\}$ **25.** $\{a, b, l\}$ **27.** $\varnothing$ **29.** $\{2, 4\}$ **31.** $\varnothing$

33. {1, 2, 3, 4, 5, 6, 7, 8}   **35.** {5, 6, 7, 8}   **37.** {2, 4, 6, 8}   **39.** {5, 7}   **41.** {5, 7}   **43.** ∅, {1}, {2}, {1, 2}

45. 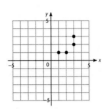   **47.**   **49.**

EXERCISE B

1. Function   **3.** Not a function   **5.** Function   **7.** Function   **9.** Not a function   **11.** Function   **13.** 4

15. -8 **17.** -2 **19.** -2 **21.** -12 **23.** -6 **25.** Function   **27.** Function   **29.** Not a function

31. Not a function   **33.** Function   **35.** Function

37. Domain = {1, 2, 3}, Range = {1, 2, 3}; not a function

39. Domain = {−1, 0, 1, 2, 3, 4}, Range = {−2, −1, 0, 1, 2}; a function

41. -27 **43.** 2   **45.** 6   **47.** 25   **49.** 22   **51.** -91 **53.** 48   **55.** 0   **57.** -7

59. $\frac{1}{5}$, $-\frac{3}{5}$, not defined

61. Domain = {0, 1, 2, 3, 4}, Range = {−3, −1, 1, 3, 5}; a function

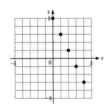

63. Domain = {0, 1, 4}, Range = {−2, 0, 2}; not a function

65. Domain = {−2, 0, 2}, Range = {−2, 0, 2}; not a function

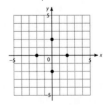

67. $C(x) = 5x$ **69.** $C(F) = \frac{5}{9}(F - 32)$

Index

471